Flora of Radnorshire

The Plants of Stanner Rocks - by *C. Port*.
Top left : Sticky catchfly. *Bottom left* : Bloody cranesbill. *Centre left*: Spiked speedwell.
Centre bottom:*Cladonia pyxidata* - a cup lichen. *Centre right*: Navelwort. *Right* : Rock stonecrop.

Flora of Radnorshire

R. G. WOODS

Produced with the assistance of

D. P. Hargreaves, Dr. D. R. Humphreys and P. J. Port

Published by the National Museum of Wales in association with the

Bentham - Moxon Trust

1993

Published in 1993

© National Museum of Wales,

Cathays Park,

Cardiff.

ISBN 0 7200 0386 5

Line Drawings

The line drawings were produced by :-

Title page Radnor lily by Mr H. Humphries

Page 5 Mr. H. Humphries

   25 Mr. H. Humphries

   26 The author

   47 Mrs M. Griffiths

   49 The author

   54 Mrs. C. Port

   55 Mr. H. Humphries

   65 The author

   71 Mrs. C. Port

   86 Mrs. M. Griffiths

  100 The author

  101 Mrs. C. Port

  147 Mr. H. Humphries

  175 Redrawn by the author from English Botany by J.E. Smith and J. Sowerby

  247 Mr. A. Orange

Contents

Foreword by *Dr. F. Rose*
Preface and Acknowledgements

1. Radnorshire-Introduction to the Vice-County	1
2. The Physical Environment	5
A) Topography	5
B) Geology	7
C) Superficial deposits	13
D) Climate	17
3. Plant Communities of Radnorshire	26
4. The Plan of the Flora	46

The Vascular Plants

5. Biogeography of the Vascular Plants	49
6. The Changing Vascular Plant Flora	55
A) Holocene Period	55
B) Modern Times	56
C) Recent Extinctions of Native Species	59
D) Recent Extinctions of Non-native Species	60
E) Vascular Plants Showing a Notable Expansion of Range	60
7. Conservation of the Vascular Plants	61
A) International Rarities	61
B) National Rarities	61
C) Nationally Scarce and Locally Rare Plants	62
D) Conservation Action	63
8. History of Vascular Plant Recording	65
9. Fern and Fern Allies--Pteridophyta	73
Clubmosses and Quill-worts--Lycopsida	73
Horsetails--Sphenopsida	73
Ferns--Filicopsida	74
10. Seed Plants--Spermatophyta	78
Conifers--Gymnospermae	78
Flowering Plants--Angiospermae	79

Non-vascular Plants

11. Algae and Cyanobacteria	163
Stoneworts--Characeae	165

12. Hornworts, Liverworts and Mosses--The Bryophyta	166
Introduction	
A) History of Recording	166
B) Biogeography	167
C) Conservation of Hornworts, Liverworts and Mosses	171
D) Arrangement and Nomenclature	173
Hornworts--Anthocerotae	175
Liverworts--Hepaticae	175
Mosses--Musci	184
13. Lichens and Lichenicolous Fungi	213
A) History of Recording	213
B) Biogeography	214
C) Atmospheric Pollution and Lichens	217
D) Lichen Conservation	218
E) Arrangement and Nomenclature	220
Lichens and Lichenicolous Fungi	221
14. Rust and Smut Fungi--Uredinales and Ustilaginales	261
Introduction	261
Rust Fungi--Uredinales	261
Smut Fungi--Ustilaginales	265
Abbreviations	266
Bibliography	268
Synonymy of Scientific Vascular Plant Names	272
Gazetteer of Place Names in the Introductory Chapters	274
Glossary of Welsh Place Names in Radnorshire with a Botanical or Ecological Connotation by *E. Gwynn*	276
Index	278
Addendum	292

Foreword

by Dr. Francis Rose, formerly Reader in Biogeography at King's College, London and now in retirement still a very active author of numerous books and papers on botanical subjects.

It is a great pleasure for me to write a foreword to the Flora of Radnorshire for my old friend Ray Woods.

Radnorshire is not well known to most British botanists, still less to European ones. Perhaps the main reason why amateur botanists visit the vice-county is to see and photograph the Radnor lily (*Gagea bohemica*), in what is so far, its only known locality in the British Isles.

But Radnorshire offers far more than that. Though generally lacking calcareous habitats, it is a delightful, very rural and varied vice-county, with much open grassland, heath and moorland. Although the extent of semi-natural woodland is small, there are woods rich in plants, especially mosses, liverworts and lichens.

My knowledge of Radnorshire has been largely acquired during a long series of most enjoyable stays with Ray, his wife and family. To a native of South East England the places in Radnorshire that I have found most interesting include bogs such as Gorsgoch with bog rosemary (*Andromeda polifolia*); agriculturally unimproved meadows with wood bitter-vetch (*Vicia orobus*); ancient woods such as those in the Elan Valley with lichens such as tree lungwort (*Lobaria pulmonaria*) and dry, rocky grassland sites for species such as maiden pink (*Dianthus deltoides*) for which Radnorshire is nationally important.

Ray Woods is an excellent naturalist and a remarkable botanist, with expert knowledge of mosses, liverworts and lichens, a rare ability these days. He tackles many other groups, studying in addition to vascular plants, micro fungi, invertebrates and birds. It is a pleasure to be in his informative company in the field.

I regard this Flora as exceptionally thorough and comprehensive. As well as being compelling reading, it will be a standard reference work for years to come

Francis Rose

1 March 1993

Preface

Without doubt "necessity is the mother of invention". As an officer of the Nature Conservancy Council and active member of the Herefordshire and Radnorshire Nature Trust, a compendium of information on the distribution and abundance of Radnorshire's plants was required in order to offer sensible advice on, for example, the merits of establishing a botanical nature reserve on one site as opposed to another.

Plant distribution records were held by a number of people and though all were willing to share their information it never seemed possible to quickly get enough information together in one place at one time to draw any sensible conclusions. A concise account of the flora of Radnorshire was required. In 1980 I decided that a Radnorshire county flora had to be produced and set about drawing together records made mostly within the last 100 years. To better understand the distribution of these plants extensive introductory chapters on such factors as the weather, geology and land-use of Radnorshire were also required. The results of these labours are presented here.

If the above was the only explanation for this book I suspect it would never have been completed. The driving force has undoubtedly come from a deep interest in plants and, in particular, a love for Radnorshire and its people.

Some explanation is perhaps required as to how an outsider from Derbyshire should become so attached to mid-Wales and so interested in plants. Animals and plants had always been an interest and membership of the North-East Derbyshire Field Club from an early age introduced me to a number of competent botanists, always willing to help. In particular Miss M.C. Hewett who lived locally, guided my early botanical steps.

With a mother born near Hay on Wye and brought up around Builth Wells, mid-Wales holidays were a regular part of life. East Derbyshire was at that time a lichen desert due to atmospheric pollution, so the discovery in Radnorshire of a whole group of organisms, sufficiently abundant on rocks, trees and buildings to colour the landscape, left a lasting impression. This, together with the luxuriance of mosses, particularly in the Elan Valley, sparked off an interest in lower plants.

It was an interest that took me through University at Manchester, where I benefited from the knowledge of lecturers in the Botany Department such as Drs. Alsop, Tallis and a youthful Clive Stace, all guided by the late Prof. David Valentine. Following a period teaching biology in Manchester Grammar School, I was fortunate to secure, in 1973, the post of Assistant Regional Officer for Radnorshire with the Nature Conservancy. The rest is recounted in chapter 8 below.

It is hoped that this work will be of wide interest. I have endeavoured to keep to a minimum the number of technical terms and latin names used, to the extent that the scientific reader may raise an eyebrow here and there. So be it. It is a small price to pay if by so doing an enthusiasm for the study of plants can be engendered in a wider public.

Acknowledgements

Personal Support

Without the unfailing help of a large number of people this book would never have been completed. Those who have contributed plant records are referred to elsewhere in the text. Without the support and interest of many landowners, occupiers, voluntary and statutory bodies who so willingly allowed their land to be searched for plants, few of these records could have been made.

Dr. F.M. Slater at the **Llysdinam Field Centre** and his staff helped in many ways. They provided records, developed the first draft computer-generated vascular plant distribution maps and Dr. Slater kindly made illustrations and maps of the geology and geomorphology of Radnorshire available.

Dr. Barry Thomas and his staff in the Botany Department of the **National Museum of Wales** have been unfailingly helpful. **Mr. R.G. Ellis** and **Mr. G. Hutchinson** have offered considerable assistance with higher plant records, whilst **Mr. A. Orange** and **Dr. R. Perry** have made available lower plant records and advice, including comments on drafts of the lichen and bryophyte sections respectively.

Over a number of years the referees of the **Botanical Society of the British Isles**, the

British Bryological Society and the **British Lichen Society** have patiently determined material. In particular I would like to thank **Dr. B.J. Coppins** of the **Royal Botanic Gardens, Edinburgh** for checking or determining countless lichens, assisting with the field work and for his comments on a draft of the chapter on lichens. **Mr. P. James** guided and encouraged my early enthusiasm for lichens and provided many useful records. **Mrs. J.A. Paton** determined a number of liverworts and offered useful comments on the bryophyte chapter.

Miss. A.C. Powell made available her substantial collection of higher plant records and has offered considerable moral support throughout the project.

Mr. and Mrs. P.J. Port have offered help in numerous ways. They contributed records and in particular drafted out the first set of bryophyte distribution maps and checked early drafts of the text. Mrs. Port's drawings considerably enhance the attractiveness of this book.

Mercy Griffiths also provided drawings of Radnor plants for which I am grateful.

Mr. M. Porter, as well as contributing much useful general advice and many useful records, has offered considerable help in the drafting of the sections on dandelions, roses and brambles.

Dr. D.P. Humphreys is to be particularly thanked for undertaking the onerous task of transcribing all the assembled records from cards to computer to create the distribution maps. Thanks should also be recorded here to **Dr. A. Morton** for making available his DMAP program which was used to generate the maps. Dr. Humpreys also checked the manuscript and created a large part of the index.

Mrs. M.C. Thomas and **Mrs. M. Lewis** helped with the typing and transcription to computer of the manuscript text.

Staff of the **Welsh Office Agricultural Department**, the **Forestry Commission**, Radnorshire District Council, Powys County Council, the **Nature Conservancy Council** and the **Countryside Council for Wales** all assisted in my search for statistics and records.

I also wish to thank **Mr. A.O. Chater** for his wise council, particularly in the latter stages of the production of this book. His many useful comments also greatly improved the text.

One of the most onerous tasks was taken on by **Mr. David Hargreaves**; that of creating by computer a camera-ready copy of this book. Thanks to facilities made available by the **Radnorshire Wildlife Trust** and its members **Mr. and Mrs. J. Wilde**, David has persisted through a variety of technological changes and innumerable text revisions to produce a book of considerable quality. Neither of us had any experience of desk-top publishing. If we had we might never have started. Without his persistence and the considerable savings in costs which have resulted, it is doubtful that this book could have been published at an affordable price.

Dr. F. Rose has made available his wide knowledge of the British and western European flora throughout this project and has kindly contributed a foreword.

Finally I wish to thank my wife Janet and family whose patience over a decade has been sorely tried.

Financial Support

Without financial support the publication of this book would not have been possible. In addition to a generous anonymous donation the **Trustees of the Late Brigadier Sir C.M.D. Venables Llewelyn Bart's Llysdinam Trust** made a substantial contribution towards the publishing and printing costs of this book.

The **Trustees of the Bentham Moxon Trusts** also made available a generous loan to cover the remaining printing costs and undertook to co-publish the book with the **National Museum of Wales** for which I am grateful.

1. Radnorshire, Introduction to the Vice-County

The rivers Wye and Severn, rising together high on the eastern slopes of Plynlymon, as they flow to the Bristol Channel create an island - the island of Fferllys. It is a quiet and beautiful place, not given to any extravagances of topography and one which has escaped the worst ravages of the industrial revolution. At the north-west end of this fair island lies Radnorshire.

Area and Population

The old county of Radnorshire, often referred to locally as simply Radnor, covered 471 square miles and was the second-smallest county in Wales. Perhaps temporarily, it now finds itself sandwiched between Montgomeryshire and Brecknock as Radnorshire District, the central district of the County of Powys, the largest county in Wales. In the east it marches with Shropshire and Herefordshire, whilst in the west it marches with the Ceredigion District of Dyfed (formerly Cardiganshire).

Radnorshire has always been thinly populated. From a population of 23,281 in 1901, it saw a continuous decline in number to 18,262 in 1971. This trend has now been reversed, perhaps due to government support for industry or a growing appreciation of the pleasure of life in what, to some, may seem a rural backwater. Despite this increase to 23,650 in 1991 it is still one of the most thinly populated areas in England and Wales, with sheep outnumbering humans 48 to 1 and 50 humans occupying each square mile (or approximately one person to every 5 hectares).

Land-use

Agriculture and forestry are major industries. Though both have shown a trend to employ fewer and fewer people in recent years, they still are the major land users in Radnor and, in consequence, have the largest impact on the flora of the county. The farms are mostly small to medium, owner-occupied family farms which rear livestock. Large estates have never occupied much ground in the county. Howse (1949) repeats the oft-quoted verse:

Radnorsheer, poor Radnorsheer,
Never a park and never a deer.
Never a squire of five hundred a year,
But Richard Fowler of Abbey Cwmhir.

The verse is attributed to a disgruntled Commissioner sent by Parliament to collect fines from Royalists after the Civil War. Today Abbeycwmhir is still the only significant parkland or pasture-woodland site for lichens in Radnor.

What Radnor lacks in parks it makes up for in common land. The 67 registered commons cover 16478 hectares (14%) of the county and range in size from less than one hectare to 2780 hectares. Almost all of this land is intensively grazed by sheep, with smaller numbers of cattle and ponies. The biological interest of Radnor's commons is described by Penford et al. (1990). Bracken covers 36% of commonland and is still increasing in area, though many of the more level patches are cut each autumn to provide livestock bedding, this being the only other significant economic use to which this land is now put.

Dwarf shrub heath covers 21% of the commons. Recently grouse moor management has been reinstituted on the commons around Llandeilo and Llanbedr Hills. Formerly the heather of Beacon Hill and the Radnor Forest was also managed to favour grouse. Acidic grassland and mosaics of grass with heathland species occupy a further 32% of this land.

Conifers are the mainstay of the forestry industry in the county, with the majority of production centred on large State Forests around Abbeycwmhir and the Radnor Forest. Sitka spruce is the most widely planted tree on upland sites, whilst on somewhat better soil, Japanese and hybrid larch and Douglas fir are the most commonly grown species. Farming and forestry are discussed in more detail in the section on the changing vascular plant flora below.

Industry

Tourism may now earn more money than either farming or forestry. Rhayader and Llandrindod Wells are heavily dependant on income from this source, with the Elan Valley being the most popular destination for tourists.

Radnor has never sported a large factory chimney constructed from bricks and its light industry is mostly grouped around the towns, with light engineering, office equipment, furniture, clothing, glassware and carpet manufactures all employing staff. Large out-of-town employers include sawmills near Newbridge on Wye and quarries at Llanelwedd and Dolyhir. Both the District of Radnorshire and the County of Powys are administered from Llandrindod Wells, which combined are the single largest employers of labour in Radnor.

Transport

This low level of economic activity has created little traffic and, barring pelican crossings, the county lacks a set of permanent traffic lights. This is of more than passing interest to the field botanist, since this lack of activity is almost certainly the main reason for the very small number of non-native plant species so far recorded from the county.

Ellis (1983) provides a statistical summary of the higher plant floras of the Welsh vice-counties. He reports a total of 864 species from Radnor, 177 of which are considered to be aliens. These are the smallest figures reported for any Welsh vice-county. He concludes this small total is due to Radnor's lack of what are often species-rich habitats, such as sea coast and its shortage of limestone as well as this small number of alien species. Plants of this latter group, even if successfully transported to Radnor by the limited amount of commerce, tend to favour squalid corners of car parks and derelict land which are rare habitats in Radnor, so establishment is not easy.

Railway tracks and quarries provided the most extensive areas of such habitats. The Central Wales Railway, built between Knighton and Builth Road in 1865 was instrumental in adding some species to the county. Wild parsnip and sand sedge may have been transported from Swansea Bay, as the trains passed daily through the sand dunes near Sketty. Modern maintenance methods on this, the only railway still running in Radnor, are less satisfactory from a botanical point of view. Herbicide trains are very efficient and most former sidings are now covered in scrub and woodland. A few notable grasslands are maintained on embankments and in cuttings, particularly between Dolau and Llangunllo Stations.

The Mid-Wales Railway was built between Tylwch and Boughrood in 1864 and closed in 1965. It followed the Marteg and Wye valleys. It is suspected that lead mine spoil from the Fan lead mines in Montgomery was applied to the track in places, probably as a weed killer. The track bed today, such as just north of Rhayader and around Erwood, is remarkably free of plant growth except for a number of nationally scarce lichens which are frequently associated with old metal mine tips. An old cutting near Aberedw provides the last known site in the county for wild carrot, whilst the track bed supports a wealth of annuals. Some of this line, including the tunnel in the lower Marteg valley, forms part of Gilfach Farm Nature Reserve of the Radnorshire Wildlife Trust. Its damp, fern-rich cuttings and dry, clinkered embankments make an interesting walk. But from a botanical point of view the most interesting stretch of this disused railway lies between Erwood Station and Llanstephan. Most of this stretch of the line has been converted to a road and access to the flower-rich banks is easy. Primroses and cowslips are a feature in the spring, whilst rare species include rock stonecrop and southern wood-rush. This line joined the Hereford to Brecon railway at Three Cocks Junction. A small stretch of this latter railway, also closed in the early 1960's, crosses the vice-county at Glasbury. The cutting here is a nature reserve of the Brecknock Wildlife Trust. A small colony of green-winged orchid survives on the side of the cutting.

Two railways just entered east Radnor. In 1875 a railway from New Radnor to Kington was opened, as was a branch line from Titley to Presteigne. The former had been built from Kington to the limestone quarries at Dolyhir on the track of a tramway opened around 1833 (Sinclair and Fenn, 1991). Both these lines closed in the early 1960's and are now either woodland or mostly destroyed. Blue fleabane was one of the more notable plants of the railway tracks at Dolyhir. It still survives about the quarries today. Between Stanner and Dolyhir a road has been built on the track bed.

Introduction

The old lineside fence of ancient sawn sleepers survives and supports a notable lichen flora.

In the north west uplands of Radnor a large private railway system was built by Birmingham Corporation to facilitate the construction of the Elan Valley Reservoirs (Judge 1987). Work commenced in 1893 from a junction with the Mid-Wales Railway near Rhayader and the line eventually stretched to just beyond the site of the Craig Goch Dam, with possibly 33 miles of track in both Brecknock and Radnor. Eventually abandoned during the First World War, most of the track bed and associated works still survive. The deep cutting of the "Devil's Gulch" beside Penygarreg Reservoir now supports fir clubmoss whilst flattened meadow-grass grows out of cracked mortar between the stones of many of the sunny retaining walls. Wood bitter-vetch is a feature of a roadside embankment near Aber-Ceuthon.

There have never been any canals in Radnor and the Wye is only unequivocally navigable as far as Hay on Wye. The effect of this right of navigation on bankside vegetation is not known. The high value attached to the game fishery on this river has, perhaps been partly responsible for ensuring that only short stretches of the bank are accessible of right to the general public. Only below Boughrood Brest can any significant lengths be examined. The silty banks and sandy bars of this area support plants such as water chickweed, yellow loosestrife and skullcap.

A number of main roads pass through the county and all have been subject, in places, to major road improvement schemes. These works have led to the extinction of at least one plant (greater broomrape), whilst at the same time they have created much plant-rich and accessible habitat. Trunk road verges are the only known sites for Oxford ragwort and dark mullein in the county. The most botanically interesting sections of trunk road lie beside the A483 from Llanddewi Ystradenni, north to the Montgomery border, with plants such as dyer's greenweed, wood bitter-vetch, wood horsetail, mountain melick and the moss *Discelium nudum*.

Verges of minor roads throughout the county are frequently rich in flowers in contrast to the all too heavily sheep-grazed surrounding farmland. Plants principally confined to roadsides include hawkweed oxtongue, hedge-bedstraw, spreading bellflower, bladder and white campion.

Radnorshire-the Vice-County

Local government boundaries are by no means stable. Since commencing the collection of data for this flora the Radnorshire District Council boundary has been significantly altered in two places. Elan Village, formerly in the Borough of Brecknock has been transferred to Radnorshire, whilst in the extreme north west corner, land formerly in Montgomeryshire District has been added to Radnorshire. To introduce stability the area chosen for the present survey is the Watsonian vice-county of Radnorshire (VC43). The vice-county system was created in 1859 by H.C. Watson to stabilise county boundaries and even out the discrepancies in size by splitting up large counties and amalgamating smaller ones. Since local floras have been, or are being produced for adjacent vice-counties, to prevent overlap and avoid gaps being left in the record, the Radnorshire vice-county appeared to be the best unit to adopt.

A set of 1 inch to 1 mile maps, marked up with the vice-county boundaries is held in the Department of Botany of the Natural History Museum in London. The Radnorshire boundary follows the boundary of the old County of Radnorshire as shown on the 1st series, 1:25,000 Ordnance Survey (OS) maps except in one area, described below. This series of maps have therefore been used to define the study area. It should, however, be noted that where the boundary follows watercourses, the centre of the main channel at the time of survey was taken as the county boundary unless there had been such a dramatic change that the OS was likely to conserve the old boundary.

The departure of the vice-county boundary from the Radnorshire County boundary on the 1st series, 1:25,000 maps occurs in the Glasbury area. In Watson's time the Radnorshire Parish of Glasbury straddled the R. Wye and the cost of maintaining the Wye bridge fell entirely on the inhabitants of Radnorshire. The cost was later shared, though somewhat reluctantly, with Brecknock when the Wye became the boundary and that land on the Brecknock bank was taken into the Parish of Tregoed and Felindre. Watson's Radnorshire vice-county still takes in land which is now in the Borough of Brecknock. The vice-county boundary follows the Tregoed and Felindre Parish boundary from the R. Wye, upstream of Glasbury in an east south east direction to Little Lodge. It then turns north along the western edge of Allt y Fran Wood, adopting a straight line north from the northern edge of the wood. Striking the lane running along the east side of the field to the east of Tir-uched Farm, the boundary follows the lane eastwards to Hoel-y-gaer. Here it rejoins the Tregoed and Felindre Parish boundary and follows it north west to the old R. Wye ox-bow at Ffordd-fawr. The boundary then strikes east to follow the old river channel down to rejoin the existing Radnorshire boundary in the middle of the R. Wye.

Map 1.1 Radnorshire Vice-County Boundary and Main Communication Routes

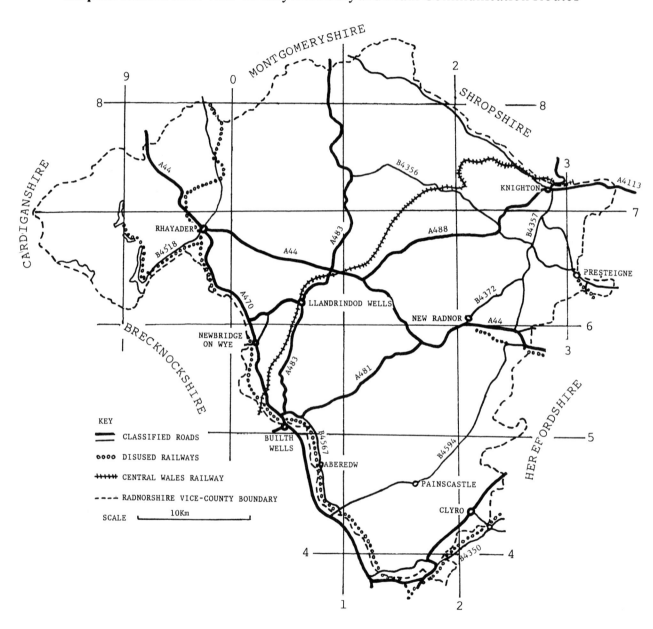

2. THE PHYSICAL ENVIRONMENT

A. TOPOGRAPHY

Radnor is perhaps the most upland county in Wales, yet it has little topography which could be unequivocably described as mountainous. There are no distinctive mountain ranges to rival the Brecon Beacons, the Berwyns, Cadair Idris or even Plynlimon. Instead, an extensive gently undulating upland plateau, lying mostly between 400m. (1312ft.) and 600m. (1968ft.), has been cut by steep sided valleys.

The mechanisms which have given mid Wales this distinctive topography are still largely unknown. Did wave action plane what must have been a jumbled assortment of hills following various mountain building events, down to roughly the same height? Rivers like the Wye flow with apparently little or no heed for rock type or fault pattern. South of Rhayader the Wye punches a gorge by Rhayader Quarries through some of the hardest rocks in the area, crossing at right angles the major structural disturbance in the area - the Twyi anticline. Was this drainage pattern formed in younger, more uniform deposits and the Wye let down on to these older rocks by the erosion away of these younger rocks? If so, nothing remains of these younger strata.

The Radnor Forest at 660m. is Radnor's highest elevation. The summit imperceptably rises out of a plateau of rolling moorland, 4 sq. km. of which are over 600m. high. Yet northern and montane species are scarce, with only mossy saxifrage, wood craneshill and stone bramble of any note. To the west of Rhayader the hills of Llansantffraed Cwmdeuddwr offer a wilder prospect yet reach only 548m. close to the Ceredigion border at Pantllwyd. Over 4 km. of difficult walking through purple moor-grass tussocks is necessary to examine this remote area of blanket bog, rock and upland lakelets.

Of easier access and botanically more rewarding are the steep rocky valleys of the Elan and Claerwen. Large areas are now flooded by reservoirs but unlike Llyn Brianne, which destroyed some of the most interesting areas of the Upper Twyi, the Elan Reservoirs seem to have flooded mostly agricultural fields

Craig Goch Reservoir, Elan Valley

and have left the crags, woodland and most of the moorland intact.

This dissected plateau extends across most of mid and northern Radnor, cut significantly only by the north to south flowing R. Ithon. The somewhat softer and more base-rich shales of this area have proved more attractive for agricultural improvement and ryegrass and white clover set against walls of Sitka spruce are all too frequent. Common land here and there offers oases to the field botanist but they are not rich enough for this area to escape being labelled the dullest part of Radnor.

South of the Radnor Forest lie extensive commons covered in moorland and bracken. This is undoubtably the best walking country in Radnor if the bracken can be avoided. It suffers botanically from being "neither up nor down" being neither high enough to be montane nor low enough to be unequivocably described as lowland heath. The area warrants more study and should, I think, be valued more highly than is perhaps the case at present. Spring head flushes, base-rich rock outcrops and water and peat filled hollows called locally mawn pools, are important features.

Lowland Radnor comes in two types. The lowland in the hills consists of wide valleys with wet boulder clay floors and mercilessly late and early frosts. The valley of the Marteg around St Harmon, the whole of the Ithon Valley, the Wye down to Builth Wells, the Edw around Franksbridge and the Bach Howey around Painscastle are good examples. Wet pastures with meadow thistle, devil's-bit scabious and heath spotted-orchid, and banks with harebell and wood bitter-vetch characterise the more interesting parts.

Map 2.1 The Topography and Drainage Pattern of Radnorshire

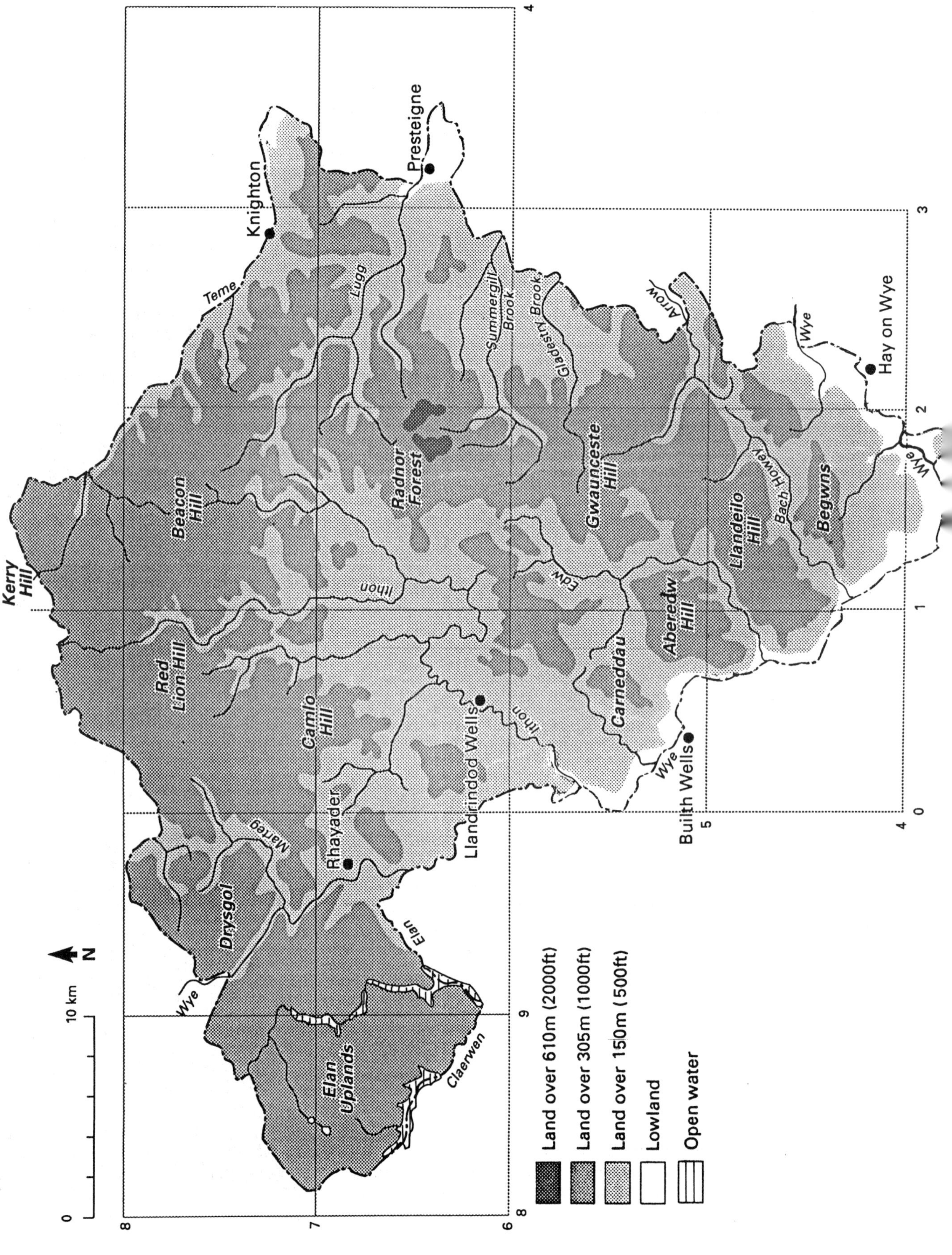

Geology

Then there is lowland Radnor proper. Fingers of the warm, fertile Hereford plain with rich alluvial soils creep in from the east. The Wye valley as far as Llanstephan, the Arrow to Newchurch, the Summergil Brook to New Radnor, the Lugg to Monaughty and the Teme to Knucklas are the core of this area. Mistletoe, white bryony, hairy St John's-wort, wood spurge, spurge laurel, southern wood-rush, travellers joy, white campion and a wealth of other arable weeds are rare but characteristic features of these valleys. Ox-bows and silty banks beside the rivers support a rich riparian vegetation and seem far removed from the sheep-bitten, wind-swept uplands.

The Wye leaves Radnor at a height of 75m. (246ft.) - the lowest point in the county, giving Radnor an altitudinal range of 585m. (1919ft.). Between these altitudes its varied topography creates a diverse range of plant habitats.

B. GEOLOGY

The 'hard' geology is considered first, geological period by period. Apart from a small outcrop of Precambrian age all Radnor's rocks were laid down in the Palaeozoic era. The origins of the rock strata, their lithology and impact on the present flora are briefly considered.

Precambrian

The oldest rocks in Radnor occur in the east of the county around Old Radnor and are of both volcanic and sedimentary origin. Volcanic activity over 1200 million years ago intruded large masses of dolerite, gabbro and acid porphyries into existing sedimentary rocks. Faulting or other weaknesses in the Earth's crust led to a depression being formed in the Welsh Marches in which over the course of around 600 million years about 7620m. (25,000ft.) of sedimentary rocks accumulated. These rocks are now exposed in the steeply rounded hills of Stanner Rocks, Worsell, Old Radnor and Hanter Hills, such distinctive features on the journey into Wales from Kington.

The complexity of their mineralogy, structure and responses to weathering have permitted one of the richest floras in the whole county to develop on the volcanic Stanner Rocks. These hard rocks weather slowly and shallow skeletal soils, rich in humus but subject to summer drought on south-facing slopes, develop in crevices and on stable screes amongst the many rock outcrops. Both the gabbro, best examined in Stanner Quarry, and the dolerite, break down in places to produce base-rich soils. These support characteristic calcicoles such as common rock-rose, bloody cranesbill, pale St John's-wort, hairy violet, the moss *Homalothecium lutescens* and the liverwort *Porella arboris-vitae*. Yet these calcicoles occur alongside acid soil-loving species such as wood sage, in a complex of calcicole and calcifuge communities.

On the steep south-facing cliffs thin soils and summer drought have prevented tree growth. Plants demanding open conditions appear to have survived here from late glacial times, squeezed out from the surrounding areas by the development of deeper soil and consequent spread of the forest. Such survivors include the Radnor lily, perennial knawel, sticky catchfly, spiked speedwell, shepherd's cress, wild onion, a range of clover species, the moss *Bartramia stricta* and many more. On the open rock a rich lichen flora has developed of over 100 species. Only on Breidden Hill in Montgomery is a similar flora found in Wales.

Worsell Wood has in recent years been almost entirely afforested with a mixture of conifers and broadleaved trees, though fragments of semi-natural woodland remain. The outcrops of the underlying mass of dolerite and smaller amounts of acid porphyries are here quite small and are all under tree cover so do not support any of Stanner's specialised flora. A thin soil has developed over scree on the lower slopes and flushing has probably introduced bases. On open south-facing ride edges calcicoles such as pale St John's-wort and calamint occur whilst under the tree canopy and perhaps betraying a long continuity of woodland cover on this site, wood spurge occurs in abundance. Open areas of all aspects on disturbed ground and scree provide a habitat for narrow-leaved bitter-cress. Rock outcrops, though shaded, on the lower slopes are sufficiently base-rich to support the mosses *Pterogonium gracile*, *Cirrophyllum crassinervium*, the liverwort *Porella playtphylla* and the lichen *Peltigera horizontalis*. More acidic rocks are characterized by the moss *Isothecium myosuroides*.

Hanter Hill in contrast has little significant tree cover. Barring the presence of one or two base-rich wet flushes on its flanks, its low dolerite, and on its southern side, gabbro outcrops support an almost entirely calcifugous vegetation. Rising to 105m (1361') it is the highest of these volcanic hills and in the presence of lichen species such as *Pseudevernia furfuracea*, *Sphaerophorus globosus* and *Ochrolechia tartarea* and the dark cushions of the moss *Andraea rothii* is more reminiscent of the hard grit and conglomerate ridges of the Elan Valley. These volcanic rocks are surrounded by and unconformably overlaid by Silurian shales. To the NW across the valley of the Gilwern Brook the Precambrian reappears as sedimentary rocks, forming much of Old Radnor Hill. A number of low rock outcrops of these hard, somewhat acidic grits and conglomerates of Wentnorian or Longmyndian age and possibly derived from the volcanic rocks described above are found on the upper, south-facing slopes. Here a notable lichen flora, somewhat akin to that of Stanner Rocks occurs, but with a greater abundance of maritime species. Old Radnor Common, now greatly eaten into by the active Gore Quarry is elsewhere covered in semi-natural acidic grassland and bracken, but an old quarry by the church and outcrops lower down on the opposite side of the road support a more species-rich flora.

Towards the SW edge of the outcropping Precambrian rocks, faulting, folding and some igneous activity has intruded base-rich materials. Hard, grit-like rocks outcrop in Yatt Wood. South-facing, summer-droughted rock ledges, free of woody species, support a notable calcicole flora with an abundance of rock stonecrop, marjoram, long-stalked cranesbill and more rarely, hound's-tongue, ploughman's-spikenard and a rich fern flora including wall-rue, maidenhair, black and forked spleenworts and rusty-back fern. Mosses found include *Pterogonium gracile* and *Homalothecium lutescens* whilst crustose lichens are rather poorly developed, though foliose species such as *Leptogium sinuatum* and *Collema* species are frequent.

Ordovician

In the absence of any rocks of Cambrian age the next youngest series are ascribed to the Ordovician period and named after a famous Welsh tribe, the Ordovices, since it was in their area of Central Wales that the stratigraphy of these rocks was first worked out. About 450 million years ago, Radnor lay on the south east edge of a shallow ocean basin. Rivers brought down sediments from the surrounding large land masses of the Welsh Geosyncline to the west and the Midland Block to the east, depositing in various times and places fine anaerobic mud, sands, grits and larger pebbles. The floor of the basin was gently subsiding so that great depths of deposits built up, but typically the ocean was never any great depth in this area. Catastrophic volcanic eruptions occurred between Llandegley and Llanelwedd depositing great quantities of volcanic ash and intruding masses of dolerite and keratopyre between the beds of sedimentary rocks.

Subsequently, buried under deposits of Silurian age amongst others, paleozoic earth-movements folded mid Wales in a north east to south west direction. The Tywi anticline is the chief fold. It is the erosion of the fractured top of this anticline and parallel features to the south east that has exposed the Ordovician rocks in an area south west from Llandegley to Llanelwedd and from Abbeycwmhir to Doldowlod.

The rocks can be examined in mostly south east facing exposures, such as those to the south of Rhayader Quarries, Dolyfan Hill and Cefn Nantmel as well as in the extensive outcrops of the Carneddau and Llandegley Rocks.

The lithology of the sedimentary rocks is very variable. Whilst mostly consisting of base-poor rocks, near Glaslyn to the south east of Rhayader Quarries, carbonate rich water seems to have carried lime salts in to mix with the siliceous silts. These lime-rich rocks support common rock-rose and spindle in an area where calcifuge species such as wood sage, wavy hair-grass, heath bedstraw and beautiful St John's-wort would otherwise be the expected species.

The Ordovician has more botanical surprises. Half a kilometre to the south of this common rock-rose site, a dry, sunny, south east facing ridge supports a plant community very reminiscent of the drier sea cliffs of West Wales. Amongst early hair-grass, sheep's-bit and English stonecrop the normally maritime lichen *Anaptychia runcinata* occurs in some abundance with the somewhat less exclusively coastal species *Ramalina subfarinacea*. This community occurs elsewhere on the Ordovician near the northern edge of its outcrop on Wenallt, Abbeycwmhir. On Dolyfan hill near Newbridge on Wye harder grits outcrop. Slow to weather, they have

Geology

Map 2.2 The Geology of Radnorshire

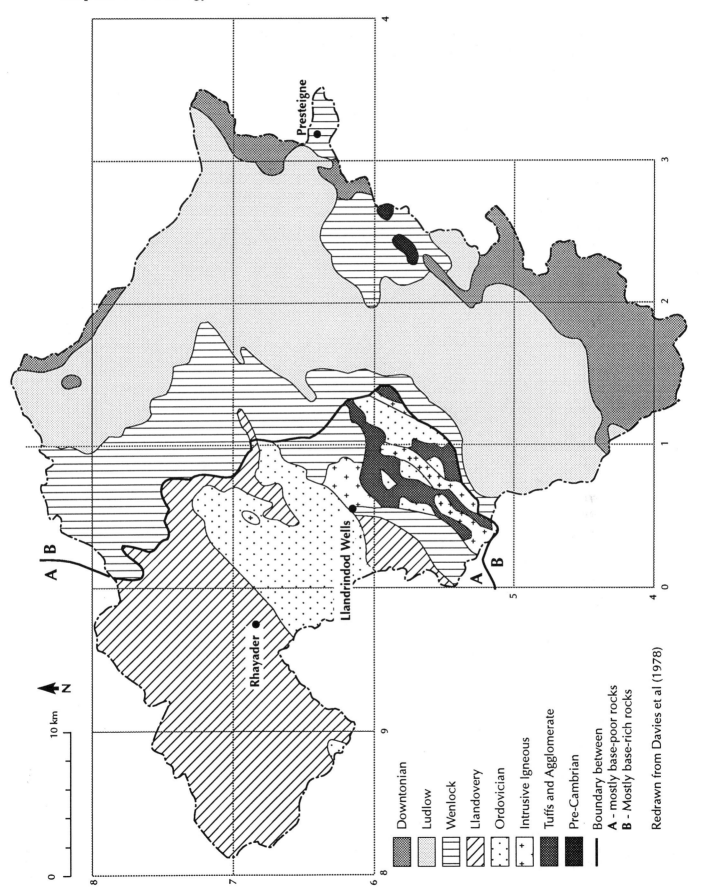

Redrawn from Davies et al (1978)

acquired a rich lichen flora but lack any maritime flavour.

This variability of flora is nothing when compared to that offered by the Llandrindod Wells Ordovician inlier. A glance at the 1:25,000 scale map of this area produced by the British Geological Survey reveals the extent and complexity of the volcanic rocks intruded and extruded amongst various sedimentary rocks between Llanelwedd and Llandegley. Within mudstones and shales, lenses or laccoliths of Albite dolerite have been intruded. Through faulting, quarrying and weathering many are now exposed with good examples visible at Llanfawr Quarries near Llandrindod Wells and in the eastern, disused quarry near Gelli-Cadwgan, Llanelwedd. Castle Bank, Llansantffraed-in-Elvel is also constructed of this material with loose blocks used to construct the rampart of an Iron age fort. It supports an almost entirely calcifugous vegetation, but the rocks, slow to weather and perhaps largely undisturbed since the Iron Age, are rich in mosses, liverworts and lichens.

Similarly intruded are massive keratophyre and platy 'andesites'. They too are slow to weather and produce characteristically crag-bound hills, best seen along the ridge of the Carneddau to the east and south east of Howey, eg. The Bower, The Banks and the ridge east of Three Wells and in the north along the ridge of Llandegley Rocks. Both the dolerite and keratophyre rocks generate only thin soils and, where subject to summer drought, open communities of ephemeral plants occur, eg. whitlow grass, upright chickweed and annual knawel in a characteristic community with others such as carline thistle, thyme, harebell and, rarely, maiden pink and storksbill.

Explosive volcanic activity shattered rock and added quantities of ash to produce beds of tuffs and ashes, agglomerates, spilites and keratophyre nearly 600m. (1968ft.) thick. Faulting and weathering exposes these various materials over wide areas in many outcrops and stream sections. Most are rich in silicates and weather to support rather botanically uninteresting acidic vegetation. The spilitic andesites are an exception as they contain calcite, which erodes to leave hollow vesicles and favours the growth of calcicoles. Good exposures occur to the east of Newmead Farm, on Caer Einion, and fragments survive below the large active quarry at Llanelwedd. Notable calcicolous bryophytes of these areas include *Ditrichum flexicaule*, *Anomodon viticulosus* and *Pterogonium gracile*, but calcicolous higher plants are sparse. Elsewhere calcicoles occur on tuffs near Alpine Bridge and The Banks.

Volcanic activity has also placed a large isolated mass of dolerite near Carmel in Ordovician rocks. Baxter's Bank and Pen-y-banc Quarry at its south end, though not yet thoroughly explored botanically, are of some interest. Under light shade of hazel and ash, twayblade occurs with primrose and the calcicolous moss *Encalypta streptocarpa* on stable scree.

Silurian

A sudden change in the fossil species of Graptolites occurring in the shales, grits and mudstones which continued to be deposited in the slowly subsiding western mid-Wales basin marks the end of the Ordovician and start of the Silurian period. Rocks of this period, named after another old Welsh tribe, are extensively exposed in Radnor. Its early stages are, however, poorly represented in the east of the county since around this time large parts of eastern Radnor were probably above sea level.

In the west, deep, submerged canyons were being infilled with coarse grits to produce the hard, weathering-resistant rocks now exposed beside Caban Coch Reservoir, the upland ridges of Blaen Restr, Craig Cwm Clyd and others in the Claerwen Valley and on Gwastedyn Hill to the SE of Rhayader. All support a relatively poor higher plant flora of wavy hair-grass, bilberry and rowan, though English whitebeam by Caban Coch is an exception. Their lichens are, however, particularly interesting and include a number of submontane species such as *Bryoria bicolor*, *Haematomma ventosum* and *Sphaerophorus fragilis*. The softer shales and mudstones seem to weather too readily and though well-covered in lichens, these are mostly widespread and common species.

The Silurian grits in places, notably along the Claerwen valley have collapsed to produce large block screes. In the humid recesses between the blocks, frequently out of reach of sheep and even perhaps largely unaffected by the all too frequent moorland fires, ferns such as Wilson's filmy fern, beech fern and the mountain male fern find a refuge amongst commoner species. Typically western Atlantic

Geology

liverworts find the humidity and shelter to their taste. *Jamesionella autumnalis* has its only Radnor station here along with other more widespread species such as *Scapania gracilis*. Lichens such as *Sphaerophorus globosus* occur in abundance, whilst higher plants include hawkweeds, goldenrod, bell heather and climbing corydalis.

The rolling uplands of the Claerwen, Elan and Upper Wye catchments with smooth summit plateaux at around 500m are all formed of materials deposited at this time. The rocks are almost all poor in bases and coupled with the high rainfall have given rise to peaty, acidic soils. A careful search of the rock outcrops however reveals here and there small lime-rich pockets picked out by calcicolous lower plants such as the lichens *Caloplaca subpallida* (on Wyloer), and *Peltigera leucophlebia* (on Cerrig Gwalch) and the mosses *Neckera complanata* (Nannerth) and *Tortella tortuosa* (Pen y Garreg). This latter species is here associated with a number of lichens more typical of base-rich tree bark eg *Nephroma laevigata, N. parile, Dimerella lutea, Pannaria conoplea* and *Peltigera horizontalis*.

Mineralization of rocks, though rare, can also create areas of botanical interest. Local iron-rich bands in rocks such as those beside Caban Coch Reservoir, carry an unusual metalophyte flora of the red-brown lichen *Acarospora sinopica* and the citrine yellow *Lecanora epanora*. Heavy metals have been intruded into rocks in scattered localities through north Radnor, their presence picked out by old mine workings. All known sites have been examined but only at the Cwm Elan Mine in the Methan Valley beside Garreg Ddu Reservoir are the tips sufficiently rich in lead or zinc or extensive enough to support a notable metalophyte flora. A preliminary survey has recorded colonies of the scarce lichens *Gyalidea subscutellaris, Placopsis lambii* and *Stereocaulon nanodes* on tipped material.

Spring-fed flushes in places show some mineralization. The most botanically interesting are rich in lime salts. A number occur between the Rhayader to Cwmystwyth mountain road and the River Elan, with mosses such as *Campylium stellatum*, *Scorpidium scorpioides* and *Cratoneuron commutatum* var. *falcatum* and rarely the bog orchid.

The harder, well-bedded shales in the Elan uplands have been quarried in places to provide slate for roofs and floors. The long abandoned tips have been colonized by a specialised flora such as parsley fern below Glog Fawr beside Garreg Ddu Reservoir. Iron-rich slate waste at the east end of Craig y Bwch in the Claerwen Valley supports a notable lichen flora.

Deposition of these rocks in the west of the county took place between 438 and 428 million years ago - the Llandovery Epoch. Towards the end of this epoch a general sinking of the land had allowed a shallow sea to creep eastwards over much of central and east Radnor and this marks the opening of the Wenlock Epoch. Fine silts and muds, often quite rich in shelly fossils and other lime salts were deposited. They are now exposed in a more or less north to south band on either side of the A483, Builth Wells to Newtown trunk road.

Recent roadworks have produced extensive exposures beside the trunk road to the north of Llanddewi Ystradenni but these are yet to be colonised by anything of note. Old cliffs to the north of Llanbadarn Ffynydd, however, support calcicoles such as marjoram, mountain melick and downy oat-grass with other more widespread species such as primrose, wild strawberry and ox-eye daisy. These contrast with nearby cliffs on which wood sage and wavy hair-grass are common. Elsewhere in this area boulder clay blankets most of the land and rock exposures are rare. Small exposures occur on Castelltinboeth around the ruins of the castle. In shallow soil pockets amongst the rocks, shepherds cress is found.

Travelling south, substantial cliffs occur beside the R. Ithon below Llandrindod Wells, some at least consisting of shales and mudstones of this age. The calcareous nature of the rocks is indicated by plants such as hairy rock-cress, stone bramble, marjoram and spindle.

Rocks of Wenlock age are mapped between New Radnor and Walton but are largely buried under late glacial deposits. The south east edge of this outlier, near Dolyhir, has been folded and faulted upwards, exposing what were former reefs produced mostly by algae in warm shallow seas and which now consist of massively bedded dark grey-white crystalline limestone. This is the only limestone in Radnor and it has been quarried to provide lime and building stone for over one hundred years. The quarrying has been so extensive that few, if any, natural outcrops now occur. Old limestone quarries have, however, been colonized by lime-loving plants. Notable species include the milk thistle, blue fleabane, hairy violet and mezereon, though the latter is almost certainly a garden escape along with the wallflower. Lichens include *Aspicilia calcarea* and *A. subcircinata*, whilst notable mosses include the tufa-forming *Eucladium verticillatum* and the distinctive fire extinguisher moss *Encalypta ciliata*. Fine basic grassland in old pastures near the quarries supports populations of meadow saffron, wild daffodil and cowslip.

During the Ludlow epoch (421 to 414 million years ago) Radnor lay on the southwest edge of a large basin in which great depths of calcareous silt and mudstones accumulated. These deposits extend today over much of east Radnor. Folded, they now form the high ground of Beacon Hill, the Radnor Forest and Gwaunceste to Aberedw Hills. The Radnor Forest is Radnor's highest point at 660m or 2166 ft. Fine heather moorland covers its large summit plateau, its flanks cut into by deep valleys or dingles. Rock outcrops, considering the height of the Forest, are not common but low cliffs occur in most dingles and are of considerable botanical interest. Davy Morgan's dingle supports stone bramble and wood crane's-bill. The valley to the west of Whinyard Rocks has a number of notable bryophytes including *Brachyodon trichodes*. Whinyard Rocks have spectacular colonies of mossy saxifrage. In Harley Dingle the Great and Little Creigiau, as with so many outcrops of these Ludlow mudstones, consist of bone dry towers of loosely piled slate, like carelessly stacked plates and threaten to collapse at any moment. Marjoram, hawkweeds, harebell and thyme occur commonly. The drought-stricken condition of these outcrops contrasts with the deep, damp ravine of Water-break-its-neck, its dripping cliffs festooned in Irish ivy and brittle bladderfern, and home of the rare moss *Philonotis rigida*. The streams of the Ludlow rock area have a curious quality about them, much resembling limestone streams. The rocks soak up water in rainstorms and release it slowly so that catastrophic or destructive floods or droughts are rare. Almost invariably sheep grazed, the margins support dense, neat lawns of bryophytes such as *Scapania irrigua*, *Philonotis calcarea* and *Cratoneuron filicinum*, and flowers such as blinks and marshwort. Their algal flora would repay investigation. A gravelly spring on Beacon Hill is the only Radnor site for the green, tissue-paper-like colonies of the freshwater alga *Monostroma* sp., which closely resembles the sea lettuce in form.

Travelling south across the heather and bracken-covered moorlands of Gwaunceste, Glascwm, Llanbedr and Llandeilo hills, outcrops increase in size until at Aberedw as Francis Kilvert so eloquently described them "the castled rock-towers and battlements and bastions of the Rocks of Aberedw" provide a botanical feast. Here shaded and sunny grey calcareous mudstone and shale crags and scree provide habitats for Welsh poppy, lesser teasel, intermediate enchanter's-nightshade, whitebeam, mossy and meadow saxifrage amongst a wealth of lower plants.

The nearby shady gorge beside the River Edw is only matched in interest by the Bach Howey gorge at Llanstephan. Physically more impressive and still not fully explored botanically, the calcareous shale cliffs there are covered in woodruff, wood melick, soft and hard shield-ferns and with alternate-leaved golden-saxifrage, wood brome and beech and oak fern. The rocks in the base of these cliffs, coloured both green and red, provide the first evidence of the dramatic changes which were taking place at the time of their deposition.

Earth movements finally after 80 million years, put an end to the Welsh basin. Uplift produced a large continent, 'St George's Land', to the north west and Radnor found itself on the edge of a Devonian sea stretching out to the south and east. Great rivers swept down iron-stained sand to form the foundation of the Old Red Sandstone.

The Downtonian beds, which outcrop in many of the wooded dingles of south east Radnor and give rise to the red soil of the lower Wye valley, provide evidence of this transitional period between the Silurian and Devonian and have been placed in both periods by different geologists. The rocks consist of flaggy sandstones, siltstones and thin limestone and cornstone beds. Notable plants of this area include hart's-tongue fern, goldilocks, wood stitchwort, giant bellflower and on cliffs by the Wye near Clyro, spurge laurel, ploughman's-spikenard and traveller's-joy. Springs on rock faces in the dingles deposit tufa around the mosses *Cratoneuron commutatum*, *C. filicinum* and *Eucladium verticillatum*.

Generally lime-rich and producing neutral to slightly base-rich soils on the high ground of the Begwns and Clyro hill, rock outcrops are sparse and tend, perhaps due to acidic air pollution, to support acidophilous lichens and mosses. Most springs are, however, lime-rich and calcareous peaty flushes have an interesting flora with marsh arrowgrass, knotted pearl-wort and the mosses *Campylium stellatum*, *Cratoneuron commutatum* var. *falcatum* and *Dicranum bonjeani*, all recorded from this area.

Downtonian outliers occur near Presteigne, between Norton and Stanage and in the Teme Valley, but seem largely to be buried by boulder clay and alluvium.

C. SUPERFICIAL DEPOSITS

Large areas of Radnor are covered in superficial deposits which isolate the flora from the influence of the underlying rock strata. This material has been created or transported by a number of processes. Active weathering is still taking place along a number of cliff lines, notably in the Elan and Claerwen Valleys, the acidic rocks producing mobile scree slopes largely devoid of vegetation. Once they reach a reasonably stable angle of repose dwarf shrubs such as bilberry, heather and bell heather, gorse and grasses colonise the material. Weathering of rocks was more active in periglacial times when extremes of temperature were more frequent. Large quantities of talus, the flaked off material, built up below most steep escarpments. Surface weathering of this material produced thin, freely-draining ranker soils. Perhaps best developed on south and east-facing slopes, they now support dwarf shrub heaths and scrub modified by burning and grazing to produce fescue and bent grasslands with heath bedstraw and tormentil. The wood and sticky groundsel are frequent in naturally unstable sites as well as those disturbed by rabbits, badgers and sheep. Woodland is undoubtedly the climax vegetation of stable talus slopes and many of the valley-side oak woods have developed on this material. On the more base-rich strata ash-oak woods form, often with a very species-rich ground flora. Leaching, however, freely occurs on the upper slopes removing lime salts, so acid soil loving plants tend to occur in the upper parts of these woods, eg. goldenrod, woodrush, woodsage, foxglove and oak fern. Lower slopes receive fragmented materials from the cliffs above and are more nutrient-rich, supporting, for example, yellow archangel, woodruff, false brome and shining crane's-bill. In glades and on scree, specialists such as the narrow-leaved bitter-cress, Welsh poppy and lesser teasel occur.

The very base-rich end of the range of skeletal soils are the Rendzinas which probably only occur on the Wenlock limestone at Dolyhir and perhaps on the igneous rocks of Stanner and Worsell Hills. They support plants such as the hairy violet, common rock-rose, marjoram, pale St John's-wort, the blue fleabane and mosses such as *Ditrichum flexicaule* and *Encalypta streptocarpa*.

The conditions provided by the screes and talus slopes are mimicked by gravel and sand deposits. Flood waters from the melting of the ice cap in late glacial times carried and sorted vast quantities of gravel and sand. In places these were piled up and recent excavations or river erosion has revealed their structure. Good exposures occur near Nantmel and below Dol y Fan near Newbridge on Wye in sand and gravel quarries. Disturbance and the open, freely-drained conditions favour plants such as broom. Rabbits enjoy the ease of burrowing and so short cropped turf with a wealth of yellow-flowered members of the Compositae such as hawkbits, dandelions and ragwort are often associated with these areas.

Deep gravel deposits are probably responsible for Radnor's vanishing stream - the Summergil Brook. It sinks into its bed near New Radnor during normal summer flows, leaving a dry river bed for some distance. Many of the small streams draining the Radnor Forest similarly disappear for short parts of their course in the summer months. Agricultural weeds are the major colonizers of the dry river channels. These disappearing streams may be the origin of the Welsh name of Maesyfed for Radnor which may be translated as "the drinking land".

Sand and gravel is being deposited today in places beside most streams and rivers. The Wye has produced the largest deposits with an outstanding shingle bed extending downstream from Glasbury bridge. Winter flooding and the constant deposition of fresh gravel provide areas free of scrub and tree growth allowing a wealth of herbaceous species to develop including yellow loosestrife, hemp agrimony and chives.

Superficial Deposits

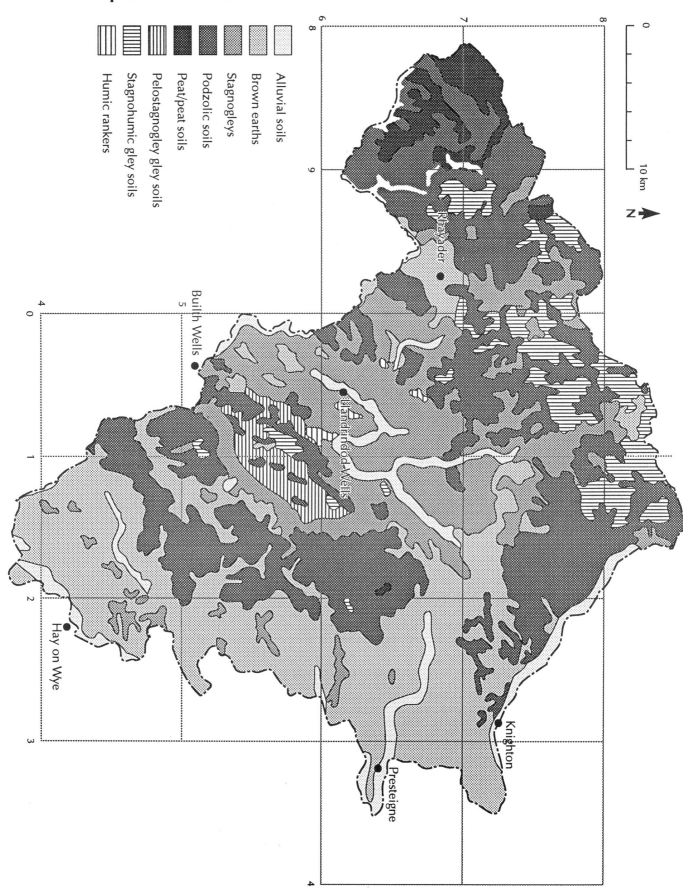

Map 2.3 The soils of Radnorshire

Superficial Deposits

The vegetation of these lowland river gravels can be contrasted with that of the extensive gravel beds beside the meanders of the upper Elan above Craig Goch Reservoir. In this latter site sheep grazing prevents any significant growth of herbs, except the sheep's sorrel and the resilient rosettes of a few composites. Bent and fescue grasses form small tussocks and lichens, notably *Coelocaulon*, *Cladonia* and *Baeomyces* species and the moss *Polytrichum piliferum* colonise the sandier parts.

The retreating ice at the end of the last ice age deposited quantities of boulder clay, particularly over the floors of many of the broader upland valleys. The middle and lower parts of the Ithon valley, the Marteg above Pant y Dwr and the Edw around Painscastle have deep deposits. The high clay and silt content results in poor drainage in parts and the formation of stagnogley and stagnohumic gley soils. The former soil types are mostly included in the Cegin soil association, whilst the latter, with its peaty surface layer, may be referred to the Wilcocks 2 association. (For greater detail on the distribution of soil types see the Soil Survey of England and Wales' 1:250,000 scale map sheet No. 2, Wales). Birch and alder woods develop on these soils, or where woodland has been lost, wet heaths, frequently dominated by purple moor-grass may form. Leaching of the upper horizons followed by the formation of peat at the surface creates stagnopodzols. Areas of this soil type typically support wet grass heath known locally as a rhos, eg. Llandegley Rhos with heath rush, cross-leaved heath, heath milkwort and mat-grass.

The uneven deposition of boulder clay drift as the ice melted created surface irregularities, many of which held water to form lakes. Llyn Gwyn, Gwyn Llyn, Llyn Heilyn and Llanbwchllyn are good examples. The growth of plants in some of the more acidic of these late-glacial lakes and the failure of their remains to decay filled them up, in part, with peat. The recent completion of this process is recorded in the name of Llyn, meaning a lake, near Newbridge on Wye. There is no open water there now, peat having partially filled the basin and a skin or "schwingmoor" of vegetation having grown to form a thin floating raft over deep water. The raft consists of bogmosses such as *Sphagnum recurvum*, *S. capillifolium* and *S. cuspidatum* with common hare's-tail cottongrass, cross-leaved heath, heather, white-beaked sedge and cranberry.

These mires, called basin mires also occur in thick glacial drift around Newbridge on Wye and Llandrindod Wells. The peat they contained has been cut in the recent past for fuel, this activity recorded in some of the place names such as Aberithon Turbary and Fedw Turbary near Newbridge on Wye. The cut over surfaces now support willow carr with bogbean, marsh cinquefoil and bottle sedge. Whilst these steep-sided basins were probably formed by the melting of large blocks of ice left buried in the drift other processes operating in these late glacial times formed another sort of basin - the pingo. Water welling up through the drift in tundra-like conditions froze. Expanding as it did so, it forced the boulder clay upwards and outwards. As the seasons changed, freezing and thawing added to the lens of ice and the quantity of material thrust out, until, when the climate grew warmer and the ice finally completely melted, a water-filled hollow had formed, surrounded by a low clay rampart. What appear to be pingos occur to the south east of Gwystre and in the Bachell Brook Valley, south east of Tyn-y-berth. Their vegetation is similar to that of the basin mires described above.

Dwerryhouse and Miller (1930) review the evidence of glacial activity in Radnor. They conclude from topographical studies and the disposition of identifiable boulders carried and later deposited by ice action that a large ice cap covered the Cambrian Mountains, filling the Wye, Irfon and Ithon Valleys. The ice spilled over the Sugar Loaf beyond Llanwrtyd Wells to the south west, over the Eppynt to the south, down the Wye and Teme Valleys and even east over the higher shoulders of the Radnor Forest to a height of at least 530m (1750 feet). The ice and its melt water cut a number of high altitude channels in these eastern Radnor uplands in a generally easterly to north easterly direction since the main valleys were still blocked with ice. The scooped out floors and unevenly deposited boulder clay created hollows which became filled with water. Llynheilyn lies in the floor of an overflow channel which once carried ice and water from the Edw down into the Summergill Valley. Its western margin sits on a rock rim whilst its eastern side is dammed by deep glacial clay deposits. Beilibedw Mawn Pool and the series of mawn pools to the south all owe their origins to similar processes, though many have been subsequently partially filled with peat. Most of Radnor's upland pools were probably created at this time. Further west, with the upper Wye valley blocked with ice, melt water was forced to flow east south east from Rhayader towards Nantmel. This broad valley is now occupied by streams far too small to have created it. Parts of the floor of the Black Brook were until recently ill-drained and peat-covered and botanically interesting. Similarly between the Carneddau Hills and Gilwern Hill the infant Colwyn Brook attempts to drain the bottom of a broad overflow channel to the north of Carneddau Farm. Peaty pastures have formed here with notable plants such as the greater butterfly-orchid and early marsh-orchid.

As the main valley glaciers melted they deposited the large quantities of clays, sand and gravel that they were carrying to form valleyside terraces. Often spread enevenly, this unsorted deposit of rounded to somewhat angular stones in a matrix of loam or clay has in the major river valleys been resorted by the rivers, with the finer silts being deposited in the bottoms of the more level valleys. These deep, stone-free silty soils are refered to the Teme, Lugwardine and Conway associations.

Wherever the speed of melting back of the glacier end matched the down valley flow of the ice, the front of the glacier remained stationary and large quantities of material formerly carried by the ice could be deposited to form moraines. In the Wye Valley below Clyro quantities of such glacial debris were deposited by the near stationary front of the Wye glacier to produce a large area of irregular hummocks and hollows. The seasonally damp hollows, in places, now support a wetland flora of great willowherb, bittersweet and comfrey. The soils here are coarse reddish loams of the Escrick 2 association.

Moraines formed temporary dams creating lakes in many of east Radnor's broad valleys. The level floor of the Gilwern Brook Valley from Kington to Dolyhir was probably a temporary lake bed as was that of the Summergil Brook below New Radnor. Wet depressions in the generally slightly seasonally waterlogged loamy soils of these valleys, (referable to the East Keswick 1 soil association) still retain an interesting wetland flora. This is perhaps best seen in the Radnorshire Wildlife Trust's Reserve at Burfa.

As streams cut through the damming moraines the lakes were drained. In the Bach Howey Valley the large shallow lake that had formed above Rhosgoch persisted longer and gradually became filled in with plant remains to form Rhosgoch Common. Studies there by D.D. Bartley (1960) demonstrate the presence of aquatic plant remains such as water-lily seeds embedded in lake sediments below the peat. This lake gradually filled up with peat. The rainfall has been enough to keep the vegetation and peat sufficiently wet to allow the surface of the bog to grow above the previous lake water level and form a raised mire. The dome of the mire shows a complex topography with bog moss filled pools, in part probably created by peat digging, set in wet dwarf-shrub heath of heather, cross-leaved heath, cottongrass and occasionally royal fern. The margin of the peat dome, called the lagg, is in places covered in willow carr with marsh cinquefoil, bogbean and narrow buckler-fern. Tussock sedge with tufts up to 1m. high occupy the more open areas, whilst seasonally open water at the south west end still supports a remnant lake margin flora of reed mace, water plantain, greater spearwort and the aquatic liverwort *Riccia fluitans*.

The most extensive areas of peatland in Radnor are the blanket bogs which first formed, probably in the Atlantic period, some 6000 years ago. Caseldine, A. (1990) and Moore, P.D. and Chater, E.H. (1969) review their history. Waterlogging of the soil on the high summit plateaux prevented plant remains decaying and peat formation began. This process may have been brought about by Neolithic inhabitants clearing the natural forest cover and establishing a heath-like vegetation. Forest brown earth soils lost their nutrients, becoming mor soils and a deep organic layer built up on the surface. The climate appears also to have become wetter and waterlogging of this layer initiated peat formation. Bogmosses, bogcotton, heather, bilberry and crowberry all contributed their remains to the peat in a process which still continues today in some sites. The best examples of blanket bog of the Crowdy soil association, occur with over 1m of peat on Ffos Trosol, Trumau and Gorsgoch in the Elan-Claerwen uplands and in the hills to the north west of Pant y Dwr near Drysgol. Notable plants include bog rosemary, lesser bladderwort, the liverwort *Mylia taylori* and the dung moss *Splachnum sphaericum* which colonizes damp sheep dung in the shallow peaty pools.

Blanket bog probably used to occur on the summit plateaux of many of the central and eastern hills such as Radnor Forest and Beacon Hill. Deep peat occurs on Black Mixen in the Radnor Forest as isolated baulks, looking very like the remnant of a former widespread deep peat cover. There is written evidence that this cover might have survived until fairly recent times. Mr James W. Lloyd reports in *British History Chronologically Arranged*, printed by John Wade that on 10 August 1800 - a particularly dry year, a person "digging for pitmar" set fire to the heath on Radnor Forest. The fire raged for 5 weeks, burning an area 30 miles in circumference, destroying, it is said, thousands of sheep and many cottages and was visible from Ludlow. This and other fires and peat cutting for fuel probably removed most of the peat. Disturbance of the bogs throughout Radnor by natural erosion and recently by all forms of human damage from ponies to motorcycles and heavy sheep grazing exposes bare peat. This now seems to lead to irreversible erosion. Naturally the growth of bogmosses should repair the damage but there is now ample evidence that many bogmoss species, which would normally recolonise the peat, can no longer do so because of their intolerance of the present levels of oxides of nitrogen and sulphur which pollute the rain, snow and mist.

Peat bogs less affected by pollution, drought and fire develop around springs and wet flushes. These soligenous peats, being more nutrient-rich, in general support a more diverse flora than the blanket bog peat. Jointed rush, blinks, lesser spearwort, white sedge, ragged robin, marsh valerian, marsh thistle, marsh pennywort and marsh St John's-wort are all widespread species of this habitat. Where the peat is flushed with base-rich water plants such as bog pimpernel, lesser skullcap, marsh arrowgrass and broad-leaved cottongrass can be found. In the absence of grazing many of these flushes would develop into alder carr. A few examples occur with smooth-stalked sedge, remote sedge, marsh-marigold, common valerian, meadow-sweet and reed canary-grass as frequent associates.

The high rainfall and low evaporation rates, so favorable to the formation of peat, have had a profound influence on the soils of Radnor. Avery, B.W., Findley, D.C. and Mackney, D. who produced a soil map of England and Wales in 1974 (published by the Director General of the Ordnance Survey, Southampton) map large areas of Radnor as stagnogley and brown podzolic soils. These acidic, ill-drained clay soils are regularly waterlogged and have been widely subjected to artificial drainage to improve their agricultural capabilities. Extreme examples of stagnopodzols with peaty surface horizons of the Hafren soil association are common in the Elan uplands and occur above around 500m on the drier Radnor Forest.

Stagnogleys verging on brown earth soils are more frequent on the boulder clay of the Ithon Valley. These brown pozolic soils, mainly of the Cegin association, are probably widespread in the wetter parts of Radnor. On the steeper slopes of the western uplands the soils are better-drained brown earths of the Manod association. Bracken favours these areas to be replaced by western gorse on the shallower, loamy ranker soils where bedrock lies close to the surface. In contrast, the favoured lowlands of east Radnor provide the more agriculturally valuable, deep, acidic to calcareous brown earth soils. In the lower Wye valley and around Presteigne the ORS rocks have produced the only significant areas of base-rich, red, sandy brown earth soils of the Milford, Escrick 1 and Bromyard associations. These, together with the riverside silty soils of the Lugwardine series support the only extensive areas of arable farmland in Radnor with a rich weed flora, including field bindweed, field woundwort and fumitory species.

D. Climate

Little has been written on Radnor's climate. Though Radnor covers a comparatively small area, the interactions of weather systems with its complex topography probably results in nearby localities experiencing different climatic conditions. The paucity of recording stations within Radnor makes quantification of these effects difficult, but the inhabitants are left in no doubt. One place may frequently be referred to as "one overcoat warmer" than another. There is a wry humour in naming clearly exposed upland farms "Treboeth" - the hot place. Yet at certain times of year, as will be demonstrated below, they may be significantly warmer than their apparently cosy neighbours in the valley below.

Temperature

The seasonal pattern of both air and soil temperatures varies from east to west across the county as well as with altitude. In the winter months mean daily temperatures, adjusted to sea level (at 0.6 degrees C for each 100m.) reveal the eastern part of the county to be colder than the west. In January 3.5 to 4 degrees C is the mean temperature in the east, whilst 4 to 4.5 degrees C prevails in the west. The ameliorating effect of the Irish Sea, some 25km. west of Radnor at its nearest point, probably accounts for this difference. By the summer this trend is reversed, with a mean daily July temperature of 16.5 to 17 degrees C in the east and 16 to 16.5 degrees C in the west of the county. Species which appear to favour the more oceanic climate of west Radnor are described in the biogeography sections below.

When the means of daily maxima and minima are considered, this probable oceanicity effect is seen in the minimum temperatures throughout the year, being around 1 degree C lower in the east compared to the west. In contrast the maxima figures are consistently higher by about 1 degree C in the east compared to the west, reflecting the greater fluctuations of temperature experienced by plants in east Radnor, in what must be a more continental type of climate.

The reduction of temperature with altitude also affects plant growth. Smith (1976) quotes mean air temperatures in Central Wales in January falling by 0.65 degrees C for each 100m. of altitude on western slopes, whilst on eastern slopes a figure of 0.45 degrees C might be used. By July the adjustment required is 0.55 degrees C for western and 0.8 degrees C for eastern slopes. Using these figures, in winter the top of the Radnor Forest would appear to be 2.6 degrees C cooler than Llandegley at its western foot. Not apparently a large amount, when taken over a year, plant growth is severely limited on hill tops. Most plants only start into growth when the air temperature reaches 5.6 degrees C. By multiplying the number of days the temperature rises above this level by the number of degrees C reached above 5.6 gives a measure of the ability of a site to support plant growth. These accumulated temperatures in degree days only reach between 825 and 1,100 in upland Radnor. The middle ground and upland valleys record accumulated temperatures of between 1,100 and 1,375, whilst the favoured lowest parts of the Wye and Teme valleys reach between 1375 and 1650 degree days. These figures might be contrasted with those from Gwent and the south of Wales where more than 1952 degree days may be accumulated per annum.

Plants which seem to demand high accumulated temperatures include white bryony and mistletoe. They are almost confined to the lowest parts of the county.

In the uplands the growing season can be very short. At Pant y Dwr in the Marteg Valley, north of Rhayader at 305m. (1000ft.), records kept by the Welsh Plant Breeding Station (WPBS) between 1967 and 1983 show an average of 1235 degree days accumulated above 6 degrees C per annum. The months June to September accounted for 70% of these degree days. April only provided on average 47 degree days. Spring comes late to upland Radnor and as the gardener well knows, is often accompanied by late spring frosts which can damage early plant growth, whilst autumn frost can prevent fruiting.

Frostiness typically increases with altitude but upland valley bottoms can suffer from radiation frosts. Heavier cold air rolls down the valley sides to flow along the floor of the valley. Like water this cold air can be ponded up by constrictions in valleys. If this air has drained off extensive areas of very elevated land it can cool valleys sufficiently to turn them into frost pockets. St Harmon, in the Marteg Valley is a well known example. Close to its confluence with the Wye the Marteg valley becomes constricted and cold air draining from the extensive areas of upland at over 400m. (1312ft.) finds difficulty escaping. WPBS records between 1967 and 1984 reveal that ground frosts can afflict this area every month of the year. July is the most frost-free month, yet there was a 60% chance of a ground frost even in this month. Air frosts occur in every month except July. In contrast at Gogerddan near Aberystwyth on the Ceredigion coast from 1954 to 1984 air frosts were unknown between the end of May and October.

The Ithon Valley is another frost pocket. The narrowing of the valley below Llandrindod Wells holds back cold air to create the "Llandrindod frost pocket". Records kept at Llandrindod between 1959 and 1972 show the latest air frost occuring on the 1 June, with a mean last date of the 4 May. At Aberystwyth these dates fall approximately a month earlier. The first autumn frost recorded at Llandrindod was on the 22 September with a mean date of the 17 October. Aberystwyth remains totally

Climate

frost free for another month, with a mean date two and a half months later.

John Goodger maintaining daily records, initially at Gwernyfed and later at Velindre, Brecknock in the Wye Valley, close to the Radnor border between 1972 and 1988 shows the lower Wye Valley to be very similar to Llandrindod as regards the times when frost was recorded and the Llandrindod figures may be a fair estimate of conditions prevailing in the major river valleys throughout Radnor.

Damage to plant growth by late spring frost can be regularly seen in many upland valleys and has been a notable feature of recent springs. Ash trees are particularly susceptible to late frost but damage can extend to young oak and alder foliage and bracken can show spectacular damage, with large areas turning brown. This colour change frequently picks out the valley bottoms, with plants on the upper slopes escaping damage and providing evidence that the upper slopes of valleys may remain frost free for significantly longer periods. The temperature inversions which produce these conditions of cold valley bottoms and warmer upper slopes are probably frequent occurrences in Radnor's hilly terrain.

Extreme cold can also kill or severely damage plants. The exceptionally cold winter of 1981-1982 caused die back of laurel and holly and the death of gorse and heather. Record low temperatures occured along the Welsh border. Newport in Shropshire at -26.1 degrees C held the British record, but temperatures not very different probably prevailed in many Radnor Valleys. John Goodger recorded -23.0 degrees C on the 14 January in Velindre, Brecknock not far from the Radnor Border. Between 1941 and 1970 the lowest temperature recorded at Llandrindod was -21.7 degrees C.

In the Climatological Atlas (1952) absolute minimum temperatures mapped out for the period 1901-1940 show central Radnor to be the coldest part of England and Wales. Only the Eastern Highlands of Scotland experience colder temperatures. A few lichens show an interesting distribution pattern which matches well the disposition of these areas experiencing extreme winter cold. They are *Catillaria globulosa*, *Cetraria sepincola*, *Ptychographa xylographoides* and *Parmelia stygia*.

The number of days with air and/or ground frost varies from year to year. At Pant y Dwr, on average, between 1967 and 1983, there were 74 days each year with air frosts; at Velindre between 1980 and 1988, 60; at Llandrindod between 1961 and 1970, 75. This compares with only 48 days between 1954 and 1983 at Gogerddan, on the Ceredigion coast. Ground frost occurs approximately one and half times as frequently.

Great heat is unusual but not unknown in mid Wales. The highest temperature recorded at Llandrindod (1941-70) was 30.6 degrees C. John Goodger recorded 33.9 at Gwernyfed in Brecknock on the 3 July in the exceptional summer of 1976. Mean maximum temperatures are probably more significant for plant growth. The hottest month is July. At Lyonshall, near the Radnor border in Hereford, the mean maximum (1947-70) was 20.1 degrees C. At Llandrindod (1941-70) it was 19.5 degrees C and at Pant y Dwr (1967-83) 18.5. Even allowing for the variations in height of the recording stations east Radnor is clearly a hotter place in summer than west Radnor.

Soil warmth may also affect plant growth. The transfer of heat through the soil depends on a number of factors. Dark, peaty soils for example absorb more of the sun's radiation than light coloured soils. Wet soils conduct heat better than dry soils. Slope and aspect greatly affect the amount of the sun's radiation that a soil can receive. At a depth of 10cm. the soil at Pant y Dwr and at Gogerddan, as measured by WPBS staff showed similar average temperatures of 3.9 degrees C in January. But at Gogerddan the soil warmed up more quickly in spring so that by May the soil was 2.6 degrees warmer, permitting earlier plant growth. A maximum temperature of 14.3 degrees was reached at Pant y Dwr in July (16.6 at Gogerddan), cooling through the autumn, yet being sufficiently warm in November at 6.6 degrees to permit limited root growth. Table 2.1. summarizes temperature data from three stations, ranging from Pant y Dwr in the west to Lyonshall, close to the eastern border of Radnor, but in Herefordshire.

Sunshine

Between 1945 and 1970 the length of bright sunshine was measured each day at Llandrindod Wells. An average of 1249 hours per year was recorded for this period and may be compared with 1474 hours at Aberystwyth

Climate

Table 2.1 Temperatures

	Jan	Feb	Mar	Apr	May	June	July	Aug	Sept	Oct	Nov	Dec	Year
Mean Monthly													
Llandrindod Wells [4]	2.7	2.9	5.1	7.7	10.7	13.6	14.9	14.8	12.7	9.6	5.7	3.7	8.7
Lyonshall [3]	3.1	3.3	5.5	8.3	11.1	14.1	15.6	15.2	13.1	10.0	6.1	4.1	9.1
Mean Monthly													
Pant-y-dwr [1]	5.2	4.5	7.1	9.9	13.3	16.5	18.5	18.3	15.4	12.0	8.0	5.6	11.2
Llandrindod Wells [2]	5.4	5.9	9.1	12.3	15.5	18.5	19.5	19.1	16.8	13.3	8.6	6.5	12.5
Lyonshall [3]	5.7	6.2	9.1	12.5	15.7	18.9	20.1	19.6	17.2	13.5	8.9	6.8	12.9
Mean Monthly													
Pant-y-dwr [1]	0.4	-0.5	0.9	1.7	7.7	9.7	9.5	7.8	5.8	2.5	0.8	0.8	4.3
Llandrindod Wells [2]	0.0	-0.1	1.2	3.2	5.8	8.7	10.4	10.5	8.5	5.9	2.8	1.0	4.8
Lyonshall [3]	0.5	0.5	1.9	4.0	6.5	9.4	11.1	10.8	9.1	6.5	3.4	1.4	5.4
Extreme Maxima													
Llandrindod Wells [4]	13.3	15.6	20.6	23.3	26.7	28.9	29.4	30.6	29.4	23.3	17.8	14.4	
Lyonshall [6]	13.3	16.1	21.1	22.2	26.7	31.7	31.1	28.9	28.9	24.4	16.1	15.0	
Extreme Minima													
Llandrindod Wells [4]	-21.7	-16.1	-13.9	-5.6	-2.8	-0.6	1.1	0.0	-2.2	-6.7	-7.8	-13.9	
Lyonshall [6]	-15.6	-14.4	-10.6	-3.9	-2.2	1.1	2.8	3.3	-0.6	-4.4	-6.7	-11.1	
Average No of													
Pant-y-dwr [1]	11.5	14.9	11.1	9.1	3.1	0.4	0.0	0.1	1.0	2.2	8.2	12.7	73.9
Llandrindod Wells [5]	13.8	14.8	13.1	5.7	1.1	0.2	0.0	0.1	0.3	1.8	8.2	15.5	74.6
Lyonshall [7]	21.7	13.2	10.0	3.2	0.5	0.0	0.0	0.0	0.1	0.6	6.2	12.7	59.2
Av. No of													
Pant-y-dwr [1]	15.8	17.7	17.4	15.0	6.5	1.8	0.6	1.1	2.5	6.2	11.6	16.5	111.6
Lyonshall [7]	22.5	20.3	19.9	13.0	5.3	1.1	0.3	0.5	1.8	6.0	15.4	21.4	127.5

1. Pant y Dwr (SN 998754), height 305m, records from 1967 - 83.
2. Llandrindod Wells (SO 061605), height 235m, records from 1941 - 70.
3. Lyonshall (SO 339576), height 155m, records from 1941 - 70.
4. As 2 but records from 1936 - 40 and 1947 - 70.
5. As 2 but records from 1961 - 72
6. As 3 but records from 1948 - 70.
7. As 3 but records from 1961 - 70.

Climate

on the Ceredigion coast, 1432 hours at Ross on Wye, Herefordshire or 1751 at Dale Fort near Pembroke. Llandrindod Wells is significantly cloudier, only enjoying as much sun as the Pennines or the east Scottish Highlands and receiving only 25% of the amount it would receive if clouds never obscured the sun.

In comparing the WPBS data for radiation measured in megajoules per square metre per day at Gogerddan, near Aberystwyth and Pant y Dwr, the latter site received about 10% less radiation on average throughout the year. The significance of cloudiness on plant growth is difficult to assess as it is so intimately connected to temperature and humidity. Plants known to favour a Mediterranean climate such as the lichen *Anaptychia ciliaris*, the moss *Bartramia stricta* and the liverworts *Targionia hypophylla* and *Riccia beyrichiana* are confined in Radnor to the valleys in the lee of the major hill blocks, probably receiving above average amounts of sunshine. (See Slater, F.M. (1990) for a discussion of this possible lee wave effect).

Table 2.2. Average total hours of bright sunshine at Llandrindod Wells (SO 051505) 1945-1970 *

	Monthly Average	Daily Average	Monthly Highest	Monthly Lowest
January	35.9	1.15	59.7	13.5
February	57.3	2.03	92.3	20.0
March	101.8	3.49	174.8	63.5
April	144.0	4.90	202.7	76.1
May	171.8	5.54	251.4	104.7
June	175.2	5.97	277.7	113.4
July	151.9	4.90	277.1	73.0
August	145.7	4.73	277.1	97.4
September	105.0	3.53	194.9	58.8
October	78.3	2.52	122.3	47.2
November	41.1	1.38	75.8	22.5
December	32.1	1.03	54.4	15.3
Year	1249.4	3.41	1542.2	1003.8

* From *Averages of Bright Sunshine for the UK, 1941-70*, Met. Office 0884, HMSO, London, 1976.

Average daily means of sunshine hours recorded at Llandrindod Wells reflect the changing length of day through the year. Minimum amounts of sunshine at 1.03 hours are recorded in December with a maximum of 5.97 hours in June. In Radnor's hilly terrain, however, many north-facing slopes never see the sun in the winter months and in the summer a steep north facing slope may receive under half the radiation of a comparable south-facing slope.

Precipitation and Humidity

Rainfall is very much influenced by topography, with the highest rainfall associated, in the main, with the highest ground. There is, however, a decreasing gradient from west to east, such that the Radnor Forest receives only slightly more than half the rainfall of the western Elan upland block at a similar height. The driest areas lie in the valley bottoms of east Radnor. The Teme valley below Knighton receives on average 825mm. of rain per year. Contast this with the 2030mm. received at the head of the Elan Valley in the west. Map 2.4 provides a general indication of the pattern of Radnor's rainfall. Table 2.3 indicates the monthly averages for a selection of sites. Throughout the county November or December is the wettest month, whilst June is the driest. These average figures hide considerable annual fluctuations. The Rev. D. Stedman Davies (1932) reviews the variation in rainfall recorded at Llandrindod Wells between 1883 and 1925, noting 1903 as an exceptionaly wet year with half as much rain again as in an average year. Notable dry summers have occured recently in 1976, 1989 and 1990 but unfortunately records are no longer kept at Llandrindod so comparisons cannot be made.

The uplands receive more rain than the lowlands and it rains more often. Most of upland western and central Radnor receives more than 0.2mm. of rain on between 220 and 230 days each year, though the Elan uplands may exceptionally have over 300 rain days in a wet year (Slater, F.M., 1974). In the drier valleys it rains on between 190 and 200 days per year. In the cooler uplands only about a quarter of the rain received is able to sink into or evaporate from the soil or through plants. A shortage of water seldom limits plant growth. In the lowlands of eastern Radnor almost two thirds of the rain received evaporates and significant moisture deficits can build up. A useful climatic summary is provided in the Ministry of Agriculture, Fisheries and Food Technical Bulletin 35, *The Agricultural Climate of England and Wales*, by L.P. Smith, HMSO, London 1976. Radnor lies within three Agroclimatic zones. 48s includes the Elan uplands along with most of north east Dyfed. 49s consists of south Montgomery, central and

Climate

Map 2.4 The Rainfall of Radnorshire (in mm)

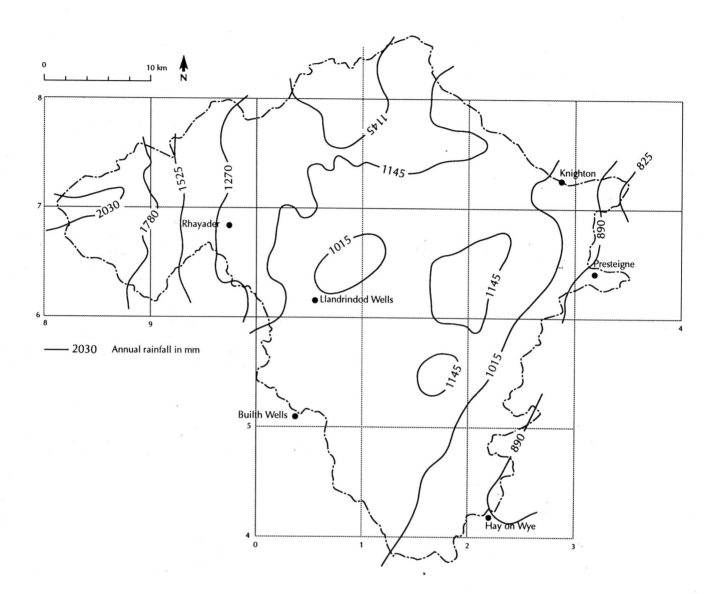

Climate

Table 2.3 Rainfall Averages 1941-70

Station	Grid Ref.	Alt (m)	Jan	Feb	Mar	Apr	May	Jun	Jul	Aug	Sep	Oct	Nov	Dec	Yearly Averag.
Cerrigllwydion	SN 843692	521	202	157	133	127	115	106	129	166	174	182	197	227	1915
Abergwngu	SN 870734	326	197	158	128	126	111	100	125	155	166	177	199	232	1874
Rhayader	SN 972680	213	134	101	83	79	89	70	71	98	110	113	137	146	1231
Bwlch y Sarnau	SO 031747	393	124	94	79	75	83	69	74	99	105	106	128	135	1171
Pen Ithon	SO 073813	366	120	94	79	77	89	71	79	102	105	105	128	131	1180
Llanddewi Ystradenny	SO 107685	234	98	76	66	65	74	64	75	91	95	91	109	105	1009
Llandrindod Wells	SO 061605	235	109	80	68	66	73	61	64	87	94	94	113	117	1026
Cefndyrys	SO 038530	229	103	77	66	63	74	63	67	90	94	100	113	110	1020
Llanbwlchllyn	SO 114475	436	109	82	73	69	84	69	72	96	100	105	122	118	1099
Llanstephan	SO 120420	134	104	78	71	67	80	66	67	91	95	101	118	114	1052
Glasbury	SO 181393	84	83	64	59	55	72	55	55	74	76	82	100	94	869
Lyonshall	SO 339576	155	82	58	59	55	71	58	66	84	77	77	95	83	865

south Radnor and north Brecknock whilst 25n covers the Hereford plain and includes east Radnor. Some of the data provided here are summarized in Table 2.4.

Table 2.4

	48s	49s	25n
Potential transpiration (mean annual in mm).	389	394	467
Rainfall (mean annual in mm)	1632	1190	746
Maximum summer soil moisture deficit (mm)	14	53	90
Frequency of irrigation need (years in 20)	0	0	17
Average date of return to soil capacity	indefinable	Sept 5th	Oct 28th
Average date of end of soil capacity	indefinable	May 27th	Apr 19th
Excess winter rain (mm)	1100	710	270

Data concerning snowfalls are sparse. At Llandrindod Wells between 1961 and 1970 snow lay at 9am on an average of 20 days per year. Snow and sleet fell on a similar number of days. At Aberystwyth snow lies on fewer than 5 days per year whilst in the Radnor uplands a figure of over 30 days might be expected. The availiable data are summarised in Table 2.5.

Table 2.5 Snow and Sleet (None recorded June to October inclusive)

Average number of days per year with :-

	J	F	M	A	M	N	D
Snow lying at 9am							
Llandrindod Wells	6.9	6.5	2.7	0.6	0	0.4	3.1
Aberystwyth	0.8	0.5	0.2	0	0	0	0.1
Snow or Sleet							
Llandrindod Wells	3.8	5.4	3.2	1.6	0.2	1.6	3.9
Aberystwyth	2.5	1.9	1.0	0.3	0	0.2	0.5

Wind

Bendelow and Hartnup (1980) in their *Bioclimatic Map of England and Wales* classify the upland two thirds of Radnor as exposed. The highest ridges probably experience mean annual windspeeds of 6.6m. per second (*National Atlas of Wales*, Board of Celtic Studies, University of Wales, 1980). Tree growth is almost entirely suppressed in these areas and the dominant vegetation is of dwarf shrubs. The middle slopes experience mean annual speeds of 4.8m. per second. Isolated trees here are deformed by the prevailing wind (usually from the south west). Only the lower valleys experience moderate shelter. Taylor (1980) reviews in detail the effects of altitude on climate. For each 100m. rise in altitude wind speed is likely to increase by between 20 and 30 percent.

Atmospheric pollution

This subject is dealt with in more detail below in the introductory section to lichens. With its sparse population and little heavy industry, local pollution sources are few. The most widespread local potential pollutants are probably nitrogen containing compounds released by agricultural operations. The increasing stock numbers, the nitrogen fixing clover leys and artificial fertilizers all leak nitrogen compounds to the air. Significant, but more distant sources of pollutants include power stations and road vehicles. They add to this mix, resulting in somewhat nutrient-rich rain and mist. The impact of these nutrients on wild plants is little understood.

Such information as exists is reviewed by Woodin and Farmer (1991). Naturally nutrient-poor systems such as acidic peatlands seem particularly at risk. The generally poor and degenerating condition of most of Radnor's blanket bogs may, in part, be caused by this form of pollution. Species-rich hay meadows and pastures are also threatened. The addition of readily soluble nutrients to such habitats tends to favour the growth of a few grass and herb species, markedly reducing species diversity.

Acidic pollutants reduce the base status of soils. Much of upland Radnor is formed of naturally base-poor rocks with a poor capacity to buffer acidic inputs. Whilst there is no direct evidence to link acidification with species loss, it is notable that a number of species which exhibit a preference for, at least, small quantities of bases in the soil seem to have declined or become extinct eg. mountain everlasting (now only common in mid Wales on limestone or Old Red Sandstone), field scabious and hoary plantain.

Climate

References

A) Geology and Superficial Deposits

Baker, J.W. and Hughes, C.P. (1973). Field meeting in Central Wales. *Proceedings of the Geologists' Association* 90, Parts 1 and 2, 65-75.

Bartley, D.D. (1960). Rhosgoch Common, Radnorshire: Stratigraphy and Pollen Analysis. *The New Phytologist*, 59, 238-262.

Caseldine, A. (1990). *Environmental Archaeology in Wales*. Dept of Archaeology, St David's University College, Lampeter.

Davies, J.H. et al. (1978). *Geology of Powys in Outcrop*. Powys County Council, Llandrindod Wells.

Davies, K.A. (1928). Contributions to the Geology of Central Wales. Notes on the Geology of the Southern Portion of Central Wales. *Proceedings of the Geologists' Association*, XXXIX, 157-168.

Earp, J.R. and Hains, B.A.(1971). *British Regional Geology. The Welsh Borderland*. HMSO, London.

Ellis, G.L. (1939). The Stratigraphy and Faunal Sucession in the Ordovician Rocks of the Builth-Llandrindod Inlier, Radnorshire. *Quarterly Journal of the Geological Society of London*, XCV, 383-442.

George, T.N. (1970). *British Regional Geology. South Wales*. HMSO. London.

Jones, O.T. and Pugh, W.J. (1941). The Ordovician Rocks of the Builth District. A Preliminary Account. *Geological Magazine*, LXXVIII, 185-191.

Kelling, G. and Wollands, M.A. (1969). The Stratigraphy and Sedimentation of the Llandoverian rocks of the Rhayader District. In *The Precambrian and Lower Paleozoic Rocks of Wales*, Wood,A. (edit). University of Wales Press.

Moore, P.D. (1978). Studies in the Vegetational History of Mid-Wales. V. Stratigraphy and Pollen Analysis of Llyn Mire in the Wye Valley. *The New Phytologist*, 80, 281-302.

Moore, P.D. and Chater, E.H. (1969). The Changing Vegetation of West-Central Wales in the Light of Human History. *Journal of Ecology*, 57, 361-379.

Owen, T.R. (1973). *Geology Explained in South Wales*. David and Charles. Newton Abbot.

Roberts, R.O. (1929). The Geology of the District Arround Abbeycwmhir (Radnorshire). *Quarterly Journal of the Geological Society*, LXXV, part 4, 651-676.

Rudeforth, C.C., Hartnup, R., Lea, J.W., Thompson, T.R.E. and Wright, P.S. (1984). *Soils and Their Use in Wales*, Soil Survey of England and Wales, Bulletin No. 11, Harpenden.

B) Climate

Bendelow, V.C. and Hartnup, R. (1980). *Climatic Classification of England and Wales*. Soil Survey Technical Monograph No. 15, Harpenden.

Climatalogical Atlas of the British Isles (1952). Meteorological Office, London.

Slater, F.M. (1974). *The Vegetation of Cors Fochno and other Welsh Peatlands*. PhD Thesis. University of Wales, Aberystwyth.

Slater, F.M. (1990). Biological Flora of the British Isles, No 168. **Gagea bohemica**. *Journal of Ecology*, 78, 535-546.

Smith, L.P. (1976). *The Agricultural Climate of England and Wales*. MAFF Technical Bulletin 35. HMSO, London.

Stedman Davies, Rev. D.(1932). The Rainfall at Llandrindod Wells. *Transactions of the Radnorshire Society*, 11, 18-21.

Taylor, J.A. (1976). Upland Climates. Ch.12, pp. 264-283. In Chandler, T.J. and Gregory, S.(Eds.). *The Climate of the British Isles*. Longman, London.

White, E.J. and Smith, R.I. (1982). *Climatological Maps of Great Britain*. Institute of Terrestrial Ecology, Cambridge.

Woodin, S.J. and Farmer, A.M. (eds.) (1991). *The effects of acid deposition on nature conservation in Great Britain*. Focus on nature conservation No. 26. Nature Conservancy Council, Peterborough.

Foel Tower, Elan Valley Reservoirs

3. The Plant Communities of Radnorshire.

No account of the flora of Radnor would be complete without some comment on the way individual plant species regularly interact with one another to form distinct communities. The Nature Conservancy Council recently commissioned Lancaster University to produce a classification system to describe the various types of higher plant-dominated vegetation found in England, Scotland and Wales. Dr. John Rodwell has now completed a draft of this National Vegetation Classification and the sections on woodland and mires have been published (Rodwell 1991a and 1991b). I am grateful to the former NCC and now Countryside Council for Wales for access to draft reports and to survey information which has enabled a very preliminary account of Radnor's higher plant dominated vegetation to be pieced together.

The NVC has yet to be extended to describe those entirely lower plant-dominated communities of, in particular, bark, wood and rock. For moss and liverwort-dominated communities Barkman (1958) has been followed, whilst for lichen-dominated communities James, Hawksworth and Rose (1977) has been followed. The following account is in no way comprehensive, but rather seeks to draw attention to the more notable vegetation types, either because of their abundance or rarity.

Woodland and Scrub

Much of the information on the woodland types of Radnor is drawn from a NCC report by Cooke and Saunders (1989). Sessile oak is the commonest woodland tree in Radnor and the **sessile oak/downy birch/ wood-sorrel woodland (W11)** is the commonest woodland type. It is widespread in Radnor on the deeper, well-drained acidic soils. Probably as a result of heavy grazing pressure, grasses such as creeping soft-grass and wavy hair-grass tend to dominate the ground layer and bluebells seldom provide the stunning displays which are a feature of this woodland type in less sheep-grazed parts of the country. Yet ferns manage to survive, especially on the steeper, rocky slopes and the **common buckler-fern**

Heather/Bell heather heath

sub-community is the commonest sub-community. Extensive stands occur in Coed yr Allt Goch and Cwm Gwynllyn Woods in the north west uplands as well as in more eastern woods such as at Timberhill, Lea Hall and in Gwernaffel Dingle. The **common stitchwort/ beautiful St John's-wort sub-community** is rare in mid Wales and may be confined to Radnor, where it occurs in areas of open canopy, on deep soil, often on more level ground above steep slopes as in Glannau, Aberedw, and Glanalder Woods. It must suffer a great range in temperature throughout the year, from sun-baked heat in the summer to considerable exposure in the winter. Perhaps best developed in this country in eastern Scotland it is another example of a continental element in Radnor's flora.

In complete contrast in the sheltered, moist valleys of north west Radnor mosses and liverworts dominate the ground layer to form the Atlantic sessile oakwood, now more precisely named in the NVC the **sessile oak/downy birch/Dicranum majus woodland (W17)**. Mosses and liverworts are most abundant in the **Isothecium/Diplophyllum albicans sub-community** which is generally developed on shallow soil, often over block scree. Fine stands occur in the Elan and Claerwen valleys, where Cwm Coel is the most extensive example. Dwarf shrubs are scarce due to sheep grazing in this sub-community, but where there is some reduction in grazing, as in small parts of Cwm Coel, bilberry and an abundance of tree saplings occur to form the **typical sub-community**. Heavier grazing pressure on somewhat deeper soils may lead to the formation of the grassier **sweet vernal-grass/common bent sub-community**, the commonest form of this woodland type in Radnor and one which merges with the sessile oak/downy birch/wood-sorrel community described above.

All the above types are confined to northern and western Britain. The southern and eastern British counterpart is the **oak/birch/wavy hair-grass woodland (W16)**, which favours acidic, seasonally dry and freely-drained soils. The **bilberry/common buckler-fern sub-community** is the most widespread sub-community in Radnor, with examples at Cefn Cennarth and Coed yr Allt Goch in the north west and Gumma, Timber Hill and Gwernaffel in the east. Climbing corydalis is a feature of this sub-community, particularly in those sites where the native canopy species have been replaced by larch. On the north slopes of Worsell Wood, near Stanner large stands of great wood-rush occur under Scot's pine in a rather interesting variant of this sub-community. The **common oak sub-community** is rare and has only been reported in a birch-dominated, probably secondary woodland at Impton.

The most frequent oak-dominated woodland type in lowland Britain is the **common oak/bracken/bramble woodland (W10)**. It is likely that the more level and now almost entirely disafforested parts of south east Radnor once supported this woodland type. Fragments of the **typical sub-community** survive along the upper edges of dingle woods such as Moity, Cilkenny, Aberedw and Gwernaffel woods. Here bramble, honeysuckle and bracken scramble over a ground layer of bluebell and wood anemone. The **sycamore/wood-sorrel sub-community**, generally developed on areas of more moist soil and with a greater abundance of dryopteroid ferns and herbs, such as common dog-violet and wood-sorrel occurs as large stands in some of the east Radnor woods such as Gumma and Cwm Gilla.

The shaded, acidic rocks of these woodlands support distinctive, if species-poor lichen-dominated communities such as the **Racodietum rupestris**, the **Opegraphetum horistico-gyrocarpae** and the **Micareetum sylvicolae**. On the trees and slowly decaying wood of these acidic woodlands are lower plant-dominated communities which show similar geographical trends to the woodland communities. Confined to the trunks and branches of trees in the wettest sessile oak/downy birch-*Dicranum majus* woodlands of the Elan and Claerwen Valleys is the **Parmelietum laevigatae**, whilst on decaying logs the **Sphenoboletum hellerianae** and **Hymenophylleto-Isothecietum myosuroidis** associations occur. On smaller branches the **Ulotetum crispae** and **Graphidietum scriptae** are common.

More widespread epiphytic woodland communities which extend into central and eastern Radnor include the **Pseudevernion furfuraceae** and the **Scoparieto-Hypnetum filiformis** on upper trunks and branches and the **Mnieto horni-Isothecietum myosuroidis** on lower trunks. Almost confined to the eastern margin of the vice-county on stumps and decaying wood is the **Dicraneto-Aulacomnietum**, whilst more widespread, but in similar habitats is the **Leucobryeto-Tetraphidetum**.

On the more base-rich soils ash becomes a common tree and the commonest woodland type is the **ash/field maple/dog's-mercury woodland (W8)**. This type is widespread in southern Britain and the finest examples in Radnor are found in the dingles of the lower Wye and Teme valleys on Downtonian rocks. The **herb-Robert sub-community** with an abundance of hart's-tongue fern, dog's-mercury and bramble is the most widespread sub-community and the one which extends the range of this woodland type into the edge of the uplands of the north and west. Fine examples occur in Cilkenny and Moity Dingle Woods where even in the depths of winter, pendulous sedge, hard and soft shield-ferns and the hart's-tongue ferns combine to produce a stunningly lush green display. In the spring herb-Paris, early-purple orchid, moschatel and goldilocks are notable plants of this sub-community.

Less abundant, though also widespread, is the **wood anemone sub-community**. It occurs in Cilkenny and Gwernaffel Dingle. Abundant bluebells and lesser celandines, as well as wood anemones mark out this sub-community. Ramsons dominates some woodland areas to form the **ramsons sub-community** such as in parts of Worsell Wood. On drier soils, particularly stable scree slopes, the **wood sage sub-community** replaces these more moisture-loving sub-communities. At Aberedw, Worsell and Stanner rare species such as pale St John's-wort, wood spurge, narrow-leaved bitter-cress and Welsh poppy enter this sub-community. Generally confined to south east Britain, the **primrose/ground ivy** sub-community has so far been noted only from Impton Wood.

Ash/rowan/dog's-mercury woodland (W9), though overlapping in part with the last woodland type, is absent from south east Britain and the Radnor examples lie at the south east edge of its British range. The **typical sub-community** is found in dry, nutrient-flushed areas in otherwise acidic oak woods or around the drier edges of alder/ash/yellow pimpernel woodland. Examples occur in many of the upland woods of north west Radnor, but also extend into the extreme east at, for example, Lower Woodhouse.

These base-loving woodland types also support a number of trees which probably only coincidentally have a base-rich bark eg. ash, hazel, wych elm, elder and field maple. Base-loving epiphytic communities tend therefore to be better developed in these woodlands. The **Lobarion** is well developed in Tyncoed Wood, in the Ithon gorge at Alpine Bridge and in the woodland beside Penygarreg Reservoir dam. The **Isothecietosum viviparari sub-association** of the **Mnieto horni-Isothecietum myosuroidis** is frequent on tree bases. Also in this habitat, but rarer following the death of wych elm, and now mostly confined to field maple, are the moss-dominated **Anomodontetum-Isothecietum** and lichen-dominated **Opegraphetum fuscellae**.

Of the woodland types of wetter ground **alder/ash/yellow pimpernel woodland (W7)** is widespread, but of no great extent, occupying the small wet-flushed areas within a range of other woodland types. Its three sub-communities are equally widespread. The **common nettle sub-community** with an abundance of golden saxifrage and creeping buttercup is well developed at Burfa Boglands and Pen y Berth Wood, whilst the **remote-flowered sedge/marsh thistle** and **tufted hair-grass sub-communities** can be found in woods such as Lakeside Wood, Glanalders and Fron Barn Dingles and on many stream and river banks. In this latter habitat the silt-encrusted tree bases may support one or more of the following charactoristic moss-dominated communities ie. the **Didymodonteto-Homalietum**, the **Tortuletum latifoliae** or an association of trunks in faster-flowing water, with an abundance of *Orthotrichum rivulare* and/or *O. sprucei*, this still requires formal recognition. These communities are much scarcer than formerly due to tree clearance work to improve agricultural drainage. Ash trees in these often warm, sheltered and relatively atmospheric pollution-free sites support the best developed examples of the **Parmelietum revolutae** on their well-lit upper trunks.

The **rusty willow/marsh bedstraw woodland (W1)** also frequently carries a distinctive lichen-dominated epiphytic community for similar reasons. The **Usneetum subfloridanae** in which *Usnea* species, often called beard lichens, clothe the twigs and trunks of the willows, is best developed where this woodland type occurs in sheltered sites such as on basin mires. Beside upland flushes in more exposed sites the willow bark carries the crustose lichen-dominated **Lecanoretum subfuscae**. This association extends onto downy birch in the **downy birch/purple moor-grass woodland (W4)**, a type which is widespread on wet, peaty soils in upland Radnor. The **soft rush sub-community** is the most common sub-community around peaty flushes, with marsh violet, soft rush, marsh valerian and tufted hair-grass. The **bog-moss sub-community** is more scarce but with good examples in Cwm Coel and at Cors y Llyn. Downy birch twigs in this latter site provide one of the best examples in Wales of the *Cetraria sepincola*-rich facies of the **Parmeliopsidetum ambiguae**, a community best-developed in the eastern Scottish Highlands.

Other wet woodland types include the **alder/tussock sedge woodland (W5)**, which is generally scarce and confined to some of the larger wet flushes and mire edges such as at Rhosgoch and Burfa Boglands, and the **rusty willow/downy birch/common reed woodland (W2)**. The **alder/meadowsweet sub-community** of this latter woodland type is also found at Rhosgoch, where the sub-community name species occur with bittersweet, marsh bedstraw and guelder-rose and show a transition, in places, towards the **Sphagnum sub-community** where bog-moss species become abundant and herbs become more scarce. The **alder/common nettle woodland (W6)** is scarce and is confined to the flood zone of the major river valleys. The **osier/almond-leaved willow sub-community** is perhaps best developed at Ffordd-fawr Mire near Glasbury, whilst the **crack willow sub-community** occurs as fragmented stands beside both the lower Wye and Teme. Nutrient-rich bark on the ancient trunks of these crack willows supports a rich

Plant Communities

epiphytic flora including the **Buellietum punctiformis**, the **Physcietum ascendentis** and the **Bryeto-Aulacomnietum**.

Scrubland.

In such a heavily grazed vice-county as Radnor scrub is confined to a few lightly, or ungrazed areas such as roadsides, derelict ground and steep banks. Hedges are common and are frequently well-maintained in the traditional Radnorshire crop and pleach style. Many of these linear patches of scrub can be refered to the **hawthorn/ivy scrub community (W21)**. On the more nutrient-rich soils of old farm dumps and around stockyards the **ivy/nettle sub-community** is found, with, in addition to ivy and nettle, cleavers and Yorkshire-fog. Elder is often frequent in these areas and its bark supports species-rich epiphytic communities such as the **Phyllantheto-Tortuletum laevipilae** and, rarely, the **Cryphaeetum arboreae**.

On base-rich soils in east Radnor the **Dog's-mercury sub-community** is frequent with flowers such as lords-and-ladies and mosses such as *Eurhynchium praelongum* and *Brachythecium rutabulum*. A sub-community transitional between the **wayfaring-tree** and the **slender false-brome sub-communities** occurs only on sunny, base-rich and thin-soiled banks such as at Dolyhir, Stanner Rocks and beside the trunk road north of Llanbadarn Fynydd. It is characterized by the presence of marjoram and wild strawberry with wood sage and common dog-violet.

Blackthorn/bramble scrub (W22) is widespread but uncommon, on unmown road verges, unused green lanes and ungrazed wood edges, where both bramble and blackthorn spread out vegetatively from existing colonies in adjacent hedges or woods. Further work is required to positively identify the sub-communities in Radnor.

The **bramble/Yorkshire-fog underscrub (W24)** develops similarly across ungrazed or unmanaged grassland. The **false oat-grass/hogweed sub-community** is found on well-drained railway and road banks, whilst the **creeping thistle/spear-thistle sub-community** is frequent on old tips and disturbed roadside banks. **Bracken/bramble underscrub (W25)** develops along woodland rides and around the edges of conifer plantations on the deeper soils of lowland Radnor. It includes species such as nettle, cleavers, bluebell, ground-ivy and wood soft-grass in the **bluebell sub-community** and wood sage, Yorkshire-fog and foxglove in the **wood sage sub-community**.

Steep banks and rocky outcrops are now some of the only areas left which support the **common gorse/bramble scrub (W23)**. It is, perhaps, too easily destroyed by burning and sheep become entangled all too easily in brambles for sheep farmers to tolerate its presence. The **sheep's sorrel sub-community** survives on sunny, freely-drained slopes, such as below Caban Coch Reservoir dam in the Elan Valley, where broom dominates ungrazed areas with early hair-grass and sheep's-bit. The **wood sage sub-community** is found on sunny and rocky slopes such as around Llanelwedd Quarries and at Stanner. In the former site spear-leaved willowherb occurs in its only Radnor locality. The **sweet vernal-grass sub-community** is of scattered occurrence on banks, particularly in south and east Radnor.

Heathland, Mires and Upland Rock Communities

The last scrub community grades into heathland on sites such as the south-facing slopes above Caban Coch Reservoir. Here **heather/western gorse heath (H8)** occurs amongst the scree and rock outcrops. Throughout the uplands on well-drained, sunny slopes grazing tends to reduce the abundance of heather and initially favours bilberry to form the **bilberry sub-community**. Short-cropped, almost turf-like growths of bilberry are common in such situations throughout Radnor. On commons such as Llanbister Common, grazing has all but eliminated bilberry to leave extensive stands of pure western gorse which, in the autumn, produces an unrivalled golden spectacle of flowers.

On limited areas of south-facing rock outcrops in the Elan and Marteg Valleys **heather/bell heather heath (H10)** occurs. Though, compared to more easily accessible areas, they are lightly grazed, the grazing is sufficient to prevent the dwarf shrubs from dominating and the **sheep's-fescue/sweet vernal-grass sub-community** is created. Annual hair-grasses and the electric-blue flowers of sheep's-bit are a feature of this sub-community.

Plant Communities

Considerable uncertainty surrounds the identity of many of Radnor's remaining heathland types. Naturally somewhat species-poor, the relative abundance of species is easily altered by management practices. Since in many cases it is only this variation in species abundance which determines the identity of a community, there appears to be much intergradation of types. The commonest heathland type on mineral soil is the **heather/bilberry heath (H12)**. The **heather sub-community** is the most widespread sub-community. Extensive stands, which colour the landscape purple in autumn, occur on Carreg Bica in the Elan uplands and on Aberedw, Llandeilo, Gwaunceste and Caety Traylow Hills. This sub-community is replaced by the **heath bedstraw/sheep's fescue sub-community** in places where grazing reduces the abundance of heather. On the lightly grazed summit plateau of the Radnor Forest the **cowberry/Cladonia impexa sub-community** occurs over wide areas. This attractive mixture of heather, cowberry, crowberry and bilberry can be found on the limited areas of deep peat as well as on mineral soils.

Natural succession following fires creates associations of species which vary with time. During the early stages lichens and bryophytes tend to dominate with scattered clumps of grass. Areas of the Radnor Forest burnt in 1976 show affinities with the **bilberry/Cladonia sub-community** of the **heather/wavy hair-grass heath (H9)**, a type common in the southern Pennines and English midlands. In the Elan-Claerwen uplands, in particular, fires followed by grazing tend to eliminate heather, whilst bilberry and in places, crowberry persist. This treatment has probably allowed **bilberry/wavy hair-grass heath (H18)** to extend onto the tops of low rocky ridges which would otherwise carry **heather/bilberry heath**. Its more natural habitat is probably north-facing slopes where bilberry naturally grows better than heather and this community persists in the absence of fires. Indeed the finest examples of the **Hylocomium splendens/Rhytidiadelphus loreus sub-community** with an abundance of hypnoid mosses and occasionally rarities such as alpine clubmoss are probably rarely, if ever, subject to fires. Particularly fine stands of this community with an abundance of the bog-moss **Sphagnum capillifolium** occur in the Marcheini uplands and show distinct affinities with the northern **heather/bilberry/Sphagnum capillifolium heath (H21)**.

Acidic rock outcrops amongst these heathlands can support a range of lichen-dominated communities. Most upland in their range of distribution are the **Fuscideetum kockianae** and the **Umbilicarion cylindricae** which occur in the Elan-Claerwen and Marcheini uplands. A notable *Umbilicaria deusta* nodum of this latter association is confined to south-facing rocks below Wyloer in the Marteg valley. Boulder scree and low rock outcrops amongst well-developed heath support both the **Parmelietum omplalodis** and the **Pertusarietum corallinae**, associations favouring well-lit, but humid sites. Rock outcrops can also offer some protection from sheep grazing. On ledges, around the edges of screes and on the sides of banks the **mountain fern/hard fern community (U19)** is frequent. Other acidic ledge communities await description, including those in the Elan Valley which support notable species such as English whitebeam and the dwarf male-fern.

The dwarf shrub-dominated communities of blanket bog, which have developed on the gently sloping summit plateaux of the wettest uplands in Radnor, show affinities with the heath vegetation. **Heather/hare's-tail cottongrass blanket bog (M19)** is scarce with scattered examples of the **cross-leaved heath sub-community** occuring in sites such as Carreg Bica in the Elan Valley and on the Radnor Forest. Intact and actively growing blanket bog is rare in Radnor. Most bogs are dissected by rapidly eroding drainage channels and are subject to regular burning, heavy grazing and the effects of atmospheric pollution. Intact areas remain in sites such as Gorsgoch, Trumau and Ffos Trosol where bog-mosses still make up a large part of the vegetation. These areas can be referred to the **Sphagnum magellanicum/bog rosemary sub-community** of the **cross-leaved heath/ Sphagnum papillosum raised and blanket mire community (M18)**. This is also the community which occurs on the central dome of the raised mire at Rhosgoch. Though *S. magellanicum* is found on only one site (see addendum), bog rosemary is known from a number of sites in the north west uplands where it occurs with leafy liverworts such as *Mylia taylori*, *M. anomala* and *Kurzia pauciflora*.

Other areas of blanket bog, in places grading into wet heath, are dominated by various proportions of deergrass, heath rush, hare's-tail cottongrass and cross-leaved heath.

Further work is required to describe the range of variability encompassed within **deergrass/ hare's-tail cottongrass blanket mire (M17), cross-leaved heath wet heath (M16)** and **deergrass/ cross-leaved heath wet heath (M15)**.

Pools in the bog surface support a species-poor bog-moss dominated vegetation. The commonest type is the **Sphagnum cuspidatum/ S. recurvum bog pool community (M2)**. The **Sphagnum recurvum sub-community** can form the even green, single species, floating lawns, which are a treacherous feature for the unwary traveller over blanket bog, or, as at Cors y Llyn, can develop a more species-rich form with, for example, cross-leaved heath, heather and cranberry. Small areas of this basin mire also support the **white-beaked sedge sub-community**. A less frequent, but well-developed bog pool community recorded, for example, at Gorsgoch, is the **Sphagnum auriculatum bog pool community (M1)**. Also species-poor, it may consist of little more than lawns of this bog-moss and scattered plants of bogbean and common cotton-grass. Further work is required on floating bog lawn communities. Turner et al. (1990) report small stands of **bog asphodel/Sphagnum papillosum valley mire (M21)** from a pingo at Crychell. Here bog moss species occur with common and hare's-tail cotton-grass, bog asphodel, cross-leaved heath and purple-moor grass to form a floating lawn of vegetation. The **cranberry/Sphagnum recurvum sub-community** is reported in the NVC from SO15, possibly from the mawn pools of Gwaunceste Hill.

Where peat, following erosion, has been deposited in wet or water-filled hollows or forms the substrate of shallow upland lakes the **common cotton-grass bog pool community (M3)** is frequent. Fine examples of this community can be found amongst the blanket bogs on the ridges between the rivers Elan and Claerwen. Sheep dung, deposited onto the wet peat of this community, can become the substrate for the dung moss *Splachnum sphaericum*. On lower ground the common cotton-grass bog pool community occurs in the water-filled pits of former peat workings as at Cors y Llyn and on Rhosgoch Common. In many sites it intergrades with the **bottle sedge/Sphagnum recurvum mire (M4)**. This is also a frequent community of the deeper seepage tracks in blanket bogs. Other seepage tracks receiving some groundwater input develop the somewhat more species-rich **marsh St John's-wort/bog pondweed soakway community (M29)**.

In the shallower seepage tracks of blanket bogs and in acidic wet flushes throughout the uplands the **star sedge/Sphagnum recurvum/ S. auriculatum mire (M6)** is common. A number of sub-communities and variants can be recognised. The commonest is the **soft rush sub-community** in which soft rush and the bog mosses *Sphagnum recurvum* and *S. palustre* frequently dominate the vegetation. The **jointed rush sub-community** is also widespread, as is the **star sedge sub-community**. In this latter type, rush species are rare, whilst a range of the smaller sedges, such as common sedge, carnation sedge and common yellow-sedge are frequent.

Where these shallow peaty flushes are fed with more lime-rich water the **dioecious sedge/common butterwort mire (M10)** is found. Dioecious sedge is only known from two sites in Radnor but the **common yellow-sedge/ bulbous rush sub-community** is widespread, occuring on many of the commons of central Radnor as well as in the north west uplands. Rare species found in this community include bog orchid and knotted pearlwort. It frequently grades into the **round-leaved crowfoot/blinks rill community (M35)** around spring-heads or on gravel or mineral soil beside spring-fed rills. Particularly well-developed examples occur on commons such as The Begwns and at Coxhead and Penybont. Where these often seasonal rills flow over rock outcrops the **Ephebetum lanatae** occasionally occurs, with such lichens as *Ephebe lanata* and *Massalongia carnosa*.

Of all the mire communities the most extensive is the **purple moor-grass/tormentil mire (M25)**. Abundant in the north west uplands, where its tussocky nature makes walking a miserable or comical operation (the gyrations induced in the stumbling walker have earned purple moor-grass its local name of disco grass), it becomes scarce in central and eastern Radnor. The **cross-leaved heath sub-community** is frequent on shallow peat on the more gently sloping ground, whilst the **sweet vernal-grass sub-community** is widespread on wet-flushed slopes. More lightly grazed flush areas such as around the edges of upland hay fields and in wet areas largely avoided by sheep in the hills support the **wild angelica sub-community**.

Plant Communities

Grasslands and Associated Communities.

Turner et al. (1990), working for the then NCC, examined in 1987 and 1988 over 40 enclosed grassland sites in Radnor. This account draws heavily on their work. **Sheep's-fescue/common bent/sheep's sorrel grassland (U1)** was not, however, sampled by them and nor were any samples collected by the NVC team. This is unfortunate since this interesting type undoubtably occurs in Radnor on banks with freely-drained, shallow, acidic to neutral soil. The **heath bedstraw/tormentil sub-community** is frequent on sunny banks, often around the edges of sessile oak-dominated woodlands or on banks with patches of common gorse. It also occurs amongst the drier stands of heather in the central and eastern uplands. It is at times separated with difficulty from the sheep's-fescue/common bent/heath bedstraw grassland (U4), which is considered in more detail below. This sub-community tends to support at least a few annual species such as early hair-grass or herbs which prefer somewhat mesotrophic conditions such as mouse-ear hawkweed and ribwort plantain. The **sweet vernal-grass/bird's-foot trefoil sub-community** is frequent on road verges and in churchyards where regular mowing keeps coarser species in check. It can also occur on shallower soil on rock outcrops, where drought resistant species such as lady's bedstraw and yarrow may be frequent.

Definition of the remaining sub-communities presents some problems. The **typical sub-community** is frequently very species-poor and most of the very heavily grazed, agriculturally unimproved grassland of the commons of central and east Radnor which consists of little else but sheep's-fescue, common bent, and sheep's sorrel is placed here. Grazing can be so heavy that even these hardy species decline and mosses such as *Brachythecium albicans* become an important component of the turf. Where bare soil appears, particularly in areas subject to summer drought, winter annuals such as common whitlow-grass, parsley piert, upright chickweed, small and common cudweed and shepherd's cress can develop. These summer-droughted communities are well-developed on the volcanic rocks between Llanelwedd and Llandegley and at Stanner Rocks. They are clearly related to the **common stork's-bill/shepherd's cress sub-community** described in the NVC from the Brecklands of East Anglia. Notable rarities of this sub-community in Radnor include perennial and annual knawel, maiden pink, upright and knotted clover, the Radnor lily and the moss *Bartramia stricta*. In pockets of slightly deeper soil at Stanner perennial species such as common rock-rose, rock stonecrop, sticky catchfly, bloody cranesbill and spiked speedwell occur in a sub-community that is possibly best placed for the time being in the **cat's-ear sub-community**. A more typical form of this sub-community, which appears to have its headquarters around the south western seaboard of Britain, consists of early hair-grass, English stonecrop, sheep's and red fescue and cat's-ear and is found on sunny rock outcrops in the north west uplands of Radnor. It occurs amongst heather/bell heather heath and intergrades with its sheep's-fescue/sweet vernal-grass sub-community. The otherwise maritime lichens *Anaptychia runcinata*, *Trapeliopsis wallrothii* and *Parmelia britannica* are confined to this habitat in Radnor.

Lower plants are an important component of the sheep's-fescue/common bent/sheep's sorrel grassland, particularly where it occurs in the many crevices of upland, acidic shale rock faces. The **Cornicularia aculeata/Cladonia arbuscula sub-community** has been created to receive a lichen-dominated sub-community of Breckland. Found on sand, the composition particularly of its higher plants is sufficiently different to prevent it being used to accommodate the lower plant-dominated ledges of Radnor. On these acidic ledges sheep's-fescue and sheep's sorrel occur with lichens such as *Cladonia subcervicornis*, *C. portentosa*, *C. furcata* and *Coelocaulon* species and mosses such as *Polytrichum piliferum*, *Racomitrium* species and *Pohlia nutans*. This community awaits formal description. A recently described (James et al., 1977) lichen-dominated community is found on sunny volcanic rocks, which in Radnor are often associated with sheep's-fescue/common bent/sheep's sorrel grassland. It is the **Lecanoretum sordidae**. Perhaps best-developed in the Welsh Marches, fine examples occur in Radnor below Cefnllys Castle, at Llanelwedd, on Old Radnor Hill and on Stanner Rocks. Crustose lichens such as *Lecanora rupicola*, *L. grumosa*, *Lecidea orosthea*, *L. sulphurea* are frequent, whilst the rock tripe, *Lasallia pustulata* can cover large areas of rock with its puppodum-like thalli.

Plant Communities

Sheep's-fescue/common bent/heath bedstraw grassland (U4) is common on the freely-drained soils of the heavily-grazed upland sheepwalks. The **typical sub-community** is commonest in the uplands and in many places may consist of little more than sheep's-fescue, common bent, heath bedstraw, tormentil and golden mosses such as *Pleurozium schreberi* and *Hylocomium splendens*. The attractive yellow to purple-brown, feather-like colonies of the leafy liverwort *Ptilidium ciliare* are frequent in this sub-community. On steeper, often less intensively grazed ground and around rock outcrops the **bilberry/wavy hair-grass sub-community** is found. On north-facing and shady sites mosses such as *Hypnum jutlandicum*, *Dicranum majus* and *Rhytidiadelphus* species may become a significant component of this vegetation.

The **Yorkshire fog-grass/white clover sub-community** occurs around flushes and in the more nutrient-rich hollows. On poorly-drained clay soils near Llanyre whorled caraway is frequent in this type of pasture. This sub-community also extends into the lowlands in permanent pastures on acidic soil. As well as the species which give this sub-community its name, sweet vernal-grass, red fescue and bird's-foot trefoil are frequent in the sward. On slightly more base-rich soils the **bitter vetch/betony sub-community** occurs rarely. In Radnor it lacks strict calcicoles such as salad burnet, supporting instead species, such as, mountain pansy, harebell and devil's-bit. plantain. Examples of the sub-community occur on slightly calcareous drift in pastures north of Llaithdu and around Llanbister and on a number of roadside banks, often associated with sessile oak woodland.

In many acidic upland sites sheep's-fescue/common bent/heath bedstraw grassland is separated with difficulty from **matgrass/heath bedstraw grassland (U5)**. The **species/poor sub-community** in particular may differ only in its greater abundance of matgrass. This latter species, being somewhat unpalatable to sheep, tends to replace more palatable grasses such as common bent in heavily-grazed sites. Other associated species include tormentil, wavy hair-grass and the mosses *Pleurozium schreberi* and *Rhytidiadelphus squarrosus*. In seasonally wet-flushed sites the hair moss *Polytrichum commune* becomes common, together with brown bent and heath wood-rush to form the **brown bent/Polytrichum commune sub-community**. It, too, is widespread in the uplands of Radnor. A usually more species-rich form of this grassland called the **heather/heath grass sub-community** with, in addition to these species, heath milkwort, sweet vernal-grass, tormentil and mosses such as *Pseudoscleropodium purum* is found on flushed ground around the edges of the heather moorland of central and eastern Radnor.

These uplands of central and eastern Radnor are built mainly of somewhat base-rich shale and mudstone rocks. Where they outcrop, as around the Radnor Forest and particularly around Aberedw, rock ledges support the base-loving **sheep's-fescue/common bent/wild thyme grassland (CG10)**. Most stands in Radnor are probably referable to the **white clover/field wood-rush sub-community**. Notable species found in this sub-community include mossy saxifrage and rock stonecrop, whilst calcicoles include lady's bedstraw, fairy flax, flea sedge and the mosses *Neckera crispa* and *Ctenidium molluscum*.

In complete contrast, on peat in the uplands, where grazing has been heavy **heath rush/sheep's fescue grassland (U6)** occurs. Only for a short time in the late spring do grazing stock nibble the heath rush. It remains largely uneaten for the rest of the year. Where it occurs in some abundance with bilberry it forms the **bilberry sub-community**. This sub-community is frequent on shallow peat on the more level ground. The **bog moss sub-community**, which tends to occur on deeper peat, is much rarer and is confined to the more intact areas of blanket bog in the Elan-Claerwen uplands. Also in this area on peaty mineral soils, often in hollows seasonally flushed with water, the **common bent/heath wood-rush sub-community** is found. In late summer the china-blue flowers of the creeping bellflower are a feature of this sub-community where it occurs beside the upper Elan and Claerwen rivers.

Bracken, with its widely-ranging rhizomes is a vigorous colonizer of many of the drier grassland communities on the deeper soils and finer talus slopes. Large bracken-dominated areas, particularly on the hill slopes of the commons of central and southern Radnor can, however, be separately placed in the **bracken/heath bedstraw community (U20)** of the NVC. Three sub-communities occur in the vice-county. On the more fertile soils in lowland areas and in more nutrient-rich hollows elsewhere the **sweet vernal-grass sub-community** is widespread. Less fertile soil on the middle ground of the hills supports the **species-poor sub-community**. Normally of no great interest to the field botanist, this sub-community, near Henllyn Mawr, Llandeilo Graban supports a population of both adder's tongue and moonwort. Towards the upper limit of bracken as a dominant species (around 450 metres) the **bilberry/Dicranum scoparium sub-community** may replace the species-poor sub-community before itself being replaced by heath, or where dwarf shrubs are kept in check by grazing or fire **wavy hair-grass grassland (U2)**. This latter community is far less common in Radnor than it is on the ORS of upland Brecknock, but stands of it occur particularly around the edges of heathland in central and east Radnor. Where grazing pressure has suppressed heather, yet bilberry survives, the **bilberry sub-community** is found, whilst where bilberry has also been eliminated the **sheep's-fescue/common bent sub-community** occurs.

Grass production forms the mainstay of Radnor's agricultural industry. To maximise production most fields have in the recent past been ploughed and the native grassland species have been replaced by ryegrass. Categorized in the NVC as **perennial ryegrass leys and related grasslands (MG7)** they are of little interest to the field botanist and are not considered further here. Similarly **perennial ryegrass/crested dog's-tail pasture (MG6)** tends to be of little botanical interest. The **typical sub-community** is found in most of the permanent and semi-permanent pastures, in churchyards, on mown roadside verges, lawns and sports fields throughout Radnor. The dominant species vary somewhat. The **meadow foxtail variant** is frequent, particularly on some of the deeper-soiled road verges in the east, where its orange-coloured flowers are a roadside feature in early summer. The **marsh foxtail variant** is frequent in damp gateways on clay soils and in seasonally wet hollows in fields, particularly on the flood plains of the major rivers. The **tufted hair-grass variant** occurs in wet-flushed hollows, most especially on the heavy clay soils. The **sweet vernal-grass sub-community** also occurs widely on shallower soil. It differs mainly in the greater abundance of sweet vernal-grass and fine-leaved fescues with field woodrush and cat's-ear as preferential species. It is found in pastures, churchyards, on hedgebanks and about rock outcrops and on steep banks in fields.

False oatgrass coarse grassland (MG1) is widespread in the lowlands on sunny, freely-drained and ungrazed places such as hedgebanks, roadverges and especially beside the railway. In these places scrub is kept in check by regular cutting. The sub-communities have not been studied in detail. The **common nettle sub-community** is found on the nutrient-rich soil of old tips and former farmyard manure storage areas and in nutrient-receiving sites such as old sunken lanes. The **red-fescue sub-community** is found on banks, such as the alluvial banks of the lower Wye and where road improvement schemes have created new embankments. Red-fescue often occurs in abundance with species such as white clover, creeping thistle, cock's-foot and yarrow. Where one particular species attains dominance above all others the NVC has recognised a number of variants, such as the rosebay willowherb variant and the meadow cranesbill variant. Others such as the colt's-foot variant and Japanese knotweed variant might be recognised in Radnor. The **common knapweed sub-community** is the most species-rich sub-community. Fine examples occur on sunny railway cuttings and embankments on the Central Wales Railway were it passes through the somewhat base-rich Ludlow beds, between Knighton and Dolau. Characteristic species include burnet saxifrage, yellow oat-grass, musk mallow, harebell and oxeye daisy as well as common knapweed and false oatgrass. Where rock outcrops amongst this vegetation it supports a number of uncommon calcicolous lower plants such as the mosses *Aloina aloides* and *Encalypta ciliata* and the lichen *Catapyrenium pilosellum*.

It is to be doubted that more than 200 botanically interesting agriculturally unimproved fields survive in Radnor. Many that have escaped reseeding have received artificial fertilizers or slurry, materials that stimulate the growth of grasses at the expense of many herbs and sedges. Turner et al. (1990) describe the following communities from a sample of 44 sites which were already known to be of some interest.

Crested dog's-tail/knapweed grassland (MG5) was the major hay-producing grassland on slightly acidic to neutral soil in Britain before being replaced by rye-grass ley in the post war period. Twenty-three MG5 grassland sites were found in Radnor.

Constantly occuring species included red-fescue, crested dog's-tail, ribwort plantain, Yorkshire fog and sweet vernal-grass with red and white clover, bulbous buttercup and autumn hawkbit being of almost equally frequent occurence. Compared to Britain as a whole cock's-foot and common bird's-foot-trefoil were far less frequent, whilst common mouse-ear was much more frequently encountered. The **meadow vetchling sub-community**, with stands recorded from 19 sites, is the most frequent sub-community. Whilst oxeye daisy and meadow vetchling tend to set this sub-community apart nationally, they are rather uncommon in Radnor and instead common sorrel and common mouse-ear provide a more reliable means of separating it from the next sub-community. Notable species include meadow saffron and adder's tongue. A group of 5 upland hay meadows in the Elan Valley support a distinctive type of vegetation which for the time being is placed in the meadow vetchling sub-community. The presence of species such as bluebell, pignut, wood bitter-vetch and lesser butterfly-orchid set them apart. Other species, such as tormentil and eyebright, provide a link with the **heath-grass sub-community**, which, though rarer, is also found in some of these fields. It is more easily characterized by the presence of species such as devil's-bit, carnation sedge and lousewort which prefer damp and acidic soils. The heavy boulder clay soils to the north west of Llandrindod support a number of examples of this sub-community.

The damp, alluvial soils of 3 sites in the Dolau-Penybont area of central Radnor supported what, on balance, were considered to be examples of **meadow foxtail/greater burnet flood-meadow (MG4)**. Though most of the typical MG5 grassland species were present, greater burnet, meadowsweet, meadow foxtail, meadow buttercup and common ryegrass were also frequent. These are all species which are characteristic of seasonally flooded hay meadows of the southern midlands of England, from where this community was described.

Wet, agriculturally unimproved pastures on peaty soils frequently support the **purple moor-grass/meadow thistle fen-meadow (M24)**. Turner et al. (1990) recorded this vegetation from 11 sites, mostly in north west Radnor. Grasses such as sweet vernal-grass, sheep's-fescue and matgrass were reported to be more frequent in Radnor sites than were generally present nationally, whilst herbs such as hemp-agrimony, wild angelica and common knapweed were more rare. Many stands seemed to be intermediate in character between the **typical sub-community** and the **jointed rush/cross-leaved heath sub-community**. Species which were preferential to the latter sub-community were sharp-flowered rush, marsh bedstraw heath spotted-orchid and sneezewort, whilst species more frequently encountered in the former sub-community included quaking-grass, marsh valerian and fen bedstraw. Notable species recorded from M24 fen-meadow include lesser and greater butterfly-orchid, southern marsh-orchid, petty whin and dyer's greenweed.

A distinctive and during the summer, colourful tall herb-rich vegetation develops in sunny, ungrazed places where the soil is moist and fairly fertile. Called the **meadowsweet/wild angelica mire (M27)** it occurs on damp road verges, beside mires and watercourses and in sunny glades in wet woodlands. The **common valerian/common sorrel sub-community** in addition to the species from which its name is derived, may, amongst others, include ragged robin, marsh-marigold and occasionally globeflower. It is widespread, particularly on the heavy glacial-drift soils of the Ithon and upper Bach Howey valleys. The **soft rush/Yorkshire fog sub-community** also occurs occasionally and favours, in particular, wet flushes in agriculturally unimproved pastures and damp silt beside rivers. Meadowsweet is less abundant here, perhaps kept in check by light grazing, whilst less palatable species such as soft rush and water mint are more abundant. Most of the flushes in which yellow flag is abundant might be placed here. Streamside stands support, in addition, hemlock water-dropwort, reed-canary grass and branched bur-reed.

Ill-drained, heavy clay soils also support another, frequently species-poor, community, the **soft rush/sharp-flowered rush/marsh bedstraw rush-pasture (M23)**. Of most botanical interest is the **sharp-flowered rush sub-community** which occurs in wet flushes in agriculturally unimproved pastures and on the lower slopes of the unenclosed hills, particularly in the north and west of Radnor. Characteristic species include, in addition to sharp-flowered rush, creeping bellflower, lesser skullcap, bog pimpernel, whorled caraway, devil's-bit scabious, greater bird's-foot-trefoil and marsh thistle. Many of these species show an Oceanic Western distribution pattern and this sub-community is characteristic of western and northern Britain. Attempts to drain this community and soil poaching by livestock leads to soil disturbance and an increase in the abundance of the more aggressive soft rush and a shift of this sub-community towards the **soft rush sub-community**.

Many once agriculturally improved fields are colonised by soft rush. Heavy clay soils, a wet autumn and large numbers of livestock create the sort of poached conditions which provide an ideal seedbed for soft rush. Amongst the soft rush clumps species such as cuckooflower, bog stitchwort, Yorkshire fog, brown bent, sweet vernal-grass and creeping buttercup grow to form the **typical sub-community** of the **Yorkshire fog/soft rush pasture (MG10)**. It is one of the few types of semi-natural vegetation that may at present be increasing in extent in Radnor following a reduction in agricultural drainage subsidies.

Swamps and tall herb fens

These habitats are widespread but uncommon in Radnor. Agricultural drainage and grazing have much reduced their extent, leaving fragments beside the major rivers and around the few remaining lowland mires.

Tussock sedge swamp (S3) with, in addition to this sedge, such flowers as wild angelica and marsh violet and the lady fern, survives as fragments in wet and peaty valley-bottoms and around the edges of mires, alder carrs and peaty lakes. The most extensive stand occurs along the north west edge of Rhosgoch Common, with individual plants of the tussock sedge growing to over one metre in height. The moss *Plagiothecium latebricola* preferentially grows on the fibrous bases of this sedge. The most accessible example of this vegetation type is found around the southern margin of Llandrindod Lake, where fishing platforms allow the examination of the vegetation from both the lake and landward sides. This lake was cut out of a peat bog in Victorian times and this tussock sedge swamp may be a remnant of its former vegetation.

Common reed swamp and reed-beds (S4) are very rare and no extensive examples are found in Radnor. Since they are also somewhat species-poor it has proved difficult to assign examples with certainty to sub-communities. Rusty willow is also colonizing examples at The Bog, Newbridge on Wye, at Rhosgoch and at Llanbwchllyn, creating stands of vegetation transitional between reed-bed and rusty willow/downy birch/common reed woodland (W2). Where common reed occurs at Llanbwchllyn with bogbean and greater spearwort stands might be assigned to the **bogbean sub-community**. This sub-community may also occur at Gwynllyn, Rhayader and on Rhosgoch Common.

Beds of the larger sedges occur sporadically through the vice-county. Lesser pond-sedge is the most frequent species, occuring in damp, former river channels in the major river valleys as well as occasionally in swamps and wet woodland. It is the major component of **lesser pond-sedge swamp (S7)**, where it occurs often with wild angelica, common valerian, meadowsweet and, particularly in the Ithon Valley, wood clubrush. At Llanbwchllyn lesser pond-sedge is reported to occur with greater pond-sedge. Further work is required to determine whether stands should be assigned to this community or the **greater pond-sedge swamp community (S6)**. In quiet reaches of the major rivers slender pond-sedge occurs with water mint, greater bird's-foot-trefoil, gypsywort and occasionally wood clubrush to form a distinctive community not formally recognised in the NVC.

The most widespread sedge-dominated community is the **bottle sedge swamp (S9)**. Almost pure stands of bottle sedge occupy the margins of peaty pools and may form floating lawns in hollows in the north west uplands. In somewhat more nutrient-rich sites it occurs with varying proportions of bogbean, marsh cinquefoil and water horsetail to form the **bogbean/marsh cinquefoil sub-community**. At sites such as Rhosgoch Common and Llanbwchllyn this sub-community intergrades with the **bottle sedge sub-community** of the **water horsetail swamp (S10)**. Only those stands consisting of little else but water horsetail can be safely separated as the **water horsetail sub-community** of this latter type of swamp. Good examples are found around the margins of Llyn Heilyn.

The **bladder sedge swamp community (S11)** (sub-community not determined) was reported from a wet field near Michaelchurch-on-Arrow by Turner et al. (1990). This field has recently been drained. The community hopefully survives elsewhere in wet hollows in east Radnor.

Bulrush swamp (S12) is found around the margins of many lowland pools. Often extending into deep water and demanding the use of a boat for detailed examination, a description of its sub-communities cannot yet be attempted. It seems probable that all four sub-communities descibed in the NVC will be found to occur in Radnor. Notable sites for bulrush swamp include Pentrosfa Bog, Pencerrig Lake and Rhosgoch Common.

Branched bur-reed swamp (S14) is frequent around the edges of ponds and slow-flowing streams in lowland Radnor. Almost pure stands of branched bur-reed, such as occur in the pond at Walton, may be referred to the **branched bur-reed sub-community**. The water-plantain sub-community occurs in shallow pools beside the Wye at Llanelwedd, where scattered plants of water-plantain and branched bur-reed fringe pools containing starwort and pondweed species. Other parts of these silty pools and elsewhere beside the slower reaches of the Wye support the more species-rich **water mint sub-community**, with plants such as gypsywort, creeping forget-me-not, bittersweet and common club-rush. In sites where this latter species becomes abundant the vegetation shows a transition towards the **common club-rush swamp (S19)**. Beside the Wye, for example, the **common club-rush sub-community** of this type of swamp vegetation replaces branched bur-reed swamp on silt encrusted stable shingle at the water's edge. Ephemeral pools occur on commons such as the Begwns and Llandeilo Hill. On their muddy bottoms common club-rush occurs with other species such as shoreweed, creeping bent, silverweed and water-purslane, showing affinities with both the **shoreweed sub-community** and the **creeping bent sub- community**.

Grant aid has been made available to landowners from a number of Government agencies in recent years to create ponds. Many of these, often shallow, newly-created clay-bottomed ponds have, to the disgust of their owners, been rapidly colonised by common sweet-grass and common duckweed to form **common sweet-grass swamp (S22)**. The species-poor **common sweet-grass sub-community** of this type of swamp is not a pretty sight and far from the flower-rich, limpid pool usually envisaged by its owner. In time its banks may become more attractive as marginal plants such as water mint, common bur-reed and water-plantain invade the site to form the **branched bur-reed/water mint sub-community**. Both of these sub-communities also occur beside slow-flowing streams and in muddy ditches in the lowlands. The **marsh foxtail sub-community** is frequent in seasonally flooded gateways and wet hollows in pastures.

The botanical exploration of **bottle sedge/marsh cinquefoil fen (S27)** is not for the faint-hearted. The interwoven rhizomes of these species, together with such common associates as bogbean, water horsetail and common cotton-grass create a floating raft of vegetation which grows out over water surfaces and may, or may not support the weight of the observer. Good examples occur around Gwynllyn, near Rhayader, whilst at The Bog, near Newbridge on Wye and at Rhosgoch Common vegetation now almost completely covers all the water. All stands may be referred to the **bottle sedge/water horsetail sub-community**, though they may differ somewhat in their associated species. At Rhosgoch, for example, ragged-robin forms a spectacular component of the vegetation in high summer.

Reed canary-grass fen (S28) occurs as fragmentary stands on silt banks beside streams, rivers and pools in ungrazed, or lightly grazed parts of the lowlands of Radnor. The **reed canary-grass sub-community** may consist of little but stands of this grass, though occasionally plants such as yellow loosestrife, bittersweet, common skullcap and meadowsweet occur as associates. Reed canary-grass, with its wiry rhizomes also colonises rock crevices and the spaces between stable cobbles in the flood zone of the major rivers such as the Ithon and Wye. In the Wye the wild chives frequently grows associated with this grass on rock outcrops.

Aquatic Communities

Whilst species lists have been prepared for many of Radnor's lakes, ponds and streams (eg. Lowther, 1987), detailed sampling with quadrats has yet to be undertaken. In consequence this account can only be of the most preliminary nature. However there can be no doubt that one of Britain's simplest plant communities-**the common duckweed** community (A2) occurs in Radnor. The **typical sub-community** may consist of little else but common duckweed and it occurs on a number of ponds in east Radnor. The ivy-leaved duckweed is rare in Radnor, occuring only in Llanbwchllyn and on Rhosgoch Common. In this former site it is possible that from time to time it occurs in sufficient abundance to form the **ivy-leaved duckweed sub-community**. In the latter site it occurs with the aquatic liverwort *Riccia fluitans* in a sub-community that may be transitional between the ivy-leaved duckweed sub-community and the **R. fluitans/Ricciocarpus natans sub-community**. The only other community in Radnor of unrooted flowering aquatic plants is the **rigid hornwort community (A5)**. Doubtfully native anywhere in Radnor, rigid hornwort has recently only been reported from an ornamental lake in Stanage Park, where it occurs with common duckweed to form the **common duckweed sub-community**.

Water-lilies are scarce in Radnor and may be native in only three lakes. The largest stands of the **white water-lily community (A7)** occur in an arc along the edge of deep water around the south west side of Llanbwchllyn. It is the only floating-leaved aquatic plant present in this area and this stand may be refered to the **species poor sub-community**. Yellow water-lily also occurs mixed with white water-lily in patches in this lake to form the **white water-lily sub-community** of the **yellow water-lily community (A8)**. This community also occurs in Gwynllyn, near Rhayader, where yellow water-lily occurs in some abundance in deep water to form the **species-poor sub-community**.

The **broad-leaved pondweed community (A9)** is uncommon in lakes and the deeper pools of clear, fast-flowing streams. In the former habitat it occurs as the **species-poor sub-community**. In lakes such as Llyn Gwyn, near Ysfa this community consists of little more than stands of broad-leaved pondweed. The **bulbous rush/alternate water-milfoil sub-community** favours the clearer water of headwater streams, particularly those flowing through agriculturally unimproved land. As well as the species from which its name is derived, it may also include small quantities of starwort species and occasionally bog pondweed.

The **amphibious bistort community (A10)**, consisting usually of little more than the somewhat rampant amphibious bistort with a little water horsetail and common spike-rush, dominates many of the quieter stretches of the River Ithon and is abundant around the margin of Llanbwchllyn. It produces in the summer a spectacular display of pink flowers with plants both floating on the water, and extending, particularly in a dry season, up the banks to become, as its name suggests, one of Britain's few truely amphibious plants.

The **fennel pondweed community (A12)** as described in the NVC has probably only been reported from Hindwell Pool in Radnor. This nutrient-rich farm pond supported stands of fennel pondweed and horned pondweed. Other stands of fine-leaved pondweed species, such as those of small pondweed in Llanbwchllyn and Llyn Heilyn and blunt-leaved pondweed in the pool north west of Gogia on The Begwns and in a pond near Upper Weston, Llangunllo, tend to occur in less nutrient-rich water. They may better be accommodated in the **small pondweed sub-community** of the **perfoliate pondweed/alternate water-milfoil community (A13)**.

The **alternate water-milfoil community (A14)** occurs frequently in pools in many of the upland streams flowing through agriculturally unimproved ground, such as on many of the commons in Radnor and occasionally around the edges of pools on the larger rivers. Alternate water-milfoil is frequently the only species present, though in the upland sites bulbous rush and intermediate water-starwort may be present. Slower-flowing streams may, in contrast, support the **common water-starwort community (A16)**. Common water-starwort is a very adaptable plant, growing as happily in water up to one metre deep as on damp mud. It is the commonest aquatic plant in wet farm ditches and shallow ponds as well as in the meandering streams which cut through silty banks in the lowlands of Radnor. It is often accompanied on the banks by fool's water-cress, hemlock water-dropwort and water-cress.

The white-flowered rafts of water-crowfoot species trailing in the current are a feature of the lower reaches of the larger rivers in Radnor. Until a detailed study has been undertaken to establish the abundance of both common water-crowfoot and stream water-crowfoot it is not possible to determine the relative abundance of the **stream water-crowfoot community (A18)** and the common water-crowfoot community (A19). A range of aquatic mosses such as the willow moss *Fontinalis antipyretica* are frequently the only associated plant species, apart from algae.

Shallow, mud-bottomed pools, many of which in a dry summer may almost completely dry up, occur on a number of the heavily grazed commons in Radnor. They support a distinctive vegetation of pond water-crowfoot, pillwort, lesser marshwort, bulbous rush, water purslane, starwort species, creeping bent and sweet-grass species. Slater, Helmsley and Wilkinson (1991) have recently described a new sub-association-the **Apietosum inundati** of the Association **Pilularietum globuliferi** of Tuxen to accomodate this vegetation. It appears to differ significantly from the **pond water-crowfoot community (A20)** of the NVC and further work is required to establish the relationship of this vegetation to a described NVC type.

The **shoreweed/water lobelia community (A22)** is very rare and possibly only occurs on the pebble bottoms of two upland acidic lakes between the Claerwen and Elan Valleys at almost 500 metres. In both Llyncerrigllwydion Isaf and Llyncerrigllwydion Uchaf water lobelia occurs with the aquatic form of bulbous rush. Quill-wort is washed up on the shore of this latter lake from time to time and it seems likely that in deep water this site supports the **quill-wort community (A23)**, which is well-developed and easily observed elsewhere in Dolymynach Reservoir in the Claerwen Valley.

The final NVC community considered here is the **bulbous rush community (A24)**. It is frequent in peaty pools, runnels and ditches in the uplands where, associated with the bog moss *Sphagnum auriculatum* var. *auriculatum* and occasionally bog pondweed, it forms the **S. auriculatum sub-community**.

Ruderal Communities

At the time of writing this account the NVC chapter on ruderal communities was not available. A brief account is given here of the more common or notable plants of such man-made habitats as walls, gardens and arable land. The plants of roadsides, railways, quarries and lead mines are discussed elsewhere in the introduction to this flora.

Walls

Drystone walls, built of local stone generally support a flora closely similar to that found on native rock outcrops of the same rock. Since field boundary walls are rare in Radnor, whilst rock outcrops are relatively common, drystone walls make a small contribution to the floral diversity of the vice-county. A few notable species have, however, only been found on drystone walls or worked stone. They include the lichens *Lecanactis dilleniana* and *Candellariella medians*.

Old lime-mortared walls can, in contrast, support a scarce and distinctive flora. Such walls tend to be more frequent in towns and villages. Associated often with gardens they provide a habitat for naturalized cultivated species which are scarcely ever found in any other habitat eg. snow-in-summer, garden arabis, wallflower, red valerian, ivy-leaved toadflax, yellow corydalis, aubretia and white, biting and reflexed stonecrop. They occur alongside more widespread native species such as wall rue, maidenhair spleenwort, herb robert, navelwort, barren strawberry, mouse-ear halkweed, ivy and wall lettuce, the moss *Tortula muralis* and lichens *Lecanora dispersa*, *Caloplaca holocarpa*, *C. citrina* and *Diplotomma alboatrum*.

A few native species, such as flattened meadow-grass, rusty-back fern and the moss *Encalypta streptocarpa*, are more common on lime-mortared walls than in natural habitats. The mosses *Barbula revoluta*, *B. tophacea* and *Bryum radiculosum* are only known from mortar. Other calcicoles recorded rarely from walls which are generally scarce in natural habitats in Radnor include hairy rock-cress and rue-leaved saxifrage. They occur on crumbling mortar between limestone boulders used to cap mudstone walls, a combination which appears to be a unique feature of walls in New Radnor. This town, together with Presteigne and Knighton have the most species-rich and interesting walls in Radnor. Towns such as Llandrindod, in contrast, are largely built of brick and though shaded walls, such as cellar recesses support a well-developed fern flora of eg. wall-rue and hart's-tongue fern, other walls support a very impoverished flora.

The use of Portland cement to repoint once lime-mortared joints in walls is a source of concern. This harder material prevents the growth of most higher plants until at least a few cracks develop. It is also a less satisfactory substrate for lower plants, though is commonly exploited by mosses such as *Tortula muralis*, *Grimmia pulvinata* and *Schistidium apocarpum* and the lichens *Lecanora dispersa*, *Aspicilia contorta*, *Protoblastenia rupestris* and *Caloplaca* species. These species are also common on solid concrete structures such as bridges, kerbing, fence posts etc. Despite its widespread use, only the moss *Orthotrichum cupulatum* and the lichen *Bacidia caligans* are known exclusively from concrete in Radnor.

Gardens

No long-cultivated garden is free of weeds. They may have been eliminated on the surface, but under the soil resides a bank of seed awaiting favourable conditions for germination. The commonest annuals in Radnor are red dead-nettle, common bitter-cress, thale cress, redshank, common field-speedwell, green field-speedwell, cleavers, annual meadow-grass, rough-stalked meadow-grass, and groundsel. No two gardens ever seem to be alike and the relative abundance of species probably changes from year to year. New species become weeds from time to time and occasionally, and for the gardener not often enough, some species decline in abundance. Slender speedwell is one plant that has greatly expanded its range recently whilst grey field-speedwell has not been refound in this survey and the only garden in which annual mercury was known has been turned into a carpark.

Perennial weeds are fewer in number. Ground elder and common couch-grass occur throughout the county, the latter being given the local name of scutch. Hedge bindweed is widespread but not common, whilst the beautiful, but insidious hairy bindweed is rare. In the wetter parts of Radnor creeping buttercup, dandelion and broad-leaved dock are frequent.

Arable fields

There is much overlap between the weed flora of gardens and fields. All the common garden weeds occur in fields. Some of the arable field weeds are, however, less frequent in gardens eg. white campion, sun spurge, long-headed poppy, common ramping fumitory, field pansy, wild pansy, scented mayweed, scentless mayweed and field woundwort, whilst some eg. corn marigold and large-flowered hemp-nettle have only been found as arable weeds. Fields in which root crops are grown for stock feeding are often rich in weed species, the use of herbicides on cereals greatly reducing the value of that crop for weeds. But perhaps the richest places for weeds are where soil dug from some considerable depth has been spread. Notable weed floras with such uncommon species in Radnor as annual nettle and small-flowered catchfly have been found on recently constructed roadside verges built up with soil from deep excavations elsewhere.

A specialised flora develops on compacted soil in gateways. Typical species include pineapple weed, knotgrass, broad-leaved plantain and lesser swine-cress.

A Conspectus of the Plant Communities of Radnorshire with their English and Latinized Names, Numbers and Lettering derived from the National Vegetation Classification.

Woodlands

W1 **Salix cinerea-Galium palustre** woodland
　　(Rusty willow/marsh bedstraw woodland)

W2 **Salix cinerea-Betula pubescens-Phragmites australis** woodland
　　(Rusty willow/downy birch/common reed woodland)
　　　　a: **Alnus glutinosa-Filipendula ulmaria** sub-com.
　　　　　(Alder/meadowsweet sub-community)

　　　　b: **Sphagnum** sub-community
　　　　　(Bogmoss sub-community)

W4 **Betula pubescens-Molinea caerulea** woodland
　　(Downy birch/purple moor-grass woodland)
　　　　b: **Juncus effusus** sub-community
　　　　　(Soft rush sub-community)

　　　　c: **Sphagnum** sub-community
　　　　　(Bogmoss sub-community)

W5 **Alnus glutinosa-Carex paniculata** woodland
　　(Alder/tussock sedge woodland)
　　　　Sub-communities not determined

W6 **Alnus glutinosa-Urtica dioica** woodland
　　(Alder/common nettle woodland)
　　　　b: **Salix fragilis** sub-community
　　　　　(Crack willow sub-community)

　　　　c: **Salix viminalis/triandra** sub-community
　　　　　(Osier/almond-leaved willow sub-com.)

Plant Communities

W7 **Alnus glutinosa-Fraxinus excelsior-Lysimachia nemorum** woodland
(Alder/ash/yellow pimpernel woodland)

   a: **Urtica dioica** sub-community
      (Common nettle sub-community)

   b: **Carex remota-Cirsium palustre** sub-com.
      (Remote-flowered sedge/marsh thistle sub-com.)

   c: **Deschampsia cespitosa** sub-community
      (Tufted hair-grass sub-community)

W8 **Fraxinus excelsior-Acer campestris-Mercurialis perennis** wood
(Ash/field maple/dog's-mercury woodland)

   a: **Primula vulgaris-Glechoma hederacea** sub-community
      (Primrose/ground ivy sub-community)

   b: **Anemone nemorosa** sub-community
      (Wood anemone sub-community)

   e: **Geranium robertianum** sub-community
      (Herb-Robert sub-community)

   f: **Allium ursinum** sub-community
      (Ramsons sub-community)

   g: **Teucrium scorodonia** sub-community
      (Wood sage sub-community)

W9 **Fraxinus excelsior-Sorbus aucuparia-Mercurialis perennis** wood
(Ash/rowan/dog's-mercury woodland)

   a: Typical sub-community

W10 **Quercus robor-Pteridium aquilinum-Rubus fruticosus** woodland
(Common oak/bracken/bramble woodland)

   a: Typical sub-community

   e: **Acer pseudoplatanus-Oxalis acetosella** sub-community
      (Sycamore/wood-sorrel sub-community)

W11 **Quercus petraea-Betula pubescens-Oxalis acetosella** woodland
(Sessile oak/downy birch/wood-sorrel woodland)

   a: **Dryopteris dilatata** sub-community
      (Common buckler-fern sub-community)

   d: **Stellaria holostea-Hypericum pulchrum** sub-community.
      (Common stitchwort/beautiful St John's-wort sub-community.)

W16 **Quercus** spp.-**Betula** spp.-**Deschampsia flexuosa** woodland.
(Oak/birch/wavy hair-grass woodland)

   a: **Quercus robur** sub-community
      (Common oak sub-community)

   b: **Vaccinium myrtillus-Dryopteris dilatata** sub-community
      (Bilberry/common buckler-fern sub-com.)

W17 **Quercus petraea-Betula pubescens-Dicranum majus** woodland
(Sessile oak/downy birch/**Dicranum majus** wood)

   a: **Isothecium myosuroides-Diplophyllum albicans** sub-community

   b: Typical sub-community

   c: **Anthoxanthum odoratum-Agrostis capillaris** sub-community
      (Sweet vernal-grass/common bent sub-com.)

Scrubland

W21 **Crataegus monogyna-Hedera helix** scrub
(Hawthorn/ivy scrub)

   a: **Hedera helix-Urtica dioica** sub-community
      (Ivy/common nettle sub-community)

   b: **Mercurialis perennis** sub-community
      (Dog's-mercury sub-community)

   c: **Brachypodium sylvaticum** sub-community
      (Slender false-brome sub-community)

W22 **Prunus spinosa-Rubus fruticosus** scrub
(Blackthorn/bramble scrub)
   Sub-communities not determined

W23 **Ulex europaeus-Rubus fruticosus** scrub
(Common gorse/bramble scrub)

   a: **Anthoxanthum odoratum** sub-community
      (Sweet vernal-grass sub-community)

   b: **Rumex acetosella** sub-community
      (Sheep's sorrel sub-community)

   c: **Teucrium scorodonium** sub-community
      (Wood sage sub-community)

W24 **Rubus fruticosus-Holcus lanatus** underscrub
(Bramble/Yorkshire-fog underscrub)

   a: **Cirsium arvense-Cirsium vulgare** sub-com.
      (Creeping thistle/spear-thistle sub-community)

   b: **Arrhenatherum elatius-Heracleum sphondylium** sub-community
      (False oat-grass/hogweed sub-community)

W25 **Pteridium aquilinum-Rubus fruticosus** underscrub
(Bracken/bramble underscrub)

   a: **Hyacinthoides non-scripta** sub-community
      (Bluebell sub-community)

   b: **Teucrium scorodonium** sub-community
      (Woodsage sub-community)

Mires and Heaths

M1 Sphagnum auriculatum bog pool community

M2 Sphagnum cuspidatum/recurvum bog pool community

    a: **Rhynchospora alba** sub-community
      (White-beaked sedge sub-community)

    b: **Sphagnum recurvum** sub-community

M3 Eriophorum angustifolium bog pool community
    (Common cotton-grass bog pool community)

M4 Carex rostrata-Sphagnum recurvum mire
    (Bottle sedge/**Sphagnum recurvum** mire)

M6 Carex echinata-Sphagnum recurvum/auriculatum mire
    (Star sedge/**Sphagnum recurvum/auriculatum** mire)

    a: **Carex echinata** sub-community
      (Star sedge sub-community)

    c: **Juncus effusus** sub-community
      (Soft rush sub-community)

    d: **Juncus acutiflorus** sub-community
      (Jointed rush sub-community)

M10 Carex dioica-Pinguicula vulgaris mire
    (Dioecious sedge/common butterwort mire)

    a: **Carex demissa-Juncus bulbosus/kochii** sub-community
      (Common yellow-sedge/bulbous rush sub-com.)

M15 Scirpus cespitosus-Erica tetralix wet heath
    (Deergrass/cross-leaved heath wet heath)
    Sub-communities not determined

M16 Ericetum tetralicis wet heath
    (Cross-leaved heath wet heath)
    Sub-communities not determined

M17 Scirpus cespitosus-Eriophorum vaginatum blanket mire
    (Deergrass/hare's-tail cottongrass blanket mire)
    Sub-communities not determined

M18 Erica tetralix-Sphagnum papillosum raised and blanket mire
    (Cross-leaved heath/**Sphagnum papillosum** raised and blanket mire)

    a: **Sphagnum magellanicum-Andromeda polifolia** sub-community
      (**Sphagnum magellanicum**/bog rosemary sub-community)

M19 Calluna vulgaris-Eriophorum vaginatum blanket mire
    (Heather/hare's-tail cottongrass blanket bog)

    a: **Erica tetralix** sub-community
      (Cross-leaved heath sub-community)

M21 Narthecium ossifragum-Sphagnum papillosum valley mire.
    (Bog asphodel/**Spahagnum papillosum** valley mire)

    a: **Rhynchospora alba-Sphagnum auriculatum** sub-community
      (White-beaked sedge/**Sphagnum auriculatum** sub-community)

    b: **Vaccinuum oxycoccos-Sphagnum recurvum** sub-community
      (Cranberry/ **Sphagnum recurvum** sub-com.)

M23 Juncus effusus/acutiflorus-Galium palustre rush pasture
    (Soft rush/sharp-flowered rush/marsh bedstraw rush pasture)

    a: **Juncus acutiflorus** sub-community
      (Sharp-flowered rush sub-community)

    b: **Juncus effusus** sub-community
      (Soft rush sub-community)

M24 Molinia caerulea-Cirsium dissectum fen-meadow
    (Purple moor-grass/meadow thistle fen-meadow)

    b: Typical sub-community

    c: **Juncus acutiflorus-Erica tetralix** sub-com.
      (Jointed rush/cross-leaved heath sub-com.)

M25 Molinia caerulea-Potentilla erecta mire
    (Purple moor-grass/tormentil mire)

    a: **Erica tetralix** sub-community
      (Cross-leaved heath sub-community)

    b: **Anthoxanthum odoratum** sub-community
      (Sweet vernal-grass sub-community)

    c: **Angelica sylvestris** sub-community
      (Wild angelica sub-community)

M27 Filipendula ulmaria-Angelica sylvestris mire
    (Meadowsweet/wild angelica mire)

    a: **Valeriana officinalis-Rumex acetosa** sub-com.
      (Common valerian/common sorrel sub-com.)

    c: **Juncus effusus-Holcus lanatus** sub-community
      (Soft rush/Yorkshire fog sub-community)

M29 Hyperico-Potamogetum polygonifolii soakway
    (Marsh St John's-wort/bog pondweed soakway)

M35 Ranunculus omiophyllus-Montia fontana rill
    (Round-leaved crowfoot/blinks rill)

Heaths

H8 Calluna vulgaris-Ulex gallii heath
    (Heather/western gorse heath)

    e: **Vaccinium myrtillus** sub-community
      (Bilberry sub-community)

H9 Calluna vulgaris-Descampsia flexuosa heath
    (Heather/wavy hair-grass heath)

    b: **Vaccinium myrtillus-Cladonia** spp. sub-com.
      (Bilberry/**Cladonia** spp. sub-community)

H10 Calluna vulgaris-Erica cinerea heath
    (Heather/bell heather heath)

Plant Communities

   c: **Festuca ovina-Anthoxanthum odoratum**
   sub-community
   (Sheep's-fescue/sweet vernal-grass sub-com.)

H12 **Calluna vulgaris-Vaccinium myrtillus** heath
   (Heather/bilberry heath)
   a: **Calluna vulgaris** sub-community
   (Heather sub-community)

   b: **Vaccinium vitis-idaea-Cladonia impexa**
   sub-community
   (Cowberry/**Cladonia impexa** sub-community)

   c: **Galium saxatilis-Festuca ovina** sub-community
   (Heath bedstraw/sheep's-fescue sub-com.)

H18 **Vaccinium myrtillus-Descampsia flexuosa** heath
   (Bilberry/wavy hair-grass heath)
   a: **Hylocomium splendens-Rhytidiadelphus**
   loreus sub-community

Acidic Grasslands and Miscellaneous Upland Communities

U1 **Festuca ovina-Agrostis capillaris-Rumex acetosella**
   grassland
   (Sheep's-fescue/common bent/sheep's sorrel grassland
   b: Typical sub-community

   d: **Anthoxanthum odoratum-Lotus corniculatus**
   sub-community
   (Sweet vernal-grass/bird's-foot trefoil sub-com.)

   e: **Galium saxatile-Potentilla erecta** sub-com.
   (Heath bedstraw/tormentil sub-community)

   f: **Hypochaeris radicata** sub-community
   (Cat's-ear sub-community)

U2 **Deschampsia flexuosa** grassland
   (Wavy hair-grass grassland)
   a: **Festuca ovina-Agrostis capillaris** sub-com.
   (Sheep's-fescue/common bent sub-community)

   b: **Vaccinium myrtillus** sub-community
   (Bilberry sub-community)

U4 **Festuca ovina-Agrostis capillaris-Galium saxatile**
   grassland
   (Sheep's-fescue/common bent/heath bedstraw grassland)
   a: Typical sub-community

   b: **Holcus lanatus-Trifolium repens** sub-com.
   (Yorkshire fog/white clover sub-community)

   c: **Lathyrus montanus-Stachys betonica** sub-com.
   (Bitter vetch/betony sub-community)

   e: **Vaccinium myrtillus-Deshampsia flexuosa**
   sub-community
   (Bilberry/wavy hair-grass sub-community)

U5 **Nardus stricta-Galium saxatile** grassland
   (Mat-grass/heath bedstraw grassland)
   a: Species-poor sub-community

   b: **Agrostis canina-Polytrichum commune**
   sub-community
   (Brown bent/**Polytrichum commune** sub-com.)

   d: **Calluna vulgaris-Danthonia decumbens**
   sub-community)
   (Heather/heath grass sub-community)

U6 **Juncus squarrosus-Festuca ovina** grassland
   (Heath rush/sheep's-fescue grassland)
   a: **Sphagnum** sub-community
   (Bogmoss sub-community)

   c: **Vaccinium myrtillus** sub-community
   (Bilberry sub-community)

   d: **Agrostis capillaris-Luzula multiflora** sub-com.
   (Common bent/heath woodrush sub-community)

U19 **Thelypteris limbosperma-Blechnum spicant**
   community
   (Mountain fern/hard fern community)

U20 **Pteridium aquilinum-Galium saxatile** community
   (Bracken/heath bedstraw community)
   a:: **Anthoxanthum** sub-community
   (Sweet vernal-grass sub-community)

   b: **Vaccinium myrtillus-Dicranum scoparium**
   sub-community
   (Bilberry/**Dicranum scoparium** sub-com.)

   c: Species-poor sub-community

Mesotrophic Grasslands

MG1 **Arrhenatherum elatius** coarse grassland
   (False oat-grass coarse grassland)
   a: **Festuca rubra** sub-community
   (Red-fescue sub-community)

   b: **Urtica dioica** sub-community
   (Common nettle sub-community)

   e: **Centaurea nigra** sub-community
   (Common knapweed sub-community)

MG4 **Alopecurus pratensis-Sanguisorba officinalis**
   flood-meadow
   (Meadow foxtail-greater burnet flood-meadow)

MG5 **Cynosurus cristatus-Centaurea nigra** meadow and
   pasture
   (Crested dog's-tail/common knapweed meadow and pasture)
   a: **Lathyrus pratensis** sub-community
   (Meadow vetchling sub-community)

   c: **Danthonia decumbens** sub-community
   (Heath-grass sub-community)

MG6 **Lolium perenne-Cynosurus cristatus** pasture
(Perennial ryegrass/crested dog's-tail pasture)
    a: Typical sub-community

    b: **Anthoxanthum odoratum** sub-community
    (Sweet vernal-grass sub-community)

MG7 **Lolium perenne** and related grasslands
(Perennial ryegrass and related grasslands)
Sub-communities not determined

MG10 **Holcus lanatus-Juncus effusus** rush pasture
(Yorkshire fog/soft rush pasture)
    a: Typical sub-community

Calcareous Grassland

CG10 **Festuca ovina-Agrostis capillaris-Thymus praecox** grassland
(Sheep's-fescue/common bent/wild thyme grassland)
    a: **Trifolium repens-Luzula campestre** sub-com.
    (White clover/field woodrush sub-community)

Swamps and Tall Herb Fens

S3 **Carex paniculata** sedge swamp
(Tussock sedge swamp)

S4 **Phragmites australis** swamp and reed-bed
(Common reed swamp and reed-beds)
Sub-communities not determined

S7 **Carex acutiformis** swamp
(Lesser pond-sedge swamp)

S9 **Carex rostrata** swamp
(Bottle sedge swamp)
    b: **Menyanthes trifoliata-Equisetum fluvialile** sub-community
    (Bogbean/water horsetail sub-community)

S10 **Equisetum fluviatile** swamp
(Water horsetail swamp)
    a: **Equisetum fluviatile** sub-community
    (Water horsetail sub-community)

    b: **Carex rostrata** sub-community
    (Bottle sedge sub-community)

S11 **Carex vesicaria** swamp
(Bladder-sedge swamp)
Sub-communities not determined

S12 **Typha latifolia** swamp
(Bulrush swamp)
Sub-communities not determined

S14 **Sparganium erectum** swamp
(Branched bur-reed swamp)
    a: **Sparganium erectum** sub-community
    (Branched bur-reed sub-community)

    b: **Alisma plantago-aquatica** sub-community
    (Water-plantain sub-community)

    c: **Mentha aquatica** sub-community
    (Water mint sub-community)

S19 **Eleocharis palustris** swamp
(Common club-rush swamp)
    a: **Eleocharis palustris** sub-community
    (Common club-rush sub-community)

    b: **Littorella uniflora** sub-community
    (Shoreweed sub-community)

    c: **Agrostis stolonifera** sub-community
    (Creeping bent sub-community)

S22 **Glyceria fluitans** swamp
(Common sweet-grass swamp)
    a: **Glyceria fluitans** sub-community
    (Common sweet-grass sub-community)

    b: **Sparganium erectum-Mentha aquatica** sub-community
    (Branched bur-reed/water mint sub-community)

    c: **Alopecurus geniculatus** sub-community
    (Marsh foxtail sub-community)

S27 **Carex rostrata-Potentilla palustris** fen
(Bottle sedge/marsh cinquefoil fen)
    a: **Carex rostrata-Equisetum fluviatile** sub-com.
    (Bottle sedge/water horsetail sub-community)

S28 **Phalaris arundinacea** fen
(Reed canary-grass fen)
    a: **Phalaris arundinacea** sub-community
    (Reed canary-grass sub-community)

Aquatic Communities

A2 **Lemna minor** community
(Common duckweed community)
    a: Typical sub-community

    b: **Lemna trisulca** sub-community
    (Ivy-leaved duckweed)

A5 **Ceratophyllum demersum** community
(Rigid hornwort community)
    b: **Lemna minor** sub-community
    (Common duckweed sub-community)

A7 **Nymphaea alba** community
(White water-lily community)
    a: Species-poor sub-community

A8 **Nuphar lutea** community
(Yellow water-lily community)
    a: Species-poor sub-community

    c: **Nymphaea alba** sub-community
    (White water-lily sub-community)

Plant Communities

A9 **Potamogeton natans** community
    (Broad-leaved pondweed community)
        a: Species-poor sub-community
        c: **Juncus bulbosus-Myriophyllum alterniflorum** sub-community
        (Bulbous rush/alternate water-milfoil sub-com.)

A10 **Polygonum amphibium** community
    (Amphibious bistort community)

A12 **Potamogeton pectinatus** community
    (Fennel pondweed community)

A13 **Potamogeton perfoliatus-Myriophyllum** alterniflorum community
    (Perfoliate pondweed/alternate water-milfoil community)
        a: **Potamogeton berchtoldii** sub-community
        (Small pondweed sub-community)

A14 **Myriophyllum alterniflorum** community
    (Alternate water-milfoil community)

A16 **Callitriche stagnalis** community
    (Common starwort community)
        a: **Callitriche** spp. sub-community
        (Starwort spp. sub-community)

A18 **Ranunculus fluitans** community
    (Stream water-crowfoot community)

A19 **Ranunculus aquatilis** community
    (Common water-crowfoot community)

A22 **Littorella uniflora-Lobelia dortmanna** community
    (Shoreweed/water lobelia community)
    Sub-communities not determined

A23 **Isoetes lacustris/setacea** community
    (Quill-wort community)

A24 **Juncus bulbosus** community
    (Bulbous rush community)
        b: **Sphagnum auriculatum** sub-community

4. The Plan of the Flora

Layout and nomenclature

The flora is split into two main sections, the vascular plants and the non-vascular plants. The major plant groups studied are introduced and, where possible placed in their national context by sections on their biogeography, conservation requirements and history of recording. The nomenclature of the flowering plants (English and Latin names) follows Ellis (1983). Where a local Radnor name has been found it is also mentioned. Welsh names are omitted, since despite an extensive search, I have failed to locate any Welsh names in current usage and Ellis (1983) provides a readily available reference. Rarely Ellis fails to provide an English name. In such cases Dony, Jury and Perring (1986) has been employed to provide an English name.

The nomenclature of ferns and their allies follows Clapham et al (1981). The flora accounts were completed before Stace (1991) became available. This was unfortunate as Stace introduces a number of name changes which are likely to be widely adopted and used for years to come. To overcome this source of potential confusion Stace's new names have been introduced to the index and an appendix is provided in which the names employed here are compared with Stace's new names. The sources of nomenclature for the non-vascular plant groups are described in their respective introductory sections.

The Species Accounts

All living or sub-fossil native and naturalized vascular plant species known to occur or to have once occured in Radnorshire since the end of the last ice age are included. The definition of a naturalized species is a vexed question. I have included here all species which have established themselves by seed or vegetative propagation in vegetation largely dominated by native species, even though these native species might themselves be maintained only by man's intervention eg. the vegetation of a roadside verge. This rule is broken to include arable weed species, plants naturalized on walls and a few introduced species, such as red oak, now widely planted, which it is believed are likely to become naturalized in the near future. That they have not yet become naturalized may be of considerable interest to future botanists. Omitted are species confined to long-abandoned ornamental plantings which have shown signs of but feeble vegetative spread and those species only recorded as spreading locally by vegetative means from adjacent gardens. An asterisk preceeds the Latin name of all non-native species.

Whilst most of the plant records have been accumulated by field work in the last 20 years, a comprehensive, though not exhaustive search of published and archived material has also been made. Inevitably errors in recording and identification are made from time to time. Where a doubtful record appears on a recent field card or in a manuscript and a search of the site subsequently has failed to reveal the species, the record has been omitted. On the other hand all published records have been included, even though the author may harbour equally strong doubts as to the correctness of the record.

No comprehensive search of herbaria has been made. I have instead relied for records of the vascular plants on searches made by National Museum of Wales (NMW) staff in the course of preparing such publications as *Flowering Plants of Wales* and *Welsh Flowering Plants* (op cit). The herbaria of the National Museum of Wales and the Botany Department of the University College of Wales, Aberystwyth have been searched for lichen and moss and liverwort records. Herbaria both of the NMW and the Royal Botanical Gardens at Edinburgh have been searched for rust and smut fungi records. Many other herbaria have been searched in pursuit of records of specific species.

For each species a description is provided of the habitat(s) from which it has been recorded, together with, in most case, a description of its frequency and/or abundance. Detailed records are provided for the rarer species (generally those found in 6 or fewer sites). The locality given usually includes the name of the civil parish. Unfortunately these parish boundaries are not shown on modern maps. Map 4.1 below is provided to assist in the location of these parishes. A 4 figure grid reference (or if not known a 10 kilometre or quadrant reference) is then provided.

The 100 kilometre grid square prefixes have been omitted. All grid references commencing with an 8 or 9 lie in 100km. square SN(22),

Plan of the Flora

The 100 kilometre grid square prefixes have been omitted. All grid references commencing with an 8 or 9 lie in 100km. square SN(22), whilst those commencing with an 0, 1, 2 or 3 lie in square SO(32).

The grid reference is followed by either the name or initials of the recorder. Generally all recorders who have contributed more the 12 records are referred to by their initials. The abbreviations section at the end of the book cross references these initials to names. Where no name is given the record was made by the author. Where the author made the record with others the author is abbreviated to an exclamation mark. All records were made post 1970 unless a date is then given. If no further information is provided the record may be assumed to be a field record, usually communicated directly to the author. Where records have been gleaned from published sources or herbaria a reference is provided to identify the source of the information. Where such a reference appears in the bibliography it is reduced to the recorder(s) name and the year of publication.

The distribution of somewhat commoner plants is illustrated by an outline map of Radnorshire on which is superimposed the 10km. grid squares of the Ordnance Survey. Each of these squares was divided for recording purposes into 4 quadrants. They are the Ordnance Survey's 1:10,000 map sheet areas which are 5x5km. in extent and are referred to by the 10km. grid square number followed by the suffixes NE, SE, SW and NW. Where only a small amount of the vice-county lies in a peripheral square the records from that square have been amalgamated with those from an adjacent square, though always one which lies in the same 10km. square. Map 9.1 provides details of the 63 quadrants which formed the recording units. A solid dot in a quadrant indicates a record of that species made in that quadrant in the period 1970-1992. An open circle indicates a record made pre-1970.

Common species, that is those recorded from 50 or more quadrants, have no distribution map. Instead the quadrants without records are listed.

The treatment as described for vascular plants is followed for the stoneworts, hornworts, liverworts and mosses. Lichens differ slightly in the treatment of common species (see section E of Chapter 13 for details).

The coverage of other non-vascular plant groups is less complete. Amongst the fungi some lichenicolous fungi are reported in the section on lichens. Of the remaining fungi only the rust and the smut fungi have been considered, since insufficient information is available on other groups. A review is provided of the limited amount of information available on algae, other than stoneworts. This latter group receives the same treatment as the higher plants.

The Record Archive

Copies of all record cards refering to specific sites have been lodged with the Countryside Council for Wales. Arrangements have been made to deposit record cards which refer to quadrants, individual species records, and all other relevant manuscript material with National Museum of Wales, Cardiff. Herbarium material, mostly of non-vascular plants will be deposited in National Museum of Wales. Some lichen and rust specimens have been deposited in the herbarium of the Royal Botanic Gardens, Edinburgh.

Melancholy Thistle x 1/6

Map 4.1 Administrative Areas of Radnorshire

----- Boundaries of the administrative areas (civil parishes etc.) used to localize plant records.

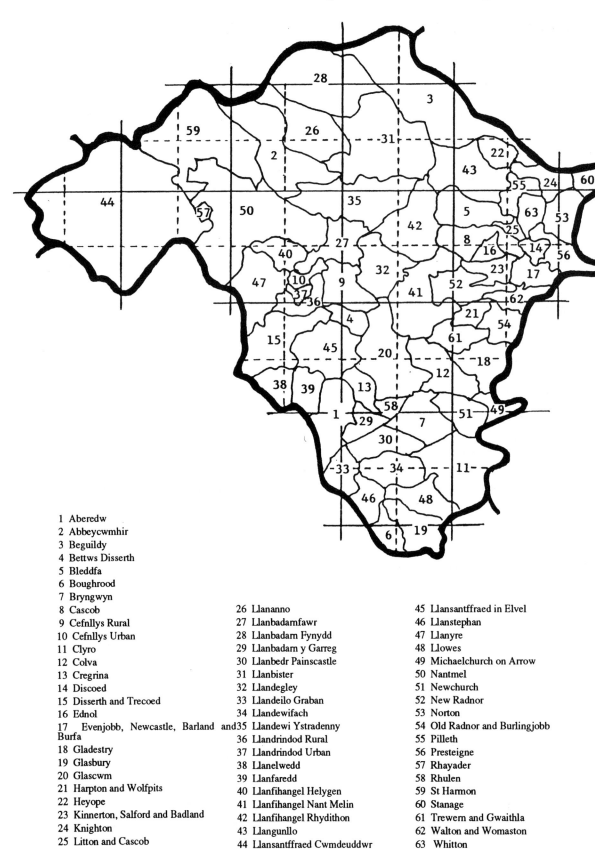

1 Aberedw
2 Abbeycwmhir
3 Beguildy
4 Bettws Disserth
5 Bleddfa
6 Boughrood
7 Bryngwyn
8 Cascob
9 Cefnllys Rural
10 Cefnllys Urban
11 Clyro
12 Colva
13 Cregrina
14 Discoed
15 Disserth and Trecoed
16 Ednol
17 Evenjobb, Newcastle, Barland and Burfa
18 Gladestry
19 Glasbury
20 Glascwm
21 Harpton and Wolfpits
22 Heyope
23 Kinnerton, Salford and Badland
24 Knighton
25 Litton and Cascob
26 Llananno
27 Llanbadarnfawr
28 Llanbadarn Fynydd
29 Llanbadarn y Garreg
30 Llanbedr Painscastle
31 Llanbister
32 Llandegley
33 Llandeilo Graban
34 Llandewifach
35 Llandewi Ystradenny
36 Llandrindod Rural
37 Llandrindod Urban
38 Llanelwedd
39 Llanfaredd
40 Llanfihangel Helygen
41 Llanfihangel Nant Melin
42 Llanfihangel Rhydithon
43 Llangunllo
44 Llansantffraed Cwmdeuddwr
45 Llansantffraed in Elvel
46 Llanstephan
47 Llanyre
48 Llowes
49 Michaelchurch on Arrow
50 Nantmel
51 Newchurch
52 New Radnor
53 Norton
54 Old Radnor and Burlingjobb
55 Pilleth
56 Presteigne
57 Rhayader
58 Rhulen
59 St Harmon
60 Stanage
61 Trewern and Gwaithla
62 Walton and Womaston
63 Whitton

The Vascular Plants

5. Biogeography of the Vascular Plants

This section seeks to evaluate the significance of Radnor in terms of the British and World distribution of higher plants. The lower plants are dealt with separately in introductory sections to the species accounts chapters 12 and 13.

British Distribution

Perring and Walters (Eds.) (1983) provide British distribution maps, recording the presence or absence of most species in the 10km squares of the Ordnance Survey's national grid or Irish equivalent. Radnor, occupying a southern-central position in the British Isles, can be seen to lie close to the edge of the range of a number of higher plant species.

Northern species, here at, or close to the southernmost edge of their range include wood cranesbill, melancholy thistle, marsh hawk's-beard, northern marsh-orchid and dioecious sedge. With much of upland Britain being confined to the north and west of the country Radnor's hills provide an outpost for a number of submontane species close to the southernmost edge of their range eg. alpine clubmoss, quill-wort, mountain male-fern, parsley fern, globe flower, mossy saxifrage, stone bramble, cowberry and crowberry.

Western species, with Radnor at the easternmost edge of their range, include Wilson's filmy fern, Welsh poppy and whorled caraway. Plants with a more south-westerly distribution and here close to the north-easternmost edge of their range include meadow thistle, western gorse, ivy-leaved bellflower, rock stonecrop and navelwort.

The largest group of all with over 60 species are those plants which occur widely in central and southern England, penetrate into eastern Radnor, but possibly find the wetter and more acidic soils of upland Radnor not to be to their liking. Whatever the reason, a number of plants strikingly common in, for example, Herefordshire are scarce in Radnorshire eg. field convolvulus, wood spurge, black bryony, white bryony, mistletoe, traveller's-joy and nettle-

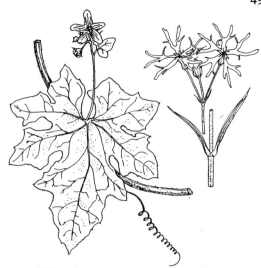

White bryony - continental southern
Ragged-robin - continental northern

leaved bellflower. Other widespread south-easterly species for which Radnor becomes a western barrier include hairy violet, maiden pink, field maple, spurge laurel, vervain, toothwort, black nightshade, meadow saffron, spreading bellflower, giant bellflower and thin-spiked wood-sedge.

World Distribution

Following Matthews, J.R., (1937) and (1955), Ellis, R.G., (1983) and Sinker et al. (1985) each of the native plant species of Radnor can be assigned to one of 15 biogeographical elements. It should be noted that Sinker et al. (1985) combined the Eurasian and European elements of Matthews and reassigned some species to elements which differed from those of Matthews, in the light of more recent information. These changes have been followed here adding also a few species not previosly considered.

The native species (excepting blackberry and dandelion species) recorded in Radnor have been assigned to 10 biogeographic elements. They are generally listed below. To save space, however, those species assigned to the two largest elements ie. Wide and Eurasian are omitted (see Ellis op cit for a complete list). No plants have been recorded from the county which might be assigned to the North American, Arctic-Subarctic or Arctic-Alpine elements.

In the lists of names which follow those plant names enclosed in brackets are thought to be extinct, whilst those marked with a * are assigned to an element which differs from that of Matthews (1955). The distribution of some of the elements is mapped below. The number of species recorded in each 5 x 5 km square of the national grid is represented by a dot. The smaller the dot the fewer the number of species recorded per square in each element.

1 Wide Element

Included here are 83 species which occur in Europe, Asia and North America, or may extend into North Africa or even be cosmopolitan. They range from such common plants in the county as red fescue and herb Robert to such rarities as ivy-leaved duckweed and nodding bur-marigold.

2 Eurasian Element

Following Sinker et al., 1985 both the Eurasian and European elements of Matthews are combined to include species which may occur just in Europe as well as those which extend into Asia. They considered the division of Asiatic from European elements to be artificial. This combined Eurasian Element accounts for nearly half of the flora of Radnor ie. 328 species and like the last element includes common as well as rare species. Examples of common species include many weeds such as nettle, creeping thistle and broad-leaved dock as well as more choice species such as devil's-bit, quaking grass and yellow flag. Rarites include lily-of-the-valley, fragrant orchid and lesser meadow-rue.

3 Mediterranean Element

Species in this element have their centres of distribution around the Mediterranean. The frost-ridden climate of Radnor appears, not unsurprisingly, to be unconducive to their survival. Only the pale flax with but a single ancient record of a probable casual occurrence provides any evidence that this element has any part to play in Radnor's flora.

4 Oceanic Southern Element

Species in this element have their chief centres of distribution in south-west Europe. Twenty eight species are assigned to this element. Almost half of these tend to occur in sheltered lowland sites or on sunny rock outcrops eg. bird's-foot, southern woodrush, upright clover, knotted clover, English elm and prickly sedge. Others, such as holly, honeysuckle, woodsage and lesser knapweed are widespread species.

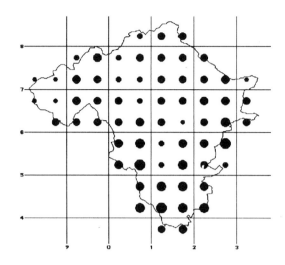

Bog pimpernel
*Lesser marshwort
*Pedunculate water-starwort
Blunt-fruited water-starwort
*Prickly sedge
*Thin-spiked wood-sedge
*Lesser knapweed
*Foxglove
*Many-stalked spike-rush
*Heath bedstraw
Tutsan
Holly
*Sharp-leaved fluellen
*Lesser hawkbit

*Honeysuckle
Southern woodrush
*Bird's foot
*(Greater broomrape)
*(Pellitory-of-the-wall)
Small-flowered buttercup
Water figwort
*Field woundwort
Slender trefoil
*Rough clover
Upright clover
*English elm
Navelwort
*Woodsage

5 Oceanic West European Element

Species which occur almost exclusively along the western edge of Europe in areas where the climate is influenced by the Atlantic Ocean make up this element. Of the 39 species assigned here 2 show a westerly distribution within Radnor ie. whorled caraway and Wilson's filmy fern. Since peatlands are such a feature of our oceanic climate it is not surprising that nearly half of the species in this element occur in peaty pastures, blanket bogs or peaty flushes eg. ivy-leaved bellflower, lesser skullcap, meadow thistle, petty whin, marsh St. John's-wort and cross-leaved heath.

Vascular Plant Biogeography

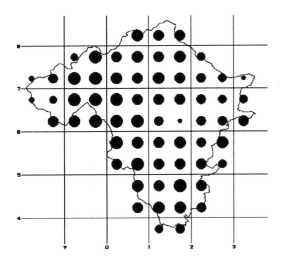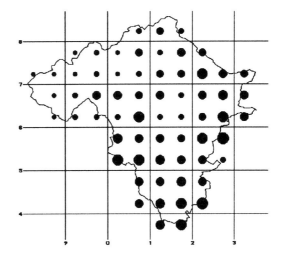

Green-ribbed sedge	Welsh poppy	*Silver hair-grass	*(Annual mercury)
*Smooth-stalked sedge	*Creeping forget-me-not	Fool's water-cress	*Upright chickweed
*Sea mouse-ear	*Wild daffodil	*Slender sandwort	*Green-winged orchid
*Opposite-leaved golden-saxifrage	Hemlock water-dropwort	Wild arum	Prickly poppy
	*Pillwort	*(Meadow brome)	*Common fleabane
Meadow thistle	*Heath milkwort	*Upright brome	*Wall barley
Pignut	*Trailing tormentil	*Smooth brome	Small-flowered sweetbriar
Climbing corydalis	*Barren strawberry	*Sterile brome	Purple willow
(Southern marsh-orchid)	*Primrose	White bryony	*Salad burnet
*Northern marsh-orchid	Ivy-leaved crowfoot	Pendulous sedge	Small-flowered catchfly
*Floating club-rush	Round-leaved crowfoot	*Rusty-back fern	Black bryony
Bell heather	Rusty-leaved willow	Traveller's-joy	*Wild thyme
Cross-leaved heath	Lesser skullcap	Meadow saffron	Large-leaved lime
Whorled caraway	English stonecrop	Spurge laurel	(Spreading hedge-parsley)
*Common ramping-fumitory	Common gorse	Fern-grass	Knotted hedge-parsley
	Western gorse	Teasel	Keeled-fruited cornsalad
Petty whin	Wood bitter-vetch	Wood spurge	(Narrow-fruited cornsalad)
Bluebell	Ivy-leaved bellflower	Common cudweed	*Wood speedwell
*Wilson's filmy-fern	Smith's pepperwort	*Red hemp-nettle	*Mistletoe
Marsh St. John's-wort	Rock Stonecrop	*Crosswort	*Squirreltail fescue
*Flax-leaved St. John's-wort		*Sharp-flowered rush	*Rat's-tail fescue
		*(Blunt-flowered rush)	*Radnor lily

6 Continental Southern Element

Species which occur chiefly in central and southern Europe make up this element. Many extend into North Africa. Of the 44 species in this group reported from Radnor 7 are thought to be extinct. Many of these were arable weeds eg. annual mercury, meadow brome and narrow-fruited cornsalad. Thirteen species favour dry, south-facing rock outcrops eg. common cudweed, upright chickweed, wild thyme and squirreltail fescue. A further group of 5 are confined to the south or east of Radnor ie. black and white bryony, mistletoe, wood spurge and traveller's-joy.

7 Continental Element

This element comprises species which are characteristic of central Europe, with some extending east into Asia and Russia. They become less common in western Europe. The 52 species of this element in Radnor show a wide range of habitat preference and include rarities such as rock cinquefoil, perennial knawel and spiked speedwell. Eighteen species have a British distribution tending to the south and east with Radnor at the western edge of their range eg. pale St. John's-wort, spreading, nettle-leaved and giant bellflowers, maiden pink and hound's-tongue.

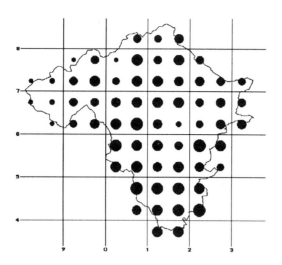

cottongrass. Reflecting, perhaps, the abundance of peatlands and open waters in northern Europe, 49 of the species in this element favour peatlands or open water in Radnor eg. the 10 sedges, common sundew, butterwort, bogbean and marsh cinquefoil.

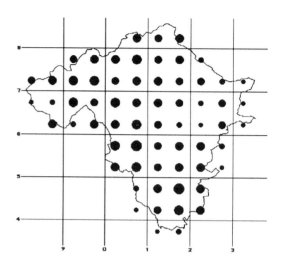

Intermediate lady's-mantle
Chaffweed
*Parsley-piert
*(Lesser hairy-brome)
*Nettle-leaved bellflower
*Spreading bellflower
*Giant bellflower
*Lesser pond-sedge
*Tawny sedge
*(Woolly thistle)
Hound's tongue
Maiden pink
Lesser teasel
*(Marsh helleborine)
*Sticky groundsel
(Pepper saxifrage)
*Greater chickweed
*Small-leaved lime
*Marsh valerian
Spiked speedwell
*Wood fescue
*Small sweet-grass
Pale St John's-wort
Ploughman's spikenard
Yellow archangel
*Great wood-rush
*Wood forget-me-not
*Lousewort

*Rock cinquefoil
Sessile oak
Common oak
Greater spearwort
*Black poplar
*Field rose
*Felt-leaved rose
Burnet rose
Sweet briar
Sherard's rose
Creeping willow
*Osier
Perennial knawel
Saw-wort
Wild service-tree
Shepherd's cress
*Knotted clover
*Dark mullein
Euphrasia rostkoviana
   -an eyebright
Dyer's greenweed
The hawkweeds-
Hieracium acuminatum
H. diaphanum
H.lasiophyllum
H.perproquinquum
*Broad Buckler-fern

8 Continental Northern Element

Species which are most abundant in central and northern Europe are included here. At the southern edge of their range, they tend to become confined to the uplands. This is somewhat reflected in the 64 species noted from Radnor in this element, since 21 occur more commonly in the hills eg. downy birch, eared willow, bog-rosemary, heath rush, cowberry, crowberry, bird cherry and common and hare's-tail

Field maple
*Sneezewort
Hairy lady's-mantle
Smooth lady's-mantle
Hare's-tail cottongrass
Bog rosemary
Wild angelica
*(Mountain everlasting)
*Downy oat-grass
Downy birch
*Intermediate starwort
White sedge
*Common yellow sedge
Lesser tussock-sedge
Dioecious sedge
Broom sedge
Star sedge
*Long-stalked yellow-
   sedge
*Greater tussock-sedge
Flea sedge
*Bladder sedge
(Cowbane)
(Alpine enchanter's-
   nightshade)
Intermediate enchanter's-
   nightshade
(Frog orchid)
Marsh hawk's-beard
*Common spotted-orchid
Common sundew
Few-flowered spike-rush
*Crowberry
Common cottongrass
Broad-leaved cottongrass
Melancholy thistle

Euphrasia micrantha
   -an eyebright
E. nemorosa
   -an eyebright
*Sheep's-fescue
*Large-flowered hemp-
   nettle
Fen bedstraw
(Field gentian)
Hieracium vulgatum
   -a hawkweed
Hairy St. John's-wort
Imperforate St. John's-
   wort
*Heath-rush
Mudwort
Bog orchid
Shoreweed
Ragged-robin
Bogbean
*Marsh lousewort
Common butterwort
*Spreading meadow-grass
Blunt-leaved pondweed
Long-stalked pondweed
Marsh cinquefoil
*Bird cherry
*White-beaked sedge
Knotted pearlwort
Eared willow
Deergrass
Lesser bladderwort
Cranberry
*Cowberry
(Wood vetch)
Marsh violet

Vascular Plant Biogeography

9 Northern-Montane Element

Many of these species are confined to the Boreal zone, only extending south into montane areas. Not surprisingly perhaps only 5 species out of the 30 or so recorded from Britain are known from Radnor ie. quill-wort, globeflower, mossy saxifrage, stone bramble and the northern subspecies of downy birch. Whilst all occur in the hills, with the exception of quill-wort, the remaining 4 are found at widely differing altitudes.

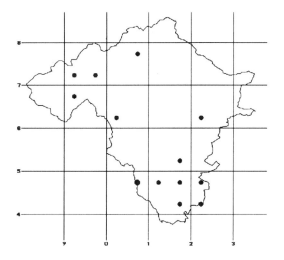

*Quill-wort Stone bramble
Mossy saxifrage Globeflower
Betula pubescens ssp **carpatica** - a birch

10 Oceanic Northern Element

The species in this element are confined mostly to north-west Europe, though some extend into north-east America. Three species are included here ie. an eyebright *Euphrasia confusa*, water lobelia and bog asphodel. They are confined to, or are more common in, the uplands of Radnor.

*Euphrasia confusa Water lobelia
-an eyebright Bog asphodel

11 Alpine Element

This element comprises species with distributions centred on west, central or south-east Europe. Only alpine penny-cress, noted once on Stanner Rocks and probably long extinct belongs here.

12 Endemic Element

The species which make up this element are not known outside the British Isles. Most are in critical genera. Of those genera only the eyebrights, hawkweeds and whitebeams have been considered here. Undoubtably a number of dandelions and brambles could be added to the 13 species reported here. *Euphrasia anglica* is a widespread eyebright of southern Britain. Of the hawkweeds *Hieracium carneddorum, H. cinderella, H. submutabile, H. subamplifolium, H. vagense* and *H. substrigosum* have their headquarters in Wales and/or the Marches. The whitebeams are both scarce trees of Wales and south-west England.

*Euphrasia anglica	***H. calcaricola**
-an eyebright	***H. placerophylloides**
*(Purple ramping-fumitory)	***H. submutabile**
*English whitebeam	***H. subampifolium**
*Sorbus porrigentiformis	***H. subcrocatum**
- a whitebeam	***H. substrigosum**
Hawkweeds-	***H. uiginsky**
*Hieracium carneddorum	***H. vagense**

Biogeographical Elements in a National Context.

Table 5.1 allows comparison of the differing proportions of native species, (excluding microspecies except of eyebrights and whitebeams) assigned to the biogeographical elements of Matthews (1955) in Radnor, the Shropshire area, Wales and the British Isles.

By comparison with the Shropshire area Radnor supports a larger proportion of Oceanic Southern and Oceanic Western species, reflecting its geographical location. This is further endorsed by the slightly lower proportion of Continental and Continental Southern species it supports.

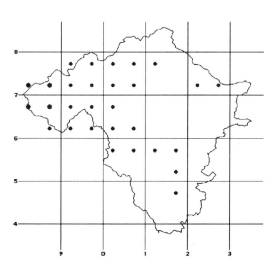

Western species such as whorled caraway, Wilson's filmy fern and southern woodrush are not known from the Shropshire area whilst Radnor does not support southern species such as marsh gentian, bee orchid or deadly nightshade which do occur in Shropshire.

A similar relationship exists between Wales and Radnor. As highlighted in chapter 2, eastern Radnor has a fairly continental climate. It supports a higher proportion of Continental and Continental Northern species and a lower proportion of Oceanic Southern and Oceanic Western species than Wales as a whole. The absence of any truely montane habitat is also reflected in the lower proportion of Northern-Montane, Arctic-Alpine and Alpine species in Radnor.

Compared to the British Isles Radnor supports a disproportionately large number of Wide and Eurasian species. Its somewhat central position may account for this, the county failing to provide suitable conditions for the most demanding oceanic, southern, northern, montane or continental species. The only other element in greater proportion to the total for the British Isles is the Continental Northern Element. This is perhaps the heart of Radnor's flora. Plants such as sneezewort, angelica, downy birch, star sedge, common spotted-orchid, sundew, cottongrass, sheep's-fescue, butterwort and eared-willow typify so much of Radnor's upland.

Table 5.1 A comparison between the modified biogeographical elements (Matthews, 1955) of the native flora of Radnor (R), Shropshire Region (SR), Wales (W), and the British Isles (BI), (omitting all microspecies except eyebrights and whitebeams).

Percentage of flora in given element

Element	R	SR*	W*	BI
Wide	12.5	18.0	10.9	10.4
Eurasian	49.3	44.7	41.8	39.8
Mediterranean	0.15	0.1	1.6	2.4
Oceanic southern	4.2	3.8	6.0	5.9
Oceanic W European	5.9	4.4	8.6	6.4
Continental southern	6.6	7.5	9.3	9.8
Continental	7.4	9.9	6.4	8.6
Continental northern	9.5	9.5	8.4	7.6
Northern-montane	0.75	0.7	1.2	1.9
Oceanic northern	0.5	0.7	2.1	1.6
North American	0	0	0.1	0.4
Arctic-subarctic	0	0	0.5	1.8
Arctic-alpine	0	0.1	2.0	4.7
Alpine	0.1	0	0.8	0.6
Endemic	0.5	0.5	1.5	1.3

* Data for Shropshire taken from Sinker et al. (1985) and for Wales modified from Ellis (1983).

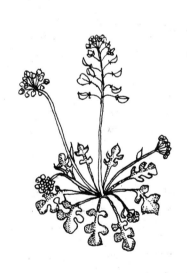

X 2

Shepherd's cress

- A Continental species -

X 2

Upright chickweed

- A Continental Southern species -

Plate 1 Stanner Rocks National Nature Reserve near Burlingjobb. A dolorite rock outcrop with an exceptional flora.

Sun-baked rock ledges viewed from the south. (*The author-CCW*)

Spiked speedwell, wood sage, rock stonecrop and associated plants. (*A.S. Ferguson-CCW*)

Plate 2 Plants of Stanner Rocks.

Perennial knawel.
(*A.S. Ferguson-CCW*)

Radnor lily.
(*P. Russell-NMW*)

The nationally rare apple moss,
Bartramia stricta. (*P. Russell-NMW*)

Marcheini uplands near St. Harmon. Heavy sheep-grazing to the left of the fence has suppressed the growth of heather. (*D.C. Boyce-CCW*)

6. The Changing Vascular Plant Flora

A. Holocene Period

At the height of the last ice age, some 18,000 years ago, it seems likely that little, if any, of Radnor escaped the icy grip of glaciers. As the ice retreated plant life, having most probably been totally exterminated, had to recolonise the area. We know a little about this process and the subsequent changes that took place thanks to plant remains being preserved from these times in peat. Moore (1978) investigated the pollen and macroscopic remains of plants preserved in a basin mire at Cors y Llyn, (or The Llyn) near Newbridge on Wye. Unfortunately no radiocarbon dating was carried out at this site, so absolute dates for when the plant remains were laid down cannot be given. From the earliest pollen grains, probably deposited in early Flandrian times, (10,000 years ago) Moore deduces that an open, herb-rich grassland, with plants such as common sorrel, meadowsweet, moonwort, a wormwood species, a rue species and sea plantain grew amongst shrubs of sea buckthorn and juniper on the land around the lake. In the lake a rich aquatic flora had developed of pondweed, water milfoil, horsetail and water-lily species.

Later an open scrubland of birch and willow with a little pine gradually spread over the grassland. This, in turn was gradually replaced by an open woodland or scrub of oak, elm, alder and hazel and/or bog myrtle (the pollen grains of the latter two species were not differentiated). There was also a decline in aquatic species in the lake at this time and gravel was washed in to form a distinct layer.

Birch then continued to decline as oak became more common and ivy appeared with bracken and royal fern. Plants of open habitats now became rare, but juniper still survived. As time passed oak and alder continued to increase in quantity as birch, pine and elm declined. Both small and large-leaved lime and their hybrid and ash appeared at this time and there may have been some evidence of human disturbance in the presence of pollen of the weed fumitory. It was not long, however, before a decline in elm and lime and a rise in the abundance of grass, ribwort plantain and the appearance of probable cereal

Llynheilyn - In a changing landscape

pollen confirmed the presence of Mesolithic or early Neolithic man.

The pattern of disturbance-indicating plant species remained fairly constant as peat accumulated through the Bronze and Iron ages and into Roman times. There was some evidence of tree clearance episodes, with localized increases in the abundance of bracken and herbaceous plant pollen grain. A major change then occured with the appearance of cannabis pollen. French and Moore (1986) discuss its presence at length. It may well have been cultivated close to the Llyn in forest clearings to provide fibers for rope, perhaps as a medieval occupation. When its cultivation declined, much of the woodland surrounding the Llyn was cleared. This possibly had a dramatic effect on the lake, encouraging sedges and bog mosses to grow over the surface of the lake to form a floating raft. The Llyn, meaning in Welsh a lake, no longer had any open water. Scots pines planted close to the lake, probably in the late 1800's began to colonise the site. By 1960 they had grown so well that, from a distance, the Llyn appeared to be a pine wood.

Bartley (1960a) studied the changes recorded in the peat of Rhosgoch Common near Painscastle. In just a small part of this complex site he describes a late-glacial lake with stoneworts gradually being colonized by reed and great fen-sedge. They, in turn, were replaced by bottle sedge and horsetail to be followed by bog mosses, cottongrass, heather and cross-leaved heath. This heath-like vegetation itself was colonized by birch and willow to form the wet woodland we see there now. Wilkinson (1988) traces a similar succession at Beilibedw Mawn Pools near Hundred House, though here the woodland was later cleared.

The basin and blanket peat of the central Wales uplands, has been examined by Moore and Chater (1969) and Wiltshire and Moore (1983). The blanket peat appears to have begun to form in Atlantic times (7000-5000 before the present day) following a widespread decline in forest cover and a spread of heathland. This change may have been induced or assisted by Mesolithic pastoralists burning the forest to improve the grazing. Valleyside woodlands persisted only to be felled in Roman and medieval times. Arable agriculture was most widespread in the Napoleonic period.

A very useful and complete summary of published data on the interactions of vegetation with man in Wales since late-glacial times is provided by Caseldine (1990).

B. Modern Times

This final section traces the more recent losses and gains of higher plant species to the county. Ratcliffe (1984) reviews at length the likely causes for post-medieval changes in the flora of Britain. He concluded that anthropogenic, rather than climatic factors were overwhelmingly responsible for most of the measured changes. Since agriculture now affects about 80% of the land surface it is this industry that has caused much of the change.

i) Agriculture

L. K. Redford contributed the Radnor section to the monumental report of the Land Utilization Survey of Britain, edited by L. Dudley Stamp and published in 1940. This provides a detailed land-use survey of the county, undertaken by its teachers and school children at a scale of 6 inches to the mile during the years 1932-3. In the introductory section Redford reviews the state of agriculture, comparing the statistics obtained in this survey with earlier surveys.

He notes that land utilisation has changed little in hundreds of years. Sheep had been the mainstay of farming, yet methods of production had changed. Instead of the sale of finished mutton, store lambs had largely taken over the trade. These were sold for fattening on farms in England. Hereford store cattle were the second most important enterprise, whilst dairy shorthorns were the latest breed for milk production. Arable land was widely scattered through the county and devoted almost entirely to the production of oats, roots and fodder crops to feed livestock, particularly horses. They still provided the major motive power on farms. There were 10,500 in Radnor in 1913 and 7,500 in 1937. The rough pastures they were turned out onto and helped to create were almost certainly botanically rich. It is sad that no contemporary record of their species composition seems to have been made.

Moorland and rough grazing covered 43% of the county whilst permanent pasture covered a further 39%. The area of rough grazing was increasing as bracken, gorse and rushes spread into grasslands. Some agricultural improvement work was, however, being carried out. He reports the best pasture seen was on land at 427m. (1,400 ft.) which was previously heather and bracken. But this improvement did not match the "thousands of acres of pasture ... deteriorating into rough grazing land". A chief cause of neglect seemed to be the widespread practice on the poorer land of earning a living purely from the judicious purchase and sale of livestock. Only enough feed was required to keep the stock alive until the price had risen enough to make a profit. This continued decline in agricultural output and consequent increased reliance on imported food began to cause the Government concern. In the late 1930's it introduced a subsidy to cheapen the cost of lime to farmers. This stimulated its use and increased the effort put into pasture improvement work.

The second world war brought a turn around in farming fortunes as thousands of acres were put to the plough to provide food to replace the lost imports. Following the war successive governments provided grants and subsidies to stimulate agricultural production and reduce our reliance on imports. In Radnor generous grants to plough, drain and reseed pastures in all but the lowlands, botanically transformed the meadows and pastures. Ryegrass and white clover were widely planted which, together with the use of artificial fertilizers, completed the destruction of the flower-filled meadows and pastures. The increased stocking levels that were possible on the improved land also permitted larger numbers of stock to be kept on the remaining areas of rough grazing and in woodlands. This has, in turn, led to the reduction in extent of these areas and the loss of species sensitive to high grazing levels.

The great reduction in the area of arable land, with cereals now only grown in the east of the county and the introduction of weedkillers has greatly reduced the weed flora. Fields of root crops until recently provided an exception, but the recent introduction of pre-emergence weedkillers has reduced their weed floras too.

Table 6.1 provides a summary of agricultural statistics, which though not all collected in precisely the same way, indicates the major trends. The total sheep number has risen from 284,576 in 1870 to 1,066,171 in 1988, whilst cattle have increased from 28,967 to 60,080. This has been achieved on a reduced area of land, since forestry has expanded in this period. The extra grass has been grown at the expense of the native flora.

Table 6.1 Agricultural Statistics 1870-1988 (Area in hectares)

	1870	1913	1937	1982	1988
Rough grazing	------	51,313	46,766	20,403	19,947
Permanent grass*	46,719	65,415	61,422	54,228	58,740
Grass ley	6,293	5,487	7,997	10,979	10,939
Arable	20,931	15,116	15,313	5,675	3,668
Wheat	2,923	787	296	792	575
Barley	2,102	1,432	232	2,015	1,770
Oats	4,991	4,664	4,271	374	267
Turnips & swedes	2,740	2,070	1,395	859	375
Total sheep	285K	278K	351K	898K	1066K
Total cattle	29K	33k	171K	63K	60K

*Permanent grass is here defined as a grass sward which has not been ploughed and reseeded for 5 or more years. It should not be equated with unimproved grassland.

ii) Forestry

Our need for wood and other products of the forest has affected the native flora of Radnor in a number of ways. The wholesale felling of woodlands and subsequent supression of woody species by grazing with domestic stock has gone on for centuries and has dramatically reduced the extent of native woodland. Where existing woodlands have been replanted, they have largely been stocked with non-native conifers to the exclusion of native plant species. The need for smallwood products such as tan bark, charcoal and mining timber has encouraged coppice systems at the expense of mature to overmature timber. This has reduced the number of ancient trees which are particularly valuable as hosts for epiphytic plant species and led to the impoverishment of moisture and shade demanding species, particularly in the western sessile oakwoods.

New plantations, largely of alien conifers, have been planted, in the main on semi-natural vegetation. Whilst rides and recently felled areas may be rich in flowers, they seldom compensate for the habitats lost during the original planting, though this may be a preferable option to agricultural improvement which leaves few areas untouched.

Linnard (1982) estimates that by the 16th and 17th century woodland probably covered less than 10 percent of Wales. He notes that even areas like the Radnor Forest had few trees by this time. An Elizabethan Inquisition of 1564 described this area as 809 hectares of "wast heath and wild, foggy and marish ground", 324 hectares of "lowe shrubbes and bushe of smalle hazill and thornes utterly destroied by reason the same been hewen and cutt down by th' inhabitants dwelling there about all waies owt of season, and al the spring tyme eaten and consumed with wild bests and goats".

The Rev. J. Williams in his *A General History of the County of Radnor* written in the early years of the 1800's (published 1905 by Davies and Co., Brecon) remarks that "until very lately no county was much better stored with woods than Radnor, particularly with groves of oak....So great has been the demand for this article for the use of the Royal Navy and...other purposes that oak timber has now become as scarce and dear in this district where formerly it abounded." Oaks of immense size had apparently been felled at Abbeycwmhir in living memory.

There seem to be no accurate early statistics of woodland cover until the Board of Agriculture summarized the agricultural returns for England and Wales in the last quarter of the 19th century (cited in Linnard, 1982). Woods and plantations in Radnor were reported to cover 3449 hectares in 1871, 4126 hectares in 1888 and 5077 hectares in 1905. Linnard suspects the early returns were underestimates of the true area of woodland and the later figures may be more accurate.

In 1928 the newly created Forestry Commission reported on its first full census of woodlands of 2 acres (0.81 hectares) or more in extent, carried out between 1921 and 1926 and called the 1924 census *(Report on Census of Woodlands and Census of Production of Home Grown Timber*, 1924, HMSO, London). The voluntary surveyors reported a total of 5887 hectares of woodland in Radnor (4.9% of the county). 988 hectares were of conifer high forest, 1603 hectares of hardwood high forest, 262 hectares of mixed high forest, 183 hectares of coppice with standards and 148 hectares of coppice. Non-productive areas included 1337 hectares of scrub and 1221 hectares of felled or devastated woodland, largely the legacy of fellings in the first world war.

The next census took place after the second world war (Forestry Commission Census Report No. 1.(1952). *Census of Woodlands 1947- 1949*, HMSO, London). Only woodlands of 5 acres (2.023 hectares) or more in extent were censused, so a strict comparison with the results of the 1924 survey cannot be made. The total extent of woodlands had, however, increased to 6550 hectares (5.4% of the county) due largely to an increase of 614 hectares to a total of 1602 hectares in the area of coniferous high forest. The first large scale plantings of conifers took place on the Elan Estate of the Corporation of Birmingham shortly after its reservoirs were completed in the early years of this century. Hardwood high forest had marginally increased to 1729 hectares, whilst mixed high forest had declined to 108 hectares. The area of scrub had decreased to 1017 hectares, whilst the second world war fellings had cleared 1761 hectares of woodland. Coppicing was reported from 252 hectares and coppice with standards had shrunk to but 82 hectares. The surveyors also reported that 246 hectares of woodland marked on the 1905 edition of the 6 inches to 1 mile OS map was no longer woodland.

The most recent Forestry Commission census took place between 1979 and 1982. Unfortunately by this stage Radnor had become part of Powys and all published data refers to Powys.

The Nature Conservancy Council in its *Provisional Radnor Inventory of Ancient Woodlands* (Peterborough, 1991) provisionally identified woodlands of 2 hectares or more in extent which were considered to be ancient in origin. Comparing the extent of such woods in 1986-88 with those considered ancient which were marked on the O.S.1.25,000 (First Series) maps, surveyed between 1885 and 1887 (the 1st edition 1 inch to the mile maps could not be used as dense hachureing obscured woodland features), it was noted that in this approximately one hundred year period 49 percent of these ancient woodlands had been destroyed by conversion to grassland or conifer crops. Only 2829 hectares of ancient, semi-natural woodland remained. Over 60% of this destruction seems to have taken place since the Forestry Commission's 1947-49 census. Government grant aid in the uplands in post-war years to drain and agriculturally improve land certainly provided a stimulus to clear woodland and the financial benefits of converting broadleaved woodlands to shorter rotation conifer plantations led to further broadleaved woodland loss.

Concerned at the rapid decline in our woodland heritage the Forestry Commission dramatically altered its woodland policy in 1985 to favour the planting of native broadleaved species and conserve ancient semi-natural woodlands. In Radnor this has greatly reduced the felling and grubbing out of semi-natural woodlands, though it has stimulated few new plantings.

Most Radnor woodlands are grazed for at least part of the year and provide valuable shelter for livestock. A recent trend to feed and overwinter greater densities of livestock in woodlands to prevent damage to agriculturally improved grassland has been a worrying development. In the worst cases it has led to almost total destruction of the natural ground-layer species.

Less intensive grazing causes the suppression of natural regeneration of woody species and threatens the long-term survival of woods. The Welsh Office Agriculture Department has recognised this problem in the Cambrian Mountains Environmentally Sensitive Area and offers annual payments to farmers prepared to exclude grazing stock from their woodlands. Of equal concern is the fact that, with but one or two notable exceptions, most broadleaved woodlands seem to have been silviculturally abandoned as profitless by their owners. This must place their future in question. Funded by both central and local government the Coed Cymru initiative has been established to look for ways to renew an interest in woodlands, particularly amongst the farming community. On the outcome of this complex interaction of government policy, market forces and personal whim hangs the long term future of most of Radnor's woodland flora.

C. Recent Extinctions of Native Species

By comparing the information provided by published plant records and specimens held in herbaria dating back about 100 years with our knowledge of the present flora of Radnor, it appears likely that 46 species of native British higher plants have become extinct in the county during this period. This is 6% of the flora of the county. However, 13 of these species may never have been more than accidental introductions, being recorded only from short-lived, man-created habitats such as arable land. A few others may possibly have been erroneously reported in the first place. Cowbane, for instance, is a distinctive plant, probably quite long-lived, which was described from two well-botanized localities by the same observer in one summer. It has never been reported again in over a 100 years, yet suitable-looking habitat persists in both sites. Bristol rock-cress remains an enigma, having been collected from an unspecified ruined fort near Llandrindod Wells in 1894. Its identity has never been challenged, nor has it ever been found again away from the Avon Gorge, near Bristol. Still other species, notably annuals, regularly spend long periods as buried seed. Their absence above ground is no proof of their extinction. The cornflower is a good example. If a viable seed bank seems likely still to be present the plant has not been considered to be extinct. Notwithstanding these problems, a number of distinctive, attractive and well-documented species have become extinct.

Table 6.2 lists the probably extinct species, with the date they were last recorded and the type of habitat it is believed they occupied. That nearly half the species show a preference for lime-rich soils in Britain may be of note. Acidic pollutants may have reduced the base status of Radnor's mostly base-deficient soils to a level at which they are no longer capable of supporting some species eg. kidney vetch, mountain everlasting and hoary plantain.

Roadside verges prove to be a habitat that has suffered a number of extinctions. Of the 5 species listed from this habitat the greater broomrape is known to have been exterminated by road widening. Others may have been affected by mechanical trimming of verges and hedges. Biennials and short-lived perennials which flower late such as spreading bellflower, greater burdock and field scabious could suffer a serious reduction in seed production if cut in August or early September.

Ploughing, reseeding and/or the use of artificial fertilizers may have contributed to the loss of frog orchid and pepper saxifrage. Housing development exterminated the last colony of southern marsh-orchid.

Eight woodland species may be extinct. An increase in grazing stock, particularly sheep, may, in part, be to blame. Limestone fern and wood vetch are most likely to have been affected in this way.

Table 6.2 Native vascular plants believed to have become extinct since 1880.

Name	Year* last reported	Habitat
Alder buckthorn	1957	Woodland
Alpine enchanter's-nightshade	1890	Woodland
Alpine penny-cress	1900	Rock outcrop
Bird's-nest orchid	1880	Woodland
Blunt-flowered rush	c.1954	Wetland
Bristol rock-cress	1894	Ruins of fort
Buckthorn	1956	Woodland
Corn buttercup	1880	Cornfield
Cowbane	1880	Wetland
Dwarf spurge	1956	Arable
Early forget-me-not	1953	Rock outcrop
Field gentian	1938	Upland pasture?
Field mouse-ear	1956	Hedge and wall
Field pepperwort	1948	Roadside
Field scabious	1969	Grassy bank
Frog orchid	1923	Grassland
Great fen-sedge	1964	Lake margin
Greater broomrape	c.1958	Roadside
Greater burdock	c.1954	Roadside
Green spleenwort	1986	Rock outcrop
Grey-flowered speedwell	c.1954	Arable
Heath cudweed	1947	Grassland?
Hoary plantain	c.1960	Grasslands
Kidney vetch	1977	Grassy banks
Knotted hedge-parsley	1938	Rock outcrop
Large-leaved lime	c.1954	Shale banks
Lesser hairy-brome	1886	Woodland
Limestone fern	1883	Woodland
Marsh fern	1879	Peatland
Marsh helleborine	1968	Wetland
Mountain everlasting	c.1954	Upland grassland
Pellitory-of-the-wall	1938	Walls
Pepper saxifrage	1890	Meadow
Purple ramping-fumitory	1920	Arable?
Ramping fumitory	1929	Arable?
Six-stamened starwort	1951	Lake margin
Southern marsh-orchid	1981	Meadow
Spreading bellflower	1965	Roadsides
Toadflax-leaved St. John's-wort	1945 1955	Rock outcrop
Tor grass	1973	Grassland
Vervain	1973	Shingle
Wild mignonette	1956	Roadside
Wild parsnip	1934	Railway
Wood vetch	c.1954	Woodland
Wooley thistle	1990	Quarry spoil
Yellow-juiced poppy	1956	Railway

* The date given in this column should be treated with caution. For some species it is the last year it was observed to grow in Radnor. For others it is the date of the last published record and the plant may have survived for many years after this time. For yet others, notably those with a last report in c.1954, they may have become extinct before this date since they are recorded without a date in a manuscript *Flora of Radnor* by Webb, written c.1954. For details see the individual species accounts below.

The causes for the extinction of 4 species reported from rock outcrops and scree can only be guessed at. In these naturally open communities an introduced seed may grow and produce flowers but might fail to to set seed eg. alpine penny-cress or early forget-me-not. Other known or suspected causes of extinction in other habitats include raising of lake water level (great fen-sedge) and changes in railway maintainance (wild parsnip and possibly yellow-juiced poppy).

For many species the cause of extinction is unknown. This is particularly true for those species where apparently suitable habitat persists, yet the plant can no longer be found eg. marsh helleborine, pellitory-of-the-wall and toadflaxed-leaved St John's-wort.

D. Recent Extinctions of Non-Native Species

Twenty-five species, not native to Britain, but considered by at least one recorder to have become naturalized in Radnor seem to have failed to maintain a foothold and have become extinct. The decision as to whether a plant has become naturalized is sometimes a difficult one. Scotch laburnum is a good example. There is no evidence that the trees which grew in Radnor were anything other than grown and planted by man, albeit in rural hedgerows.

Table 6.3 lists these probably extinct non-native species. All the provisos made above with respect to the date of the last record and the possibility of a viable seed bank remaining apply here. The overwhelming majority of these species favoured disturbed ground. In Radnor this habitat is generally scarce and is seldom maintained for long in any one place. Even Oxford ragwort is only able to maintain a tenuous hold in the county.

E. Vascular Plants Showing a Notable Expansion of Range.

In contrast to the last group of species considered, this group seem to have either successfully colonized the county in the last 100 years or to be expanding their range. The open conditions of river banks and disturbed ground are the favoured habitats of these species. Table 6.4 lists the most successful non-native species and indicates when they were first recorded from the county. As so little recording was carried out in the first half of this century it is likely that first colonization took place some time before at least some of these records were made. Few native species seem to have shown such an expansion. The many-seeded goosefoot and red goosefoot have both been found in new localities recently, favouring disturbed, nutrient-rich sites. Henry Ridley makes an interesting remark regarding rosebay willowherb in 1880. Writing in the *Journal of Botany*, **19**, 171-172 he records it as native on Rhos Common (probably Rhosgoch Common), but as being planted or escaped in many other places, especially about railway banks, suggesting it had recently expanded its range.

Table 6.3 Extinct non-native vascular plants

Name	Year last reported	Habitat
Annual mercury	1984	Garden
Annual wall-rocket	1973	Waste ground
Black mustard	c.1954	Clay bank
Chicory	c.1975	Roadside/grass ley
Common amaranth	1983	Arable
Dwarf cherry	1982	Woodland edge
False thorow-wax	1976	Footpath
Flax	1881	?
Grey field-speedwell	c.1954	Arable
Himalayan cotoneaster	c.1954	?
Hoary mustard	c.1954	Waste ground?
Irish saxifrage	1938	Waste ground?
Lucerne	1931	Railway bank
Meadow brome	1886	Field border
Narrow-fruited cornsalad	1890	Arable
Pearly everlasting	1947	Field margin
Perfoliate honeysuckle	1951	Wall
Poor-man's pepper	1957	?
Procumbent yellow-sorrel	1945	Pathside Arable
Rye brome	1880	
Scotch laburnum	c.1954	Hedges
Shepherd's needle	1874	Arable
Slender soft-brome	1956	?
Velvetleaf	1983	Arable
White mustard	c.1954	Arable?

Table 6.4 Non-native vascular plant species which have recently successfully colonized Radnor.

Name	Year first reported	Habitat
American willowherb	1983	Various
Indian balsam	1920	River bank
Japanese knotweed	1945	River bank
Lesser swine-cress	1945	Arable
Monkeyflower	1945	River
New Zealand willowherb	1961	Wet rock
Pineapple weed	1909	Roadside
Pink purslane	1971	River bank
Slender rush	1938	Paths
Slender speedwell	1956	River
Variegated yellow-archangel	1988	Road verge

7. Conservation of Vascular Plant Species

The maintainance of a diverse wild plant flora can no longer be left to chance. If there has been a single innovation in forestry or farming in recent times that has created opportunities for anything other than ubiquitous weeds it has passed my attention. The loss of species-rich grasslands and the spread of conifer plantations has been documented above. To conserve our flora requires positive action. To help determine priorities this section seeks to value the many elements of the flora on an international and local scale. Since the rarer a species is the more likely it is to face extinction, considerable weighting is given to rarity.

A. International Rarities

The British and even more particularly the Radnor vascular plant flora, when considered on a European scale, has relatively few species. This is probably because only a limited number managed to return following the retreat of the ice at the end of the last ice age, the English Channel soon becoming an effective barrier to further immigration. In the short time since this event, there has been little opportunity for new species to evolve. Hence there are few British endemic species, other than in apomictic genera such as **Hieracium**, **Rubus** and **Taraxacum**. On an international scale endemic species, since they occur nowhere else, are of considerable importance.

In Radnor all the extant British endemic species are in apomictic genera. They are listed (except for species in **Rubus** and **Taraxacum** which require further study) in the section on biogeography above. Only for the two whitebeam species from Aberedw and the Elan Valley do we have any clear knowledge of their present status in the county. The remainder, an eyebright and 9 hawkweeds, all require survey to establish their current status.

The importance of endemic taxa was recognised by the International Union for Conservation of Nature and Natural Resources (IUCN) when in 1976 they drew up a list entitled Rare, Threatened and Endemic Plants for the Countries of Europe. Of Europaean endemic species they placed the pillwort in the vulnerable species category. Throughout its range the shallow, mud-bottomed pools it

COUNTRYSIDE COUNCIL FOR WALES

requires have become rarer due to drainage. Radnor's commonland is now probably an international centre for this species.

Also listed by IUCN as vulnerable is the bog orchid. Again drainage threatens its existance on mainland Europe, whilst Britain still supports a significant population. In Radnor it is known from a single bog in the Elan Valley. The status of Killarney fern may need to be reviewed following the recent discovery that its prothallial stage is widespread in western Britain. Considered vulnerable in Europe by IUCN, it is, however, never likely to be found widely in Europe and care should be taken to conserve Radnor's prothallial population.

Other flowers, though common in Radnor, have a surprisingly limited world distribution. These include bluebell, western and common gorse. They are confined as native species to western Europe, with probably their largest populations in the British Isles. Many of the Oceanic West Europaen element of Radnor's flora are internationally scarce and also warrant careful conservation. They are listed in the section on biogeography above.

B. National Rarities

Perring and Farrell (1983) in the British Red Data Book of Vascular Plants list those native plant species which have been recorded in 15 or fewer 10 kilometre squares of the national grid since 1930. These are regarded as nationally rare species. Eight species from this list have been reported from Radnor and are listed in Table 7.1. All have been reported from a single Radnor locality. Since 1983 further survey has shown that the cornflower is now so scarce it should be added to this list.

Table 7.1 Nationally rare vascular plant species

Species	Number of recent British Sites	In Radnor Present On NNR	SSSI
Bristol rock-cress	1	No	No
English whitebeam	12	No	Yes
+Killarney fern	*	No	No
+Perennial knawel ssp.	1	Yes	Yes
+Radnor lily	1	Yes	Yes
+Rock cinquefoil	4	No	Yes
Sticky catchfly	12	Yes	Yes
Upright clover	3	Yes	Yes

* The recent discovery of many new sites for the prothallial stage and great secrecy surrounding records for the sporophyte stage prevents the calculation of an accurate figure.

+ Protected by Schedule 8 of the Wildlife and Countryside Act 1981.

From Table 7.1 it can be seen that the sites for half of these nationally rare species are protected by National Nature Reserve status. The Bristol rock-cress was only reported once in Radnor and the record requires confirmation. Cornflower, mentioned above but not included in the table, appears never to have been more than a casual species in Radnor. Of the remaining species only the Killarney fern is not protected by having its site designated as a Site of Special Scientific Interest. Its status as a rare species needs to be reviewed.

C. Nationally Scarce and Locally Rare Species.

i) Nationally Scarce Species

The Nature Conservancy Council in its "Guidelines for the Selection of Sites of Special Scientific Interest"(1989) provides a lists of higher plant species which occur in between 16 and 100 ten kilometre squares of the national grid. To these have been added a few species which, whilst occuring slightly more commonly than this, have shown a significant decline in abundance recently and might by now fall into this category. Forty-three of these species have been recorded from Radnor, though 15 are believed to be extinct and 3 may be introductions.

Of the extant species, notable populations are found in this county of chives, bog rosemary, narrow-leaved bitter-cress, maiden pink, upright chickweed, rock stonecrop, wood bitter-vetch and pillwort (the latter also considered above in the internationally scarce category). Some form of statutory protection is afforded 19 out of the 25 species of extant, native higher plants included in this category. The remainder consist of 3 weed species of sporadic occurrence, whilst 2 occur in proposed SSSI's and 1 in a Wildlife Trust nature reserve. A complete list is provided in Table 7.2.

Table 7.2 Nationally scarce vascular plant species in Radnor

Probably Extinct

Alpine enchanter's-nightshade	Lesser hairy brome
Alpine penny-cress	Limestone fern
Corn buttercup	Narrow-fruited cornsalad
Cowbane	Needle spike-rush
Greater broomrape	Purple ramping-fumitory
Heath cudweed	Shepherd's needle
Large-leaved lime	Six-stamened waterwort
	Spreading bellflower

Extant

Bog orchid	Rock stonecrop
Bog rosemary	Shepherd's cress
Chives	Small-flowered buttercup
Euphrasia rostkoviana -an eyebright	Small-flowered catchfly
Fenugreek	Spear-leaved willowherb
Forked spleenwort	Spiked speedwell
Green-winged orchid	Upright chickweed
Long-stalked pondweed	Wall whitlow-grass
Maiden pink	Welsh poppy
Mudwort	**Sorbus porrigentiformis** -a whitebeam
Narrow-leaved bitter-cress	Wood bitter-vetch
Pale St. John's-wort	Wood fescue
Pillwort	

Probable introductions

Mezereon	Stinking hellebore
Mountain currant	

ii) Locally Rare Species

Some additional 138 species of vascular plant occur in 5 or fewer localities in Radnor. They may be identified from the species accounts below since, where known, detailed site information is provided for all plant species found in six or fewer localities. In many cases their existance in Radnor is precarious. Every effort should be made to safeguard their sites. The Countryside Council for Wales maintains a register of these species and their sites, in addition to the categories considered above and would welcome any information on the current status of these rare species.

D. Conservation Action

The selection and notification of the best examples of wild plant habitat as **Sites of Special Scientific Interest** was the responsibility of the Nature Conservancy Council, now the Countryside Council for Wales. SSSIs form the backbone of the statutory conservation system. It is a legal designation applied to land of special biological or geological interest. By notifying all parties that have a legal interest in the land and all agencies such as local authorities, the National Rivers Authority etc. who may, at some time affect the land, the importance of its scientific interest is made widely known. Through a formal consultation system potentially damaging activities are drawn to the attention of CCW, allowing them to seek ways of resolving conflicting interests. Resolution is, sadly, not always possible and this system cannot protect all such sites forever. The land remains in private ownership and unless an access agreement has been negotiated, the public (and CCW) have no rights of access to the land. It is, perhaps, a typical British compromise and one which, given goodwill on all sides, is moderately successful. Without this system, it is doubtful that many species-rich hay meadows, for example, would have survived.

By the end of 1991 there were 69 SSSIs in Radnor, occupying an area of 15,204 hectares. The largest was Elenydd, an upland site of which 6,620 hectares occur in Radnor, the longest was the R. Wye, whilst the smallest was a tiny roadside quarry of 0.04 hectares of special geological interest. There are, in all, 8 geological sites and 61 sites of varying botanical importance in Radnor.

The rationale behind the selection of biological sites is detailed in Guidelines for Selection of Biological SSSIs produced by the NCC at Peterborough in 1989. Botanical sites are selected as the best examples of habitat types as defined by the National Vegetation Classification (Rodwell 1991a and 1991b) or as supporting notable populations of nationally rare or scarce species. On this basis in Radnor 17 woodland, 31 grassland, 8 open water, 8 upland, 12 wetland and 2 higher plant sites have been scheduled as SSSIs. Note that, as some SSSIs are of multiple interest, there are more "habitat sites" than SSSIs.

The further designation of SSSI's as **National Nature Reserves** (NNR) may allow extra control to be excercised. The amount of control depends on the form of tenure or terms of an agreement the CCW has been able to secure over the land. In Radnor the CCW manages 3 NNRs. One, (Cors y Llyn) is owned outright with considerable freedom to manage; one, (Stanner Rocks) is held on a long lease, providing almost equal freedom, whilst a third, (Rhosgoch Common) though owned, is subject to commoners' rights and effective management may only proceed if unopposed by the commoners. All three sites have been threatened by man-induced changes and as a result require active management if their interest is to be retained. The interest of these sites is discussed in the section on plant communities below.

The adverse effects of modern forestry practices and agriculture has been recognised by both the Government agencies reponsible for their promotion and control. The Forestry Commission has announced the establishment of a **Forest Nature Reserve** at Stanner Rocks to complement the CCW NNR. The Welsh Office Agricultural Department (WOAD) has established two "**Environmentally Sensitive Areas**" (ESA's) in Wales. As a result of a British Government initiative in 1985 the European Council of Ministers approved a scheme whereby farmers might be paid to conserve the best features of our countryside. West Radnor was included in the Cambrian Mountains ESA designated in 1986, with an extension in 1988.

Farmers in ESA's are offered a five year agreement which provides them with an annual payment in return for following a prescribed set of farming practices. Within the Cambrian Mountains ESA payments are made to conserve agriculturally unimproved grasslands, semi-natural rough grazings and native broadleaved woodlands. Whilst the system is voluntary, the takeup of all except the provisions for woodland has been good. It is to be hoped that ESA status is soon extended to the rest of Radnor before agricultural improvement work destroys what few plant-rich sites are left.

Voluntary nature conservation bodies have established nature reserves in Radnor. The most notable is the **Radnorshire Wildlife Trust**. It manages 10 reserves, some of which are owned whilst others are managed by agreement with their owners. Its largest reserve of 153 hectares is Gilfach Farm, north of Rhayader. Purchased in 1988, the Trust manages it to maximise the interest of its meadows, rough grazings, woods and wetlands. For a visitor to Radnor wishing to see its upland flowers this is, perhaps, the best site to visit.

Other notable Wildlife Trust reserves include Burfa Bog, near Walton with wet pasture and woodland, Bailey Einon, a species-rich wood at Cefnllys and Cefn Cennarth, a sessile oakwood north of St Harmon.

The **Royal Society for the Protection of Birds** manages two woodlands, Dyffryn Wood and Gwynllyn Wood, near Rhayader. These are good examples of sessile oakwoods. The **Woodland Trust** owns two fine dingle woodlands in the lower Wye Valley. Cilkenny Dingle Wood is one of the most species-rich woodlands in Radnor.

The map below illustrates the distribution of the areas considered above.

Map 7.1 Areas of Radnorshire protected by statutory designations and/or managed to conserve wildlife by various organisations as of 1st January 1992

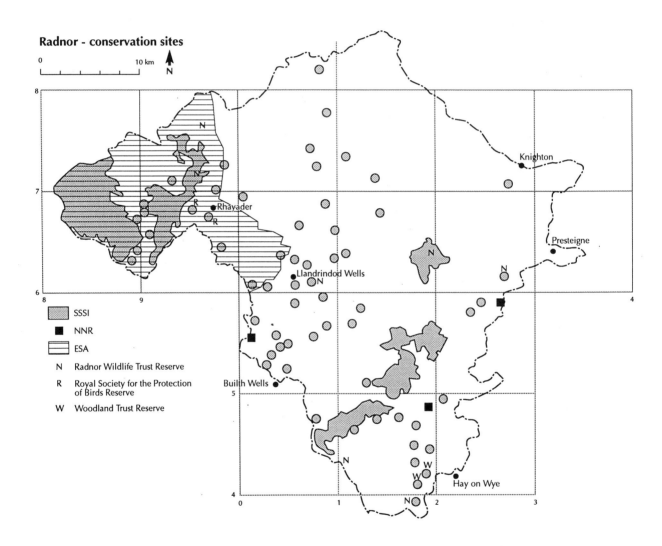

Footnote: Since the above map was prepared a new Environmentally Sensitive area covering the rest of Radnorshire has been announced.

8. History of Vascular Plant Recording

A brief history of higher plant recording is provided here. Whilst it is hoped that reference is made to all major published works and manuscripts no claim is made to completeness. In particular a survey of major herbaria has not been undertaken. **P.W. Carter** in Vol. 20, 42-86 of the *Radnorshire Society's Transactions* for 1950 provides a detailed and readable account of the history of botanical exploration of Radnor up to that time. He refers to Radnor as the "Cinderella" of Welsh vice counties, noting the paucity of early records. The earliest records he was able to trace were made by the **Rev. Littleton Brown**, a Shropshire man, born in Bishop's Castle and residing there as a clergyman until 1731. He visited Rhayader in 1726 and in a letter to Dillenius he wrote of the quantity of butterwort and bulbous rush in the bogs and found mountain pansy "in ye mountains". This latter plant must have been a feature of the hills since he notes it as being plentiful in the mountains on his way to New Radnor. A specimen of brittle bladder-fern which he collected from Water-break-its-neck is preserved in the Dillenius herbarium at Oxford.

Most travellors in these early years bypassed Radnor. **John Evans**, a Bristol school teacher, however, seemed somehow to contrive to visit it on a journey from Cardigan to Tregaron in 1804. In his *Tour through South Wales* 10 species of plant were noted from Radnor. Carter (1950) selects *Swertia perennis* for mention. Quite what this plant was I am not certain. There is an alpine plant of this name today but it has never been reported from Britain. *Allium boreale* is similarly a mystery whilst the cloudberry (*Rubus chamaemorus*), also reported by Evans, has never again been reported south of the Berwyn Mountains.

With **T. Westcombe** we are on firmer ground. In a letter dated October, 1843, written from Worcester to *The Phytologist* 1, 781, the first useful list of higher plants for the county is found. He visited Clyro, finding enchanter's-nightshade, herb-paris, and giant bellflower amongst others. At Rhosgoch he reports bog asphodel, bogbean, marsh cinquefoil, petty whin, shoreweed, early marsh-orchid, lesser marshwort and broad-leaved cottongrass. At Llanbwchllyn were white water-lily, greater spearwort, marsh

Aberedw Rocks

willowherb and shoreweed. It is gratifying to note that most of these species can still be found in these sites over a century and a half later.

In 1851 the **Woolhope Naturalists' Field Club** was formed. Taking its name from a village near Hereford it, in fact, drew members from a wide area of Mid Wales and the Marches. Excursions were held on Tuesdays, the society not catering for the working class. For many years it was also a male preserve, with occasional meetings admitting ladies "by special ticket". With one of its objectives being "to form and publish correct lists of the various natural productions of the County of Hereford" Offa's Dyke presented no barrier and its *Transactions* became a major source of published plant lists for Radnor as well.

It strayed over the border into Radnor on its third field meeting. The staying power of its newly recruited membership can be but wondered at. This meeting commenced at 6am at the Green Dragon Hotel in Hereford, took breakfast at Kington, then via Presteigne and Nash Quarry, entered Radnor at Stanner Rocks. It returned to Kington for dinner, after which papers on various subjects were read. Tea was taken at 8pm and the party regained Hereford at 11pm. On Stanner Rocks, a place known locally as "The Devil's Garden", were noted the now familiar flora of shepherd's cress, sticky catchfly, spiked speedwell, stork's-bill, crow garlic, perennial knawel, hare's-foot clover, bloody crane's-bill, bird's-foot etc..

In 1867 the club visited Radnor twice. On June 28th, taking advantage of the recently opened Central Wales Railway, they visited Llandrindod Wells. The "bog at the upper edge of the common" was examined. It yielded records of common cottongrass, marsh cinquefoil, bog St.

John's-wort, bogbean, early marsh- orchid, fragrant orchid, marsh lousewort and meadow thistle. The majority of this peatland was almost certainly excavated shortly afterwards to create Llandrindod Lake. A small but interesting fragment of bog survives across the road, to the south west of the lake, beside a housing estate. The common itself has changed also. The Woolhope Club members recorded heath rush, cross-leaved heath, lousewort, heath bedstraw and, most interestingly of all, mountain everlasting. None of these species survive today. In this account there is also a whimsical note on the apparent absence of Llandrindod Wells. Stepping from the train no town could be seen. Apparently at that time Llandrindod consisted of little more than a few large hotels, so buried in vegetation they could not easily be seen.

On the 18th of July the Club held an extra meeting, visiting the Bach Howey gorge near Llanstephan. Considered then to be one of the scenic wonders of mid Wales, it had recently become more accessible following the construction of the Mid-Wales Railway. The railway company had cooperated further by providing a temporary station close to the viaduct over the gorge. Ladies were also admitted and a successful party reported finding a couple of fronds of Wilson's filmy fern together with oak, beech and brittle-bladder ferns. All but the filmy fern still survive. The party noted with displeasure the spider-like web of new wire fences around the recently enclosed Trewern Hill. This is the only reference in a botanical account I have found to the enclosure of common land, an activity which was proceeding apace at that time. The effect on the flora must have been as dramatic as it was on the landscape.

From 1874 there appear in the literature many records made by the **Rev. Augustin Ley** (1842-1911). He had been born in Hereford in 1842 and educated at Oxford, became the incumbent of Sellack with King's Capel, near Ross-on-Wye. This was to be his home for 30 years. He was an outstanding botanist whose first notable contribution to the flora of Radnor was the publication of a catalogue of the flowering plants and ferns of Radnor in the *Report of the Botanical Locality Club* of 1874, pp. 80-86.

The 303 species listed here, without any information as to locality, represent the first substantial published account of Radnor's higher plants. Whilst most of the species noted are still found in Radnor today, this list provides the only evidence for the occurrence of shepherd's needle in the county. Additional records, with, in most cases a brief description of locality, made in 1881 were published in the *Report of the Botanical Record Club*, 1883, pp. 246-247. The 88 species recorded here included greater butterfly-orchid and twayblade from the Elan Valley.

But perhaps his most interesting list of plants appears in the report of this club for 1884-1886, pp. 144-146. Published here was his exciting discovery on the 13th July 1886 of rock cinquefoil from a "rocky ridge, in one station in some plenty". The exact locality, in what was then its second known British site, was withheld to protect the plant and speculation continues as to whether this was the population known today from the bank of the R. Wye. His other notable finds from this period include chives and lesser meadow rue from beside the Wye at Llanstephan and lesser teasel from the gorge at Aberedw.

The Rev. Ley's botanising in the Aberedw area produced one of the few detailed species lists ever published for a circumscribed area of Radnor. His *First Contribution towards a Flora of Aberedw, Radnorshire* appeared in the *Transactions of the Woolhope Naturalists' Field Club* for 1891, pp. 180-198. It had been first read on the lawn of Mr. Mynor's fishing cottage on the bank of the R. Edw upstream of the church during a visit of the Woolhope Club to Aberedw on the 30th June 1891. It lists, with brief details of locality, 340 flowering plant, 18 fern and 120 moss species. (The Rev. Ley never seems to have taken to liverworts). The area encompassed includes such botanically interesting sites as Llanbwchllyn and the Wye Valley as far down as Boughrood. Amongst first county records which appear here were wild service tree- "a small bush at one station in the rocks," welsh poppy- "plentiful and clearly native in the wooded gorge" and both meadow and mossy saxifrage. All are still present today. The losses have been surprisingly few. Most notable are wood bitter-vetch, which grew near the castle and hound's-tongue, which was described as being abundant near the bottom of the rocks. With Ley's death in 1911, the Welsh Marches lost its most gifted botanist. Many obituaries appeared, written by friends who had also taken an interest in the plants of Radnor.

In the *Woolhope Club's Transactions* published in 1914, pp. 195-204 both the **Revs. W. Moyle Rogers and HJ Riddlesdell** mourned his death. The former had published in 1899 in the *Journal* *of Botany* 37, 17-25 an account of his visit in 1898 to Brecknock and Radnor, part of the time accompanied by his son and the Rev. Ley. As with so many visitors to Radnor, this account commences with a list of those species the visitor was surprised not to find. Fortunately most of them, eg. cross-leaved heath, wood spurge and hemp agrimony, were probably not truly absent, but were merely overlooked. Nevertheless it is notable that accounts of the flora of Radnor seem to dwell frustratingly on that which was not found, whilst telling us all too little about what had been discovered. The **Rev. W. Moyle Rogers** achieves a far more satisfactory balance by listing over 370 species as having been found, whilst noting the rarity of such common southern British plants as white dead-nettle, bittersweet and purple loosestrife and the abundance of musk mallow, still a feature of Radnor. Most notable of his records are pale St. John's wort and ploughman's spikenard from around Stanner Rocks, dwarf spurge from Boughrood and lesser teasel from Llowes.

The **Rev. H.J. Riddelsdell** visited the county in July, 1909 with a party from the **Cotteswold Club**, reporting his findings in the *Proceedings of the Cotteswold Naturalists' Field Club* of 1910, pp. 57-61. He notes the abundant, if rather species-poor heathland vegetation, so different from the Cotteswold Hills and Vale of Gloucester. They explored the Carneddau Hills, Llanelwedd, producing a list of species which, with but a few exceptions, might be made today. Pineappleweed, then a recent colonizer of Radnor was found on roadsides at both Llanelwedd and Boughrood. Penddol rocks on the Wye above Llanelwedd were visited and a white-flowered form of chives was remarked upon. A visit to Llyn Gwyn, near Rhayader, produced a very interesting record of water lobelia, never reported from here again, whilst in the hills about the Elan Valley the yellow-flowered form of mountain pansy was found to be abundant. Finally Aberedw Wood was visited, providing records of Welsh poppy, nodding melick and mossy saxifrage amongst others.

These accounts had built on a well-established tradition by then, of publishing records, particularly of plants previously unreported in a county. **H.C. Watson** had in 1873-1874 produced his *Topographical Botany* which listed the occurrence of plants in each of his British vice counties. A correspondent, who along with the Rev. Ley probably supplied Watson with many of his records for Radnor was **E.H. Jones** of Cwmithig, near Nantmel. If I am correct in my assumption, he was actually Edward W.H. Jones, who was buried in Nantmel Churchyard in 1900, aged 74 years. Kent and Allen (1984) state that his herbarium was donated to H.C. Watson. Since Watson's herbarium is in Kew, a search there, a task I have been unable to undertake, may reveal some early plant specimens from Radnor. An interesting manuscript in Edward Jones' hand, written in 1879 is preserved in the Welsh Folk Museum. Entitled *A Pebble Rescued From the Waves of Time*, it had, in fact, been rescued from a fire in Llangeitho, Cardiganshire, though how it got there is unknown. Sadly, though containing a little local geography he confines himself to a historical account of his own family and a number of local households without a reference to a plant. Cwmithig was unoccupied for many years and though a few relatives remain, any plant records made by that exceedingly rare species, the native Radnor botanist, have still to be found.

Despite all the above-mentioned activity, Radnor vice-county had not been well-studied and it was perhaps the chance of filling in some of the gaps in Watson's list that brought **H.N. Ridley** to the county in 1880. He published his records in the *Journal of Botany* of 1881, pp. 170-174. Commenting that the "Radnor Forest consists of a considerable area of low hills, none more than 2166 feet in height", he was clearly unimpressed with the uplands of the district. Plants he considered common in Herefordshire which he failed to detect in Radnor included comfrey, greater knapweed, mistletoe and small toadflax. Amongst the 111 species he lists are the narrow-leaved bitter-cress at Aberedw, burnet rose at Stanner, white bryony near Radnor (an area in which it has never been reported again), cowbane from "Rhos Common" and "in a stream on Mount Carneddau" and bird's-nest orchid from a wood above the Edw, near Aberedw. Neither of these latter two species have ever been reported in the county again. Further notes of H.N. Ridley appear in the *Journal of Botany*, 1884, p. 378, including the only record ever of limestone fern from Radnor, which he discovered in Aberedw Wood.

Further gaps in *Topographical Botany* were filled by a visit to Wales by **G. Claridge Druce**, reported in the *Journal of Botany* of 1908, pp.335-336. Visiting Boughrood, he notes vervain, hemp agrimony, sessile oak and white willow, amongst others, as new county records. Not new, but notable was soapwort beside the Wye. This plant must have escaped from cultivation at a fairly early date. He makes no mention of Indian balsam, although he recorded it from beside the Usk at Crickhowell in Brecknock. It, presumably, had not yet become established beside the Wye.

Other contributions to our knowlege made about this time arose from a visit in 1896 by the Woolhope Club to the site of the construction of Birmingham Corporation's waterworks in the Elan Valley. Reported in their *Transactions* published in 1898 are the discovery of lily-of-the-valley, wood bitter-vetch, greater butterfly-orchid and mountain pansy, amongst other plants.

W.A. Clarke visited Llandrindod Wells, providing a brief account in the *Journal of Botany*, 1901, pp. 279-280. He found whorled caraway to be abundant beside the lake and maiden pink in several places near the quarries. The latter still persists, but the former appears to have gone. The botanical contents of local guidebooks he notes are generally "feeble and disappointing". Bufton's guide to Llandrindod Wells he claims as an exception. Produced by W.J. and J.O. Bufton and entitled *Illustrated Guide to Llandrindod Wells* it was presumably to the 1906 edition that he referred. Fortunately the **Rev. John B. Lloyd** of Liverpool had stayed in the town in late June and early July, 1896 and contributed a list of over 300 plant species to this revised edition. All were apparently recorded within 2 miles, or thereabouts, of the Emporium in the town, though no information on habitat or locality was provided. Whilst almost all these species can be still found in this area today, a few notable species appear to have been lost eg. knotted pearlwort, tutsan, field scabious, melancholy thistle and parsley fern.

The site for the latter species was cryptically hinted at by the **Rev. A. Wentworth Powell**, one-time rector of Disserth. He had contributed to a chapter on flora in the first edition of Bufton's guide. This appears to have been reproduced in its entirety in the 1906 edition. Though lacking the detail of Lloyd's list it provides some useful insights. He notes the lamentable loss of ferns from around Llandrindod Wells. Their destruction had been so great he refused to provide any information as to their whereabouts. The fern trowel had clearly been used to effect. There was a hill, he notes, almost in sight of his house, where the parsley fern grew so abundantly a farmer mowed it for stable litter. As he wrote, only a few straggling plants were left and there have been no subsequent records from the Llandrindod Wells area. He mentions the more interesting flowers of more distant places such as the mountain pansy on the Radnor Forest; the hillsides in places white with the wild everlasting; the woods near Erwood Station in which the "great blue lampankula" was abundant.

After the Victorian era there was a rapid decline in the number of published plant records for Radnor. This was not peculiar to Radnor. Sinker et al (1985) refer to Shropshire botany in the first half of the twentieth century as entering a "dark age". The staff of the Botany Department of the National Museum of Wales, however, continued to collect material. From the early 1920's both **H.A. Hyde** and **A.E. Wade**, who for nearly 40 years were colleagues, visited the county and their findings and the records they were able to glean from the NMW herbarium and elsewhere were compiled to produce the book *Welsh Flowering Plants* in 1934. Radnor was found to support fewer species of flowering plant (573 were recorded), than any other Welsh vice county. This was attributed to its lack of large areas of limestone and limited amounts of open water. The annotated catalogue in this book provides a complete list of all flowering plants reported from Radnor, the first since Watson (1883). A second edition of *Welsh Flowering Plants* appeared in 1957, by which time 595 native species had been reported from Radnor.

The same authors compiled records of ferns to produce in 1945 *Welsh Ferns*. This edition did not deal with the clubmosses, quillworts or horsetails, but reported 25 species of true fern from Radnor. Locality information was provided for only 4 species, the most notable being pillwort from near Llanbwchllyn. A.E. Wade had found it there in 1928.

Wade had collaborated with **J.A. Webb**, a Swansea schoolteacher, for many years. The latter seems to have spent some time in Radnor, probably on regular annual visits. He was an active member of the **Swansea Field Naturalists' Society**, a group which seems to have made a number of char-a-banc trips into mid Wales, visiting such far-flung places from Swansea as Llanbadarn Fynydd and the Teme Valley.

Records were published regularly in the *Transactions* of the Society from 1929 to the mid 1950's. A collation of notable plant records (183 species) was published by Wade and Webb in 1945 in *The North Western Naturalist*, pp.

156-160. (It was later republished in the *Transactions of the Radnorshire Society* in 1947, vol. **XVII**, pp. 3-12). Webb contributed over three quarters of these records, many of which were of naturalized species. From the valuable information in this paper it has been possible to deduce a number of alterations in the distribution of plant species in subsequent years. These are detailed in the chapter on the changing vascular plant flora.

Perhaps the most valuable document of this period is a manuscript prepared by Webb with Wade's assistance entitled *Materials for a Flora of Radnor*. Written in now rather faded ink in fairly poor quality notebooks, it is preserved in the Botany Department library of the National Museum of Wales at Cardiff. To give an indication of the scope of this work a few sample entries dealing with species of varying degrees of rarity might be considered.

i) Common species eg. toad rush "Common and abundant". Bluebell "Common and abundant. White-flowered individuals not rare".

ii) Widespread but less common species eg. common duckweed "Thinly dispersed through the Shire". Green-ribbed sedge "Well dispersed on heaths but not very plentiful."

iii) For more local species the county was split into its major river basins ie. Wye, Teme and Lugg. Locality information was provided, if possible, for each. For example the entry for green-winged orchid states "Locally fairly frequent. **Wye.** Builth Road towards Llandrindod by the direct road in pastures: Builth to Builth Road. **Teme**. Beguildy 1931."

iv) Species previously reported from the county by others received the entry "N.M.W. List," being recorded on a card index held in the National Museum of Wales.

In the species accounts below all notable records from this work are reproduced. It is interesting to contrast this manuscript with a paper entitled *The Flora of Radnor* which appeared in the *Transactions of the Radnorshire Society* for 1937, pp. 48-58. Written by **O. Gibbin**, a Presteigne schoolteacher, he provides a useful description of the general vegetation of Radnor at a time when even the better quality (from an agricultural point of view) grasslands were flower-rich. Over 200 species of flowering plant are listed by the months of the year in which they flower. For only the 30 rarest species he noted is any indication of locality given. Beware that 8 of these are listed from Nash Rocks, a site just over the border in Herefordshire.

In the 1950's the **Botanical Society of the British Isles** (BSBI) began the enormous task of mapping the distribution of the vascular plants of the British Isles. The mapping unit chosen was the 10 kilometre square of the Ordnance Survey's national grid. A special meeting of the Society was held in 1956, based at Llandrindod Wells, to record in Radnor's 10 kilometre squares. Led by **F.H. Perring**, his report in the *Proceedings of the BSBI*, **2**, 416-418 follows a not unfamiliar pattern of plant reports for Radnor by providing almost as full an account of what was not found as that which was discovered. Thirty three species were recorded from the county for the first time. These included a number of calcicoles such as blue fleabane, milk thistle and houndstongue, which had been noted on the limestone at Dolyhir.

Much of the remainder of the work had to be shouldered by the voluntary BSBI county recorder. **Miss A. C. Powell** held this post from 1965 to 1988, in her field work contributing much to our knowledge of Radnor's flora. She compiled annually lists of notable records to be published in *Watsonia* and the *Welsh Bulletin* of the BSBI.

All this survey work was drawn together by F.H. Perring and S.M. Walters and in 1962 the BSBI and Nelson published an *Atlas of the British Flora*. For the first time it was possible to picture the distribution of the majority of Radnor's vascular plants and place it in a national context. By 1968 enough work had been carried out on a range of the more difficult-to-identify taxa to allow F.H. Perring to edit the records and the BSBI and Nelson to publish a *Critical Supplement to the Atlas of the British Flora*.

In 1969 a fifth edition of *Welsh Ferns* was produced. Expanded by **S.G. Harrison**, the Keeper of Botany at the National Museum of Wales, it now included clubmosses, horsetails and quillworts. Details were provided of specimens held of fir, alpine and stag's horn clubmoss from Rhayader, with additional specimens of the latter from Aberedw. But perhaps the most significant publication of the National Museum of Wales came in 1983 with the production of *Flowering Plants of Wales*. This monumental work was produced by **R.G. Ellis**, assistant keeper of botany. It was more than an update of Welsh Flowering Plants. For the rarer species details of localities and recorders are provided, whilst the commoner species are mapped at a 10 kilometre square level. It considerably updates the *Atlas of the British Flora* since post-1962 recording effort had greatly increased in mid Wales.

In 1970 the Department of Applied Biology of the University of Wales established a field station close to the Radnor border at Llysdinam in Breconshire. **Dr. F.M. Slater**, its curator, reviewed its history in the *Radnorshire Society Transactions* of 1987, pp. 92-100. The Llysdinam Estate had been chosen due to the long interest of the Venables Llewelyn family in natural history and the offer of land and buildings by the Llysdinam Trustees. The field centre has provided a focus for numerous projects which have documented the plant life of mid Wales. Early work was funded by the **Nature Conservancy Council** to allow D.G. Merry to carry out studies on the vegetation of the R. Wye (see Merry, D.G., Slater, F.M. and Randerson, P.F., 1981).

Dr. Slater made good use of finances from the Manpower Services Commission in conjunction with the then Hereford and Radnor Wildlife Trust, to survey Radnor's meadows and woodlands. The results of this work, together with additional surveys carried out by Sarah Mason and Anne Helmsley in 1984 and 1985 were edited by Dr. Slater and D.G. Moncur to produce *The Flowering Plants of Radnorshire*. This work, published by Llysdinam Field Centre in 1985, includes chapters on all the major plant habitats and an appendix listing over 600 species assigned to habitats.

Other notable botanical papers of Dr. Slater include *The rarer plants of Radnorshire* published in the *Transactions of the Radnorshire Society*, 1980, pp. 73-77 and *The biological flora of the British Isles*, No. 168, **Gagea bohemica** *(Zauschner) J.A. & J.H. Schultes* in *Journal of Ecology*, **78**, 535-546.

At the invitation of the University of Wales and the Llysdinam Trustees the Nature Conservancy Council (NCC) was able to occupy an office at the Llysdinam Field Centre from 1976 until it eventually outgrew the accommodation in 1985. I had joined the then **Nature Conservancy** (NC) in 1973 as the officer responsible for Radnor, though soon after finding myself also caring for Brecknock and Montgomery. By 1979 the workload was such that additional staff had become necessary and **I.D. Soane**, also a keen botanist, took over responsibility for Radnor, allowing me to pursue an interest in its flora as a pleasant recreation.

The NC and later the NCC, with a statutory duty to declare National Nature Reserves and notify Sites of Special Scientific Interest, had, since its creation in 1949, been carrying out botanical surveys and had collated the information supplied by others in order to identify the most important botanical sites. This work is now continued by **D.R. Drewett** and **D.C. Boyce** of the **Countryside Council for Wales**. Thanks to the generosity of these organisations, the University of Wales, the Botanical Society of the British Isles members and recorders (currently **Dr. D.R. Humphreys**) and the Radnorshire Wildlife Trust (Conservation Officer **D.P. Hargreaves**), all of whom have willingly shared their botanical records with me, it was possible to consider the collation of data leading to the production of a county flora for Radnor.

At the invitation of the Radnorshire Society and to stimulate an interest in Radnor's flora in the hopes of acquiring new records, two papers were prepared by the author for their *Transactions*. In 1987 a paper *The ferns and fern allies of Radnor* was published, to be followed in 1989 by one entitled *The trees and shrubs of Radnor*. So completes this somewhat subjective and necessarily selective account of the history of higher plant recording in Radnor. For the "Cinderella amongst the Welsh counties" described by Carter (1950), she has clearly known a number of recorders. I trust I have done justice to their labours, though fear I have not recaptured that incomparable pleasure that so many of these recorders must have felt on the discovery of a previously unreported plant. Of all the vice-counties in England and Wales, Radnor still probably offers the greatest opportunity to discover a new vice-county record.

To date 1105 vascular plant taxa have been recorded from the vice-county, 905 of which are considered to be British native species.

The total number of vascular plant taxa recorded from each 5 x 5 km square within Radnor is shown on Map 9.2 below.

Plate 3 Rhosgoch Common National Nature Reserve, near Painscastle. A dome-shaped or raised mire.

The acidic central area of the mire in August with heather and hare's-tail cottongrass. Downy birch is invading this area. (*A.S. Ferguson-CCW*)

The fen-meadow in June with ragged robin, common cottongrass, bulrush and many species of sedge. (*A.S. Ferguson-CCW*)

Plate 4

Aberedw Rocks. These lime-rich mudstone cliffs support a notable flora, including mossy saxifrage, whitebeam and Welsh poppy. (*RWT*)

Cwm Gwynllyn, near Rhayader. The finest nutrient-poor lake in Radnorshire with fringing reed-swamp. Sessile oakwoods clothe the valley sides, supporting a rich moss and lichen flora. (*D.C.Boyce-CCW*)

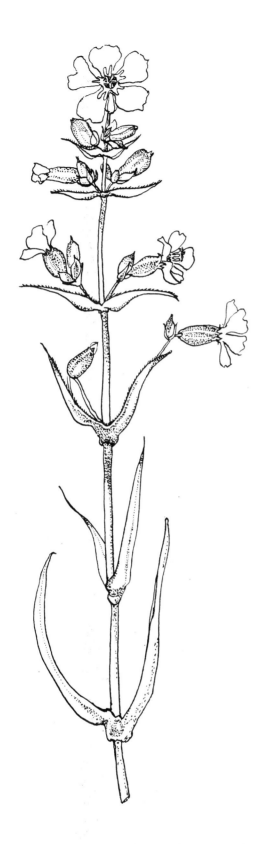

Sticky catchfly

THE VASCULAR PLANT FLORA

Map 9.1 Radnorshire vice-county - the recording units

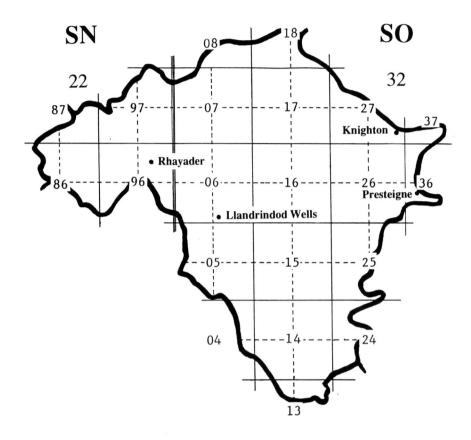

Map 9.2 The number of vascular plant taxa recorded from each recording quadrant (excluding microspecies).

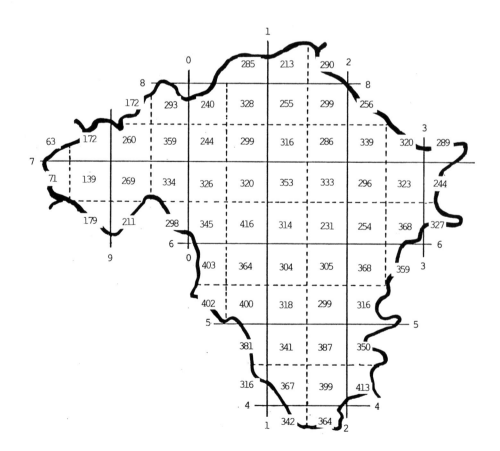

9. Fern and Fern Allies- Pteridophyta

LYCOPSIDA

Clubmoss Family
Lycopodiaceae

Fir clubmoss

Huperzia selago

A scarce sp. of peat-covered ledges and banks in the NW uplands. Browsed by sheep, well-developed specimens only occur on ledges inaccessible to grazing stock. Rocks to S of Llyn Cerrigllwydion Uchaf 8469; rocks to S of Craig Goch Resr. dam 8968; streamside, Gororion 8773 and Aberhenllan 8972, all in Llansantffraed Cwmdeuddwr; noted from near St Harmon 97SE by NMW staff(internal report); damp rocks near Cwm-berwyn, Llansantffraed-in-Elvel 0754; Llanbadarn Ffynydd c07 or 17 JS Stephens (in MsFlR, no date given).

Stag's-horn clubmoss

Lycopodium clavatum

An uncommon or overlooked sp. of short, acidic grassland and dwarf shrub heath, particularly on the top of banks or on ledges. Intolerant of heavy grazing or competition from other spp. it may be a ready colonizer, surviving only briefly in some sites such as the sides of forestry tracks. Stony woodland bank, Cwm Coel, Elan Valley 8964 RK et al and on trackside nearby 9063; hillside, Cefn Rhydoldog, Llansantffraed Cwmdeuddwr 9468 JH; acidic grassland above Aberedw Rocks 0846 ACP; Aberedw Wood c.0747 AJ Davy (NCC records 1968); acidic grassland on edge of forestry track, Little Park, Abbeycwmhir 0472; near Bwlch y Sarnau c.07(MsFlR); hills around Llanbadarn Fynydd c.07 J Stephens (in AEW and JAW *Trans Rad Soc* (1947) XVII,12); edge of forest track, Fforest-fach, Radnor Forest 1868; in forestry, Cascob 2367 GL Powell det. ACP; Gwernaffel Dingle, Knighton 2770 PR (1966). Not noted with sporangia.

Alpine clubmoss

Diphasiastrum alpinum

Very rare in acidic grassland and dwarf shrub heath in the western hills. SW of Glanhirin, Elan Valley 8470 HLJ Drewett and DRD; Y Gym, Elan Village 9366; N slopes, Treheslog Bank 9369; Cefn Rhydoldog 9468 JH, all in Llansantffraed Cwmdeuddwr; hills around Llanbadarn Fynydd c07 J Stephens (in AEW and JAW, *Trans Rad Soc* (1947) XVIII, 12).

Quill-wort family
Isoetaceae

Quill-wort   *Isoetes lacustris*

Rare in upland lakes. Llyn Cerrigllwydion Uchaf 8369, CCW survey; Dolymynanch Resr., Claerwen Valley 9061; Llynheilyn 1658 W.O. Wait.

SPHENOPSIDA

Horsetail family
Equisetaceae

Water horsetail

Equisetum fluviatile

Frequent and widespread arround the edges of pools, lakes and ponds from the acidic uplands of the NW to the oxbow lakes of the lower Wye valley. Also present in the wettest of bogs and ditches.

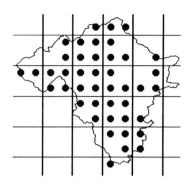

Marsh horsetail

Equisetum palustre

Absent from the most acidic upland areas, it is frequent in damp woodland, grassland, marsh, stream and lakesides elsewhere in the county.

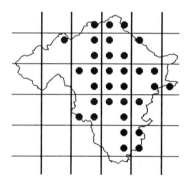

Wood horsetail

Equisetum sylvaticum

Widespread on damp roadside verges, in unimproved pastures, upland hay meadows, flushes and occasionally on hedgebanks, river banks and in damp woodland.

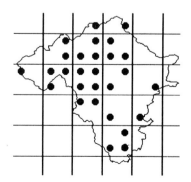

Common horsetail

Equisetum arvense

Common throughout on river banks, as a garden weed, in hedgerows, churchyards, road and railway banks except in the NW uplands where it is confined to areas arround habitations. In all squares except 86, 87SW, 96SE, SW & NE, 97NW, 06NE & NW, 26SW and 27NW.

Great horsetail

Equisetum telmateia

A scarce sp. of wet clay flushes in or on the edge of woodlands. Bryn Person Wood, Llananno 0872 MEM; Cilkenny Dingle Wood, Llowes 1741; Cilbigyn Farm Wood, Gladestry 2155 JSB and DEG. Reported in MsFIR as being locally frequent, with records from Llanstephan 14, Doldowlod 96 or 06, Rhayader 96, frequent on clays in the Upper Ithon Valley, Llanddewi Ystradenny 17, near Abbeycwmhir 07 and Lloyney 27, this sp. appears to have declined in abundance.

FILICOPSIDA

Adder's tongue family Ophioglossaceae

Moonwort

Botrychium lunaria

Rare, in unimproved, acidic grassland and occasionally with the next sp. in old hay meadows, on common grazings and rarely roadverges. 9 sites.

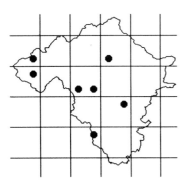

Osmundaceae

Adder's tongue

Ophioglossum vulgatum

A scarce fern found in the short turf of unimproved acidic grassland, such as upland hay meadows and pastures, churchyards, common land and roadside verges. Possibly overlooked and varying greatly in abundance from year to year. 11 sites.

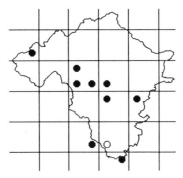

Royal fern family Osmundaceae

Royal fern   *Osmunda regalis*

A rare fern of long undisturbed peatlands and upland cliffs. On rocks in waterfall, Cwm Coel, Elan Valley 8964 T Inskip and M Davies (*Nat Wales* 15, 143(1977)); wooded basin mire, Cors y Llyn, Newbridge on Wye 0155 PD Moore et al (NCC internal report); wooded basin mire, Dreavour, Colva 1852 ACP; raised mire and lagg woodland, Rhosgoch Common 2048. A record from Llanbwchllyn 1146 by Lewis Davies in *Radnorshire, Cambridge County Geography* series 1920 has never been confirmed by other visitors to the lake.

Parsley fern family Cryptogrammaceae

Parsley fern

Cryptogramma crispa

Rare on rock outcrops and scree. One plant on a heap of waste slate from an old quarry S of Glog Fawr, Elan Valley 9166 C Parry and DEG; scree below Cerrig Gwalch, Rhayader 9370 DRD et. al.; rock outcrop, Wyloer, Gilfach, St.Harmon 9571 IDS and DPH; a single small plant on a cliff SE of Cwm-berwyn, Llansantffraed-in-Elvel 0854 and Stanner Rocks 2658 BSBI excursion (1953). In the absence of a specimen and with no other records from this well botanized locality the Stanner record requires confirmation. The earliest record traced was of the Rev. A Wentworth Powell (in Bufton,WJ and JO,1906) from the Llandrindod Wells 06 area where he remarks "there is a hill, almost in sight of my house where a few years ago the parsley fern was so abundant that the farmer mowed it for stable litter. Now there are only a few straggling specimens left". This somewhat improbable story should perhaps be followed up since there are no other records from this area.

Bracken family Hypolepidaceae

Bracken

Pteridium aquilinum

Common in acidic freely-drained woodlands. Abundant on open, well drained, uncultivated sites throughout the county. Extensive, expanding stands occur on many of the upland commons, though large areas of inbye land have recently been cleared by ploughing and reseeding and the use of herbicides. It is still cut and baled in some areas to provide winter bedding for stock. Common in acidic, freely drained woodlands. Recorded from all squares.

Filmy fern family Hymenophyllaceae

Killarney fern

Trichomanes speciosum

The prothallus only occurs in some abundance in dark recesses in acidic mudstone cliffs of the Bach Howey gorge, Llanstephan 1143 and 1243 (British Pteridological Soc. exe. 1991).

Wilson's filmy fern

Hymenophyllum wilsonii

A rare sp. of damp, shaded and sheltered rocks in ravines. By Afon Claerwen below Craig Cwm-clyd 8862 MH Rickard; wooded gorge, Cwm Coel, Elan Valley 8963; wooded gorge N of Y Foel, Elan Valley 9166 ACP; gorge of the Cwmnant, Rhodoldog, Llansantffraed Cwmdeuddwr 9367; cliffs in forestry E of Esgair Dderw, Llansantffraed Cwmdeuddwr 9469; waterfall, Nant y Sarn, Pont Marteg, St.Harmon 9571 AJ Worland (Brit Pterid Soc exc 1963); damp rocks in gully near Aberedw 04NE M Taylor (1961, *Nat Wales* **7**, 170), and known from this area for c100 years (AL *Trans Woolhope Club* 1890-1892, 194); a couple of fronds were noted "in the glen above the falls" on the Bach Howey below Trewern Hill, Llanstephan 1243 by AL (*Trans Woolhope Club* 1867). It has never been reported there since.

Asplenium

habitat persists there in the extensive willow carr and this fern may yet be rediscovered.

Beech fern

Phegopteris connectilis

In wet, peaty crevices in shaded rock faces and on stream banks. Frequent only in the NW uplands but with outposts in the Radnor Forest 16, upper Teme Valley 1283 and Bach Howey gorge 1143.

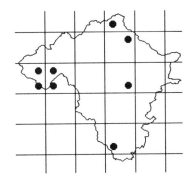

Marsh fern family
Thelypteridaceae

Mountain fern

Oreopteris limbosperma

Common in the uplands on stream banks, woodland edges and damp valley sides, especially where humus accumulates.

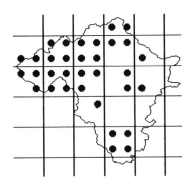

Marsh fern

Thelypteris thelypteroides

Only a single record made last century by E Armitage et al. from Rhosgoch Common c1948 (*Trans Woolhope Club* 1879, 181) provides evidence of its occurrence in the county. Suitable

Spleenwort family
Aspleniaceae

Black spleenwort

Asplenium adiantum-nigrum

Widespread, but rarely abundant in well-drained crevices in rockfaces, hedgebanks and amongst block scree. Avoids the most acidic sites and appears to favour warm, humid and sheltered localities.

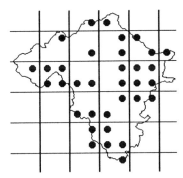

Maidenhair spleenwort

Asplenium trichomanes

Common on well-drained rock outcrops, walls and stable scree throughout the county. Recorded in all squares except 86, 87, 06NW & NW, 07SW & NW, 07SE & SW, 15NW, 16SW, 17SW and 24NE. The sspp. have not been separated in this survey. Ssp. *quadrivalens* is probably the most abundant.

Green spleenwort

Asplenium viride

Reported by Woolhope Club members from rocks in the wooded gorge of the Edw at Aberedw 04 during an excursion there in 1871 (*Trans Woolhope Club* 1871, 6). JM Roper reports seeing this fern on a calcareous shale cliff in a wood to the E of Knucklas 2673. It could not be relocated a few weeks later and subsequent searches have failed to find it. There are no other records of this somewhat montane sp., though it occurs in abundance nearby in the Black Mountains in Brecknock.

Wall-rue

Asplenium ruta-muraria

Common in crevices in walls and rock outcrops, particlarly favouring old lime-mortar and calcareous rock outcrops.

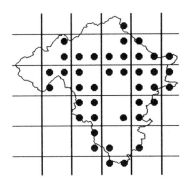

Forked spleenwort

Aplenium septentrionale

Very rare. On S facing acidic gritstone rock outcrop, Yatt Wood, Dolyhir 2458 DC Boyce, DRD & !.

Dryopteridaceae

Rusty-back fern
Ceterach officinarum

Scarce on base-rich natural rock outcrops, but of occasional occurence on old mortared walls in the lowlands and in disused quarries on the limestone and dolerite at Dolyhir 2458 and Stanner 2658 respectively.

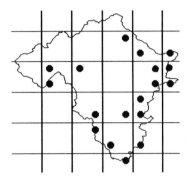

Hart's-tongue fern
Phyllitis scolopendrium

Frequent on moist, calcareous, shaded woodland banks, rock outcrops and walls. Most abundant in the dingle woodlands on the ORS in the lower Wye valley. Extends into naturally base-poor areas by colonizing damp mortar of drains and other shaded walls.

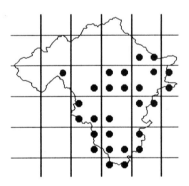

Lady fern family Athyriaceae

Lady-fern
Athyrium filix-femina

Common in woodlands and scree and on rock ledges and hedge and stream banks throughout Radnor. Recorded from all squares.

Brittle bladder-fern
Cystopteris fragilis

Scattered on damp, shaded, calcareous rock outcrops, particularly those of the Wenlock and Ludlow beds, in dingles and woodlands. Rare on old mortared walls and bridges. Carter, PW, (1950) reports a specimen of this fern collected by the Rev Littleton Brown from Water-break-its-neck, Llanfihangel-Nant-Melan c1860 in 1726 as being included in the Dillenius herbarium, Oxford. This is the earliest higher plant specimen traced from Radnor. 24 sites.

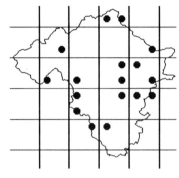

Male fern family Dryopteridaceae

Male fern *Dryopteris filix-mas*

Common in woodlands, on stream, river and hedge banks throughout the county. Recorded from all squares except 86NW and 87.

Scaly male fern
Dryopteris affinis

In similar habitats to the last sp. and often with it, though perhaps a less successful colonizer, being scarce in hedgerows and plantation woodlands, particularly in the lowlands. Further

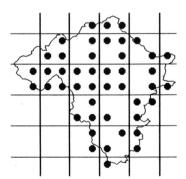

work is required to determine the distribution of the sspp..

Dwarf male fern
Dryopteris oreades

A scarce or overlooked fern of cliffs in the uplands. Craig y Bwch, Claerwen Valley 9061; Craig y Foel, Elan Valley 9264; track-side cliff in valley W of Tyfaenor Park, Abbeycwmhir 0671. (This latter site has been destroyed by track widening).

Broad buckler-fern
Dryopteris dilatata

Common in damp woodlands, in hedgerows, on old tree stumps, shady stream and river banks, the edge of bogs, particularly in the uplands and in block scree and on rock ledges, always in acidic sites. In all squares except 86NW, 04SE and 25NE.

Narrow buckler-fern
Dryopteris carthusiana

An uncommon fern of peatlands, being most abundant in wet alder woods and in willow carr around the larger peatlands. Less frequent amongst rush spp. in flushes in the uplands.

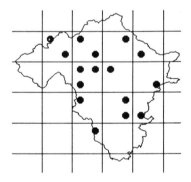

Soft shield-fern
Polystichum setiferum

Frequent on the basic, clay soils and cliffs of damp, shaded dingle woodlands in the S and E of Radnor on the ORS and Ludlow beds. Elsewhere scattered and confined to the more lime-rich woods on volcanic soils or boulder clay. 20 sites.

Polypodium

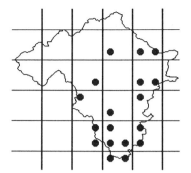

Hard shield-fern
Polystichum aculeatum

Over twice as abundant as the last species and occuring in similar, though often drier habitats. 45 sites.

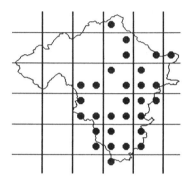

Oak fern
Gymnocarpium dryopteris

A scarce fern of damp, acidic rock ledges and scree in humid and shaded sites. Unable to survive heavy grazing, in the uplands it is confined to block scree and ledges out of reach of sheep, whilst on the lower ground it is confined to rocky gorges and screes in ungrazed woods. 16 sites.

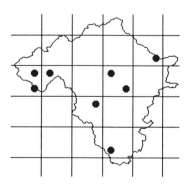

Limestone fern
Gymnocarpium robertianum

A single record of HN Ridley published in *J Bot*, 1884, 378 from Aberedw Woods c04NE is the only evidence of the occurrence of this fern in the county. The habitat there today appears suitable and the fern may yet be refound.

Hard-fern family
Blechnaceae

Hard-fern *Blechnum spicant*

A common species of acidic soil in woodlands and on shady upland banks amongst heather and bilberry. In all squares except 04SE, 05SW, 13, 15NW, 25SW & NE, 26NW and 36NW.

Polypody family
Polypodiaceae

Polypody
Polypodium vulgare group

Polypody is a common fern of tree trunks, hedgebanks, walltops and rock outcrops. Further work is required to determine the distribution of the three spp. reported from the county. The group is recorded from all squares except 87SW, 07 and 13NE. *P. vulgare* s.s. is widespread, particularly in acidic sites.

P. interjectum is probably widespread on trees with a base-rich bark such as ash, elm and ancient oak and common on base-rich rock outcrops and mortared walls, especially in the S and E of the county.

P. cambricum is rare on basic rock outcrops. Recorded on ledges by the R.Wye, Boughrood 1240 by ACP and on dolerite rock ledges at Stanner Rocks 2658 by PM Benoit.

Pillwort family
Marsileaceae

Pillwort
Pilularia globulifera

This strange fern, endemic to Europe, is possibly Radnor's most notable plant. Threatened throughout most of its range by the drainage of its habitat, the Radnor population with 28 separate sites known is of international importance. It inhabits mud-bottomed pools which generally loose their water in summer, though often remain damp. Competition is reduced by intensive grazing and poaching of the soil. Associated spp. include water purslane, lesser marshwort, starwort spp, blinks and marsh cudweed. Sorocarps appear to be rare and have only been seen in three sites. The commons of Aberedw Hill 05SE, Llandeilo Hill 04NE, Maelienydd 17SW and the Begwns 14SE are the headquarters of this fern.

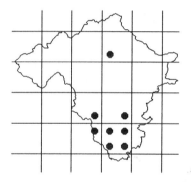

Pillwort

10. Seed Plants - Spermatophyta

Conifers
GYMNOSPERMAE

Pine family
Pinaceae

*European silver-fir
Abies alba

Occasionally grown as a forest and onamental tree, it seeds into adjacent woodland. Noble fir *A. procera* and Grand fir *A. grandis* are occasionally grown in plantations such as on the Radnor Forest 26 but no regeneration has been detected.

*Douglas fir
Pseudotsuga menziesii

Commonly planted as both an ornamental and timber tree with mature specimens abundantly seeding onto adjacent road verges and into nearby woodland.

*Western hemlock
Tsuga heterophylla

Was fashionably planted in the late 1950's to early 1960's, on the better soils or underplanted in broadleaved woods. Strangely, despite abundant natural regeneration in other parts of Powys, it has yet to be detected regenerating in Radnor.

*Norway spruce   *Picea abies*
Widely planted to supply Christmas trees and timber. Frequently regenerating along nearby tracksides.

*Sitka spruce
Picea sitchensis

Large plantations have been created recently on the wetter soils with mature specimens seeding abundantly into adjacent moorland and cleared areas of plantations.

*Larch   *Larix* spp.
Japanese larch, *L. kaempferi,* European larch, *L. decidua* and their hybrid are all planted for timber and ornament and regularly regenerate along forest tracks and on roadsides.

*Lodge-pole pine
Pinus contorta

Planted on rocky outcrops and around the edge of exposed conifer plantations, it seeds readily to regenerate along firebreaks and forest tracks.

Scots pine
Pinus sylvestris

Reported as subfossil pollen grains from late glacial (Flandrian) peat deposits in the basin mire at Cors y Llyn, Newbridge on Wye 1055 (Moore, P.D., 1978). This tree occurs abundantly on the site today, but a long break in the pollen record and information from local sources suggests the tree recolonised the site around the 1900's from nearby planted specimens. Subfossil pollen is also reported from the peat of Rhosgoch Common 1948 (Bartley, D. D., 1960). The tree occurs frequntly in ornamental woodlands and is occasionally grown in plantations, about which it regularly seeds.

A few other species of pine have been planted on a small scale eg. Corscian pine *P. nigra* ssp. *laricio,* Bhutan pine *P. wallichiana,* Weymouth or white pine *P. strobus* and Macedonian pine *P. peuce*. There is no evidence of them naturally regenerating.

Redwood family
Taxodiaceae

Three sp. are occaisionally includedin ornamental plantings ie, Coast redwood *Sequoia semperivens,* Wellingtonia *Sequoiadendron giganteum* and Japanese cedar, *Cryptomeria japonica*. The latter has also been used in experimental forestry plots. all set seed but none have been found naturalised.

Juniper family
Cupressaceae.

*Lawson cypress
Chamaecyparis lawsoniana

Commonly planted as an ornamental tree in gardens and churchyards and occasionally grown in small plantations for landscape purposes and its foliage sold to the florists trade, it occasionally seeds itself onto nearby banks.

*Western red-cedar
Thuja plicata

Like the last sp. it is regularly planted in gardens but is also grown on a small scale for timber and foliage and occasionally naturalizes in forest rides and on disturbed roadside banks.

Juniper
Juniperus communis

Reported as subfossil pollen from Flandrian age peat deposits in the basin mire at Cors y Llyn, Newbridge on Wye 1055 (Moore, P.D., 1978) the plant appears to have become extinct in the county.

Yew family
Taxaceae

Yew   *Taxus baccata*
Its status as a wild tree is uncertain. Large specimens occur in many churchyards and ornamental woodlands. It is rare in the wild, perhaps having been exterminated to prevent livestock poisoning. Yew typically survives in roadside hedges, on inaccessible cliffs such as at Cefnllys Castle 0861 and in the ungrazed dingle woodlands on the ORS in the lower Wye valley.

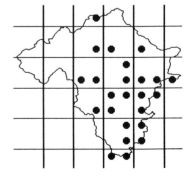

Flowering Plants

ANGIOSPERMAE
DICOTYLEDONES
Willow family
Salicacaeae

Crack willow *Salix fragilis*

In the lowlands, common on stream and river banks and beside pools and occasional in damp woodlands and hedgerows. Very scarce in upland districts and probably planted in some sites.

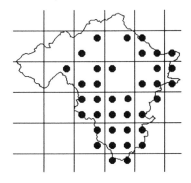

White willow *Salix alba*

Probably planted in most, if not all its Radnor localities on river banks and lakesides in the lowlands. Beside the lower Wye, where it might possibly be native, it hybridises freely with the crack willow making certain identification difficult.

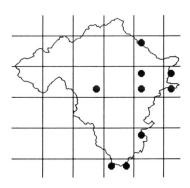

Almond willow *Salix triandra*

Rare and probably native in old ox-bow lake on the Wye flood plain, Fforddfawr, Glasbury 1840 MP & MEM (NCC records). Also noted from 26 on a BSBI exc. in 1956. The record in Woods (1990) from 08 should be referred to *S. purpurea*.

Grey and rusty willow
Salix cinerea

The rusty willow, ssp. *oleifolia*, is the commonest willow in wet woodlands, marshes, bogs, upland hedgerows, damp road verges, on moorland, wet acidic cliffs, damp disused quarries, railways and abandoned, once disturbed ground. Recorded from all squares except 86NW. The grey willow, ssp. *cinerea* is very rare or overlooked on more base-rich soils. By small stream SE of Esgairdraenllwyn, Llanbadarn Fynydd 0881 conf. D Meikle.

Eared willow *Salix aurita*

Frequent around the edges of wet, peaty upland woods and hedgerows, on wet acidic cliffs and moorland and by streams in the uplands. Hybrids with the rusty willow are frequent.

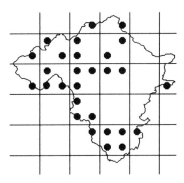

Goat willow *Salix caprea*

Common in damp woodlands, especially on the more base-rich soils of the S & E and frequent in hedgerows, on roadsides and on disturbed ground. Where it occurs with the grey willow, hybrids usually can be found. Recorded in all squares except 86, 87 and 18SW.

Creeping willow *Salix repens*

Very rare. The 97 record in FPW is based on a doubtful specimen. The Woolhope Club reported seeing this species on an excursion from Hay, through Rhosgoch to Painscastle in their *Transactions* for 1911, 207. They may have seen it in its only known site in herb rich, damp pasture at the SW end of Rhosgoch Common 1948 GC Cundale. Occasional in vice-counties to the S and W, its rarity in Radnor is difficult to explain.

Osier *Salix viminalis*

Common on the banks of the River Wye from Llanelwedd to Hay. Less frequent in the lowlands elsewhere, where it occurs on streambanks and beside ponds. Possibly planted to supply materials for baskets in some sites. The majority of upland populations some of which appear to have been of hybrid origin may have arisen this way.

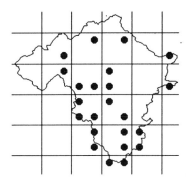

Purple willow *Salix purpurea*

A scarce sp. of wet ground. Hedge N of Llyn Gwyn, Nantmel 0065; Penithon 08 JAW (1949 WFP) and seen recently on stream banks and in a roadside hedge to the NE, below Esgairdraenllwyn, Llanbadarn Fynydd 0881; old ox-bows in Wye Valley, Fforddfawr, Glasbury 1840 MP & MEM (NCC files, but not refound in survey in 1990); pondside, Stanage Park 3371. Also noted from 16 (*Atlas Brit Fl*).

*White poplar *Populus alba*

Occasionally planted and spreads by means of suckers in gardens, parks and about farms. Not mapped.

Aspen *Populus tremula*

Widespread, especially on the heavy boulder clay soils of the Ithon Valley. Found in woodlands, but more frequently seen alongside streams and as hedgerow groups.

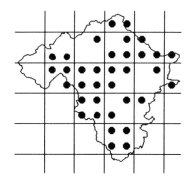

Black poplar *Populus nigra*

A rare tree around farms and on the flood plains of the larger rivers which has probably always been planted. Two trees in 1840 near Maesllwch show the characteristic features of the native type (conf. E Milne-Redhead), but occurring in hedgerows, are also probably planted. Two large moribund trees on a remote stream bank in Ddol Wood, Llanbister 1272 occur in an otherwise natural looking wood. recorded from 06SE, 08SE, 14SE, 15SW, 16SW and 17SW.

Other introduced poplar spp. noted include *P. gileadensis* and *P. trichocarpa* spreading by means of suckers, in for example Llandrindod Wells 0561 and 0660 and the grey poplar *P. x canescens* in 06 and 2350. Hybrid poplars, (probably *P. balsamifera* x *P. trichocarpa*), are widely planted in the lower Wye Valley as a source of timber.

Birch family
Betulaceae

Silver birch *Betula pendula*

A common pioneer tree sp. of disturbed woodlands, conifer plantations, disused quarries, track and roadsides. Commoner than the downy birch on the low ground and more common on drier sites. In all squares except 86, 87, 96SW, 07SW, 15SE, 18 and 26SW.

Downy birch

Betula pubescens

This sp. has been recorded from all squares except 86NW, 87, 13SE and 26SW. Ssp. *pubescens* is a widespread tree of wet to dry acidic soils, occurring commonly in disturbed woodlands and hedgerows, particularly in the uplands. The ssp. *carpatica* has been rarely distinguished. In Cwmfaerdy Wood, Abbeycwmhir 0769 and Bryn Person Wood, Llananno 0872, both MEM (NCC records).

Alder *Alnus glutinosa*

Common on river and stream banks, by pools and in wet woodland and hedgerows throughout the county. In all squares except 86NW and 87.

*Grey alder *Alnus incana*

An alien species planted as an ornamental tree and occasionally used in forestry plantings. It has sometimes been accidentally introduced in mistake for common alder. Plantings beside forestry roads in the Radnor Forest 1966 are now naturally regenerating along the road banks in damp places, the first time regeneration has been noted in the county.

Hazel family
Corylaceae

Hornbeam *Carpinus betulus*

Probably not native in Radnor, though pollen is present in the pre-Roman peat deposits at Cors y Llyn, Newbridge on Wye 0155 (Moore PD, 1978). Planted in a few hedgerows and woodlands eg Old Radnor Hill 2558 DRD; Presteigne Castle Wood 3164 CC and possibly naturalising at Silia, Presteigne 3064.

Hazel *Corylus avellana*

Common in all woodlands except in heavily grazed, acidic upland woods and one of the commonest hedgerow shrubs. In all squares except 86NE & NW and 87.

Beech family
Fagaceae

Beech *Fagus sylvatica*

As with hornbeam the status of this tree is uncertain. It has been widely planted on freely drained sites in woodlands and as specimen trees in hedgerows. Pollen is present in peat deposits to at least pre-Roman times (Moore PD, 1978). Unlike hornbeam it regularly regenerates, and in some woods is an aggressive colonizer.

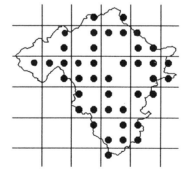

*Sweet chestnut

Castanea sativa

Occasionally planted in woodland and parkland. Rarely forming seeds, it does not appear to have naturalized in Radnor.

*Red oak *Quercus rubra*

Planted in recent years in forestry plantations, especially to form a hardwood edge to conifer crops. Regeneration not yet noted.

*Turkey oak *Quercus cerris*

Generally rare and confined to the large estates. Fruits regularly, but no significant natural regeneration noted.

Sessile oak *Quercus petraea*

The dominant tree of steeply sloping, upland valley-side woodlands and in hedgerows throughout the county. Mast production is erratic and with most woods grazed, regeneration is sparse. The tree is seldom managed for timber production, many woodlands having been abandoned silviculturally, following major fellings in the two world wars. Few large specimens remain and the majority of trees in woodlands appear to be of coppice origin. In all squares except 86NW, 87, 07NW, 08SE, 15SW and 16SE.

Hybrid oak

Q. petraea x *robur* = *Q.* x *rosacea*

Many oaks, especially in the S of the county appear to be intermediate in character between the sessile and pedunculate oak. No attempt, in view of the gradual variation of characters, has been made to map this possible hybrid.

Pedunculate oak

Quercus robur

The dominant tree on the somewhat acidic, heavy boulder clay soils, it is frequent in the wetter lowland sites in woods and hedgerows. Regularly producing acorns and apparently preferred as a timber tree to sessile oak, it consequently is more frequently planted. In all squares except 86, 87, 96SW, NE & NW, 97SW & NW, 07SW & NW, 08SE, 15NE, 17SW & SE and 18SW.

Urtica

*Southern beech

Nothofagus spp.

Both Roblé beech *N. obliqua* and Rauli beech *N. procera* have been planted in forestry trials, particularly in the Radnor Forest. There is no evidence of any natural regeneration yet.

Elm family
Ulmacae

Wych elm   *Ulmus glabra*
Once common on base-rich soils in the S and E of the county in woodlands and hedgerows, and with scattered individuals by dry, base-rich flushes, on cliffs and roadsides in the uplands. Dutch elm disease, affecting the whole of the county throughout the 1970's, has killed almost all the large trees. Even isolated trees such as the large specimen near Ciloerwynt in the Claerwen Valley 8862 have died. Some regrowth has occurred in places and a few trees may yet prove resistant. Large clearings have appeared in some of the ORS dingle woodlands and on lime-rich Wenlock and Ludlow age cliffs such as around Aberedw 04NE as a result of tree death. In all squares except 86NE & NW, 87, 97SE, SW & NW, 06SW & NW, 07NW, 17SW & SE and 18SW.

English elm   *Ulmus procera*
Probably native, but noted mostly in hedges, where it suckers freely. Common only in the lower Wye Valley and recently much affected by Dutch elm disease, so few trees of any stature remain.

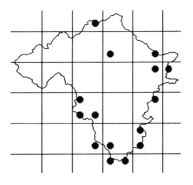

Hemp family
Cannabaceae

Hop   *Humulus lupulus*
Widespread but infrequent in sheltered roadside hedgerows. It is not cultivated in Radnor.

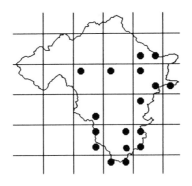

*Indian hemp
Cannabis sativa

Formerly cultivated for its fibres, pollen grains remain preserved probably from medieval times in peat at Cors y Llyn, Newbridge on Wye 0155. (French & Moore, 1986).

Nettle family
Urticaceae

Common nettle   *Urtica dioica*
Common on nutrient-rich soils in woodlands, hedgerows, farmyards and gardens. Recently it has become a common weed of grass leys following the use of slurry and larger quantities of fertilizers. In all squares except 86NW and 87SW.

Small nettle   *Urtica urens*
A very scarce annual weed of cultivated and disturbed soil. On stony waste on roadside tip, Pentre Caeau, Llandeilo Graban 0844 ! & AO; silt banks by Wye, Fforddfawr, Glasbury 1840 NCC survey; yard of Forest Inn, Llanfihangel-Nant-Melan 1758 JAW (1945, *NW Nat* **20,** 160); site of recently buried pipe, near Presteigne Withybeds 3164 RK. Recorded in FPW from 25 as well as 06 and 07, most of which squares occur in Radnor.

Pellitory-of-the-wall
Parietaria judaica

Probably extinct. Noted in Presteigne 3164 in 1932 and Knighton 2872 in 1931 and 1938 in MsFlR. A search of both towns has failed to find it recently.

Mistletoe family
Loranthaceae

Mistletoe   *Viscum album*
Common only in sheltered, sunny sites in the Wye Valley from Yr Allt, Llandeilo Graban 0844 to the English border, where it most frequently occurs on hawthorn, apple cultivars and crab apple. Noted on field maple in Fron wood, Llowes 1942 DEG and JSB, between Clyro and Hay 2143 ACP and near Cwm, Cabalva 2345; on hazel, Cwrt-y-graban, Llanstephan 1141 and near Cwm, Cabalva 2345; on poplar cultivar near Skynlais, Glasbury 1639; on common lime and a *Crataegus* cultivar, Llanstephan 1141; on false acacia *Robinia pseudacacia*, Brynrhydd, Llowes 1841 ACP; on ash, near Neuadd farm, Llanstephan 1141. Outposts for the sp. include an apple in an orchard Merry Hall, Newbridge on Wye 0158 FMS, apple in a garden at Howey 0558 C Parry and a crab apple at Discoed 2764 D Smith.

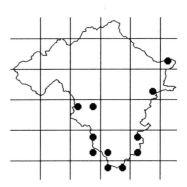

Knotweed family
Polygonaceae

Knotgrass

Polygonum aviculare agg.

Common on unsurfaced trackways, road and pathsides, in field gateways and as a weed of cultivated ground. Both *P.*

Polygonaceae

aviculare s.s. and *P. arenastrum* are present in the county, the latter being more common on track sides. They were not mapped separately. In all squares except 86NE & NW, 87, 96SW, 97NW, 07SW, 18SW and 26NW.

Water-pepper

Polygonum hydropiper

Common on silt and mud on the sides of rivers, streams, ponds and ditches, on reservoir margins and in wetlands subject to disturbance. A pioneer colonist.

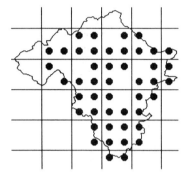

Redshank

Polygonum persicaria

Common in cultivated ground of fields and gardens, on disturbed roadsides, silt and detritus on riverbanks, tips and about farmyards. In all squares except 86NE & NW, 87, 96SW, 16SE, 18SW, 26NW & SE and 36NW.

Pale persciaria

Polygonum lapathifolium

In similar habitats to the last sp. but less abundant.

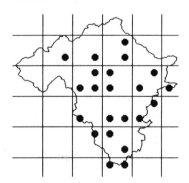

Amphibious bistort

Polygonum amphibium

Widespread, but not common, in and on the margins of slow-flowing reaches of rivers and in lakes, ponds and marshes, generally in the lowlands. The River Ithon supports fine stands in places.

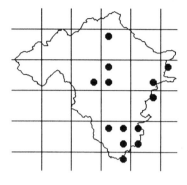

Common bistort

Polygonum bistorta

Widespread on the damp clay soils of agriculturally unimproved pastures and haymeadows, streamsides, roadside verges and occasionally churchyards. 19 sites.

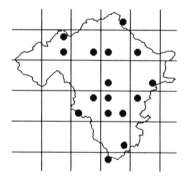

*Himalayan knotweed

Polygonum polystachyum

This rarely grown alien is scarcely naturalized. Roadside, Norton Manor, Norton 2966. Noted from 24 in WFP, a record which may be from Radnor.

Black bindweed

Fallopia convolvulus

An uncommon weed of arable fields, gardens and rubbish tips.

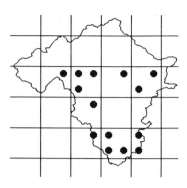

*Japanese knotweed

Reynoutria japonica

Still a fairly scarce introduction on roadsides, waste ground and river banks. Suffering regularly from frost damage in late spring, many of the upland valleys may have too extreme a climate to permit its establishment.

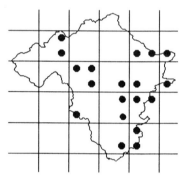

*Giant knotweed

Reynoutria sachalinensis

A scarce introduction. Between railway and road, Howey 0558; by Llandrindod Wells lake 06SE JAW (1954, NMW); by sewage works Llandrindod Wells 0460 and Hotel Commodore 0561; waste ground, Penybont village 1164.

Sheep's sorrel

Rumex acetosella agg.

Common and in all squares except 96NE, 97SE, 14NW, 15SW and 37SW on well-drained banks, dry pastures and sunny rocky outcrops throughout. No attempt has been made to subdivide this aggregate of species. *R. angiocarpus* is recorded from Llanelwedd 05 AEW (1927 NMW) and a form intermediate with *R. tenuifolius* from Stanner Rocks 2658 HA Hyde (1956 NMW).

Rumex

Common sorrel
Rumex acetosa

Common in grasslands, marshes, open woodlands and streambanks. Recorded from all squares.

Water dock
Rumex hydrolapathum

Beside the ornamental lakes in Stanage Park 3271 ACP, but almost certainly introduced here.

Curled dock   *Rumex crispus*
A widespread weed of roadside verges, tips and occasionally an invader of pastures.

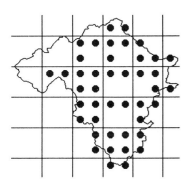

Clustered dock
Rumex conglomeratus

An uncommon species of wet hollows in field margins and roadsides, marsh and lake margins.

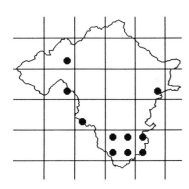

Wood dock
Rumex sanguineus

Common in woodlands and hedgerows throughout the county.

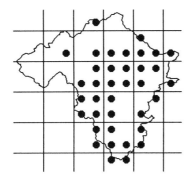

Broad-leaved dock
Rumex obtusifolius

A common weed in pastures, on roadsides, riverbanks and waste-ground. Of late it has become a very significant agricultural weed, favoured by the use of slurry as a fertilizer. In all squares except 86NW, 87SW and 96SW.

Goosefoot family
Chenopodiaceae

*Good-King-Henry
Chenopodium bonus-henricus

Once grown as a pot herb, this plant is now apparently more rare than formerly. Seen recently only in a farmyard, Cwmbach, Rhayader 9569; Maengowan, Llanelwedd 0552; in a farmyard in 26SW ACP and at Llanshay, Knighton 2971 PR (1964). MsFIR considered it to be "not very frequent" but recorded it from Rhayader 96, Llandrindod 06, New Radnor 26, Presteigne 36 and Knighton 27.

Red goosefoot
Chenopodium rubrum

Rare. On dried-out margins of Llanbwchllyn 1246 DRD & IDS; roadside dung heap, New Radnor 2160 and on dung on farmyard waste scattered in roadside quarry near Upper Pitts Farm, Stanage 3171.

Many-seeded goosefoot
Chenopodium polyspermum

A rare colonist of open ground. Car park, Llanelwedd 0551; swede field E of Boughrood Castle 1338; muddy margin of Llanbwchllyn 1246 BR Fowler det. JPM Brenan (1976, *Nat Wales* **15**, 28); rubbish tip, Clyro 2143.

Fat-hen
Chenopodium album ssp. *album*

A common weed of arable fields, particularly swede fields, gardens, tips and disturbed waste places.

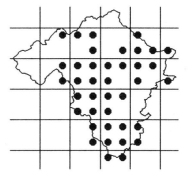

Common orache
Atriplex patula

Often associated with the last sp. and in similar habitats but less common.

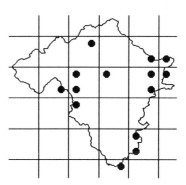

Spear-leaved orache
Atriplex prostrata

Rare or overlooked on waste ground and as a weed of arable fields. Reported from 07SE, 24SW and 36SW. There are, in addition, records in FPW from 04, 05, 07 and 08 which may be from Radnor.

Pigweed family
Amaranthaceae

*Common amaranth
Amaranthus retroflexus

A single record of this N American sp. from a cultivated field at Llanfaredd 0650 J Jones det. !.

Purslane family
Portulacaceae

Blinks *Montia fontana* agg.
Common around springs, in wet runnels, in hollows of seasonally wet rocks, on the exposed mud of reservoirs and pools, in ditches and wet field gateways. The sspp. *variabilis*, *amporitana* and *chondrosperma* are all reported from Radnor in FPW. A recent sample survey suggests all are widespread.

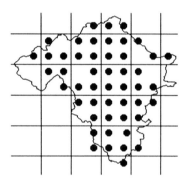

*Pink purslane
Montia sibirica

It is uncertain when this sp. first established itself on the banks of the Wye. Not mentioned in MsFIR (1954), by the early 1970's it was widespread by the Wye, especially between Builth Wells 05SW and the English border.

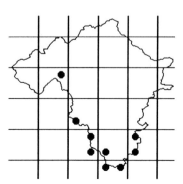

Pink family
Caryophyllaceae

*Mossy sandwort
Arenaria balearica

Naturalized on Dolymynach dam, Clerwen valley 9061 and on shady base of stone wall, Maesgwynne, Howey 0656.

Thyme-leaved sandwort
Arenaria serpyllifolia ssp. *serpyllifolia*

In dry, sunny, open habitats such as on rock ledges, scree, old quarry workings, railway ballast and gravelled tracks. Common in a number of widely scattered localities.

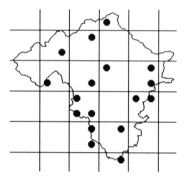

Ssp. *leptoclados*

Perhaps overlooked in this survey, it was noted by AEW and JAW in MsFIR from Cregrina 1252, Stanner 2658, Old Radnor 2559 and New Radnor 2160 in 1938 but has recently only been seen on the bed of the disused railway near Aberedw 0747.

Three-nerved sandwort
Moehringia trinervia

Common in ungrazed, shady, freely drained woodlands and hedgerows on the less acidic soils.

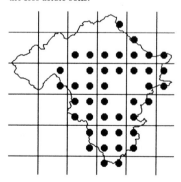

Wood stitchwort
Stellaria nemorum ssp. *glochidisperma*

A rare sp. of damp, shady ungrazed woodland. Fron Barn Dingle, Dutlas 1976 DEG & JSB; Bettws Dingle, Cabalva 2346 ! & IDS.

Common chickweed
Stellaria media

A common weed of arable land, gardens, road verges, riverbanks, tips, etc. where the nutrient status is fairly high. In all squares except 86NE & NW, 87SW and 18SW. *S. neglecta*, the greater chickweed, has not always been separated with certainty and cannot be accurately mapped. It appears to be widespread, but uncommon in hedgerows, on streambanks and by woodlands. *S. pallida*, the lesser chickweed, has been sought in dry, sunny places but has never been found.

Greater stitchwort
Stellaria holostea

A common flower of hedgerow, open woodland and scrub, recorded from all squares except 86NE & NW, 87SW, 96SW and 16SE.

Bog stitchwort
Stellaria uliginosa

A common sp. of springs, flushes, marshes, wet ditches and lakesides throughout. In all squares except 86NW, 87SW and 96NW.

Lesser stitchwort
Stellaria graminea

Common in undisturbed grassland, and scrub, on roadside banks, stream banks and in woodland clearings. In all squares except 86SE & NW, 87SW and 17NW.

*Snow-in-summer
Cerastium tomentosum

Often present in tipped garden rubbish, it has rarely been found naturalized. Old railway bank St. Harmon 97SE (NMW record); rock outcrops, formerly a garden, by lake, Llandrindod Wells 0660; walls, New Radnor 2160; Knighton c2872 MsFIR; walls Presteigne 3164.

Cerastium

Field mouse-ear
Cerastium arvense

On walls and in hedges just outside Presteigne c3164 JAW (MsFlR), it was first noted in 1938 and last seen by AEW in 1952. JAW also collected a specimen from hedges at nearby Norton c3067 in 1956 (NMW).

Common mouse-ear
Cerastium fontanum ssp. *glabrescens*

A widespread sp. of open soil in grassland, on tracksides, stream and river banks, in scrub and woodland glades. In all squares except 87SW.

Sticky mouse-ear
Cerastium glomeratum

Frequent as a weed of arable fields and along farm tracks, in farmyards and gateways.

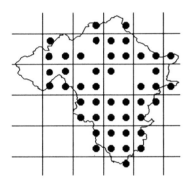

Little mouse-ear
Cerastium semidecandrum

A scarce or overlooked sp. of freely drained soil and gravel. Noted in MsFlR from Cregrina c1252, Colwyn Castle c1053, near Glascwm c1553 and Stanner c2658 and in *Atlas Brit Fl* from 26. It has recently been seen only on gravel near the old gas works, Knighton 2972.

Sea mouse-ear
Cerastium diffusum

A scarce sp. of open, freely-drained ground. Dry bank N of Llanerch-y-cawr, Claerwen Valley 9061; bank above Visitor Centre, Elan Village 9264; old railway, N of Rhayader 9668; gravel by disused track, Llanelwedd Quarries 0551; soil by road, near Bach Howey viaduct, Llanstephan 1042.

Upright chickweed
Moenchia erecta

A plant of the short, open turf of unimproved, dry, summer-droughted grassland, on thin soils. The best populations are found on the volcanic Carneddau hills between Llanelwedd 0552 and Llandrindod Wells 0761. Record from Presteigne (Lewis Davies, 1920) was from Stapleton Castle, Herefordshire. Recorded from 04SE, 05SE & NE, 06SE, 14SW, 15NW & SW, 25NE and 27SW.

Water chickweed
Myosoton aquaticum

Today a sp. of the Lower Wye. Damp hollows, flooded in winter, Fforddfawr mire, Glasbury 1840 MP and MEM (NCC records); silt banks beside the R.Wye near Hay 2242 ! and ACP, where it is abundant. It was, however, recorded from a stream between Dolyhir and New Radnor c25? by HNR in 1880 (*J Bot* 1881,171).

Knotted pearlwort
Sagina nodosa

A rare sp. of base-rich, peaty flushes, seen recently only on the SE slopes of the Begwns, Llowes 1643 where its habitat is subject to poaching and heavy grazing. Noted in MsFlR from the lower Elan Valley, and near Rhayader c96; from near Llandrindod c06 JB Lloyd (Bufton WJ and JO,1906); from 16, site unknown, RCLH and FK Johnston (post 1950, BSBI recording scheme).

Procumbent pearlwort
Sagina procumbens

A common and widespread plant of heavily grazed, seasonally damp grassland, track and path sides and open, disturbed soil. In all squares except 87SW.

Fringed pearlwort
Sagina apetala

Of occasional occurrence on the edges of gravel and old broken tarmacadam paths and tracks, and rarely on dry and sunny rock outcrops. The sspp. *apetala* and *erecta* have not been separated. Both occur in the county.

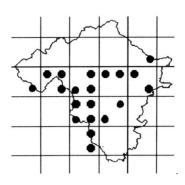

Perennial knawel
Scleranthus perennis

This nationally rare sp. has been known from Stanner Rocks 2658 for over one hundred years. The earliest specimen traced is in BM, collected by Anne Gardner in 1850. It grows in shallow pockets of summer-droughted soil on these dolorite and gabbro cliffs. A biennial, its population varies from year to year, but is always small and despite some protection by National Nature Reserve status for the rocks, its existence still appears to be precarious. From experience in cultivation it is very sensitive to slug and snail grazing and may be confined to a small area of the reserve by grazing pressure. Being small and inconspicuous, all intended visitors should contact the CCW first for guidance. A misplaced boot could exterminate it, in this, the only British locality for the ssp. *perennis*.

Annual knawel
Scleranthus annuus

Known from two distinct habitats: summer-droughted, S facing soil pockets on rock outcrops and from disturbed soil. Root field, St Teilo's barn, Llandeilo Graban 0945 ACP (1968); recently constructed roadside verge, Nantmel 0266; rock outcrop, Llandrindod Hill Farm 0660; rock outcrop by Bach Howey, N of Trewern Hill, Llanstephan 1244; rock outcrops, Hirllwyn Bank, Llansantffraed-in-Elvel 1055; rocky bank, Cwm, Frank's Bridge 1056; Stanner Rocks 2658 AL (*Trans Woolhope Club*, 1853, 91); oat field, Cold Oak, Presteigne 2963 ACP (1969).

Caryophyllaceae

Corn spurrey
Spergula arvensis

A widespread weed of cultivated ground, tips and farmyards.

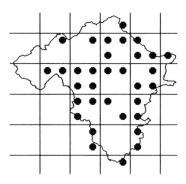

Sand spurrey
Spergularia rubra

An uncommon sp. of open, dry stony habitats. Roadside by Garreg Ddu Resr. dam, Elan Valley 9163; Rhayader Quarries 9765; site of former Builth Wells Station. Llanelwedd 0451 (BSBI exc.); stony ground, Llanelwedd Quarries 0452 (BSBI exc.); S facing rock outcrop, Yr Allt, Doldowlod 0062; rail bank, Rhayader to Doldowlod 06 &/or 96 AL (1881, *Bot Loc Rec Club* 1882, 246); gravel of forestry track, Cwmhir bank, St Harmon 0172; forestry track, Abbeycwmhir 0570 ACP; gravel drive, Evancoyd, Evenjobb 2663 ! and ACP.

Ragged-robin
Lychnis flos-cuculi

Widespread in wet hollows in unimproved pasture, flushes, marshes, ditches and wet roadside verges. In all squares except 86, 87SW, 96SW, 97SW & NW, 04, 18SW, 13NW, 16SE, 26SW & NW and 27SE.

Sticky catchfly
Lychnis viscaria

Frequent in crevices on the S facing, sunny, dolerite cliffs of Stanner Rocks 2658. Known here for over 130 years (*Trans Woolhope Club*, 1853, 91).

*Corncockle
Agrostemma githago

Probably extinct. Noted from 06 (possibly Radnor) and 15 in *Atlas Brit Fl*. The source of these records is not known.

Bladder campion
Silene vulgaris ssp. *vulgaris*

Described as "fairly common especially on railway banks" in MsFlR, this sp. is now far from common. It still survives near the remains of Aberedw Station 0747 (closed along with the Mid-Wales Railway in 1962); on a roadside nearby at 0748; on the track bed of the railway near Llanfaredd 0650! and J Jones and on the site of the old station at Glasbury 1839 MP & H Soan. Elsewhere it is reported from the edge of a hay field, Llanfadog Lower, Elan Valley 9366 IHS and DRD; Nantgwyn churchyard 9776 DRD and RK and a roadside verge, Howey 0558.

White campion
Silene pratensis ssp. *pratensis*

As with the last sp. apparently much decreased of late. Reported in MsFlR as "rather frequent around Rhayader, Cwmdeuddwr, Crossgates, Boughrood, New Radnor" etc. Recently seen only on verge S of Rhayader 9767; near the old railway station, Aberedw 0747; garden, Newhouse, Llanbister Road 1669, DP Humphries; roadside verge NE of Llanfihangel Rhydithon 1767; roadside verge, Bleddfa 2167; roadside verge, Dolley Green, Whitton 2865.

Red campion
Silene dioica

Common in hedgebanks, scrub, woodlands, shady stream banks and occasionally on disturbed ground throughout. In all squares except 86, 87, 96SW & NW, 97NW & NE and 18SW.

Small-flowered catchfly
Silene gallica

A very scarce weed of disturbed ground. On recently reconstructed road verge to the S of Rhayader 9767 in 1987 and in a field near New Radnor c26SW (1947, MsFlR).

*Soapwort
Saponaria officinalis

Naturalized mostly on river gravels and silt banks. Frequent by the Wye from Boughrood 1338 to the English border.

*Sweet William
Dianthus barbatus

Naturalised on railway bank near Dolau Jenkin, Penybont 1065 (ITE railway survey for NCC).

Maiden pink
Dianthus deltoides

A plant of sunny and dry, summer droughted grassland on rocky banks. Benefiting from some grazing, it is doubtful if this species can cope with today's high density of sheep, which in some sites now appear to prevent flowering. Maengowan, Llanelwedd 0552 J Gethin (Reported from this area of the Carneddau in 1880 by HNR in *J Bot* 1881, 171 and in 1909 by HJ Riddelsdell in *Proc Cotteswold Club* XVII, 57-61); "in greater profusion on a sunny bank a few hundred yards above the Baptist Chapel at Howey" c0558 A Wentworth-Powell (c1900 Bufton WJ & JO 1906); collected from dry banks in the lane above Howey by a Miss Matthews det. AEW (1954 NMW), this may be the same site. No suitable habitat seems to be present now; S facing bank, N of Brynhir, Howey 0558 CAC; by church, Llandrindod Wells Hall Farm 0660 (known here for over 80 years); Llanfair Quarry, Llandrindod Wells 0661 LMP & J Langford; old quarry Tanygraig, Llandrindod Wells 0662 AO; rocky bank Cefncoed, Llandrindod 0762 IDS; roadside bank, Cwm, Frank's Bridge 1056 ACP; Hirllwyn bank, Llansantffraed-in-Elvel 1055 ACP; Graig Fawr, Hundred House 1358 ACP.

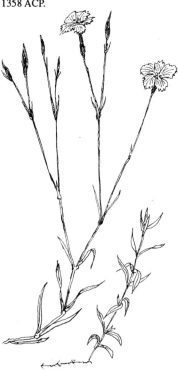

Maiden pink

Water-lily family
Nymphaeaceae

White water-lily
Nymphaea alba

A rare aquatic plant probably introduced to some sites. Mill ponds, Cennarth Mill, Nantgwyn 9777 FE & DRD; *Nymphaea* pollen, probably this sp., is reported from the lower peat deposit of Cors y Llyn, Newbridge on Wye 0155 (Moore PD, 1978); lake, Pentrosfa, Llandrindod Wells 0559; Llyn Gwyn, Nantmel 0165 FE & DRD; Llanbwchllyn 1146 (noted here in 1843 by T Westcombe in *The Phytologist* 1, 781); Nyth-grug pool, Llanfihangel-Nant-Melan 1660 NCC survey. Noted in MsFlR from Penybont pool 1164 and Kinnerton c2463.

Yellow water-lily
Nuphar lutea

Scarce in lakes and ponds. Gwynllyn, Llansantffraed Cwmdeuddwr 9469; Llyn Gwyn, Nantmel 0165 FE & DRD; Llanbwchllyn 1146 AL (*Trans Woolhope Club*, 1891, 179 and still present).

Hornwort family
Ceratophyllaceae

Rigid hornwort
Ceratophyllum demersum

This plant of lakes and ponds is probably not native, being introduced deliberately or accidentally with fish. Pencerrig Lake, Llanelwedd 0454 ACP det. BA Seddon, where a single piece found in 1969 was not relocated in subsequent searches; ornamental fish pond, Old Town hall, Llandrindod Wells 0661; in quantity in lake, Stanage Park 3371.

Buttercup family
Ranunculaceae

Stinking hellebore
Helleborus foetidus

Well established on wooded cliff on the bank of the Wye below Wyecliff, Clyro 2242 ! and ACP, probably having escaped from the garden above.

Green hellebore
Helleborus viridis ssp. *occidentalis*

Possibly native in shady, calcareous wooded dingles. Wood "on the left of Erwood Wood Hill " c04 JG Williams (1939 NMW); large patch in deep gulley at Aberedw 04NE M Taylor (1961 *Nat Wales* 7, 170); in quantity in woodland near Penberth, Llanbadarn-y-garreg 1049 RK; a number of plants on roadside bank between Boughrood and Llanstephan 1240 HRNT survey.

*Winter aconite
Eranthus hyemalis

Naturalized from nearby garden on wooded cliff, by R. Wye, Wyecliff, Clyro 2242.

Globe flower
Trollius europaeus

An uncommon sp. of damp upland hay meadows and pastures, much decreased due to drainage and reseeding. It also occurs rarely on the margins of and on islands in the larger rivers such as the Wye and Ithon and in wet patches in open woodlands. 16 sites.

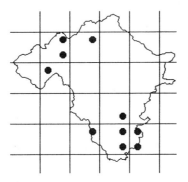

Marsh-marigold
Caltha palustris

A widespread sp. of marshes, lake and pool margins, river banks, and wet woodlands. It has much decreased in wet pastures due to drainage. In all squares except 86NW, 87SW, 96NE, 97NW, 04SE, 25SW & NE and 26NW.

Monk's-hood
Aconitum napellus

Naturalized on the bank of the R. Dulas, Builth Road 0354 E Matheson and on a disturbed roadside bank, possibly part of an old garden, Evenjobb 2662.

Wood anemone
Anemone nemorosa

Frequent in woodlands, hedgerows and occasionally in upland hay meadows, pastures, on cliff ledges and in gorges.

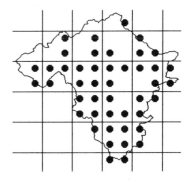

Traveller's joy
Clematis vitalba

Doubtfully native and known only as isolated plants. Wood at 1639 and by roadside wall near Glasbury 1739; ruins near Clyro Court 2042 AEW & JAW (1938 *Trans Rad Soc* (1947) **XVII**, 4); on a field fence on bank of Wye, Wyecliff, Clyro 2242 ! & ACP; walls at Knighton 27SE AEW & JAW (1938 see above); edge of old park near Boultibrooke, Norton 3065.

Ranunculaceae

Creeping buttercup
Ranunculus repens

Common in damp pastures, streamsides, edges of marshes and flushes, in wet woodlands and a weed of cultivated land. In all squares.

Meadow buttercup
Ranunculus acris

Common in meadows, pastures, roadside verges, woodland glades, stream and river banks. In all squares except 86NW and 97NW.

Bulbous buttercup
Ranunculus bulbosus

Widespread in well drained, sunny meadows, pastures and roadside banks. In all squares except 86SE & NW, 87SW, 97SW & NW, 06NW, 07SW & NW, 16NE, 17SE & NE, 18SW, 26SE and 37SW.

Corn buttercup
Ranunculus arvensis

Noted once in cornfields about Clyro c24SW HNR (1880 *J Bot*, 1881, 170).

Small-flowered buttercup
Ranunculus parviflorus

A rare sp. of disturbed ground. Garden, Glasbury 1838 SI Leitch, hollow of old limestone quarry, Yatt Wood, Dolyhir 2458 ACP.

Goldilocks buttercup
Ranunculus auricomus

A scarce sp. of shaded, base-rich soil in woodland glades and rarely on hedgebanks and in churchyards. 10 sites.

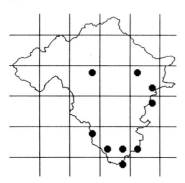

Celery-leaved buttercup
Ranunculus sceleratus

Known only from a marsh, probably on the site of a former pond, N of Barland, Evenjobb 2762 PJP & CP.

Lesser celandine
Ranunculus ficaria

The sspp. have not been separately mapped. Ssp. *ficaria* is common in woodlands and on hedge and streambanks. Ssp. *bulbifer* is frequent only as a weed of waste ground in towns, in gardens, churchyards and occasionally on river banks. The sp. is found in all squares except 86, 87 and 97NW.

Lesser spearwort
Ranunculus flammula

Common in marshes, flushes, on lake and river margins, in ditches and damp pastures. Recorded from all squares except 86NW and 27SE.

Greater spearwort
Ranunculus lingua

A rare sp. of sedge dominated fen. Llanbwchllyn 1146 (first noted by T. Westcombe in 1843 (*The Phytologist* 1, 781, and still present); abundant in small area of fen at SW end of Rhosgoch Common 1948 NCC survey.

Ivy-leaved crowfoot
Ranunculus hederaceus

In waterfilled ditches, shallow pools and slow flowing streams, mostly on mineral soils. Widespread.

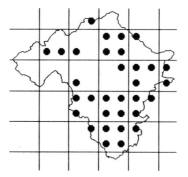

Round-leaved crowfoot
Ranunculus omiophyllus

In similar habitats to the last species but preferring peaty soils and commoner in the uplands.

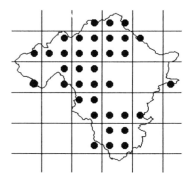

Three-lobed crowfoot
Ranunculus tripartitus

A small specimen collected from a pool on the top of Colva Hill 1953 by ACP was determined as this sp. by BA Seddon (1965 NMW). It is doubtfully this sp. Confirmation is required as to the presence of this plant in Radnor.

Water crowfoot
Ranunculus aquatilis agg.

Much further work is required on this group with populations, particularly in the R. Wye rarely setting seed and appearing to be of hybrid origin.

Pond water-crowfoot
Ranunculus peltatus

Frequent in mud-bottomed pools mostly on the upland commons of central Radnor and in slow flowing reaches of brooks and rivers. 27 sites.

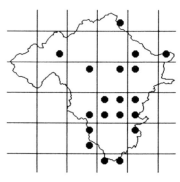

Stream water-crowfoot

Ranunculus penicillatus

Frequent in the more nutrient-rich rivers such as the lower Wye, Ithon, Arrow and Teme. The vars. have not been systematically separated. The var. *calcareus* is recorded from the Teme at Milebrook 3172 NTH Holmes.

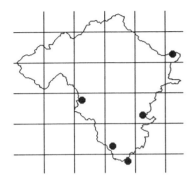

Common water-crowfoot

Ranunculus aquatilis s.s.

Plants which fail to set seed are widespread in the major rivers. These may be of hybrid origin. The true sp. has not certainly been discovered during this survey. FPW reports it from 04 and 18, parts of which occur in Radnor and from 26. Sinker et al. (1985) report it from a number of tetrads in 27 and 37. Further work is required on this taxon.

Thread-leaved water-crowfoot

Ranunculus trichophyllus

A scarce or overlooked species of ponds, lakes and rivers. In the R. Ithon near the sewer beds at Llandrindod Wells c0460 JAW (1938 MsFlR); wet floor of quarry by roadside Llanbadarn Fynydd 07 NE R Lewis, FH Perring & BM Sturdy (1956 *Proc BSBI* 2, 367 (1957); Llanbwchllyn 1146 S Thurley.

River water-crowfoot

Ranunculus fluitans

Infrequent, or overlooked, mostly in the larger rivers. R. Wye near Aberedw 04NE HJ Riddelesdell (1909, *Cotteswold Naturalists Field Club* (1910), **XVII**, 57-61); R. Dulas near Llanfihangel Helygen c06NW VG (1956 *Proc BSBI* 2, 367); R. Wye, Glasbury 13NE, J Mountford; R. Arrow, Milton Bridge, Michaelchurch on Arrow 2350 NTH Holmes; R. Teme, Milebrook 3172 NTH Holmes.

Papaver

Columbine

Aquilegia vulgaris

Probably not native but escaped from cultivation in all its Radnor sites such as disused railway banks, tips, waste ground, beside footpaths and on river banks. Recorded from 97SE, 04NE & SE, 05SW, 06SW & NW, 07SE, 08SE, 14SW, 15NE, 17SW, 25NE and 36SW.

Lesser meadow-rue

Thalictrium minus

Rare. On bank of the R. Wye between Llanstephan and Boughrood 1239. Recorded from this site first by AL in 1886 (*Bot Rec Club Rep* 1884-1886, 144). Waste ground Presteigne c3164 RG Ellis.

Barberry family
Berberidaceae

Barberry   *Berberis vulgaris*

A scarce sp., rarely naturalized and more probably planted in hedges. By R. Wye below Yr Allt, Llandeilo Graban 0744; near derelict cottage, Aberedw 0846 ACP. Noted in MsFlR from hedges Glasbury, Disserth and Newbridge.

*Oregon-grape

Mahonia aquifolium

Occasionally planted as cover for pheasants in woodlands and rarely escaping from gardens, as at Pistyll, Glasbury 1538 and Stanner Rocks 2658.

Poppy family
Papaveraceae

*Opium poppy

Papaver somniferum

A scarce weed sp. of disturbed ground associated usually with habitation and on rubbish tips. Recently seen on roadside bank, Pant-y-dwr 9874; old garden near Erwood Station 0943; waste ground, Llandrindod Wells 0561; Clyro churchyard 2143; roadside verge, Evenjobb 2662; tip, Knighton 2872.

Common poppy

Papaver rhoeas

Far from common on roadsides and waste ground and not a field weed. On builders sand, old station yard, Builth Road 0253; by recently rebuilt wall, Boughrood Church 1239; roadside, Llangunllo 2170 PJP; roadside Monaughty in many places 2369, 2470; Presteigne 36SW PJP.

Long-headed poppy

Papaver dubium

Of occasional occurrence on disturbed roadside verges, arable fields and gardens and tips, mainly on the low ground in the S and E.

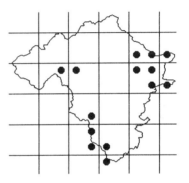

Yellow-juiced poppy

Papaver lecoqii

Rare and only seen on the railway line, Penybont c0964 RCLH (1956 *Proc BSBI* 2, 368).

*Prickly poppy

Papaver argemone

Noted by AJ Bird from 18 on a BRC card, this field and wayside weed should be sought in the E of the county.

Welsh poppy

Meconopsis cambrica

Probably native only in the rocky gorge of the Edw at Aberedw 0747 where it was first seen in 1884 (AL in *Trans Woolhope Club* 1884, 180) and on cliffs by Craig Pwll Du in the Bach Howey gorge, Llanstephan c1143, T Westcombe, 1843 (*The Phytologist* 1, 781), though not apparently recorded again in this latter site. Elsewhere it occurs naturalized by regularly escaping from cultivation eg. on walls by Caban Coch Resr., Elan Valley 9264, pavement edges, Llandrindod Wells 0561 and river banks and roadsides in various places.

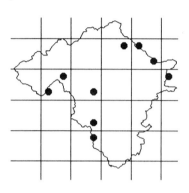

*Greater celandine

Chelidonium majus

Frequent in hedgerows and on waste ground about habitations throughout Radnor.

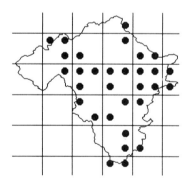

Fumitory family
Fumariaceae

Climbing corydalis

Corydalis claviculata

Frequent on acidic, peaty soil in ungrazed or lightly grazed open, upland woods and plantations, especially those with block scree and rock outcrops. Occasional in hedges, on road banks and in rough pastures, possibly as a relic of former woodland. 29 sites.

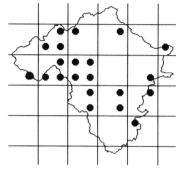

*Yellow fumitory

Corydalis lutea

A scarce sp. of lime mortar in stone walls about habitations. Rhayader 9667 Llandrindod Wells 0561; Clyro 2143; New Radnor 2160; Dutlas 2077; Knighton 2872; Presteigne 3164.

Purple ramping-fumitory

Fumaria purpurea

Known only from a single record of ES Todd from Llandrindod Wells c06, (*Rep Bot Exc Club* 1920, 110).

Ramping-fumitory

Fumaria muralis ssp. *muralis*

This rare weed of disturbed soil has been noted only once from near Boughrood c13NW by GC Druce in 1929 (*Rep Bot Exc Club* 9, 102).

Common ramping-fumitory

Fumaria muralis ssp. *boraei*

The commonest fumitory, but nowhere is it abundant. Scattered on disturbed roadside verges, in arable fields, gardens and on hedgebanks.

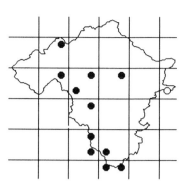

Common fumitory

Fumaria officinalis

Slightly less common than the last sp.. In gardens, arable fields and on disturbed road verges.

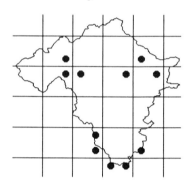

Cabbage family
Cruciferae

*London Rocket *Sisybrium irio*

Reported from the Llandrindod Wells area c06 in 1896 by the Rev JB Lloyd in Bufton, WJ & JO, 1906, 114. There are no other records.

*Tall rocket

Sisymbrium altissimum

Reported from the county by Druce, GC (1932). The source of this record has not been traced.

*Eastern rocket

Sisymbrium orientale

A single record by JAW from railway banks, Rhayader 96NE (1938 MsFlR).

Barbarea

Hedge Mustard
Sisymbrium officinale

Not common, but widespread in settlements, on path and track sides, tips and wasteground, mostly in the east.

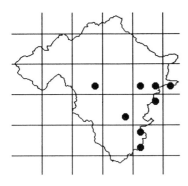

Garlic mustard
Alliaria petiolata

Common on fertile roadside banks and disturbed ground around farmyards. Less frequent in more wild places, but present about sheep lairs in woods and on calcareous rock outcrops where nutrients accumulate and on silt banks by streams and rivers.

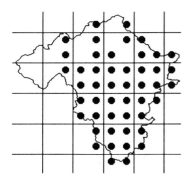

Thale cress
Arabidopsis thaliana

Frequent as a winter annual on sunny and free-draining open sites such as rock outcrops, ant hills, tracksides, wall tops and gardens.

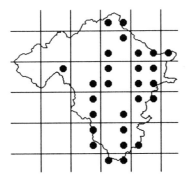

*Treacle mustard
Erysimum cheiranthoides

A local sp. of disturbed soil and arable fields. Waste ground, Erwood 0943 ACP; recently constructed roadside verge, Ridgebourne, Llandrindod Wells 0559; layby and road verge, Nantmel 0266; oatfield, Carmel c0566 VG (1956, *Proc BSBI* **2**, 369); field gateway in lane W of Llangunllo 2071 JM Roper; roadside bank, Knucklas 2474.

*Dame's-violet
Hesperis matronalis

Scarce in hedges and on tips around towns and villages. Noted in MsFlR as once being frequent but by 1954 "it appears to have largely disappeared". Recently seen near Rhayader 9768; churchyard, Llanbedr 1446; in churchyard, New Radnor 2160; hedge N of Whitton 2869 PJP and CP and on roadside banks, Knucklas 2574.

*Wallflower
Cheiranthus cheiri

Very rare and recently seen only on walls at Rhayader 9667, in an old limestone quarry S of Yatt Farm, Burlingjobb 2458 and at Presteigne 3164. Reported by AEW and JAW from quarries at Rhayader 96NE and at Cwmbach, Glasbury 1639 and on walls at Knighton 2872.

Winter-cress
Barbarea vulgaris

Occasional on stream, river and roadside banks, often in damp places.

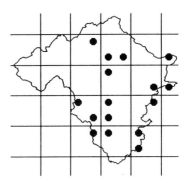

*American winter-cress
Barbarea verna

A rare sp. of waste ground. Rhayader 96NE JH; Nantmel c06NW JAW (1938, *Trans Rad Soc* (1947), **XVII**, 4); on gravel in old station yard, Stanner 2658 BSBI exc. det. TG Evans.

*Intermediate winter-cress
Barbarea intermedia

Noted by AL in a clover field and on a railway bank near Aberedw c04NE (1886-1890, *Trans Woolhope Club* 1890-92, 185), no other more recent records are known.

Creeping yellow-cress
Rorippa sylvestris

An uncommon sp. of damp stream and riverbanks, on pool margins and in ditches.

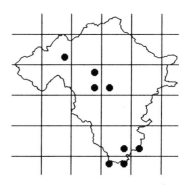

Cruciferae

Marsh yellow-cress
Rorippa palustris

Less common than the last, occurring in damp hollows on tips and in field gateways mostly from Boughrood 1338 to the English border.

Large bitter-cress
Cardamine amara

Only known as a few plants growing in a wet flush in a garden at Tan-y-Coed, St Harmon 9677; in an old ditch in unimproved pasture, Portway meadow; Rhosgoch Common 1948 NCC survey; and in a wet flush in a pasture, Church House Farm, Michaelchurch 2450 DEG (this latter site recently destroyed)

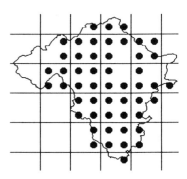

Cuckooflower
Cardamine pratensis

A common sp. of damp, unimproved grasslands. In pastures, on roadside verges, in woodland glades, marshes and flushes. Double flowered forms are frequent in a meadow near Ditchyeld Bridge, Evenjobb 2760 and occasionally elsewhere. In all squares.

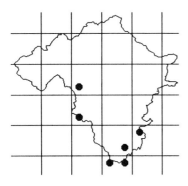

Narrow-leaved bitter-cress
Cardamine impatiens

A rare sp. of open scree in woodland glades and edges and on disturbed ground. Often common in a site following disturbance of the ground and then declining as other vegetation develops. Frequent around Aberedw 0746 for over 100 years on railway ballast and embankment; in flower beds, Pontshoni 0746 and on scree behind the house. Roadside, Smithfield Farm, Cwmbach Llechrhyd 0254 SE Gooch; Pen y Berth Wood, Llanbadarn y Garreg 1049 ACP; scree, Graig Fawr, Hundred House 1358 ACP; roadside ditch, Weythel, Dolyhir 2357 ACP; common on scree and stream bank, Worsell Wood, Stanner 2257, 2558 and 2658.

Hairy rock-cress
Arabis hirsuta

A rare sp. of dry and usually sunny calcareous rock outcrops. Cliffs by R. Ithon near Disserth 0459 CAC and near Dol-berthog, Llandrindod Wells 0460; Cwm Blithus Rocks, Radnor Forest 1661 NCC survey; limestone cliffs, Yatt Wood; Dolyhir, 2458; dolerite cliffs, Stanner 2658; wall, New Radnor 2160. Noted in MsFlR from Aberedw 04, Disserth 05, Llananno 07, Llanstephan 14, Presteigne 36, Bleddfa 26 and Knighton 27, and by HNR on walls between Builth Wells Station and Llanelwedd 05SW in 1880 *(J Bot* 1881, 170).

Bristol rock-cress
Arabis scabra

Noted "on the ruins of ancient fortifications or camp on the top of a fairly high hill in the district" at 1020 feet near Llandrindod Wells, DT Gwynne Vaughan (1894 *Sci Gossip NS* 1, 93). This sp., otherwise confined in Britain to the Avon gorge, has never been reported from Radnor again. Cefnllys Castle 0861 seems the most likely locality, though little is left but scree and low rock outcrops. Whilst a small area is base-rich most of the site is acidic and probably unsuitable.

*Horse-radish
Armoracia rusticana

Seldom grown in gardens and now rarely naturalizing on waste ground around towns and villages. Considered locally frequent in MsFlR, occurring by the railway, Howey c0458; Llandrindod Common 0660; Penybont c1164; Newbridge 0158; Llanfadog c9365; and Rhayader 96NE. In this survey only noted on waste ground or road verges in 05SW, 14SW and 24NW.

Water-cress
Nasturtium officinale agg.

Frequent on the edges of streams and rivers, about springs, in ditches and marshes on the lower ground in the S & E. Rarer elsewhere. *N. officinale* and *N. microphyllum* have not been certainly separated and are mapped together. Hybrids of these two species appear to be widespread but pure *N. microphyllum* has yet to be found.

Wavy bitter-cress
Cardamine flexuosa

Common on damp, open soil by streams and flushes, in woodlands, gardens and ditches and on hedgebanks, tracksides. In all squares except 86NW & NE, 87 and 97SW.

Hairy bitter-cress
Cardamine hirsuta

Frequent on shallow, open, freely drained soil on rock outcrops, walls, scree, tracksides and in gardens.

*Garden arabis
Arabis caucasica

Noted in MsFlR as well established as a garden escapee on walls round Llandrindod Wells 06, Presteigne 36, Knighton 27, Norton 36, Beguildy 17 or 18 and Dutlas 27. Noted recently on walls Cwmbach, Glasbury 1639; Whitton 2767; Knighton 2872 and Presteigne 3164.

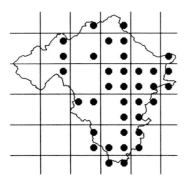

Erophila

*Aubretia *Aubrieta deltoidea*

This common garden plant has become naturalized on the S facing cliff face of an old limestone quarry at Burlingjobb 2458 and on walls in New Radnor 2160 and Presteigne 3164.

*Honesty *Lunaria annua*

An occasional colonist of roadside verges near gardens and on waste ground. Roadside, Builth Road 0253; roadside, Rhosgoch 1848; churchyard, Llanfihangel-Nant-Melan 1858; roadside Old Radnor 2458; churchyard, New Radnor 2160. Noted in MsFlR from New Radnor where it apparently grew in the same site from 1938 to 1952.

*Sweet Alison

Lobularia maritima

Rarely escaping from cultivation. On clinker and ash banks around Erwood Station 0843; tip, Nantmel 0165 and on a roadside tip Knighton 2872.

*Golden Alison

Alyssum saxatile

A few plants naturalized in the mortar of a brick wall, Spa Road, Llandrindod Wells 0561 and on walls New Radnor 2160 and Presteigne 3164.

Wall whitlow-grass

Draba muralis

A scarce probable introduction from England, seen only on a roadside tip near Ashfield, Ysfa 9763 and on a verge in Stanage Park 3271. Noted in FPW from 25, the source of this record has not been traced.

Common whitlow-grass

Erophila verna ssp. *verna*

Widespread on sunny, open, free-draining sites such as rock outcrops, scree, quarries and tracksides.

Shepherd's-purse

Capsella bursa-pastoris

A common weed of disturbed ground in gardens, fields, tips, gateways and tracksides.

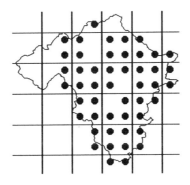

Shepherd's cress

Teesdalia nudicaulis

A scarce sp. of open turf on summer droughted rock outcrops. Llanelwedd c05SE or SW JAW (1948-51 *Proc Swansea Sci and Fld Nats Soc* **II** pt **10**, 325); Llandrindod c06SE JAW (1948-51 loc cit.), with an earlier record from the golf links Llandrindod Wells of JAW (1948 MsFlR); Castelltinboeth, Llannano 0975 ! & DS Wilson; Old Radnor Hill 2558 DR Humphreys. (Noted here by AL in 1888 in *Trans Woolhope Club* 1886-1889, 218). Stanner Rocks 2658, known here for over 130 years (AL *Trans Woolhope Club* 1853, 91); Hanter Hill 2457 PJP & CP.

Field penny-cress

Thlaspi arvense

A scarce plant of open ground. Disturbed roadside verge between Crossway and Howey 0558; kitchen garden, Ridgebourne, Llandrindod Wells 0560 AEW & JAW (1938 *Trans Rad Soc* (1947) **XVII**, 5); by R. Ithon above Dolau Jenkin, Penybont 1066 G Pell; on gravel by R. Ithon, Penybont 1164 CAC; disturbed roadside verge E of Bleddfa 2368.

Alpine penny-cress

Thlaspi alpestre

A specimen of this sp. was collected on Stanner Rocks 2658 by HJ Riddelsdell in 1900 and determined by EF Linton (*J Bot* 1903, 409). It was there in very small quantity and could not be found in 1903. The nearest extant Welsh locality is on lead mine spoil in Cardiganshire (FPW). With seed known to have little capacity to remain dormant the origin and fate of the Stanner population remains a mystery.

*Perennial candytuft

Iberis sempervirens

Planted and now naturalised on rocky bank near the N shore of Llandrindod Wells Lake 0660.

*Garden candytuft

Iberis umbellata

A rare garden escape noted only on the site of former gardens, now supermarket, in Llandrindod Wells 0561 and a roadside verge near the disused New Radnor Station 2160.

Field pepperwort

Lepidium campestre

A scarce sp. of tracks and roadsides, not seen in the county for perhaps 30 years. Recorded in MsFlR from the lower Elan Valley 96, Newbridge on Wye 05, Llandrindod Wells 06, Abbeycwmhir 07, and N of Llanelwedd 05. JAW adds Stanner Rocks 2658 (1948 *Proc Swansea Sci & Fld Nats Soc* **II**, pt **10**, 324), whilst the earliest record is by HNR from a roadside near Builth Wells 05 (1880, *J Bot* 1881, 171).

Cruciferae

Smith's pepperwort

Lepidium heterophyllum

Occasional on freely-drained roadside banks, quarry spoil and summer droughted turf on shallow soil on sunny rock outcrops.

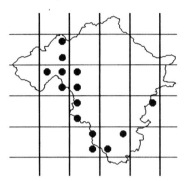

*Poor-man's pepperwort

Lepidium virginicum

Fowl-runs, Llandrindod Lake c0660 JAW (1938 NMW) and near Gladestry 25 RA Langdale-Smith det. RCL Burgess in 1957 (BRC record). The former specimen has been reported variously as *L. densiflorum* and *L. neglectum*, both of which species are treated as synonymous with *L. virginicum* by Stace (1991).

*Swine-cress

Coronopus squamatus

A scarce weed of disturbed ground. Field gateway, Upper Cabalva Farm, Clyro 2346 ACP and beside the Gilwern Brook near Stanner 2657 CAC.

*Lesser swine-cress

Coronopus didymus

An uncommon weed of track sides, field gateways and waste ground, most common on the ORS in the Wye valley.

*Annual wall-rocket

Diplotaxis muralis

Noted once on the disused railway line near Glasbury Station 1839 by SI Leitch in 1973. The site was cleared in the same year.

*Rape and Swede

Brassica napus

Commonly sown as a cover crop to aid establishment of grass leys and as a root crop for human and livestock consumption. Seed spilled on roadsides and waste ground produces short-lived naturalized populations.

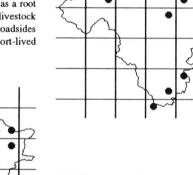

*Turnip *Brassica rapa*

A common weed of arable fields, disturbed roadsides and tips. Some of the records may also refer to naturalized populations of the tuberizing cultivated turnip.

*Black mustard *Brassica nigra*

No recent records. Scarce, near Clyro 24SW and in rocky, clayey cutting, Newchurch c2151 (both MsFlR).

Charlock *Sinapis arvensis*

Widespread but not common as a weed of arable fields, tips and waste ground.

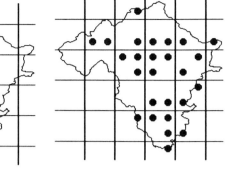

*White mustard

Sinapis alba ssp. *alba*

In similar sites to the last. Noted in MsFlR from the Wye valley between Llanfaredd 0650 and Builth Road 0253, it has not been encountered in this survey.

*Hoary mustard

Hirschfeldia incana

This alien from the Mediterranean is recorded from 06SE, the Llandrindod Wells area by AEW (NMW records).

*Wild radish

Raphanus raphanistrum ssp. *raphanistrum*

A scarce weed of disturbed ground. Noted in MsFlR from Llanbister 17SW, Llanelwedd and Newbridge (both 05); it has recently only been seen in a swede field S of Llidiarddau, St Harmon 9778 and on the bank of the R. Wye near Clyro 2242 ! & ACP.

Mignonette family Resedaceae

Weld *Reseda luteola*

An uncommon plant of sunny, freely drained open ground. Roadside gravel, Aberedw 0748 ACP; gravel, disused Erwood Station goods yard 0943; quarry spoil, Llanelwedd 0552; soil of landslip, bank of Wye near Llowes 2042; bank of Wye, Upper Cabalva 2345 ACP and on quarry tip, Dolyhir 2458 ! & IDS.

Wild mignonette

Reseda lutea

A single record from a disturbed roadside near Clyro c2143 VG (1956 *Proc BSBI* 2, 369).

Sundew family Droseraceae

Common sundew

Drosera rotundifolia

A widespread sp. of acidic, peaty flushes, blanket and raised bogs and wet heath, especially in the NW uplands.

Stonecrop family Crassulaceae

Navelwort

Umbilicus rupestris

Frequent in dry crevices of rock outcrops, in stone walls and freely drained hedgebanks, avoiding the most acidic sites. It is occasionally found as an epiphyte in pockets of humus on old tree trunks.

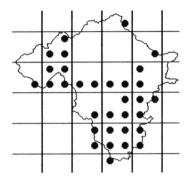

*House-leek

Sempervivum tectorum

Commonly cultivated and naturalized on walls and roofs about houses and farmyards. Planted populations proved so difficult to separate from naturalized ones that a map could not be produced.

Orpine *Sedum telephium*

An uncommon sp. of rock outcrops, particularly volcanic rocks, sunny, freely-drained hedgebanks, and scree. 18 sites. Both the ssp. *telephium* and *fabaria* are recorded from the county (FPW) but have not been separated in this survey.

*Caucasian stonecrop

Sedum spurium

This and the next sp. are aliens of vigorous growth, typically exceeding their allotted space in the garden and hence are regularly thrown out! Rarely naturalized on rock outcrops and seen recently only on the bank of the Wye near Llanstephan 1043 ACP and on an outcrop in wood E of Knucklas 2673. Recorded in MsFlR from a number of sites including railway cuttings at Builth Road 0253, Newbridge 0158 and from Rhayader 96, Nantmel 06, Presteigne 36, Llangunllo 27 and Beguildy 17 or 18.

*Reflexed stonecrop

Sedum reflexum

Uncommonly naturalized on rocks, walls, and roadsides. Disused quarry near church, Rhayader 9668; Rhayader Quarries 9665; old road, near Vulcan Arms, Ysfa 9863; roadside verge near Llowes 1942; walls, Presteigne 3164. Probably formerly more widespread as described in MsFlR as "locally frequent to abundant" with localities such as railway banks, Llandrindod 06 and Newbridge 05, Llandegley 16, Clyro 24, Boughrood 13 etc..

Rock stonecrop

Sedum forsterianum

A widespread but not common sp. of basic rock outcrops and scree. Favours the volcanic rocks of Llandrindod 06SE, Dolyhir 25NW and Stanner 25NE and the calcareous shales of the Radnor Forest 16 & 20. Unaccountably absent from Aberedw Rocks 04NE. c20 sites. The sspp. *forsterianum* and *elegans* both occur (FPW) but have not been separated in this survey. It is occasionally cultivated and escapes onto walls and stone roofs.

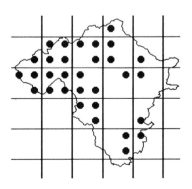
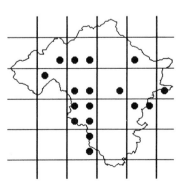
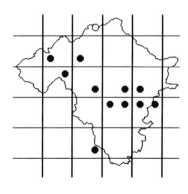

Saxifagaceae

Biting stonecrop *Sedum acre*

Frequently cultivated, it occasionally becomes naturalized on walls and gravel paths eg farm wall, Llanevan, Llandegley 1561 ACP; Old Radnor churchyard 2559 DRD & IHS; walls and roadsides Knucklas 2574. It is doubtfully native anywhere in Radnor.

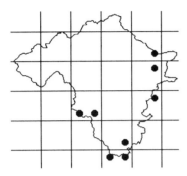

*Tasteless stonecrop

Sedum sexangulare

Rarely naturalized in open, freely draining sites. Walls, Penithon c0781 (1938 and 1948 MsFlR); river gravel by Wye, Llowes 1941 ACP.

*White stonecrop

Sedum album ssp. *album*

Widely cultivated and occasionally naturalized on rocks and walls eg bank of R Wye Llanfaredd 0651; old wall, Bryngwyn Church 1849 ACP; old wall, Bailey Bog Farm, Waun Marteg 0375 ACP; rocks by Wye, Wyecliff, Clyro 2242. Noted in MsFlR from most of the towns and larger villages on walls and probably still present.

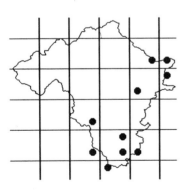

English stonecrop

Sedum anglicum

Common on dry, acidic, shallow soil in rock crevices and on rock ledges in the S & W. MsFlR reports that it had much decreased since 1918 though offered no possible reasons. Its scarcity in the NE where apparently suitable habitat occurs cannot be explained.

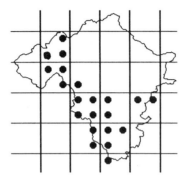

Saxifrage family Saxifragaceae

*London pride

Saxifraga x urbium

This well known garden plant has been rarely seen naturalized. Rocks by river, Disserth 0357 ACP; old quarry in wooded valley, Gwernaffel Dingle, Knighton 2770; near Llangullo 17SW (MsFlR).

*Celandine saxifrage

Saxifraga cymbalaria var. *huetiana*

A single record from a wall at Glasbury c1739 E Battiscombe (1954 NMW).

Rue-leaved saxifrage

Saxifraga tridactylites

A scarce sp. of freely-drained, open, calcareous sites. Aberedw 04NE (MsFlR); gravel near old railway station, Glasbury 1839 SI Leitch; on limestone rocks in disused quarry near Dolyhir ! and in rocky pasture nearby ACP, both 2458; Stanner 2658 (MsFlR); 26 (FPW); on mortar of stone walls, Presteigne 3164.

*Irish saxifrage

Saxifraga rosacea

Noted as a naturalized sp. from Llandrindod Wells 06SE JAW (1938, NMW).

Mossy saxifrage

Saxifraga hypnoides

On sunny, calcareous shale rock outcrops and cliffs. Aberedw Rocks and gorge 0746, and 0847 (first noted here by AL 1891); Craig y Fuddai, Llanbedr Hill 1247 ACP; many outcrops around the Radnor Forest 1661, 1761, 1862, 1963, 1964, 2062 NCC survey, ACP & ! with the earliest record from T Westcombe in 1844 at Water-break-its-neck c1860 (*The Phytologist* **1**, 781). Noted in MsFlR in addition from Llandrindod Wells 06, Beguildy 17 and Llangunllo 27, though possibly as garden escapes.

Meadow saxifrage

Saxifraga granulata

Exclusively a woodland plant, in Radnor. Frequent only on the calcareous soils in the S of the county. Habitats occupied include wooded cliff ledges, scrub and on the alluvium accumulated on the wooded banks of streams and rivers.

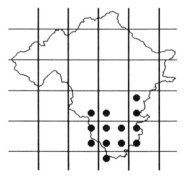

*Fringe-cups

Tellima grandiflora

This rare garden escape noted only once from a roadside verge between Nash and Walton c26SE EM Rutter (1961, NMW).

Alternate-leaved-golden-saxifrage

Chrysosplenium alternifolium

Widespread in the S and E on damp soil and decaying logs in woodlands. Favours streambanks and particularly wet and humid woodlands. Noted by AL in 1891 at Aberedw 04NE, AEW and JAW (in MsFlR) seem never to have found it in Radnor. This is curious as it is now known from over 40 sites.

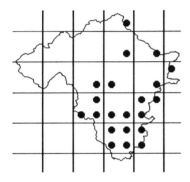

Opposite-leaved-golden-saxifrage

Chrysosplenium oppositifolium

Common in flushes, spring heads, wet woodlands and on damp stream banks throughout. In all squares except 86NE & NW, 87SW and 97NE.

Gooseberry family
Grossulariaceae

Red currant *Ribes rubrum*

Frequent in damp, shady woodlands, on stream banks and occasionally in hedgerows.

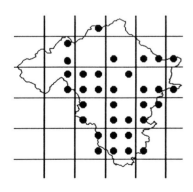

Ribes

Black currant *Ribes nigrum*

Less common than the last, but frequent on river banks and occasional in damp woodlands and hedgerows. Widely grown in gardens for its fruit, most populations probably arose from this source.

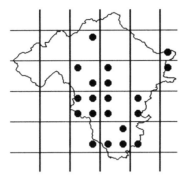

*Flowering currant

Ribes sanguineum

A rare alien noted in MsFlR from a brookside, Abermethil Llandegley 16SW; Cefnllys 06SE and Gaufron 96NE. Seen recently on roadsides in woodland near Doldowlod 0160 and SW of Clyro 2143.

Gooseberry *Ribes uva-crispa*

Common in hedgerows, woodlands, on rock outcrops and river banks. In all squares except 86, 87, 96NW, 97SW & NW, 06NE, 17NE & NW, 18SW and 26NE.

Mountain Currant

Ribes alpinum

Doubtfully native anywhere in Radnor it is an uncommon shrub of hedges, generally about habitations. Best detected in spring as one of the first shrubs to come into leaf. Possibly once cultivated for its fruits or as a hedging shrub, it no longer seems to be grown. Pant y Dwr 9875, Vaynor, Nantmel 0169; edge of Rock Park, Llandrindod Wells 0561; between Gwystre and Crossgates 0765; streambank, Aberdwynant, Llananno 0575; New Cottage, David's Well, Llananno 0578; Rhosgoch 1847; Rhonllwyn, Llandegley 1464; Newhouse, Llanfihangel Rhydithon 1669; dumped in roadside quarry near City Farm, Moelfre City, Llanbister 1275.

Rose family
Rosaceae

*Bridewort

Spiraea spp. and hybrids

Frequently spreading vegetatively in ornamental woods, hedges and on waste ground from original plantings or tipping. Widespread, no attempt has been made to separate the hybrids or spp. and not mapped due to the difficulty in determining whether naturalized or not in so many sites.

Meadowsweet

Filipendula ulmaria

Common in wet pastures and meadows, ditches, marshes and on stream and river-banks and on damp roadsides. Reported from all squares except 86NE & NW and 87.

Rubus

Subgen. Cylactis

Stone bramble *Rubus saxatilis*

On slightly calcareous, damp shale rock outcrops. Cerrig Gwalch, Rhayader 9370 CCW survey; by the Ithon near Dol-berthog, Llandrindod Wells 0460 and in Davy Morgan's Dingle, Radnor Forest 1862 ACP

Subgen. Idaeobatus

Raspberry *Rubus idaeus*

Common on roadside verges, wood and plantation clearings and edges, scrub and waste ground, with cultivated forms common in gardens. Recorded from all squares except 86, 87, 04SE, 13NW, 14SE and 26SW.

Rosaceae

Subgen. Rubus
The Brambles

No attempt has been made to map the microspecies of bramble. The following account is based on records made during BSBI excursions, the accounts given in FPW and in *Brambles of the British Isles* by Edees, ES and Newton, A, Ed. Kent, DH, Ray Soc., London 1988, supplemented by valuable records made by M. Porter, whose assistance in the preparation of this account is gratefully acknowledged.

The order and taxonomy followed is that of Edees & Newton 1988 (op cit). It has proved impossible in some cases to determine whether a 10 km grid square record was made in Radnor or in an adjacent vice-county or both. Such records are placed in ( ).

Section A Rubus

Subsect. Rubus

R. bertramii G. Braun
96, 04, 05, 06.

R. plicatus Weihe & Nees
05, (25).

Subsect. Hiemales E.H.L. Krause

Series 1 Sylvatici

R. scissus W.C.R. Watson
96, 05, 06, 14, 15.

R. vigorosus Mueller & Wirtgen
16, 26.

R. laciniatus Willd.
05.

R. lindleianus Lees
96, 04, 05, 13, 14, 25, 26.

R. ludensis W.C.R. Watson
96, 04, 05, 06, (13), 14, 16.

R. perdigitatus Newton
(04), 05, (25).

R. platyacanthus Mueller & Lef.
(05, 06), 36, 37.

R. pyramidalis Kaltenb.
(05), 06.

R. silurum (Ley) Druce
(87), 96, (97,04), 05, 06, 13, 14, 15, 16, 24, (25, 36, 37).

Series 2 Rhamnifolii

R. acclivitatum W.C.R. Watson
04, 05, 13, 14, 16, 24, (25), 26, (36,37).

R. amplificatus Lees
14.

R. cardiophyllus Lef. & Mueller
13, 24, (25).

R. dumnoniensis Bab.
96, (14), 24.

R. incurvatus Bab.
(04), 05, (37).

R. lindebergii Mueller
04, 05, 26, 27, 36, (37).

R. nemorales Mueller
05, 06, 14, (24, 36, 37).

R. polyanthemus Lindeb.
04, (05), 06, 13, 14, 16, 23, (24, 25), 26, (36, 37)

R. prolongatus Boulay & Letendre ex Corbiere
04, 05, 06.

R. rhombifolius Weihe ex Boenn.
96, (04), 05, (13), 14, 16, (36).

R. rubritinctus W.C.R. Watson
04, 05, 13, 14.

R. septentrionalis W.C.R. Watson
05.

Series 3 Sprengeliani

No records

Series 4 Discolores

R. ulmifolius Schott
04, 05, 13, 14, (24), 25.

Series 5 Vestiti

R. bartonii Newton
04, 05, 06, (25, 36).

R. longus (Rogers & Ley) Newton
04, 05, 13, 14, 24, 25.

R. vestitus Weihe
04, 05, 06, 13, 14, (24), 25, (36, 37).

Series 6 Mucronati

R. mucronatoides Ley ex Rogers
16, (24, 25), 26, 36.

R. wirralensis Newton
14.

Series 7 Micantes

R. moylei W.C. Barton & Riddelsd.
96, 04, 05, 06, (24, 25, 36).

Rubus

R. raduloides (Rogers) Sudre
96, (04), 05, 06, 13, 24, 27, (36, 37).

R. biloensis Newton & Porter
Reported from the vice-county in Newton, A. & Porter, M. (*Watsonia* **18**, 195).

Series 8 Anisacanthi

R. leyanus Rogers
05, (13), 14, 25, 36.

R. pascuorum W.C.R. Watson
26, (36, 37).

Series 9 Radulae

R. echinatus Lindley
14, 27.

R. euryanthemus W.C.R. Watson
06.

R. longithyrsiger Lees ex Focke
(24, 25), 36.

R. pallidus Weihe
05, 06, (36).

Series 10 Hystrices

R. babingtonii Salter
(04), 05, (13), 16, (24, 25, 36).

R. dasyphyllus (Rogers) E.S. Marshall
96, (97), 04, 05, 06, (08), 13, 14, 16, 18, (24), 25, 26, (36).

R. hylocharis W.C.R. Watson
(96, 04), 16, (24).

R. merlini Newton & Porter
97

R. purchasianus Rogers
14.

Series 11 Glandulosi

No records.

Section B
Corylifolii Lindley

R. pruinosus Arrh.
26.

R. tuberculatus Bab.
05, 06, 13, 14, 25.

Section C
Caesii Lej. & Courtois

Dewberry *R. caesius*
A scarce sp., though hybrids with *R. fruticosus* agg. are somewhat more widespread. On verges, railway banks and by carparks. By disused Builth Wells Station, Llanelwedd 0451; by railway, Llandrindod Wells 0560; roadside hedge between Dolau and Llanfihangel Rhydithon 1466. Reported from New Radnor 26SW in MsFlR and from 07 and 25 in FPW.

Field rose *Rosa arvensis*
Common in hedges, on waste ground, woodland edges and scrub, generally in freely drained, sunny sites.

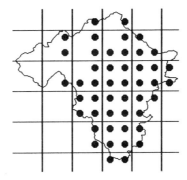

Burnet rose

Rosa pimpinellifolia

Known only from the S facing cliffs of Stanner Rocks 2658. First reported there in 1888 by AL in *Trans Woolhope Club* 1886-1889, 218.

*Japanese rose *Rosa rugosa*
This alien persists probably from tipped garden rubbish, bird-spread seed or deliberate planting in a few sites. Disused railway bank near St Harmon 97SE NMW field record and at edge of wood near layby S of Howey 0457.

Close-styled rose

Rosa stylosa

Known only from a hedge near the disused Glasbury railway cutting, Glasbury 1839 MP.

Dog Rose

Rosa canina agg.

Commonly found in hedgerows, woodland clearings, river banks and waste ground on the more fertile soils. Further work is required to elucidate the distribution of the segregated species of this group. Hybrids appear to be more frequent than the true species.

R. canina is likely to be widespread

R afzeliana is reported from Rhayader c96 by AL *(Bot Loc Rec Club* 1882, 246), roadsides at Llanelwedd 05SE and near Howey 0558 by MP and from near Llanbadarn Fynydd 1079 by IM Vaughan.

R. coriifolia is reported from 05 and 06, one or both of these records may be from Radnor (FPW).

R. obtusifolia Widespread in the lower Wye Valley. Disused quarry Llanelwedd 0551 ! & MP; Bach Howey gorge and nearby roadside hedge, Llanstephan 1041 ! & MP.

R. dumetorum Reported from old railway bank near St Harmon 97SE NMW record; disused quarry Llanelwedd 0551 ! & MP; roadside hedge Llanstephan 1042 ! & MP and old railway bank near Glasbury Station 1839 MP.

Rosaceae

Downy rose
Rosa tomentosa agg.

Members of this group seem to be commonest in the uplands when their frequently dark flowers are a conspicuous feature of hedgerows.

Further work is required to resolve the distribution of *R. tomentosa*, *R. sherardii*, and *R. villosa*, all of which have been reported from Radnor (FPW).

Sweet briar
Rosa rubiginosa agg.

Only *R. micrantha* s.s. has been reported and in small quantity. Bushy place near Llanstephan c14SW AL (*Bot Loc Rec Club*, 1886, 145) and on disused railway bank near Glasbury Station 1839 MP.

Agrimony
Agrimonia eupatoria

Mostly confined to roadside verges on sunny, freely-drained sites. Occasionally in woodland clearings and beside rides and on field margins.

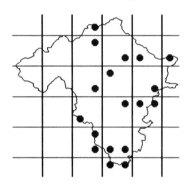

Fragrant agrimony
Agrimonia procera

Rare. On the disused railway Aberedw 0648 ACP; roadside verge near Builth Road Station 0253 MP; stream bank at Gaer, Llansantfread (sic) c05 HNR (1880, *J Bot* 1881, 171); Worsell Wood, Stanner c2557 CI and NY Sandwith (1945 *Rep Bot Exc Club*,1947, 56).

Great burnet
Sanguisorba officinalis

Frequent on damp clay soils on roadside verges, hedgebanks, unimproved pastures and haymeadows and rare on ungrazed upland cliff ledges.

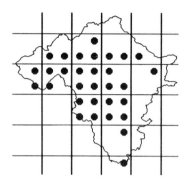

Salad burnet
Sanguisorba minor ssp. *minor*

Very rare on basic soil and possibly much decreased in recent years. Seen recently only on the gravel of a little used forestry track in Caen Wood, Presteigne 3063. Reported in MsFlR from Erwood Station, Llandeilo Graban and Aberedw Rocks, all probably in 04; Llandrindod Wells and Cefnllys in 06; near Ffynnon Gynydd c14 and Stanner 25.

Water avens *Geum rivale*
Widespread but not common on damp stream banks, by flushes in woodland, on damp roadside verges and occasionally in wet unimproved meadows and rough pastures.

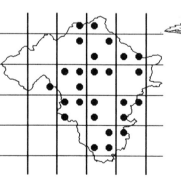

Wood avens *Geum urbanum*
A common sp. in the more nutrient rich woodlands, hedges, roadside banks and waste places. Recorded from all squares except 86, 87, 96SW & NW and 07NW.

Marsh cinquefoil
Potentilla palustris

Widespread in the wettest of marshes, on pool sides, basin mires and flushes. Known from over 60 sites, it has undoubtably declined recently due to widespread agricultural drainage.

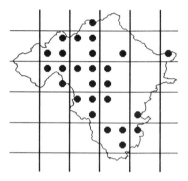

Silverweed
Potentilla anserina

Common on the open soil of gritty road margins, tracks and gateways. Colonises muddy tracks and grazed ephemeral pools which dry out in the summer. Recorded from all squares except 86, 87SW, 96SW and 18SW.

x 1/3

Marsh cinquefoil

Potentilla

Rock cinquefoil
Potentilla rupestris

This national rarity, known from only three sites in Britain, was first found in Radnor by the Rev. A Ley on the 13 August 1886 (see *Bot Loc Club Rep* 1884-1886, 145 and *J Bot* 1887, 28), He describes finding around 50 plants on the side and summit of a range of rocks of the Llandeilo series in the SW of the county. No more details were supplied, to protect the plant from unscrupulous collectors. There has been much speculation as to the exact locality of his find, with Llanelwedd Rocks, displaying similarities to the Breidden Hill, Montgomery locality of the plant, a favoured site. There are however, good reasons to believe that Ley found it in the site on rocks in the flood zone of the Wye where RA Graham collected a specimen (BM) in 1934 and where it still occurs today. It maintains a tenuous hold, kept free of competing woody species by winter floods and summer droughts, conditions which the plant itself seems ill adapted to. To protect the plant its exact locality will not be revealed here.

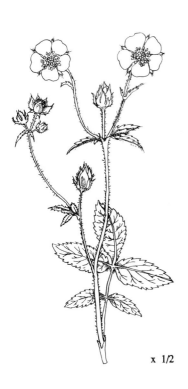

x 1/2

Rock cinquefoil

Tormentil *Potentilla erecta*

Common and widespread in acidic grassland, heathland, hedgebanks, woodland glades, scrub and road verges throughout, on acidic soil. Recorded in all squares except 13NW, 26SW and 37SW.

Trailing tormentil
Potentilla anglica

Reported in FPW as being widespread in central Radnor. This survey has failed to find many localities, but possible hybrids with *P. erecta* and/or *P. reptans* have been reported. The true species occurs in hedgebanks in dry, sunny places in 96NE, 05SW, 13NE, 14SW and 16NE.

Creeping cinquefoil
Potentilla reptans

Widespread on roadside banks, waysides, wasteground, quarry wastes etc.

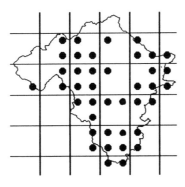

Barren strawberry
Potentilla sterilis

Common in woodland glades, on hedgebanks, rock outcrops and in short turf in dry places. In all squares except 86, 87, 97NW and 07NW.

Wild strawberry
Fragaria vesca

Widespread on woodland banks, hedgerows, rock outcrops, scree and old walls. Most abundant on the more base-rich soils in the S & E. In all squares except 86, 87, 97NW, 07NW & SW and 26SW. A yellow fruited form occurs by the roadside 1km N of Llanbadarn Fynydd 0878.

*Hautbois strawberry
Fragaria moschata

The only record is from between houses, New Radnor c2160 (1948 MsflR).

Lady's-mantle
Alchemilla vulgaris agg.

Widespread but not common in the uplands, in meadows, pastures, roadside verges and churchyards.

Intermediate lady's-mantle
Alchemilla xanthochlora

Widespread in damp agriculturally unimproved old pastures, roadsides and churchyards.

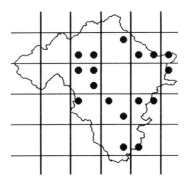

Hairy lady's-mantle
Alchemilla filicaulis ssp. *vestita*

In similar habitats to the last but generally less common except in the Teme valley where it is the commonest species.

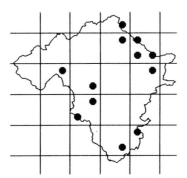

Rosaceae

Smooth lady's-mantle

Alchemilla glabra

Most frequent in the north on river, road and old railway banks.

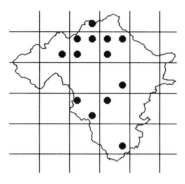

Alchemilla mollis

Naturalized in churchyard, Rhayader and on bank of R. Wye below 9668.

Parsley-piert

Aphanes arvensis agg.

Common as a weed of arable fields and on the freely-drained soil of roadside banks, rock outcrops and ant hills. The aggregate is mapped below. Further work is required to elucidate the distribution of *A. arvensis* s.s. and *A. microcarpa*, both of which appear from a preliminary survey to be equally widespread. Walters S.M. (1948) reports in "*Aphanes microcarpa* in Britain " (*Watsonia* **I**, part **III**, 168) that a specimen collected by A. J. Wilmott in 1932 from Stanner Rocks 2658 was intermediate in character between *A. microcarpa* and *A. arvensis*. The aggregate is reported from all squares except 86, 87, 97SW, NE & NW, 06NE & NW, 07NW, 14NE, 17SE and 36.

Wild pear   *Pyrus pyraster*

Scarce. A large specimen occurs beside The Bog or Aberithon turbary, Newbridge on Wye 0157. A hedge at Wernhir, N of Newbridge on Wye 0160 JH appears to have been planted with a number of trees producing very small fruits, approaching those of *P. cordata* in size.

Crab apple

Malus sylvestris agg.

Common in hedgerows, woodlands and on riverbanks and roadsides. The native *Malus sylvestris* s.s. is a sp. of woodlands whilst *M. domestica* is found in hedges, on riverbanks, and roadsides. They were not separately mapped. Various cultivars are undoubtably naturalized from discarded apple cores eg a tree bearing fruit very similar in appearance to "Golden Delicious" is naturalized on the roadside below Stanner Rocks 2658.

Rowan or Whitty

Sorbus aucuparia

Common on acidic rock outcrops, upland hillsides, woodlands, hedges and roadsides. Less frequent in the lowlands and occasionally planted as an ornamental tree on roadsides and in gardens. In all squares except 87SW.

Wild service tree

Sorbus torminalis

Very scarce and confined to a few rock outcrops. Yr Allt, Llandeilo Graban 0844 ACP; Aberedw Rocks 0746 AEW (1927 MsFlR) and still present; Edw gorge, Aberedw 0747; cliffs by the R. Ithon from its confluence with the Wye to above Disserth ie 0257 CAC, Craig Fain 0357 DEG & JSB, 0359 A Crawford; Stanner Rocks 2658 AS Fergusson.

Whitebeams *Sorbus aria* agg.

Occasionally planted as an ornamental tree but probably not naturalized. Two trees were noted from an oakwood on E side of Penygarreg Resr, Elan Valley c96NW by WM Condry (NMW). Their origin requires investigation.

Sorbus porrigentiformis

On calcareous shale cliffs near Pontshoni, Aberedw 0746 and 0845 MEM and MP. Whitebeams were first recorded at Aberedw by AL in 1886 and still occur in good quantity on the cliffs. He identified them as *Pyrus rupicola* and AEW records *S. rupicola* at Aberedw in 1927 (NMW records). These records are believed to be referrable to *S. porrigentiformis*. JAW reports *S. rupicola* as occurring in plenty near Cefnllys 0861 in 1918 and 1938 (*Proc Swansea Sci & Fld Nat Soc.* Vol **1**, pt **4**, 94). No specimens have been traced and no whitebeams have been found recently in the area.

English whitebeam

Sorbus anglica

Five trees occur in crevices in the grit and shale rocks overlooking Caban Coch Resr. dam, Elan Valley 9264.

*Wall cotoneaster

Cotoneaster horizontalis

Naturalised on stony roadside bank W of Cwmbach, Glasbury 1639; on rocks in Yatt Wood, Dolyhir 2458 ! & IDS and on walls New Radnor 2160 and Presteigne 3164.

*Himalayan cotoneaster

Cotoneaster simonsii

Not recorded naturalized in this survey but noted in MsFlR from Doldowlod 9962; Gwystre 0665; Llandrindod 06SE; and Crossgates 0864.

*Small-leaved cotoneaster

Cotoneaster microphyllus

Not recorded naturalized in this survey but noted in MsFlR from walls, Llandrindod 06SE; Beguildy 1979 and Knighton 2872.

Hawthorn or quickthorn

Crataegus monogyna

Common as isolated specimens on grazed hillsides, in open woodlands, scrub, waste ground and in hedges. It is the major hedgerow shrub planted into new hedges created after roadworks, usually from Dutch stock. In all squares except 86NW & NE and 87.

Prunus

Midland Hawthorn

Crataegus laevigata

Possibly planted in hedge above disused railway tunnel, Llansantffraed Cwmdeuddwr 9667 B Jenkins & DPH.

Blackthorn   *Prunus spinosa*

Common in hedgerows, woodland glades and edges, on stream and river banks. Recorded from all sqares except 86NE & NW, 87 and 96SW & NW.

Plum   *Prunus domestica*

Plums and damsons have been regularly planted in orchards and in hedges near cottages and farms. Any occurring as standard trees in trimmed hedges have been assumed to be planted and ignored in this survey. Specimens laid into hedges have probably been overlooked. Consequently naturalized plums appear to be scarce, and further work is required to determine the distribution of the ssp. *domestica* - wild plum and ssp. *institia* - the bullace.

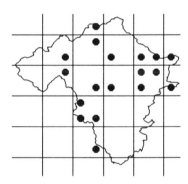

Wild cherry   *Prunus avium*

A scarce tree except in woodlands on heavy calcareous clays in the S and E. Present also on road verges, old railway banks and in hedgerows and occasionally planted as an ornamental tree or for its wood.

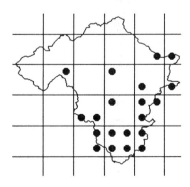

*Dwarf cherry

Prunus cerasus

Described in MsFlR as very local and infrequent and recorded from Painscastle c1646; Old Radnor c2459; New Radnor 2160; Presteigne c3164 and Michaelchurch c2450. AL notes it as being "plentiful along the rock bank of the gorge under the churchyard" Aberedw 0847 (*Trans Woolhope Fld Club* 1884, 186). R Birch found it in this latter site in 1977 but in poor condition. By 1982 it could not be found. This survey has failed to locate this alien anywhere in the county. Similarly Sinker et al 1985 record its demise in Shropshire.

Bird cherry   *Prunus padus*

A frequent sp. of damp woodland, streamsides and hedgerows on the heavy clay soils of the upland valley bottoms. Particularly common along the Ithon Valley.

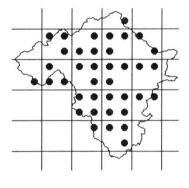

*Cherry laurel

Prunus laurocerasus

This alien has established itself in a few woodlands, having been planted as pheasant cover or to provide evergreen screens. Spreading by layering, seedlings have never been seen. The winter of 1981-82 killed it back to ground level but it has since recovered. Below Yr Allt, Doldowlod 0061; Penybont Hall 1164; Navages Wood, Stanner 2659.

Pea family
Leguminosae

*Laburnum

Laburnum anagyroides

Reported in MsFlR as "plentifully introduced on the hills between Llanbadarn Fynydd 0978 and Camnant Bridge 0882 and makes a beautiful show in June". This is hardly the case now, being very scarce except where planted in gardens. Near Gwenfron, Pant y Dwr 9476; Tylwch 9779 and Fieldstile Farm, Kinnerton 2463.

*Scotch laburnum

Laburnum alpinum

Reported in MsFlR as being with the last but much scarcer. Not seen in this survey. FPW reports it from 05 & 07. The sources of these records have not been traced.

*Hairy-fruited broom

Cytisus striatus

This broom from the Iberian Peninsula has been sown together with the native species on roadside banks to provide cover and has naturalized. Beside A470 near Doldowlod 9962 AO Chater (*Nat Wales* 1978, **16**, 57), the A438 near Bettws Dingle, Clyro 2345 and the A44 near Stanner Rocks 2658. Easily distinguished from common broom by its silver-hairy seed pods when in fruit.

Broom   *Cytisus scoparius*

Frequent on dry, sunny roadsides and field banks, rock outcrops, scree, beside gravel tracks and in disused quarries. Suppressed by heavy grazing in many suitable-looking sites.

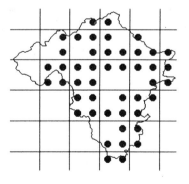

Leguminosae

Dyer's greenweed
Genista tinctoria ssp. *tinctoria*

Widespread in unimproved, dry, acidic pastures and in grassland on road and railway banks. Many sites have been lost through agricultural improvement work but it is still known in about 50 localities.

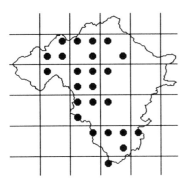

Petty whin   *Genista anglica*
Frequent only in the uplands in heathy, damp, unimproved pastures and roadside banks. Like the last, reduced in extent recently by agricultural improvement but present in c40 localities.

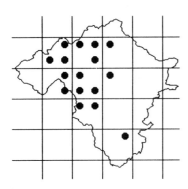

Gorse   *Ulex europaeus*
Frequent on hedgebanks, rocky hillsides, especially on the commons, favouring sunny and dry places. Known locally as French gorse, some populations may have been planted as cover for game. In all squares except 86, 87, 07SE, 14NW, 16SE, 24NW, 25NW and 27NW.

Western gorse   *Ulex gallii*
In similar sites to the last but more abundant on heathland in the hills and less common in hedgerows and disturbed sites generally. Both spp. show little sign of expansion in view of the now heavy grazing of most of their habitat, burning and the widespread death of colonies in hard winters such as 1985-86. In all squares except 87SW, 13, 15SW & NW and 16SE.

*Garden lupin
Lupinus polyphyllus

A rare colonist of roadsides & tips. Roadside tip, Knighton 2771.

Wood bitter-vetch
Vicia orobus

An uncommon sp. of unimproved upland hay meadows and occasionally pastures, hedgebanks, roadside and railway banks in the W and N. Heavy grazing and the agricultural improvement of pastures has reduced its abundance recently to c30 sites.

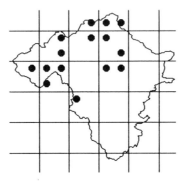

Tufted vetch   *Vicia cracca*
Common on roadside banks and in hedgebottoms, marshes, riverbanks and old shingle beds where grazing is absent or very light and occasionally in unimproved hayfields.

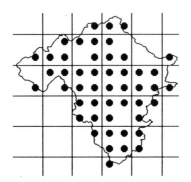

Wood vetch   *Vicia sylvatica*
No recent records. Noted in MsFlR from Cwm Ithon and Hundred House c1154 and by AL from a wood near Stanner c2658 (1886 *Bot Record Club Rep*, 1884-1886, 145). The source of the record E of Heyop c2375 in Sinker et al, 1985 could not be traced, but this sp. is frequent close by in Shropshire.

Hairy tare   *Vicia hirsuta*
Widespread but not common on roadside banks and occasionally on waste ground, about farmyards and on silt banks by the larger rivers.

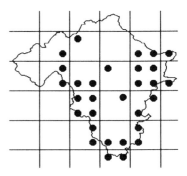

Smooth tare
Vicia tetrasperma

A rare sp. of somewhat open grassland subject to disturbance from time to time. Roadside verge near Erwood Stn. 0943 A Hemsley; road verge, Llanstephan 1042 ! & MP; disused railway bank, Boughrood 1240; layby NW of Clyro 2044; verge of A44 near New Radnor Stn. 2160.

Bush vetch   *Vicia sepium*
Common on hedgebanks, in woodland, on roadside verges and in ungrazed and irregularly cut grassland. Recorded from all squares except 86, 87SW, 96SW & NW and 97NW.

Common vetch   *Vicia sativa*
Frequent on roadverges, hedges and railway banks and occasionally in woodland glades. Sspp. *sativa* and *nigra* have not been separated but the latter is probably the most frequent ssp..

Trifolium

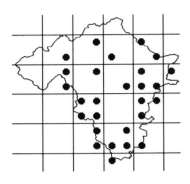

Spring vetch
Vicia lathyroides

This predominantly coastal species in Wales (FPW) has a single mid Wales record from Stanner Rocks 2658 GC Druce in 1912 (*Rep Bot Exc Club*, 1912, 210). It appears never to have been seen subsequently.

*Broad Bean   *Vicia faba*

A casual seen only on Clyro rubbish tip 2143.

Bitter vetch
Lathyrus montanus

Common in unimproved meadows and pastures on well drained soils, in woodland glades and on hedge and river banks. The var. *tenuifolius* with narrow leaflets is scarce. On roadside bank, Cwmbach Lechrhyd 0354 and at edge of forestry near Fron, Bleddfa 2168 C Parry.

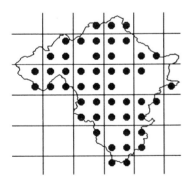

Meadow vetchling
Lathyrus pratensis

Common on roadside and railway banks, in unimproved haymeadows and on woodland edges, avoiding heavily grazed sites. In all squares except 86, 87, 96SW, 27NW and 36NW.

Narrow-leaved everlasting-pea
Lathyrus sylvestris

Very rare. In scrub woodland on Burfa camp, Evenjobb 2860 DR Humphreys and on a bank near Knucklas Station 2574 J Roper. Possibly of long standing in this latter site since JAW reports this sp. from Knucklas in 1925 (*Rep Bot Exc Club*, 1925, 871).

*Broad-leaved everlasting-pea
Lathyrus latifolius

A scarce naturalized alien or colonist. Bank of R. Wye, Boughrood 1338 and on waste ground, Frank's Bridge 15NW RCLH (1956 *Proc BSBI* 2, 371).

*Garden pea
Pisum sativum

Noted on waste ground, Presteigne 3164 RG Ellis.

Common restharrow
Ononis repens

A scarce sp. of old permanent pasture and disturbed ground. Described in MsFlR as fairly frequent in parts of the Upper Wye and Ithon Valleys, it has not been recorded recently in this area. Old railway, Aberedw 0748 ACP; by railway in old station yard, Llandrindod Wells 0561 RK; abundant on bank between old railway and cereal crop, Boughrood 1239; railway bank near Dolau 1568 ITE-NCC survey; in field, Clyro 2044 IHS & CC; common on sand and shingle banks by R. Wye, Clyro 2242 ACP & !.

*White melilot
Melilotus alba

A rare casual noted growing on a heap of builders sand in old station yard, Builth Road 0253 and on gravel in builders yard Llanelwedd 0451 (it was first reported from 05 on a BSBI exc. in 1956). Also on waste ground, Tremont Road, Llandrindod Wells 0662 DRD.

Black medick
Medicago lupulina

An infrequent sp. of waste ground and roadside verges, mostly in the lowlands.

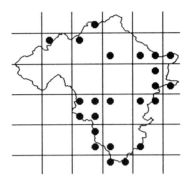

*Lucerne   *Medicago sativa*

Present on railway embankment near Llangunllo Stn. c27SW until 1931 (MsFlR).

Fenugreek
Trifolium ornithopodioides

Known only from a shingle bank by the R. Wye, Clyro 2445 D Merry det. QON Kay (1977, *Nat Wales* 15, 144).

Upright clover
Trifolium strictum

This national rarity of shallow, summer-droughted soil occurs in small quantity on rock ledges at Stanner Rocks 2658. First reported by FM Day and WH Hardaker in 1936 (NMW) it has been subsequently reported on few occasions, but as it occurs with other annual clover spp may have been overlooked. Noted most recently in 1990 and 1991.

White clover
Trifolium repens

Regularly sown with grasses to produce semi-permanent pastures and on roadside banks to provide cover. It occurs wild in unimproved meadows and pastures, in flushes, marshes and riverbanks in all squares except 86NW and 87SE.

Leguminosae

*Alsike clover

Trifolium hybridum ssp. *hybridum*

Sown to provide cover on newly created roadside banks and verges but seldom persisting long. Occasionally seen as a field weed and perhaps sometimes sown. Rare on tips and waste ground.

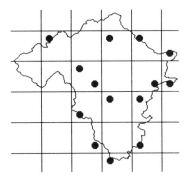

Hop trefoil

Trifolium campestre

An uncommon sp. of dry, sunny grassland, preferring the lowlands. Most frequently noted on road and railway banks, it also occurs on calcareous rock outcrops and rarely in old meadows and pastures.

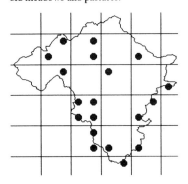

Lesser trefoil

Trifolium dubium

Common in freely drained, sunny grasslands on roadsides, lawns, meadows, pastures and rock outcrops. In all squares except 86NW, 87, 07SW and 14NW.

Slender trefoil

Trifolium micranthum

Very scarce but seen recently by MP on gravel beside road near Llanerch-y-cawr, Claerwen Valley 9061 and by BSBI members on the road verge of the A481 to the S. of Llanelwedd Rocks 0551 where it was plentiful. Noted in MsFlR from lawns at Llanbister 1073 and Knighton 2872 and from the Links, Llandrindod Wells 06SE. Recorded by HNR from a road near Clyro c24SW in 1880 (*J Bot* 1881, 171) and by FPW from 25.

Knotted clover

Trifolium striatum

A rare sp. of disturbed, freely-drained soil. Tip, Llanelwedd Quarry 0452 BSBI exc.; levelled tip, Llanfair Quarry, Llandrindod Wells 0661; disturbed roadside verge, Llanbister 1071; AL records it from Stanner Rocks 2658 (*Trans Woolhope Club* 1853, 91) but AEW and JAW reject a record from Stanner and its presence is in need of confirmation; Silia, Presteigne c 37SW WMR (1899 *J Bot* XXXVII, 19).

Hare's-foot clover

Trifolium arvense

A scarce sp. of open and dry soil and gravel. S facing bank on site of old railway bridge near former Rhayader Stn. 9667; on the railway, Aberedw c04NE AL (1881 *Bot Loc Rec Club* 1882, 246); old track bed of railway, now road verge S of Middle Hall, Aberedw 0648 ! and JD Woods; abundant on disused platform and old goods yard, Builth Road Stn. 0253; roadside of A470 NW of Penmaenau, Llanelwedd 0352 ! , CP and PJP; on roadside verge near old railway viaduct over the Bach Howey, Llandeilo Graban 1042 ACP (first seen 1973 and persisted until 1982 when site became overgrown); ashes and clinker near Dolau Stn. 16NW JAW (1938 MsFlR); Stanner Rocks 2658; Silia, Presteigne c37SW WMR (1899 *J Bot* XXXVII, 17-25).

Rough clover

Trifolium scabrum

Only known from dry, S facing, stony grassland on dolerite. Stanner Rocks 2658 AI Nock and JAW (1948 MsFlR) and still present.

Red clover *Trifolium pratense*

Common in haymeadows, pastures, and on roadsides, railway and open river banks. Occasionally included in seed mixtures used for semi-permanent pastures. In all squares except 86NW and 87SW.

Zigzag clover

Trifolium medium

Occasional in unimproved meadows, especially on boulder clay and in hedgerows and on road verges, especially those associated with the edges of woodlands.

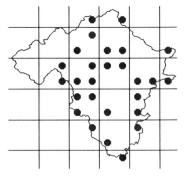

Common bird's-foot-trefoil

Lotus corniculatus

Common in well-drained grassland, especially in hedgerows and on road sides, churchyards and rocky outcrops. Often a pioneer colonist of recently disturbed shale banks by roads and in forestry. In all squares except 86NW, 87SW and 07SW.

Greater bird's-foot-trefoil

Lotus uliginosus

Common in wet places such as marshes, pond and lake margins, ditches and river banks, wet meadows and woodland glades. In all squares except 86NW, 87SW, 26SW and 37SW.

Kidney vetch

Anthyllis vulneraria ssp. *vulneraria*

A rare plant of dry, sunny banks, seen recently only on a roadside verge near Llanstephan, 14SW FMS (1977 *Nat Wales* 15, 144). Recorded from Aberedw c04, Erwood c04, Llandrindod c06 and Llanddewi Ystradenni c16 areas in MsFlR.

Bird's-foot

Ornithopus perpusillus

A scarce sp. of sunny, dry, shallow, summer-droughted soil on rock outcrops and gravel. Grassy bank above wood, Glan-llyn, Llansantffraed Cwmdeuddwr 9469 C Parry; volcanic outcrops around Llanelwedd Quarries 0551; open rocky woodland, Caer, Llansantffraed in Elvel 0854 J Langford; gravelly roadside bank, Gt Vaynor, Nantmel 0169; Graig Hill, Radnor Forest 1867 PJP and CP; rock outcrop near Lower Hanter, Burlingjobb 2557; Stanner Rocks 2658 (1st reported here by AL in 1853 in *Trans Woolhope Club*, 1853, 91); Forest Wood, Cascob 2467 PJP and CP; forest track, Caen Wood, Presteigne 2962. Noted in MsFlR in addition from Doldowlod c96SE; Cwm Elan c96 and near Forest Inn, Llanfihangel-Nant-Melan 15NE.

Poached-egg-flower family Limnanthaceae

*Poached-egg-flower

Limnanthes douglasii

This American flower, frequently grown in gardens, is reported naturalized in MsFlR from Abbey- cwmhir c07SE by JAW (1929) and roadside tip Lloyney 2475 DPH.

Wood sorrel family Oxalidaceae

*Procumbent yellow-sorrel

Oxalis corniculata

A rarely established alien. Pathside by Llandrindod Lake 0660 JAW (1945 *NW Nat* 20,157).

*Upright yellow-sorrel

Oxalis europaea

Rare as a garden weed. Argoed Mill, Doldowlod 9962; New Inn, Newbridge on Wye 0158 and Pen-y-gweirglawdd, Clyro 2146.

Geranium

Wood-sorrel

Oxalis acetosella

Common on shady woodland and hedge banks, amongst scree and on rock ledges in the uplands. Recorded in all squares except 86NW and 87.

Geranium family Geraniaceae

Bloody crane's-bill

Geranium sanguineum

A few colonies occur on sunny dolerite rock ledges at Stanner Rocks 2658. Known here by AL as early as 1853 (*Trans Woolhope Club* 1853, 91), it is undoubtedly native. Naturalized with other garden plants, it was recorded also from a railway bank near Penybont 1065 NCC/ITE survey.

Meadow crane's-bill

Geranium pratense

Of occasional occurrence on roadside and river banks, lanesides and in churchyards, mostly on boulder clay and alluvium. Considered by the authors of MsFlR to have decreased.

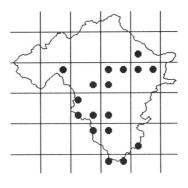

Wood crane's-bill

Geranium sylvaticum

A rare sp., close to the southernmost edge of its range. Newbridge on Wye c05NW WMR (1898, *J Bot* XXXVII, 21); small colony on shale rock outcrop, Davy Morgan's Dingle, Radnor Forest 1862 ACP; Water-break-its-neck, Llanfihangel-Nant-Melan 1860 AL (1885 BM cited FPW) and not seen since; roadside bank near Turgy, Crug y Byddar 1583 SF Bond; woodland, Gwernaffel Dingle, Knighton 2770 PR (1964 and gone by 1971).

*Dusky crane's-bill

Geranium phaeum

A rare alien of hedgerows, which might also have been planted in churchyards. Shady laneside, Llanfaredd 0651 (destroyed in 1989); churchyard, Llanstephan 1142 AP Conolly; churchyard, Franksbridge 1156 (destroyed 1986); hedge, Newchurch c2150 AM Pell (NMW); hedge near Old Radnor c2559 JA Whellan (1943 *Rep Bot Exc Club* 1943-44, 710); between Norton and Presteigne c26 D Taylor (1956 NMW) Knighton c2872 A McKensie (1924 WFP). Noted also from 36 in FPW.

Hedgerow crane's-bill

Geranium pyrenaicum

A rare denizen of old railway and roadside banks. Roadside, Tan-y-coed, Pant-y-dwr 9677; old railway, Aberedw 0748 ACP and LE Whitehead; rough grassland on site of former Builth Wells Stn., Llanelwedd 0451 BSBI exc.; hedgebank, Llanfihangel Rhydithon 1466 FMS (*Nat Wales* 16, 59). Also recorded from 15 in FPW.

Dove's-foot crane's-bill

Geranium molle

A common weed of fields, gardens, gateways, road edges and waste ground.

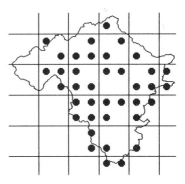

Small-flowered crane's-bill

Geranium pusillum

A scarce sp. of freely drained, disturbed ground. Old tip, Llanelwedd Quarries 0452 BSBI exc.; rock outcrops, S end Llanelwedd Rocks 0551 BSBI exc.; old quarry, Llandrindod Wells 06SE VG (1956 *Proc BSBI* 2, 370); by railway line, Llangunllo c27SW JF Hall, PCH and FH Perring (1956 op cit).

Long-stalked crane's-bill

Geranium columbinum

Rare on sunny, freely drained, somewhat basic soil. Roadside banks, formerly part of railway embankment near Aberedw 0747 ACP (recorded on bank, Aberedw by AL in 1881, *Bot Loc Rec Club* 1882, 246); on rock outcrop on bank of Wye below Yr Allt, Llandeilo Graban 0844; old railway, Glan-gwy, Disserth 0253 ACP, CP and PJP; on roadside, formerly railway bank to E of Erwood Station 1043; on rocks in Yatt Wood, Dolyhir 2458; Stanner Rocks 2658; side of forestry track, Taylor's Wood, Norton 2967 PJP; forestry track, Cold Oak, Presteigne 2863; waste ground, Presteigne 3164. MsFIR records it more widely from Llandrindod Wells c06SE; Llanbadarn Fawr c06NE; Cefnllys c06SE; New Radnor Castle 26SW and Beguildy c17NE.

Cut-leaved crane's-bill

Geranium dissectum

Widespread as a weed of gardens, arable fields, farmyards, tips, track sides and occasionally on thin soil over rock outcrops on sunny valley sides.

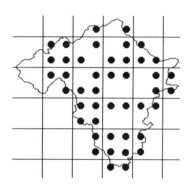

Shining crane's-bill

Geranium lucidum

Uncommon on freely-drained calcareous soil of rock outcrop and scree, particularly on the Wenlock and Ludlow shales of the Wye Valley and limestone at Dolyhir 2458. Rarer on walls and tracksides such as at Cefndyrys 0353 and Presteigne 3164. 15 sites.

Euphorbiaceae

Herb Robert

Geranium robertianum ssp. *robertianum*

Common in woodland, on hedgebanks, rock ledges, scree, tracksides, in yards, gardens and on riverbanks except in the most acidic sites. In all squares except 86NE & NW, 87 and 97NW.

Common stork's-bill

Erodium cicutarium ssp. *cicutarium*

Rare in freely-drained, sunny sites. Old track bed of railway to N of Builth Road 0253 S Mason; Llanelwedd Rocks 0551; rocks, Maengowan, Llanelwedd 0552 CAC; Llanstephan 14SW GP (1952); Stanner Rocks 2658 (1st record AL 1853 *Trans Woolhope Club* 1853, 91). Noted in MsFIR from Walton c2559 and Knucklas c2574.

Flax family
Linaceae

Pale flax *Linum bienne*

PD Moore (1978) notes a pollen grain, in all probability of this sp. from early Flandrian peat deposits at Cors y Llyn, Newbridge on Wye 0155. FPW notes its presence, pre 1930, in the county. The source of this record is unknown.

*Flax *Linum usitatissimum*

This once cultivated sp. is known from a single record of AL from Rhayader c96NE in 1881 (*Bot Rec Club* 1878-1882, 246).

Fairy flax

Linum catharticum

Widespread in short turf on somewhat basic, dry or wet flushed sites such as unimproved meadows and pastures, roadsides and upland flushes.

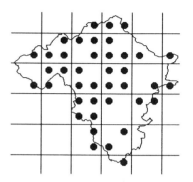

Spurge family
Euphorbiaceae

*Annual mercury

Mercurialis annua

Noted by AP Conolly, both male and female plants appeared when the gardens between Middleton Street and the railway, Llandrindod Wells 0561 were cleared in 1984 prior to building a supermarket and car park. It has not been seen since.

Dog's mercury

Mercurialis perennis

Abundant on the damp soil of woodland, hedges, shady river and stream banks and cliff ledges on base-rich rock types and calcareous boulder clay. In all squares except 86NE & NW, 87, 96SW and 07NW.

Sun spurge

Euphorbia helioscopia

Widespread as a weed of arable crops, particularly swedes. Also in gardens, tips and on disturbed roadside verges.

Polygala

Wood spurge
Euphorbia amygdaloides

Noted for the first time in Radnor by CAC on the wooded bank of the Gilwern Brook, Worsell Wood, near Stanner 2657 in 1982. On searching Worsell Hill for other colonies, this sp. has been found to be common under larch and beech plantations on all sides of the hill 2557, 2558, and 2658.

Milkwort family
Polygalaceae

Common milkwort
Polygala vulgaris

Widespread in the short turf of unimproved pastures, road verges, hedgebanks and churchyards. Avoids the wet peaty sites where it is replaced by the next sp..

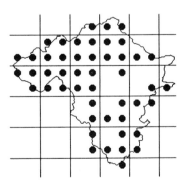

Heath milkwort
Polygala serpyllifolia

Common in the damp acidic turf of upland pastures and sheep walks and on rock ledges, heath or moorland, road verges and old quarries.

Maple family
Aceraceae

*Norway maple
Acer platanoides

Occasionally planted in gardens and parks and sometimes naturalizing, such as in the Rock Park, Llandrindod Wells 0560 and Yatt Wood, Dolyhir 2458.

Field maple   *Acer campestre*
Common as a hedgerow and woodland tree on the more base-rich soils of the S and E of the county. Rare in the NW.

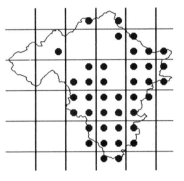

*Sycamore
Acer pseudoplatanus

Commonly planted about farmhouses and now well naturalized in all woods except the most acidic sessile oak types in the uplands. Frequent in hedgerows and an aggressive colonist of disturbed and ungrazed ground. The bark is regularly eaten by grey squirrels perhaps allowing the coral spot fungus (*Nectria cinnabarina*) to enter the tree causing a yellowing of leaves and death of the upper crown, a common disease in some years. Recorded from all squares except 87NE & NW and 87SW.

*Caper spurge
Euphorbia lathyris

A rare escape from cultivation. Old gardens, now site of supermarket, Llandrindod Wells 0561; disturbed roadside bank, Evenjobb 2662; Knighton 27SE JHA Stenart (1885 WFP).

Dwarf spurge
Euphorbia exigua

Noted on a BRC card by VG from 24 in 1956 and from Boughrood c13NW by WMR in 1898 (*J Bot* **XXVII**, 23), it should be sought as a weed of disturbed ground in the S of the county.

Petty spurge
Euphorbia peplus

Often with sun spurge in arable crops, gardens and on disturbed road verges or tips.

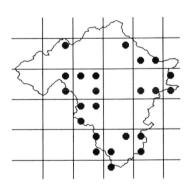

Horse chestnut family
Hippocastanaceae

*Horse chestnut
Aesculus hippocastanum

Widely planted as an ornamental tree but not common and very rarely naturalized.

Balsam family
Balsaminaceae

*Small balsam
Impatiens parviflora

An uncommon alien, well established in damp woodland edge sites in the E of Radnor. Bronydd, near Clyro at confluence of stream with Wye 2345 R Birch. It was first reported, possibly from this site (by R. Wye, Clyro) by C Goodman in 1956 (*Proc BSBI* **2**, 370); Evancoyd, Evenjobb 2563 CP; Boultibrook, Presteigne 3065 K Lomax (det. ACP 1969, *Nat Wales* **12**, 36).

*Indian balsam
Impatiens glandulifera

Frequent on silt banks beside the Wye from Llanfaredd 0650 to the English border. The earliest record traced is of JAW from by Rhayader Bridge c9667 in 1920 (*Rep Bot Exc Club* 1920, 118). It does not appear to have been recorded here in recent years. Also noted by forestry track near Cold Oak, Presteigne 2863 and by the R. Lugg below Presteigne 3264.

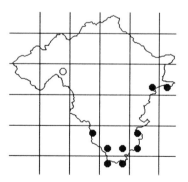

Holly family
Aquifoliaceae

Holly *Ilex aquifolium*
Common in woodlands and hedgerows except in the heavily grazed uplands, where winter browsing of the bark tends to eliminate it. In all squares except 86NE & NW, 97 and 18SE.

Spindle family
Celastraceae

Spindle
Euonymus europaeus

Occasional in woodland, scrub and hedgerows on the base-rich soil of the lower Wye Valley. Rare elsewhere and confined to locally base-rich sites. 15 localities.

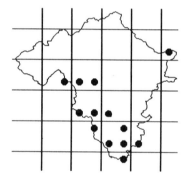

Box family
Buxaceae

Box *Buxus sempervirens*
Widely planted in churchyards, gardens and ornamental woodlands but hardly naturalizing. Rare on waste ground, probably arising from garden rubbish. A single plant on a ledge inaccessible to sheep on Craig Gigfran, Elan Village 9264 may have established from seed.

Buckthorn family
Rhamnaceae

Buckthorn
Rhamnus catharticus

Noted only from near Llanstephan 14SW by HJM Bowen in 1956 (*Proc BSBI* **4**, 283). Favouring calcareous soils, it should be sought on lime-rich rock outcrops and on the banks of the Wye.

Alder buckthorn
Frangula alnus

Like the last the present status of this sp. is uncertain. It was recorded twice on a BSBI recording weekend in 1957 from 07 and 13. Limited details of the presumably 07 record were published as near Llanbadarn, DE de Vesian (*Proc BSBI* **23**, 71). No other records are known.

Lime family
Tiliaceae

Large-leaved lime
Tilia platyphyllos

Reported from Aberedw 04NE, by the Wye below Builth c05, Llanfaredd 05SE, Monaughty 26NW, near Knighton 27 and from Presteigne 36SW in MsFlR. No certain records of this tree have been made in this survey.

Small-leaved lime
Tilia cordata

A rare tree, native in some sites, of uncertain origin elsewhere. Wood below Pen y Garreg Resr. dam, Elan Valley 9167 ! & P Jennings; hedgerow S of Vulcan Arms, Ysfa 9863; beside trunk road N of Newbridge on Wye 0159 FMS (*Nat Wales* **16**, 59); by Clywedog Brook above Crossgates 0867 AO; Abbeycwmhir c07 AEW and JAW (1938 *Trans Rad Soc* **XVII**, 6); Bryn Person Wood, Llananno 0872 NCC survey; cliffs in Bach Howey gorge, near Llanstephan 1143 and 1243; Glascwm c15SE (1954 MsFlR); Ddol Wood, Llanbister 1272 ! and PJP; Presteigne withy beds 3164 RK (possibly planted).

*Lime   *Tilia x vulgaris*

Widely planted but, seldom setting seed, it rarely, if ever, naturalizes. Not mapped.

Mallow family
Malvaceae

Musk mallow

Malva moschata

The beautiful pink flowers are a feature of sunny gravel banks, disused quarries, roadsides and rock outcrops in freely drained sites.

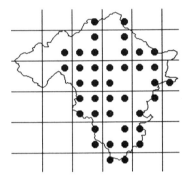

Common mallow

Malva sylvestris

Rather scarce and mostly confined to waste ground by roads, tips and farmyards in the lowlands. The leaves were used to produce poultices to treat livestock at Llanfaredd 0650 (J Jones pers comm.), until recent times.

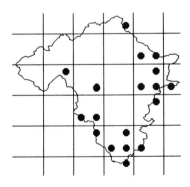

Dwarf mallow   *Malva neglecta*

The records of *M. pusilla* in MsFlR probably refer to this species. It was noted as being very local, with records from Builth Road and Cwmbach c0253, Aberedw c04NW, Glasbury c1739, New Radnor c2160 and Presteigne c3164. Recently seen only in a garden in Glasbury 1739 SI Leitch.

Velvetleaf
*Abutilon theophrasti

Two plants were seen in a field cultivated earlier in the season near Llanfaredd 0650 in autumn 1983. The origin of the seed is unknown but similarly isolated specimens turned up in a swede field at Llysdinam 95, Brecknock near the Radnor boundary in the same year. Reported by J Jones to FMS; it was det. by TG Evans.

Daphne family
Thymelaeaceae

Mezereon   *Daphne mezereum*

A single plant occurs in roadside scrub on the top of an old limestone quarry above Burlingjobb 2458 ! and IDS. Probably arising from seed spread from a nearby garden.

Spurge laurel

Daphne laureola

Doubtfully native, it occurs in a roadside hedge near Penmaenau, Llanelwedd 0352 !, CP and PJP; abundant in scrub woodland near garden, Wyecliff, Clyro 2242 ACP; a single plant on roadside verge below Yatt Wood, Dolyhir 2458 ! and IDS; frequent in old shrubbery, Womaston 2660 ! and DR Humphreys; several plants in roadside hedge, SW of Lloyney, Beguildy 2375; old ornamental woodland, Silia Wood, Presteigne 3064 ! and T Sykes.

Sea buckthorn family
Elaeagnaceae

Sea buckthorn

Hippophae rhamnoides

Reported as subfossil pollen in late glacial peat deposits at both Cors y Llyn, Newbridge on Wye 0155 (Moore PD, 1978) and Rhosgoch Common 1948 (Bartley DD, 1960).

St John's-wort family
Guttiferae

*Rose of Sharon

Hypericum calycinum

Noted in MsFlR as established on rocks and in the dingle near Llandrindod Wells c06SE. These records probably relate to clearly planted populations still present by the Lake and in the Rock Park. Also colonizing wooded road verge from ornamental planting, Llanstephan 1141.

Tutsan

Hypericum androsaemum

A scarce sp. with all recent records confined to damp calcareous soils and rocks in the SE of the county. Damp road verge, Llandeilo Graban 1043 FMS; lane bank below Llanstephan Church 1241; Cilcenni Dingle, Llowes 1741 MEM; bank of R. Wye, Clyro 2042. Formerly more widespread and noted in MsFlR from Doldowlod c96NE; Llanfaredd c05SE; Llandrindod Dingle 0560; Abbeycwmhir c07SE; Llanbadarn Fynydd c07NE and Llanbadarn y Garreg 14NW.

Hairy St John's-wort

Hypericum hirsutum

A scarce sp. of sunny, calcareous woodland edges and banks, mainly in the E and S of Radnor.

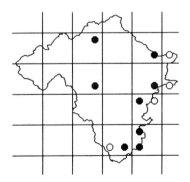

Violaceae

Slender St John's-wort
Hypericum pulchrum

A common sp. of sunny, freely-drained, acidic soil on hedge banks, rock ledges, woodland clearings and edges, heathland and unimproved upland pastures. Reported from all squares except 86NW, 87SW, 97SW, 14NW and 15SE & SW.

Pale St John's-wort
Hypericum montanum

Known only from the edges of the scrub-covered screes at Stanner Rocks 2658 from where it has been recorded for over 130 years *(Trans Woolhope Club*, 1853, 91) and on a S facing trackside at the edge of a beech plantation, Worsell Wood, Stanner 2557.

Marsh St John's-wort
Hypericum elodes

An uncommon sp. of wet, peaty flushes in unimproved pastures mostly in the hills of mid and W Radnor. Present also in basin mires and rarely in ditches, it is known from around 40 sites, but must have declined due to agricultural drainage works. Noted in MsFlR as common in the Wye Valley but scarce in the E, yet with records from Clyro c24sw, near c27NW and Presteigne c36SW.

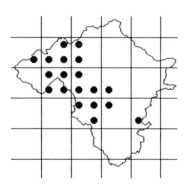

Toadflax-leaved St John's-wort
Hypericum linariifolium

This nationally scarce sp. was noted by NY Sandwith in 1945 "within a few miles of Llandrindod, in fissures of rocks at an altitude of 1400 ft. on igneous tuffs, ashes and agglomerates" (K; *NW Nat* Sept-Dec 1945, 266 and *Trans Rad Soc* 1947, **XVII**, 13). It was associated with rowan, honeysuckle, bilberry, sheep's-bit, foxglove, woodsage and orpine. It has not been found since, despite a search of many sites which fit the description eg. Llandegley Rocks 1361.

Trailing St John's-wort
Hypericum humifusum

A widespread sp. of open acidic soil on banks, tracksides, scree and rock outcrops in woodlands, hedgebanks and heath.

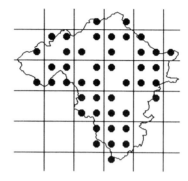

Square-stalked St John's-wort
Hypericum tetrapterum

A frequent sp. in flushes, marshes, ditches and in the reedswamp of pools and lakes and of river banks. Strangely considered to be rare in MsFlR with a single record.

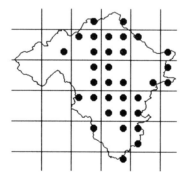

Imperforate St John's-wort
Hypericum maculatum ssp. *obtusiusculum*

Intolerant of grazing it is confined to the grasslands of roadside verge, railway embankments, churchyards, gardens, forest rides and wasteground, where it is widespread.

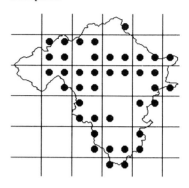

Perforate St John's-wort
Hypericum perforatum

In somewhat drier, but in other respects similar habitats to the last. Equally widespread. The hybrid between this and the last sp. is reported from the county in FPW and may have been overlooked in this survey.

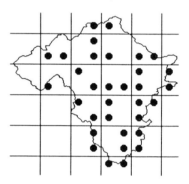

Violet family
Violaceae

Sweet Violet   *Viola odorata*

A widespread but scarce garden escape. Roadside verge below garden, The Screen, Llandeilo Graban 0943 (pink flowered); hedge near County Hall, Llandrindod Wells 0660; old railway station, Glasbury 1839 MP and H Soan; roadside bank and churchyard, Llanstephan 1241 and 1142 (white flowered); Bryngwyn churchyard 1849; wooded cliff below Wyecliff garden, Clyro 2242 ! and ACP; frequent on roadside bank near Burlingjobb 2458 (white flowered). Formerly perhaps more widespread, it is noted in MsFlR from Cwmbach c05SW; Llandrindod c06SE; Gwystre c06NE; Boughrood c13NW; Pistyll, Boughrood Brest c13 and Knighton 27SE.

Helianthemum

Hairy violet

Viola hirta ssp. *hirta*

A rare sp. noted in MsFlR from Aberedw c04NE and Stanner Rocks 2658. It still occurs in clearings amongst the scrub on the S facing screes at Stanner and on the limestone in nearby Yatt Wood, Dolyhir 2458 ! and ACP.

Early dog-violet

Viola reichenbachiana

A scarce or overlooked sp. apparently almost confined to the calcareous ORS woodlands of SE Radnor. Cilcenni Dingle, Llowes 1741; Moity Dingle, Llowes 1842 and 1942 FE and DRD; Bettws Dingle, Clyro 2246; Monaughty Dingle, Bleddfa 2368 DEG and JSB.

Common dog-violet

Viola riviniana

A common plant of woodlands, hedgerows, upland grassland, river banks, roadside verges and occasionally a garden weed. Recorded from all squares except 86NE & NW and 87SW.

Heath dog-violet *Viola canina*

Very rare or overlooked. HJ Riddlelsdell records it from the Carneddau near Llanelwedd c05SW or SE in 1909 (*Cotteswold Nat Fld Club* **XVII**, 57-61). Problems in the identification of this sp. render the record somewhat uncertain in the absence of a specimen. A record of GP (confirmed by M Walters) from Llanstephan c14SW in 1952 adds credibility to the 1909 record.

Marsh violet *Viola palustris*

A common plant in wet flushes, bogs, damp rocks and streamsides in the uplands. The sspp. have not been separated, though both are reported from the county in FPW.

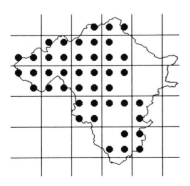

Mountain pansy *Viola lutea*

An uncommon or overlooked sp. of short, acidic turf in upland pastures, haymeadows, road verges and hill grazings. MsFlR noted a decline between 1938 and 1950. Ploughing, reseeding and increasing levels of sheep grazing have probably accelerated that decline. It is however a resilient sp. in unimproved areas and a reduction in grazing often reveals its previously unsuspected presence by a fine show of flowers. 31 sites. Flowers of yellow, purple, and mixed coloured forms are equally common.

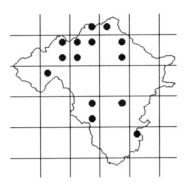

Wild pansy

Viola tricolor ssp. *tricolor*

Rather scarce but occuring in a range of open habitats from swede fields, and meadows to the edge of hayfields, gardens and roadside banks. Purple flowered forms seem to be the commonest.

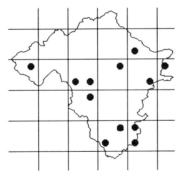

Field pansy *Viola arvensis*

The general scarcity of this sp. reflects the small amount of arable land in Radnor. Swede and cereal fields and gardens appear to be its most significant habitats.

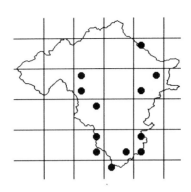

Rock-rose family
Cistaceae

Common rock-rose

Helianthemum nummularium ssp. *nummularium*

A rare sp. known from a few sunny and dry, basic rock outcrops. A few plants on rocks S of Rhayader Quarries, Nantmel 9765; noted in MsFlR from Aberedw Rocks 04NE but recorded as absent from the area by A Ley in surveys between 1886 and 1890 (*Trans Woolhope Club* 1894, 185) and not seen by any other observers; open limestone woodland, Yatt Wood, Dolyhir 2458; dolerite rock outcrops, Stanner Rocks 2658; volcanic rocks near Bleddfa c26NW MsFlR (the reference to volcanic rocks in this area is puzzling as none are shown on maps of the Geological Survey).

Waterwort family
Elatinaceae

Six-stamened waterwort

Elatine hexandra

Recorded from the margin of Llynheilyn, Llanfihangel-Nant-Melan 1658 by G Nock in 1950 (NMW) and from "pools around the margin" of Llynheilyn by AEW in 1951 (MsFlR). Not seen for many years.

Melon family
Cucurbitaceae

White bryony
Bryonia cretica ssp. *dioica*

Confined to the lowest ground in the Wye Valley from Llanstephan 1141 to the English border, it scrambles over hedges, fences and shrubs. Particularly fine specimens grow through the yew trees in Clyro churchyard 2143. 10 sites. An early record of HNR from a "roadside near Radnor" in 1880 (*J Bot* 1881,172) suggests it may have been more widespread formerly.

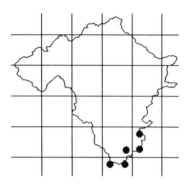

Loosestrife family
Lythraceae

Purple-loosestrife
Lythrum salicaria

The scarcity of this showy plant in Radnor was not appreciated at the outset of this survey. MsFlR reports it as "local and not very frequent. In various spots from Boughrood 1239 up the river to Llanwrthwl 9763. Scarce in the Elan Valley c96, Llandrindod 06SE and Llanbister c1073". Seen recently only from the margin of Gwynllyn, Llansantffraed Cwmdeuddur 9469, in 05SW & NE, old oxbow lake, Ffordd Fawr, Glasbury 1840 and in 24SW.

Water-purslane
Lythrum portula

Common on soil which is wet in winter and dry in summer. Favours ephemeral, mud-bottomed pools, reservoir margins, ditches, field gateways, flushes and tracksides where there is little competition.

Lythraceae

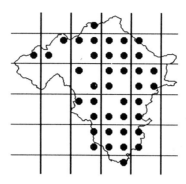

Willow-herb family
Onagraceae

Enchanter's-nightshade
Circaea lutetiana

Common in woodlands, hedgerows and on shady banks. Recorded from all squares except 86NE & NW, 87, 97SW, NE & NW, 04SE, 07NW, 16SE and 18SW.

Alpine enchanter's-nightshade
Circaea alpina

Reported with the upland enchanter's-nightshade as filling the shady part of the gorge at Aberedw c04NE by AL (1886-1890 Notes on the Flora of Aberedw. *Trans Woolhope Club* 1894, 188). Only the upland enchanter's nightshade now appears to be present and though HNR visited the area in 1880 (*J Bot* 1881, 171) he noted only the next sp..

Upland enchanter's-nightshade
Circaea x intermedia

A rather scarce sp. of damp and shady woodland banks and hedgerows. Most frequent in 06, it is known from 9 sites.

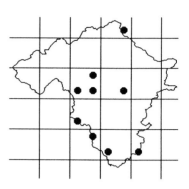

*Large-flowered evening primrose
Oenothera erythrosepala

Rare on waste ground and roadsides. Lane bank near bungalow, Llanelwedd 0652 and by footpath Ddole Road, Llandrindod Wells 0662.

Rosebay willowherb
Epilobium angustifolium

Widespread on roadside verges, railway banks, woodland rides and clearings, wasteground and occasionally on upland cliffs. Highly palatable it cannot survive in sites regularly grazed. Noted from all squares except 86SE & NW and 87SW.

Great willowherb
Epilobium hirsutum

In wet ditches by slow flowing streams and on river banks, pond and lake margins. Common in the lowlands.

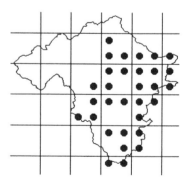

Epilobium

Hoary willowherb

Epilobium parviflorum

Uncommon in damp hollows in pastures, ditches, pond and stream margins, mostly in the lowlands.

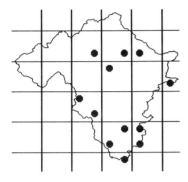

Broad-leaved willowherb

Epilobium montanum

Common on moist, shady, open soil in woods, hedgerows, stream banks and gardens. In all squares except 86, 87 and 96SW.

Spear-leaved willowherb

Epilobium lanceolatum

Only known from sunny, south facing screes of volcanic rock amongst scrub, Llanelwedd Rocks 0552 F Rose and !. Its only mid Wales locality.

Square-stalked willowherb

Epilobium tetragonum

Possibly rare and not recorded in this survey. Noted as rare in MsFlR with records from Llandrindod c06SE and Llananno c07SE and from 14, 25, and 36 in FPW. A specimen from Llandrindod Wells 06SE of JAW in NMW is doubtfully this sp..

Short-fruited willowherb

Epilobium obscurum

Widespread on moist to damp soils in pastures, woodland glades, ditches, on road verges, stream, pond, lake and river margins, mostly in the valley bottoms.

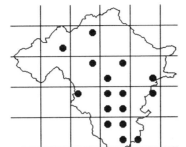

Pale willowherb

Epilobium roseum

A plant of damp hollows, ditch and streambanks, roadsides and gardens, which is scarce or has been overlooked. Noted in MsFlR from a garden at the Ridgebourne, Llandrindod Wells 06SE, it has been seen in 05NW, 06SE, 07SE and 16NE in this survey.

Marsh willowherb

Epilobium palustre

A common sp. of wet flushes and pastures, springs, pool margins and fens. Favouring peaty places, it is consequently commonest in the uplands. Recorded from all squares except 87SW, 04SE, 13NW, 17NE, 24SW, 25NE, 26NW, 36SW and 37SW.

*American willowherb

Epilobium ciliatum

A recently established alien, unrecorded in MsFlR, now widespread on damp, disturbed ground such as in ditches, on stream and roadside banks and occasionally in drier sites such as hedgebanks and gardens. Hybrids with other spp. of willowherb are common and some records may refer to hybridized material.

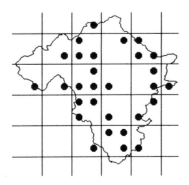

*New Zealand willowherb

Epilobium brunnescens

This naturalized alien was first noted in the county in 1961 by A Jones at the foot of Claerwen Reservoir dam c8763 (NMW). It has now colonised a wide scatter of sites, usually shaded, damp and open, such as rock outcrops, ditches, tracksides, springs and landslips.

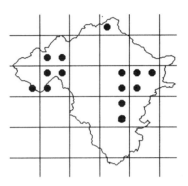

Water-milfoil family Haloragaceae

Whorled water-milfoil

Myriophyllum verticillatum

Recorded from the R. Wye at Llanelwedd 0352 and upstream of Glasbury 1638 by D Merry (NCC Wye survey). It may well prove to be frequent in the Wye. Noted "within 2 miles of Llandrindod Wells (06SE) or thereabouts" by JB Lloyd in the early 1900's (Bufton WJ and JO, 1906), probably from the R. Ithon. Subfossil seeds are reported from the peat of Rhosgoch Common 1948 over 2.5 metres below the surface (Bartley DD, 1960a).

Spiked water-milfoil

Myriophyllum spicatum

Scarce or overlooked. Recorded as subfossil pollen in late Flandrian peat deposits at Cors y Llyn, Newbridge on Wye 0155 (Moore PD, 1978); ornamental pond, Old Town Hall, Llandrindod Wells 0661, presumably introduced with fish; Llandrindod Lake 0660 (1954 MsFlR but not seen recently); R. Wye above Glasbury c13NE DC Hutt; R. Wye, Boughrood 1338 D Merry; Llanbwchllyn 1146 S Thurley; Llynheilyn 1658 (1948 MsFlR).

Alternate water-milfoil

Myriophyllum alterniflorum

Frequent in small mud-bottomed pools and in slow flowing parts of brooks, mostly in the uplands.

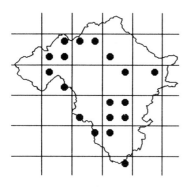

Dogwood family
Cornaceae

Dogwood *Cornus sanguinea*

Uncommon in hedges, woodland and on streambanks on the calcareous soil of SE Radnor.

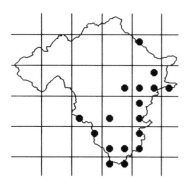

*Red-osier dogwood

Cornus sericea

A barely naturalized alien, planted as an ornamental shrub by Pencerrig Lake, Llanelwedd 0454 (MsFlR); Rock Park, Llandrindod Wells 0560 (MsFlR) and still present in both sites and on roadside below Dol-y-fan Hill, Newbridge on Wye 0160.

Ivy family
Araliaceae

Ivy *Hedera spp.*

Two wild spp. of ivy are now recognised in the British Isles (McAllister, H.A. and Rutherford, A 1990); common ivy *H. helix* L. s.s. and atlantic ivy *H. hibernica* (Kirchner) Bean. They were not generally separated in this survey. As an aggregate, ivy has been recorded from all squares except 86NW, 87SW and 18SE, being common in ungrazed woodlands, on hedgebanks and cliffs. From Radnor ivy samples collected by A.O. Chater, McAllister determined the chromosome number. Three out of the four samples from mid Radnor were atlantic ivy (from a hedgebank and tree trunk near a river in 0365 and a churchyard in 1858). The only common ivy was from a stump in a wood W of Pen-y-bont 1164. A recent sample survey by the author using trichome charateristics only, showed atlantic ivy to be the commonest species.

*Persian ivy *Hedera colchicha*

Naturalized in oak wood, Llowes Dingle Wood 1842.

Umbellifer family
Umbelliferae

Marsh pennywort

Hydrocotyle vulgaris

Widespread in wet flushes, short peaty grasslands, by springs and ditches, mostly in the uplands.

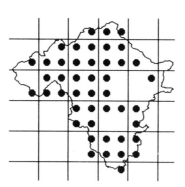

Sanicle *Sanicula europaea*

Frequent in the more base-rich woodlands and occasional on shady hedge, roadside and stream banks.

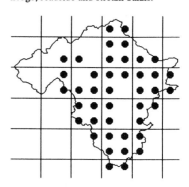

Rough chervil

Chaerophyllum temulentum

Mostly a lowland sp. of hedge banks but occasional by rivers and on waste ground.

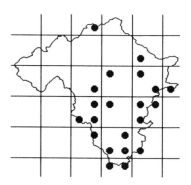

Cow parsley

Anthriscus sylvestris

Common on roadside verges and hedgebanks and extending occasionally into hayfields and along riverbanks, except in the western uplands where it is rare. Recorded from all squares except 86, 87, 96NW & SW and 97SW.

*Garden Chervil

Anthriscus cerefolium

A rare casual. Roadside verge W of Boughrood 14SW TG Evans and established for some years in a garden, Gorther, Felindre 1682.

Aegopodium

*Shepherd's needle

Scandix pecten-veneris

Recorded from Radnor in Watson's Topographical Botany ed. 2 (London 1883 presumably based on a record (unlocalized) of AL in *Bot Loc Rec Club* 1874, 82). No other records known.

*Sweet cicely

Myrrhis odorata

A scarce sp. long established near habitations. Alltgoch, St Harmon 9772 ACP; Nantgwyn 9776 AP Conolly (1965 *Nat Wales* 9, 216); in and around Glascwm churchyard 1553; Llanfihangel-Nant-Melan churchyard 1858; wet ditch, Fron Rocks, Beguildy 1976 ACP (1962); road verge, Gorther, Beguildy 1682; verge, Garth Farm, Knighton 2772 PR (1964); Folly Wood, Presteigne 3163 FE and DRD.

Pignut *Conopodium majus*

Common in unimproved pastures, churchyards, roadside verges and open woodlands in well-drained sites. Recorded from all squares except 86SE & NW, 87SW and 07SW & NW.

Burnet-saxifrage

Pimpinella saxifraga

Frequent in unimproved pastures, hedgebanks, open scrub, churchyards and road verges in well-drained, sunny sites.

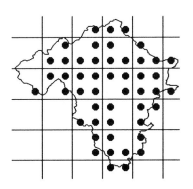

Ground-elder

Aegopodium podagraria

A common weed of gardens, churchyards, hedgerows and streambanks, most frequent about habitations.

Tubular water-dropwort

Oenanthe fistulosa

A rare marshland plant. Reedswamps, Llanbwchllyn 1146 GP(1957) and S Thurley (1977); in marsh by pool W of Gogia, The Begwns, Llowes 1643 ACP; reedswamp, Rhosgoch Common c1948 DD Bartley (1954 NCC records).

Hemlock water-dropwort

Oenanthe crocata

Common beside streams, in ditches, fens, wet pastures and glades in alder carr. Recorded from all squares except 86, 87, 96SW & NW, 97NW, 07NW, 16SE, 18, 26SW and 27NW.

Fool's parsley

Aethusa cynapium

Noted in MsFlR as "fairly common but not abundant", this weed of cultivated ground and roadsides appears to have become more rare. Soil heap by road S of Middle Hall, Aberedw 0748; road verge, Painscastle 1746 ACP; verge, Evenjobb 2662; verge, Dolley Green, Whitton 2865; verge, E of Bleddfa 2267; tip, Knighton 2872; on silt by R. Lugg below Presteigne 3264.

Pepper-saxifrage

Silaum silaus

A flower of damp unimproved pastures known from a number of sites in Brecknock but with no recent records from Radnor. Pastures between Aberedw and Builth 04 or 05 AL (1886-1890 *Trans Woolhope Club* 1891, 188) and Radnor Forest c16 or 26 HNR (1880 *J Bot* 1881, 172).

Hemlock *Conium maculatum*

A plant of disturbed ground, generally with nutrient rich soil. Recorded from roadside verges, tips, farmyards and river banks, mostly in the lowlands. Appears to have declined since the 1950's having been recorded in MsFlR from most of the villages and towns in the E. 13 recent records.

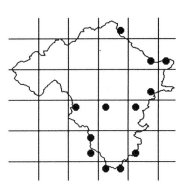

*False thorow-wax

Bupleurum subovatum

A single plant collected by FMS from beside Llandrindod Lake 0660 in the hot summer of 1976 and det. E Clement. Possibly originated in seed fed to the birds.

Fool's water-cress

Apium nodiflorum

A frequent plant of streamsides, springs, wet ditches and marshes on the more calcareous soil of the S and E.

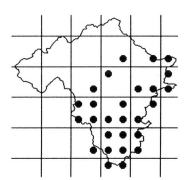

Umbelliferae

Lesser marshwort
Apium inundatum

Frequent in the mud-bottomed pools of the upland commons. Rarer in ditches, willow carr and flushes. 33 recent records.

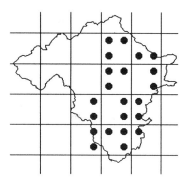

*Garden parsley
Petroselinum crispum

Noted in MsFlR from walls and waysides, Cwmbach Llechrhyd c05SW and Glasbury c13NE (in 1938). There are no other records.

Cowbane   *Cicuta virosa*

Reported by HNR from a stream on the Carneddau c05 and from "Rhos Common" (probably Rhosgoch Common c1948) in 1880 (*J Bot* **19**, 178). The habitat persists possibly in both sites, but strangely this conspicuous plant has never been reported again.

Whorled caraway
Carum verticillatum

In Radnor a scarce sp. on the eastern edge of its range. Drainage of its favoured habitat, damp pastures, reduced its extent in the 1980's. It may no longer be present in some of these sites. Peaty flushes by the R. Claerwen, S of Craig Cwm Clyd, Llansantffraed Cwmdeuddwr 8862 DRD. Noted in 1919 by Charles Waterfall (MsFlR) from near Llandrindod c06, it was not recorded again until seen by the author in a field between Llanyre and Crossgates 0564 in 1977. It proved to be more widespread in wet pastures on boulder clay in this area, with subsequent records by LMP, JL and ACP from five fields in 0463 and three in 0563 and by DRD in 0660. Outlying stations are reported by SMS Gooch from Garth Fawr, Llandeilo Graben 1143 and by IHS and CC from Painscastle 1744.

Wild angelica
Angelica sylvestris

Common in tall-herb communities of wet pastures, flushes, reedswamp, open, damp woodlands, ditches and streambanks. Recorded from all squares except 86NW, 87 and 96NE.

Wild parsnip
Pastinaca sativa

Reported from lime and rubble between the railway tracks in Dolau Station coalyard 1367 by JAW in 1934 (MsFlR). It had apparently gone by the 1950's. This sp. still reaches Llandovery along the Cental Wales line. The lifting of almost all of Radnor's sidings in recent years and use of efficient herbicides reduces its chances of recolonising the county.

Hogweed
Heracleum sphondylium

Common on roadside banks, in unimproved hayfields, waste ground and woodland clearings. Reported from all squares except 86 and 87. The narrow-leaved form (var. *angustifolium*) appears to be rare.

*Giant hogweed
Heracleum mantegazzianum

Reported in MsFlR from Llandrindod Wells c06SE and Pencerrig c0453 in 1938, possibly from ornamental plantings. It was not recorded in the county again until 1976 when ACP noted one plant on the bank of the Wye below Hay Bridge 2343 and five at Upper Cabalva 2446. Far from becoming the invasive weed which might have been feared, the Hay Bridge population in 1987 remained a single plant (now in square 2242) with another single plant on the Wye S of Clyro Court 2042 in 1982.

Knotted hedge-parsley
Torilis nodosa

Reported only from Aberedw Rocks c04NE by JAW in 1938 (MsFlR), it has not been seen since.

Spreading hedge-parsley
Torilis arvensis

Reported from the county by Watson (1883). There is no information as to the source of this record. Very rare in Wales, it is presumed extinct in Radnor.

Upright hedge-parsley
Torilis japonica

A frequent hedgerow and woodland-edge plant in freely-drained, sunny sites on the less acidic soils.

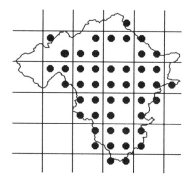

Wild carrot
Daucus carota ssp. *carota*

Reported in MsFlR as "locally abundant especially along railway banks etc". This sadly is no longer the case. A few plants survive along a bank near the old Aberedw Station 0747 and were reported in Glasbury cutting 1839 by MP in 1969. ACP noted it from roadside grass at Boughrood 1141 in 1969 and a HRNT survey reports it from a castle mound near Burfa 2761 in 1986 but not seen in 1987. No other localities are known. Reduced maintenance of railway banks and lack of fires with the coming of diesel locomotives may have adversely affected it.

Heath family
Ericaceae

Cross-leaved heath
Erica tetralix

A common sp. in the wetter areas of the various types of acidic peat bog found in the county.

Vaccinium

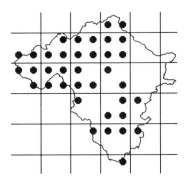
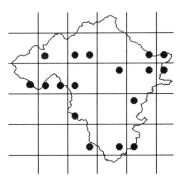
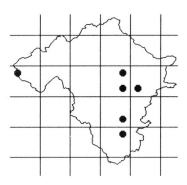

W of Blaen Rhestr 8369, unrecorded elsewhere in the Elan uplands.

Bell heather   *Erica cinerea*

Frequent on sunny, S facing, acidic shale outcrops in the Elan uplands and in a scattering of sites elsewhere.

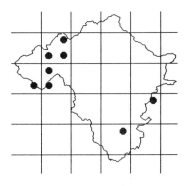

Heather   *Calluna vulgaris*

Widespread on acidic soil and peat in the uplands and occasionally in wet pastures and open woodlands. Ploughing, afforestation, burning and heavy grazing and extremes of drought and cold have much reduced the extent of heather moor in recent years. Extensive stands remain on Y Foel in the Elan Valley 96NW, Rhulen and Glascwm Hills 14, Gwaunceste Hill 15, Radnor Forest 16 and Beacon Hill 17. Recorded from all squares except 13NW, 15NW, 24SW, 25NW, 26NW, 27SW and 36NW.

*Rhododendron

Rhododendron ponticum

Planted in many woodlands on the larger estates and naturalizing in shady acidic sites. Not so invasive or as widespread as in N and W Wales.

Bog rosemary

Andromeda polifolia

A scarce plant of what few intact blanket bogs remain in the NW uplands. Trumau 8667, Gorsgoch 8963 R Birch and Cnapyn Drawsffos 8370 DRD et al. in Llansantffraed Cwmdeuddwr and Drysgol, St Harmon 9475 IDS.

Cranberry

Vaccinium oxycoccos

Widespread but not common on the wetter *Sphagnum* moss-dominated peatlands of the county. Whilst mostly in the western uplands, small populations survive on Black Mixen, Radnor Forest 1964 and around mawn pools between Colva and Caety Traylow Hills 1955 ACP. Surprisingly never recorded from Rhosgoch Common 1948. 13 sites.

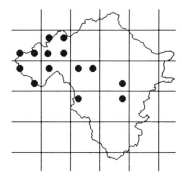

Cowberry

Vaccinium vitis-idaea

Widespread and best developed in the upland heaths of Colva and Gwaunceste Hills 15 and the Radnor Forest c16. Elsewhere rare and apart from a single record of a specimen on a boulder top

Bilberry, Wimberry or Whinberry

Vaccinium myrtillus

Common on the acidic soil and peat of the unenclosed uplands, especially about the ridges and on steep slopes. Often very dwarf as a result of grazing. Occasionally present in some of the less heavily grazed oakwoods, on steep banks, rock outcrops and roadside banks. Few sites offer sturdy enough plants with sufficient fruit to warrant the effort of picking, though it was once a popular industry. Recorded from all squares except 05SW, 13NW, 14NW, 15NW, 16NW, 24SW, 36NW and 37SW.

Crowberry   *Empetrum nigrum*

Widespread on the upland moors, particularly on areas of peat. It persists long after heather has been grazed out. Less common in the lowlands, it is found on basin mires such as Cors y Llyn, Newbridge on Wye 0155 and the raised mire of Rhosgoch Common 1948.

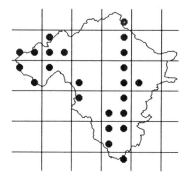

Primrose family
Primulaceae

Primrose *Primula vulgaris*

Widespread in woodlands, hedgerows and on stream banks, especially on moist calcareous soil. Rare in the uplands and largely confined to flushed rock ledges and banks out of reach of sheep. Common in churchyards and possibly planted there to decorate the graves. Francis Kilvert describes the decking of graves with primrose flowers at Easter in Clyro churchyard in 1870 (Plomer 1938). A form with pink flowers is particularly frequent in churchyards. Whether deliberately selected from the wild or the result of crossing with cultivars is not known. Recorded from all squares except 86NE & NW, 87SW, 96SW and 97NW.

Cowslip *Primula veris*

Confined to the calcareous soils of the S and E of the county it is nowhere common. Unable to cope with heavy sheep grazing, artificial fertilizers or ploughing it is now confined to a few unimproved pastures, roadside verges, railway banks, churchyards and sites such as grass-covered underground reservoirs. 22 sites.

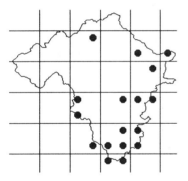

The hybrid between the cowslip and primrose, *P.* x *tommasinii*, sometimes called the false oxlip, has been seen in a number of sites where both parents are present.

*****Cyclamen**

Cyclamen hederifolium

Well established and spreading in the lightly mown grassland, amongst the yews, of Clyro churchyard 2143 ACP. (Apparently originally planted there by her grandmother in c1870, when they flourished under a, then, canopy of beech). Also established in Norton churchyard 3067 A Helmsly and S Mason (in Slater, FM and Moncur, D, 1985).

Yellow pimpernel

Lysimachia nemorum

Frequent in damp woodlands and on streambanks, especially on the more calcareous clay soils. Occasional in marshes and flushes in the uplands. Recorded from all squares except 86NW & NE and 87.

Yellow loosestrife

Lysimachia vulgaris

Mostly found in reedswamps beside the larger rivers such as the Ithon and Wye. Occasional in marshes and swamps in wet pastures and woodland. Over 30 sites.

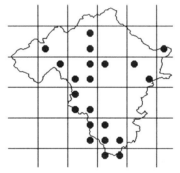

*****Fringed loosestrife**

Lysimachia ciliata

Reported by RCLH from Tyn-yr-ynn, Llanbister 17, presumably as a garden escape (1957, *Proc BSBI* **2**, 375).

Creeping-jenny

Lysimachia nummularia

Of scattered occurrence in damp places such as seasonally waterlogged meadows, marshes, by rivers, flushes in woodland and occasionally in ditches by roadsides. Frequently cultivated as a shade bearing rockery plant, some river and roadside records may be of garden escapes.

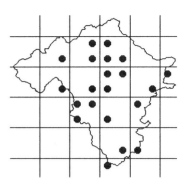

*****Dotted loosestrife**

Lysimachia punctata

An escape from cultivation persisting on damp grassland and on tips. Field record of AEW from 06SE (NMW); verge of A488 near its summit on the Radnor Forest 1868; marshy ground, Norton Manor 2966 LE Whitehead (1964); tip by R. Teme, Milebrook, Knighton 3172.

Chaffweed

Anagallis minima

This predominantly coastal sp. of minute size (usually less than 1 cm tall) is known only from the dried up mud of a pool, S of Llandeilo Graban 0943 and from soil by a small stream on common land to the W of Yr Allt, Llandeilo Graban 0844.

Bog pimpernel

Anagallis tenella

Frequent in the low turf of wet, peaty, somewhat base-rich flushes, often on heavily grazed common land. Occasional in open fen vegetation, such as on Rhosgoch Common 1948. 17 sites.

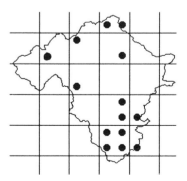

Scarlet pimpernel

Anagallis arvensis

A scarce sp. of disturbed ground, noted in arable crops, on reconstructed road verges, waste-ground and gardens.

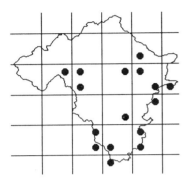

Olive family
Oleaceae

Ash   *Fraxinus excelsior*

Common in woodlands, hedgerows, on riverbanks, cliff and scree and as a colonist in open ungrazed sites such as old quarries, roadsides etc, in all except the most acidic areas. Recorded from all squares except 86NW and 87SW.

*Lilac   *Syringa vulgaris*

Doubtfully naturalized, it occurs rarely as suckers spreading along hedgerows from gardens. Not mapped.

Wild privet

Ligustrum vulgare

Frequent in hedgerows, often about gardens, its status as a wild sp. is not certain. It is recorded from most 10 km squares in the county but only in the following sites does it appear to occur wild or truly naturalized; Tyn y caeau Wood, Llanstephan 1241 LMP and JL; Coed y Marchog, Glasbury 1541 LMP and JL and on a limestone outcrop, Yatt Wood, Dolyhir 2458.

*Garden privet

Ligustrum ovalifolium

Commonly planted as a garden hedging shrub and doubtfully naturalized in hedges and about abandoned gardens and in old shrubberies. Not mapped.

Gentian family
Gentianaceae

Common centaury

Centaurium erythraea

An uncommon sp. of open grassland, often on track and roadsides and disused railway lines and in the open turf of summer-droughted grassland on shallow soil over rock. 15 sites.

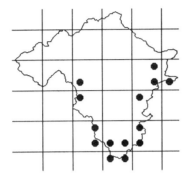

Field gentian

Gentianella campestris

Reported in MsFlR from a single site, Llanfihangel-Nant-Melan c15NE (1938 JAW). A record of "the rare blue gentian" from upland pastures above Aberedw Rocks c0846 by A Wentworth-Powell, reported in Bufton, WJ and JO, 1906, 109 may be of this species. No other records.

Bogbean family
Menyanthaceae

Bogbean

Menyanthes trifoliata

Still widespread in the wetter bogs, marshes, basin mires and on the margins of lakes and pools, mostly in acidic places. c40 sites.

Menyanthes

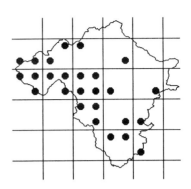

Periwinkle family
Apocynaceae

*Lesser periwinkle

Vinca minor

Rarely established in hedgerows close to gardens. Noted in MsFlR from Cwmbach Llechryd c05SW, Nantmel 06NW and Penybont c16SW. Seen recently only on a lane bank near Nant y Mynach, Nantmel 0166.

*Greater periwinkle

Vinca major

Rarely established alien. On the site of the old Aberedw railway station 0747 ACP; the old railway, now road bank, by the Skreen, Llandeilo Graban 0943; Llowes churchyard 1941; roadside bank near old level crossing, Burlingjobb 2458; churchyard, Gore, Burlingjobb 2559; and the old railway station, Stanner 2658.

Bedstraw family
Rubiaceae

Field madder

Sherardia arvensis

Described as "thinly dispersed" in MsFlR as a weed of arable crops. It has not recently been seen as a field weed, surviving mostly as a plant of summer droughted, open, rocky turf. Llanelwedd Rocks c05 HJR (1909 *Cotteswold Nat Fld Club* **XVII**, 57-61); gravel heap in farmyard, Bryn y Groes, Crossway 0557 DRD det. ! ; Llanstephan c14SW GP (1952); Dolyhir Quarry 2458 ACP (1968) and in disused quarry S of Yatt farm, nearby; Stanner Rocks 2658; near Old Radnor Church 2559.

Rubiaceae

Woodruff *Galium odoratum*

Widespread on damp, calcareous soil in woodlands, often favouring the steeper banks and rock ledges and occasionally seen in hedgerows.

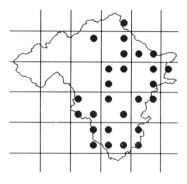

Lady's bedstraw
Galium verum

Widespread in mid Radnor on sunny, freely-drained, grassy banks by roads and railways, on rock outcrops, in unimproved pastures and churchyards.

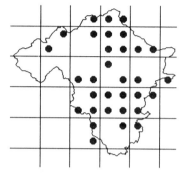

Cleavers *Galium aparine*

Common on open, nutrient-rich soil of hedgebanks, lowland woodland, gardens, wasteground, detritus by rivers and streams and rides in plantation woodland. Recorded from all squares except 86, 87 and 96SW.

Crosswort *Cruciata laevipes*

Widespread in hedgebanks, on grass verges, by roads and railways and on the edges of scrub and woodlands, generally in freely-drained, sunny sites.

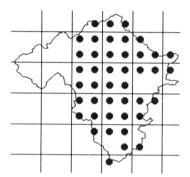

Fen bedstraw

Galium uliginosum

Widespread but scarce in damp, somewhat calcareous flushes in unimproved pastures and mires, especially on the boulder clay. 21 sites.

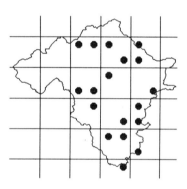

Hedge-bedstraw

Galium mollugo agg.

Rare and probably a fairly recent introduction. First reported from the county in 1918 by Druce from a bank at Aberceuthon, near Rhayader c9566 as *G. album* (MsFlR). The authors of MsFlR say the record had been queried and themselves report *G. mollugo* s.s. from banks at Aberceuthon in 1939 (*Trans Rad Soc* **XVIII**, 8). A search of this area in 1990 revealed *G. mollugo* s.s. still present in, or possibly close to its original locality. A specimen well established in mown grass behind the Clwyd Pavillion on the Royal Welsh Agricultural Society's showground at Llanelwedd 0351 could only be det. as the agg.. Plants which on balance are *G. mollugo* s.s. occur in a recently planted hawthorn hedge on a roadside between Llanbedr and Painscastle 1546. *G. mollugo* was also reported from near Presteigne c36 by WMR. (1899 *J Bot* **XXXVII**, 17-25).

Heath bedstraw

Galium saxatile

Common on acidic soil of rough upland hill grazings including heath, rock ledges and grasslands, roadside banks, woodlands and hedgerows and reported from all squares except 24 and 26NW.

Jacob's ladder family Polemoniaceae

Jacob's ladder

Polemonium caeruleum

A single pollen grain noted in late glacial peat deposits on Rhosgoch Common 1948 by Bartley, DD (1960a) provides evidence that it might have once been a native plant. Probably ousted by the spread of forest it now returns as an escape from gardens. Abandoned gardens, now super-market, Llandrindod Wells 0561; near Stanner c25NE AH Trow (1923 WFP); Cefnsuran, Llangunllo 2271 (Sinker et al. 1985).

Bindweed family Convolvulaceae

Dodder *Cuscuta epithymum*

Only noted from the county by Watson (1883), the source of this record is unknown.

Common marsh-bedstraw

Galium palustre agg.

Common in marshes, bogs, wet ditches and pastures, riverbanks and on lake margins and reported from all squares except 86NW and 87SW. No attempt has been made to separate *G. palustre* from *G. elongatum*. The latter has never been reported from Radnor but is probably present.

Myosotis

Hedge bindweed

Calystegia sepium ssp. *sepium*

Widespread but not common on hedges, scrub by rivers, woodland edges and occasionally waste ground. The hybrid between this and the next species *C.* x *lucana*, is reported by JAW from New Radnor 2160 in 1938 in 1938 (FPW).

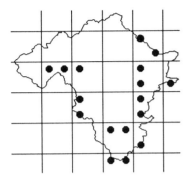

*Large bindweed

Calystegia silvatica

An uncommon sp. of wasteground, hedges and scrub, mostly associated with human habitation and most frequent in the E.

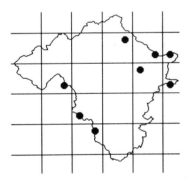

*Hairy bindweed

Calystegia pulchra

Scarce. On garden hedge, Newbridge on Wye 0158; on hedge, Gladestry 2454 ACP; wooden roadside fence, Knighton 2872 ACP.

Field bindweed

Convolvulus arvensis

Very local. Noted in MsFlR as being only associated with the railways at Llanelwedd c05SE or SW, Builth Road c0253 and Rhayader c9667. The railway closed in 1962. Apparently suitable habitat persists but this plant has only been refound at Llanelwedd 0451. It survives by the old railway in Aberedw 0648 ACP; in a garden, Llandrindod Wells 0560 RK; on roadside verge near old level crossing, Boughrood 1238; on verge between Boughrood and Boughrood Brest 1438; old railway station, Glasbury 1839 MP and H Soan; and on a roadside W of Presteigne 3065 PJP. Details of the record in 26 in FPW have not been traced.

Borage family
Boraginaceae

Viper's-bugloss

Echium vulgare

Reported by Watson (1883) from the county, the source of this record is unknown.

*Lungwort

Pulmonaria officinalis

A scarce alien long naturalized on the bank of the Ithon in Bailey Einon Wood, Cefnllys 0861. Reported naturalized from 26 in FPW, the locality is unknown. A small plant grows on the base of a wall in Presteigne 3164.

Common comfrey

Symphytum officinale and x *uplandicum*

It has not proved possible to safely separate the above two taxa with populations showing much variation. They are not common and are confined to damp roadsides, streambanks and waste ground. .

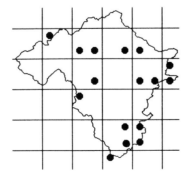

*Green alkanet

Pentaglottis sempervirens

A rarely established garden escape. For example in deciduous plantation by lane, Maesllwch 1740 DPH det. ! ; Llanfihangel-Nant-Melan churchyard 1858; roadside and walls, Knucklas 2574; naturalized in churchyard, Knighton 2872; by drive, Stanage Park 3371.

Field forget-me-not

Myosotis arvensis

Frequent on sunny, well-drained soil on banks, track sides, gardens and arable fields.

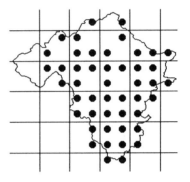

Early forget-me-not

Myosotis ramosissima

Noted at Stanner Rocks 2658 by RA Boniface in 1953 (BRC field record), it has never been reported again.

Changing forget-me-not

Myosotis discolor

Widespread on shallow, summer-droughted soil over rocks in sunny sites. Occasional on gravel by tracks, in arable fields, on ant hills and on sunny hedgebanks.

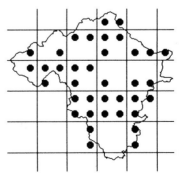

Wood forget-me-not

Myosotis sylvatica

Established on riverbanks and occasionally in hedgerows and on old railway banks. Probably under-recorded., having widely escaped from cultivation.

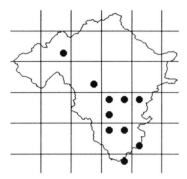

Creeping forget-me-not

Myosotis secunda

Common by springs, flushes, in marshes, by streams, rivers and pools and in wet peaty pastures. Recorded from all squares except 86, 87SE, 96SW, 97NW, 16NE, 25NW and 27SE & NW.

Tufted forget-me-not

Myosotis laxa ssp. *caespitosa*

Frequent in damp places by streams, pools, in wet pastures, ditches and marshes, mostly on the lower ground and avoiding the most acidic sites.

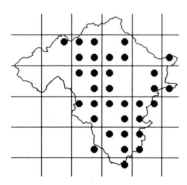

Water forget-me-not

Myosotis scorpioides

Widespread on streamsides, pool edges, wet ditches and marshes, especially in the lowlands.

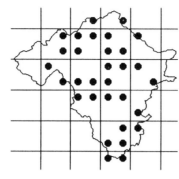

Hound's tongue

Cynoglossum officinale

Rare and seen recently only in a hedge bank, Weythel, Dolyhir 2457 ACP, about the limestone quarries there 2458 and on the rocks in Yatt Wood, Dolyhir 2458. It has been recorded from this area of limestone since 1957. Quarrying has provided much apparently suitable habitat for the plant but it remains in very small quantity. Recorded in the last century as "abundant near the bottom of the rocks, Aberedw" c04NE by AL (*Trans Woolhope Club* 1884, 191) it does not appear to have been recorded again from that area, though suitable looking habitat persists.

Vervain family
Verbenaceae

Vervain *Verbena officinalis*

Considered to be "very local" in MsFlR with records from Aberedw c04NE; Llanstephan c14SW; Clyro 24SW; New Radnor c26SW; Presteigne c36SW and Norton c36 NW. It has clearly declined, being last seen on a shingle island in the R. Wye near Glanwye, Glasbury 1537 M John (1973).

Starwort family
Callitrichaceae

Common water-starwort

Callitriche stagnalis

Common in ponds and lakes, pools in streams and rivers and in wet ditches.

Callitrichaceae

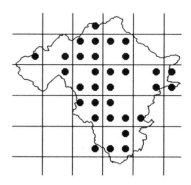

Various-leaved water-starwort

Callitriche platycarpa

Claimed for the county by A Bennet in a supplement to *Topographical Botany* Ed 2 (*J Bot* **43** supplement). The origin of this record is unknown.

Blunt-fruited water-starwort

Callitriche obtusangula

Appears amongst species recorded in 07 by D Bayliss in 1956, the exact locality and habitat are unknown.

Intermediate water-starwort

Callitriche hamulata

Frequent in pools in streams and brooks, mostly in the uplands. Rare in lakes.

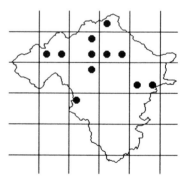

Pedunculate water-starwort

Callitriche brutia

A scarce or overlooked sp. of pools in streams and marshes. Marsh near St Harmon 97SW (NMW records); Llandeilo Hill c04NE and/or 14NW AEW (1927 MsFlR); small pool on dammed stream near Bleddfa 1869 ACP (det. JE Dandy); SE of Beacon Hill 17NE PJP; Beguildy 1980 ACP (det. RG Ellis).

Galeopsis

Mint family
Labiatae

Bugle   *Ajuga reptans*
Common in damp soil of woodlands, streamsides, pastures, roadside verges and ditches. Recorded from all squares except 86, 87SW, 96SW, 97SW & NW, 04SE, 16SE, 26SW & NW and 27NW.

Wood sage
Teucrium scorodonia

Common on the acidic, freely-drained soil of woodland banks, hedgerows, cliffs, block scree and heathy banks. In all squares except 86NW, 87SW, 15NW and 25SW.

Skullcap
Scutellaria galericulata

Widespread on riverbanks, in wet pastures, reedswamps and damp woodland edges, generally in ungrazed or lightly grazed sites. 45 sites.

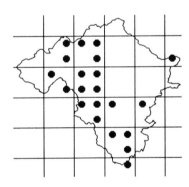

Lesser skullcap
Scutellaria minor

Frequent in unimproved, wet, peaty pastures, flushes and mires, mostly in the uplands. Drainage, ploughing and reseeding have reduced its abundance recently. 42 sites.

Red hemp-nettle
Galeopsis angustifolia

Known from a single record of HNR on a roadside near Llanelwedd c05 (1880, *J Bot* 1881, 172).

Large-flowered hemp-nettle
Galeopsis speciosa

A scarce sp. of arable fields and tips in the uplands, it has been seen in 12 sites since 1960.

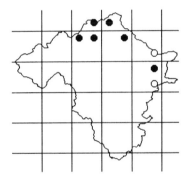

Common hemp-nettle
Galeopsis tetrahit agg.

Frequent and widespread on disturbed soil in woodlands, bracken, heath, arable fields, tips, roadsides and wasteland. Recorded from all squares except 86, 87SW, 96SW and 18SW. No attempt has been made to seperate *G. tetrahit* s.s. from *G. bifida* in this survey. No published records of this latter sp. from Radnor have been located.

*Spotted dead-nettle
Lamium maculatum

Frequently grown in gardens but rarely establishing on hedge banks. Recently seen only near old level crossing, Dolyhir 2458 BSBI exc., but reported in MsFlR from Newbridge on Wye c05NW, Llandrindod Wells c06SE, Llanbadarn Fawr c06SE and Penybont c16SW.

White dead-nettle
Lamium album

Widespread but not common on roadside banks and wasteground, mostly associated with the trunk roads through the county.

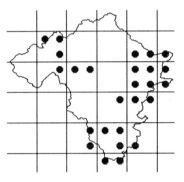

Red dead-nettle
Lamium purpureum

Frequent as a weed of arable crops and in gardens and on disturbed roadsides, hedgebanks and tips.

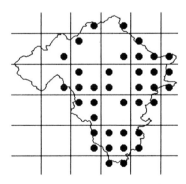

Labiatae

Yellow archangel

Galeobdolon luteum ssp. *montanum*

Common on the shady, damp, calcareous soils of woodlands, cliffs and occasionally river and hedgebanks in the S and E of Radnor. Rare and confined to calcareous flushes elsewhere.

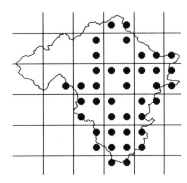

*Variegated yellow-archangel

Galeobdolon argentatum Smejkel

A recently described sp. of unknown origin now fairly widely grown in gardens. Vigorous and invasive, it escapes to the wild from gardens and tipped garden refuse and is likely to become more frequent. Wooded bank and streamside near garden by the Cnithio Brook, E of Gelli-cadwgon, Llanfaredd 0651; on roadverge near garden, Llynheilyn 1658; on steep wooded bank near Gungoed Farm, Llangullo 1971, ACP, CP and PJP and disused roadside quarry by Rock Bridge, Presteigne 2965.

Black horehound

Ballota nigra ssp. *foetida*

Reported from Knighton c27S, Presteigne c36SW and Norton c36NW by AEW and JAW (1938 in *Trans Rad Soc* (1947) **XVII**, 10) this colonist of wasteground and tracksides has not been seen during this survey.

Betony *Stachys officinalis*
Common in unimproved, well-drained, acidic pastures and meadows, hedgebanks, open woodlands and roadside verges. Recorded from all squares except 86NW, 87, 97NW, 04SE, 07NW, 16SE, 18SW and 26SE.

*Lambsear

Stachys byzantina

This common garden plant was noted growing on gravel in a disused part of Llanelwedd quarries 0551 by RG Ellis on a BSBI exc..

Hedge woundwort

Stachys sylvatica

Common, except in the most acidic sites, in woodlands, hedgerows, on streambanks, road verges and wasteground. In all squares except 86, 87 and 97NW.

Marsh woundwort

Stachys palustris

Of occasional occurrence in ditches, on riverbanks, in damp meadows and on hedgebanks where grazing is light or infrequent. *S.* x *ambigua*, the hybrid between this and the last sp., is widespread but uncommon and has not been mapped.

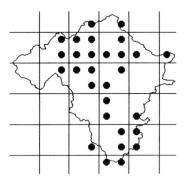

Field woundwort

Stachys arvensis

Of sporadic occurrence in arable fields. Swede field near Cilgwynfydd Farm, Cwmbach Llechrhyd 0354; swede field S of Llwynbrain, Crossway 0456; swede field N of Boughrood 1240. Noted in FPW from 15 also and reported as "fairly common" in MsFlR.

Ground-ivy

Glechoma hederacea

Common in hedgerows and on waste ground and occasional in woodland and on scree and cliff ledges, especially in calcareous sites. Recorded from all squares except 86, 87, 97SW & NW, 07SW & NW, 15SE and 16SE.

Selfheal *Prunella vulgaris*
Common in unimproved pastures and meadows, roadverges, churchyards and woodland clearings in sites ranging from freely-drained to moist. In all squares except 86NW, 97NW and 15NW.

*Balm *Melissa officinalis*
A rarely naturalized alien noted from Rhayader c96NE in MsFlR (1938); by R. Wye below the Vicarage, Boughrood 1338 PCH et al (1956 *Proc BSBI* **2**, 377); by Wye, below Wyecliff, Clyro 2242 ACP & !.

Wild Basil

Clinopodium vulgare

Occasional on sunny roadside verges and in scrub woodland on the more calcareous soils.

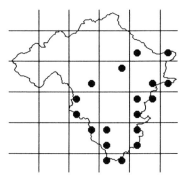

Marjoram *Origanum vulgare*
Scarce and confined to sunny, freely drained, calcareous sites such as riverbanks, roadside banks and rock outcrops.

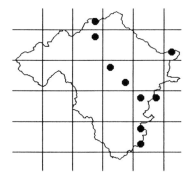

Wild thyme

Thymus praecox ssp. *arcticus*

Common amongst short turf on well-drained, sunny banks and rock outcrops throughout. Recorded from all squares except 86NW, 87, 96SW, 97SW & NW, 07SW, 24SW, 27NW, 36NW and 37.

Gipsywort

Lycopus europaeus

Of scattered occurrence in the reedswamp of lake and pond margins, riverbanks, marshes and ditches where grazing is light or absent.

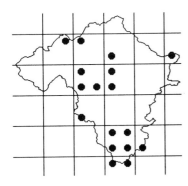

Corn mint *Mentha arvensis*

Widespread but not common on damp roadside banks, wasteground, river gravels, field margins and occasionally on arable land.

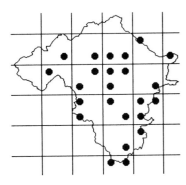

Water mint *Mentha aquatica*

Common in wet flushes, marshes, ditches, on riversides, lake and pond margins. Recorded from all squares except 86, 87, 13NE and 16SE.

Mentha

Round-leaved mint

Mentha suaveolens

Reported naturalized in 06SE by AEW (NMW records).

*Spear mint   *Mentha spicata*

Occasionally naturalized by streams or established on roadside banks and waste ground. Llandewi-fach 1244 ACP; Cwmyrafor, Rhydspence 2347 ACP; Milton Mill, Michaelchurch on Arrow 2350 ACP; Bryndraenog, Dutlas 2178 PR (1964); R. Teme, Milebrook 3172 ACP.

Much futher work is required on the mints to determine the true distribution of the species and hybrids. The following hybrids have been reported from Radnor.

M. x *gentilis* **Bushy mint**
04, 13 and 24.

M x *verticillata* **Whorled mint**
96, 04, 05, 06, 13, 14, 17, and 24.

M. x *smithiana* **Red mint**
96, 97, 13, 23 and 25.

M. x *piperata* **Peppermint**
05, 17, 24 26 and 27.

M.* x *villosa* **Large apple mint
96, 13, 14, 24 and 36

Nightshade family
Solanaceae

*Duke of Argyll's teaplant

Lycium barbarum

Rarely cultivated and doubtfully naturalized. Farm hedge, Llandeilo Graban 0844 ACP; garden hedge, Penybont 1164 ACP.

[Henbane *Hyoscyamus niger*]

Reported from Presteigne by Lewis Davies in *Radnorshire*, pg. 58 published in 1920, a volume in the Cambridge County Geography series and repeated by Carter, 1950. This may repeat a record from the *Trans Woolhope Club* for 1898. It is clear that none of the species mentioned in this report occurred in Radnor. The henbane was on nearby Stapleton Castle in Herefordshire.

Black nightshade

Solanum nigrum

Very rare on disturbed soil. Llanstephan c14SW GP (1954); roadside verge, newly constructed by A438, Clyro Court 2143; Clyro rubbish tip 2143.

Bittersweet

Solanum dulcamara

Frequent in the lowlands beside rivers and ditches, damp hedgerows, pond and lake margins and on waste ground.

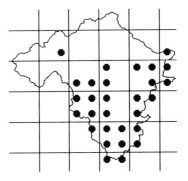

*Potato   *Solanum tuberosum*

Occasional on tips and waste ground but not persisting. Not mapped.

*Tomato

Lycopersicon esculentum

Rare on rubbish tips. Nantmel tip 0165 and Clyro tip 2143.

*Thorn apple

Datura stramonium

A few specimens have been reported from arable fields, particularly in warm summers. St Harmon 9672 MR John det. ACP (1967); Pant y Dwr 9477 Mr Evans det. R Lovegrove (1973); Little Vaynor, Nantmel 0269 Mr Price det. DRD (1989); Crangoed Farm, Llangunllo 1971 Mrs Jones det. D Smith (1975); garden, Llangunllo 2171 JM Roper; Cefnsuran, Llangunllo 2271 DP Humphries.

Butterfly-bush family
Buddlejaceae

*Butterfly-bush
Buddleja davidii

Reported naturalized from Radnor, possibly from 13 in FPW, the source of this record has not been traced. A large specimen grows from a crack in a disused area of tarmac by a car park in Presteigne 3164.

Figwort family
Scrophulariaceae

Mudwort *Limosella aquatica*
A rare plant of mud-bottomed pool margins known only from a small ephemeral pool in a rough track on common land N of Pen-y-graig, Llandeilo Graban 1045. First reported in 1988.

*Monkeyflower
Mimulus guttatus

Frequently naturalized by streams and rivers, many populations appear to be of hybrid origin. *M. luteus* and/or *M. cupreus* may cross with this sp.. The mapped records are of *M. guttatus* and hybrids.

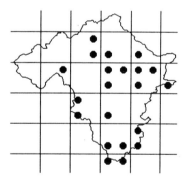

*Musk *Mimulus moschatus*
Uncommonly naturalized by ponds and streams. Streamside, Bach Howey, near Erwood c14SW JG Williams (1935 NMW); below Hanter Hill, 2457 & 2556 PJP; Llanshay Dingle, Knighton 2971 PR (1964); by pool, Stanage Park 3271 ACP.

Scrophulariaceae

Great mullein
Verbascum thapsus

Of scattered occurrence on freely-drained, sunny shale, shingle and sand on riverbanks, scree, disused quarries and roadsides.

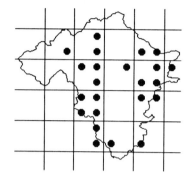

Dark mullein
Verbascum nigrum

A few plants occur on the verge of the Nantmel bypass 0266 constructed in 1981.

Common figwort
Scrophularia nodosa

Widespread on hedgebanks, ditch and riversides, open damp woodlands, wasteground and occasionally a garden weed. In all squares except 86, 87, 96SW & NW, 06SE, 15SW, 16SE, 17NW and 18SW.

Water figwort
Scrophularia auriculata

Uncommon on damp or disturbed soil. A single plant in a flower bed, North Rd, Llandrindod Wells 0562; by Afon Llynfi, Glasbury 1738; Llanstephan c14SW GP (1952); by Clyro Brook below Clyro 2143; Burfa Bog, Evenjobb 2761. Noted also from 05, 16, 17 and 36 in FPW.

*Snapdragon
Antirrhinum majus

Rarely naturalized on waste ground and walls. Rubble by old railway bridge Yr Allt, Llandeilo Graban 0844; derelict gardens, now supermarket, Llandrindod Wells 0561; walls New Radnor 2160; tip by road, W of Knighton 2872. Reported in MsFlR also from Clyro c24SW and Knighton c27SE.

*Small toadflax
Chaenorhinum minus

Rather scarce on sunny, freely-draining gravel and clinker of railway ballast, tips and quarry spoil. Noted as frequent on the railways in 1938 by AEW & JAW (1947 *Trans Rad Soc* XVII, 10), it still persists in suitable habitats on the disused Mid Wales line and in disused sidings away from the herbicide trains on the Central Wales line. Frequent at Llanelwedd Quarries 0552, it has appeared on tracks elsewhere, surfaced with scalpings from this quarry.

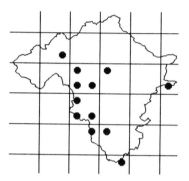

*Purple toadflax
Linaria purpurea

Not commonly grown in gardens in Radnor, it rarely becomes naturalized. Roadside, Rhayader 9668; on walls and waysides, Llandrindod Wells 0562; tip by road, W of Knighton 2872.

Pale toadflax
Linaria repens

An uncommon plant of sunny, freely-drained banks, scree and cinders. Rhayader c96NE MsFlR; scree on roadside bank by A470, Neuadd-ddu, N of Rhayader 9274; disused railway, W of Aberedw 0648 ACP; cinders of old railway station, Builth Wells 0451 JH; Builth Road railway station 0253; railway ballast, Llandrindod Wells 0662; shingle island in Wye, Boughrood Brest 1537 ACP; Llanbister c1673 JAW (1938 MsFlR); railway embankment, Llangunllo Station. 2172.

Veronica

Common toadflax

Linaria vulgaris

Frequent only in the lowlands on dry, sunny hedgebanks and on wasteground.

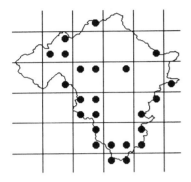

*Ivy-leaved toadflax

Cymbalaria muralis

An uncommon sp. of sunny and sheltered base-rich walls in towns and villages. Frost sensitive, hard winters may limit its spread in central and W Radnor.

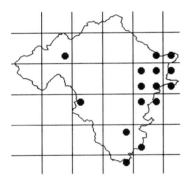

Sharp-leaved fluellen

Kickxia elatine

A rare weed of disturbed ground. On rubble used to infill old railway bridge below Yr Allt, Llandeilo Graban 0844; cornfield, Boughrood c13NW JFH & PCH (1956 *Proc BSBI* **2**, 376); Llanstephan c14SW GP (1954).

Round-leaved fluellen

Kickxia spuria

A single plant was found growing on disturbed soil in a lay-bye adjacent to forestry on a lane NW of Clyro 2044 in 1988. The only record of this south eastern species.

Foxglove *Digitalis purpurea*

Common in woodlands, hedgerows, riverbanks, tracksides, rock outcrops, banks and screes. Often abundant for a few years in plantation woodlands following felling. Recorded from all squares except 86NW and 87SW.

*Fairy foxglove

Erinus alpinus

Naturalized on stone walls around garden, Pencerrig, Llanelwedd 0453.

Thyme-leaved speedwell

Veronica serpyllifolia ssp. *serpyllifolia*

Common in open turf on seasonally damp soil. Upland pastures, banks, roadsides, lawns, stream and ditchbanks, disturbed soil of gardens and arable fields. Recorded from all squares except 86NW and 87SW.

Heath speedwell

Veronica officinalis

Common on acidic, freely-drained soil of hedgebanks, heaths, wood edges and upland pastures. Recorded from all squares except 86SE & NE, 87SW, 13NE, 24NW and 27NW.

Germander speedwell

Veronica chamaedrys

Common in woodland, hedgerows, shady streambanks, unimproved meadows and pastures, railway and road banks and churchyards. Recorded from all squares except 86SE & NW and 87SW.

Wood speedwell

Veronica montana

Frequent in woodlands, hedge and streambanks on damp clay soils.

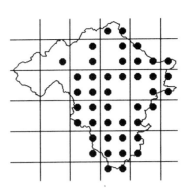

Marsh speedwell

Veronica scutellata

Widespread in wet flushes, marshes, damp pastures and on the edges of peaty pools and upland streams. 30 sites.

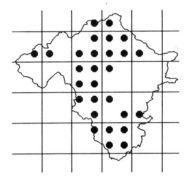

Brooklime

Veronica beccabunga

Frequent beside streams, in wet ditches and by ponds. Recorded from all squares except 86, 87, 96SW & NW & 97SW and NW.

Blue water-speedwell

Veronica anagallis-aquatica s.s.

Reported from Radnor in FPW. It is mapped from squares 14 and 24, part of which occur in Radnor and 15 which entirely lies in the county. The source of these records is unknown and the plant has not been seen in this survey.

Scrophulariaceae

Wall speedwell

Veronica arvensis

Widespread on open, seasonally droughted soil on scree, rock outcrops, walls and tracksides.

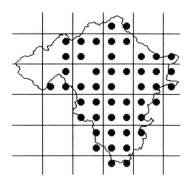

*Common field-speedwell

Veronica persica

Almost certainly under-recorded. It is a common weed of arable fields, gardens, tips and disturbed ground, developing early in the year.

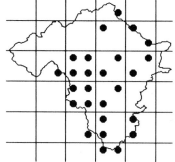

Spiked speedwell

Veronica spicata

Frequent on the sunny dolerite rock outcrops at Stanner 2658, from where it has been known since at least 1853 (AL in *Trans Woolhope Club* of that year, pg. 91).

Green field-speedwell

Veronica agrestis

Developing later in the year than common field-speedwell, it has been less frequently recorded. It is certainly likely to be more frequent than the records suggest, being a widespread weed of gardens and arable crops.

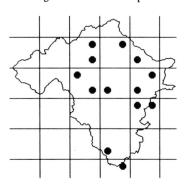

*Slender speedwell

Veronica filiformis

This alien from the Caucasus and Asia Minor was first reported from the bank of the Wye near Boughrood c13NW by PCH and JFH and from Penybont c16NW by RCLH in 1956 (Bangerter and Kent in *Proc BSBI* **2**, 197-217). Though seedpods have not been seen in Radnor, it has spread into damp grasslands, streamsides, gardens and churchyards throughout the county.

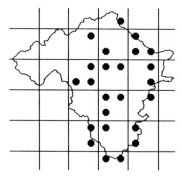

Common cow-wheat

Melampyrum pratense

Widely scattered amongst mosses in ungrazed sessile oakwoods and extending into heathland on cliff ledges and scree not reached by grazing animals.

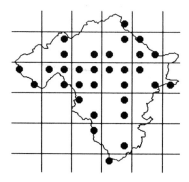

Grey field-speedwell

Veronica polita

Noted from Rhayader c96NE and Llandrindod Wells c06SE in MsFlR and from 96, 06, 16, 26, 27 and 36 in FPW, it has not been encountered in this survey. Sinker et al. 1985 note its absence from the hill country of Shropshire.

Ivy-leaved speedwell

Veronica hederifolia

Frequent on the more nutrient-rich soils of the E and S in hedgerows, woodlands, on streambanks, wasteground and as a garden weed. The sspp. have not been separated.

Eyebright

Euphrasia officinalis agg.

Frequent in unimproved upland meadows and pastures, edges of flushes and road verges on shallow soil with short grazed turf.

Euphrasia

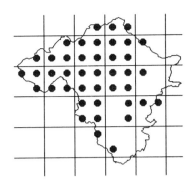

Euphrasia confusa

Near Llanbadarn Fynydd 18 FM Perring (1956 NMW conf. PF Yeo); Pen y Banc, Beacon Road, Llangunllo 2073 PR (1965 det. PF Yeo);Corner Farm, Llangunllo 2074 PR (1965 det. PF Yeo); Bailey Hill, Knighton 2373 PR (1965 det. PF Yeo); Craig y Don, Knighton 2673 PR (1965 det. PF Yeo).Reported also from 96 ,04 and 13 in FPW, part of which squares occur in Radnor and from 15 and 26 within Radnor.

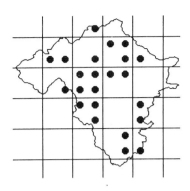

No systematic survey has been made of the individual species. The account below is a very preliminary one.

Euphrasia rostkoviana ssp. *rostkoviana*

The following records were made on a BSBI excursion in 1956 (*Proc BSBI* **2**, 376). Afon Claerwen Valley 86; Nant Brithgwm, Penygarreg Resr. Elan Valley 86; Gwynllyn, Llansantffraed Cwmdeuddwr 9468; Bachell Brook, Llananno 0872; Bettws Cottage, Bettws Disserth 1157. Also recorded from Witterley's Farm, near Knighton 2673 PR (1965 det. PF Yeo).

Euphrasia anglica

Four records were made on a BSBI exc. in 1956. Abbeycwmhir 0472 DE de Vesian; Hundred House 1154 FH Perring; near Dolbedwen, Newchurch c2049 CC Goodman; Cwm Ifor, Knighton 2971 FH Perring. Noted also by PR in 1965 from Pen y Banc, Beacon Road, Llangunllo 2073; Lower Cwmheyope Farm, Heyop 2174; Witterley's Farm, Knighton 2673; Woodhouse Farm, Knighton 2771 and Garth Hill, Knighton 2772 (all det. PF Yeo).

Euphrasia arctica ssp. *borealis*

Reported from St Harmon railway station 9972 DE de Vesian (1956 BSBI exc.) and Llangunllo 2171 by FH Perring (1956 BSBI exc.).

Euphrasia nemorosa

Claerwen Valley c86 BSBI exc. (1956 det. PF Yeo and PD Sell, reported in *Proc BSBI* **3**, 297) and streamside, Aberedw 04 VG (1956 NMW conf. PF Yeo).

Euphrasia micrantha

Reported from 15 in FPW. The source of this record has not been traced.

Euphrasia scottica

Boggy ground above Penygarreg Resr., Elan Valley 8968 WE Warren (1959 NMW conf. PF Yeo).

Red bartsia

Odontites verna

Widespread on roadverges, in field gateways and occasional in pastures. The recent revision of this group by Snogerup (1982) necessitates the aggregation of records of the sspp. *verna* and *serotina* and re-examination of the group.

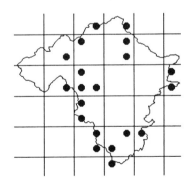

Marsh lousewort

Pedicularis palustris

In the wetter peaty flushes and marshes, mostly in the uplands.

Lousewort

Pedicularis sylvatica

More frequent than the last in damp acidic grassland and peaty flushes in unimproved pastures and hill grazings.

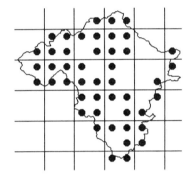

Yellow rattle

Rhinanthus minor

Common in unimproved pastures and haymeadows, on roadsides, railway banks and in churchyards. Becoming abundant in upland haymeadows following the first few applications of artificial fertilizer, but later eliminated.

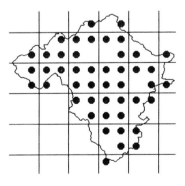

Orobanchaceae

Toothwort
Lathraea squamaria

Frequent only in the woodlands on the ORS of the lower Wye Valley with scattered records on the more base-rich soils elsewhere. The host has in most cases been difficult to determine with certainty, but it is most frequently associated with wych elm and hazel. 24 sites.

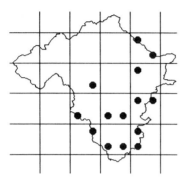

Broomrape family Orobanchaceae

Greater broomrape
Orobanche rapum-genistae

Parasitic on gorse and broom, it was last seen on broom on the side of the trunk road near Doldowlodd c9962 in the period 1955 - 1958 by J Wood. Road works have much altered this area and though the host species remains, a recent search by J Wood and T Sykes has failed to relocate it. Miss Wilding reports its presence in Llandrindod Wells c06SE in *Rep Bot Exc Club* 1925, 888 and members of the Woolhope Club report seeing it at "Silia", Presteigne 3064 and found numerous specimens on broom between the top of Harley's Hill and Offa's Dyke to the west c2864 (*Trans Woolhope Club* 1912, 32-35). T Sykes has also searched this area and though both gorse and broom remain, no broomrape was seen.

Butterwort family Lentibulareaceae

Common butterwort
Pinguicula vulgaris

Widespread in wet, peaty and stony flushes, in wet moss cushions on rock outcrops and in peaty seepage areas, particularly in the uplands.

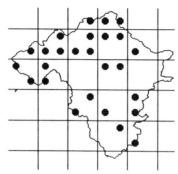

Lesser bladderwort
Utricularia minor

An uncommon sp. of peaty pools in basin and raised mires and in mawn pools, especially in the Aberedw 0845 to Glascwm Hill area 1551. 14 sites.

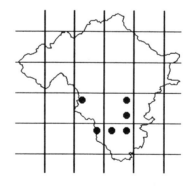

Greater bladderwort
Utricularia australis

Known only from pools in the reedswamp of Rhosgoch Common 1948. Possibly first noted here in 1938 by PM Holt (WFP), it was still present in 1982 (NCC survey). A record of *U. vulgaris* of HNR 1880 (*J Bot* 1881, 173) from Rhos Common may be of this species.

Plantain family Plantaginaceae

Greater plantain
Plantago major

Common in lawns, churchyards, tracksides, field gateways, waste ground, pastures and farmyards. In all squares except 86NW, 87SW, 16SE and 18SW.

Hoary plantain
Plantago media

This sp. of calcareous grassland is noted only from near Dolyhir Quarries c2458 MH Bigwood; near Knighton 27SE BS (1956 BRC survey) and near Bucknell, but in Radnor, 37SW BS (1956 BRC survey). There are no more recent records in 27 or 37 and the Shropshire Flora Project recorders failed to find it in this area.

Ribwort plantain
Plantago lanceolata

Common in meadows, pastures, churchyards, on road verges, lawns, path and tracksides and wasteground. In all squares except 87SW.

Shoreweed *Littorella uniflora*

Widespread on the muddy margins of upland pools and reservoirs and rarely in lowland pools. Occasionally colonising the beds of ephemeral, muddy streams on the heavily grazed upland commons. Over 30 sites.

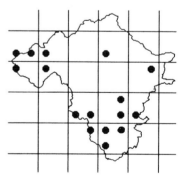

Honeysuckle family
Caprifoliaceae

*Dwarf elder
Sambucus ebulus

Known only from a roadside verge at Knucklas 2574. First reported here in 1952 by the Swansea Field Naturalists Society and still present.

Elder *Sambucus nigra*
A common shrub of hedgerows, especially near habitations, streambanks, rubbish tips, the more nutrient-rich woodlands and scrub, particularly on the sites of former rabbit warrens, many of which date from before the introduction of myxomatosis. All squares except 86SE & NW, 87, 96SW and 97NW.

*Red-berried elderberry
Sambucus racemosa

This rarely naturalized alien was reported from the Kerry Hills 08 by AJ Bird in 1957 (BRC records).

Guelder rose
Viburnum opulus

Frequent in damp woodlands and hedgerows on the more base-rich clay soils.

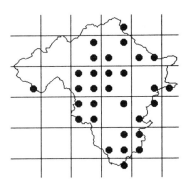

Wayfaring-tree
Viburnum lantana

A single plant grows on the shale bank of a disused railway near White Hall, Aberedw 0748 ACP. Not apparently planted, it may have originated in a nearby garden, though there is no local recollection of it ever having recently been grown.

*Snowberry
Symphoricarpos albus

Commonly planted in hedgerows, shrubberies and to provide cover for game in woodlands, it long persists and spreads vegetatively, especially on the moister soils. Due to difficulty in separating obviously planted from naturalized stands no map has been produced.

*Perfoliate honeysuckle
Lonicera caprifolium

Noted in MsFlR as "found in plenty on the Castle Walls, Presteigne 3164 in 1948 but gone by 1951 - apparently cut away".

Honeysuckle
Lonicera periclymenum

Common in woodlands, scrub, hedgerows and on cliffs out of reach of sheep in the uplands. In all squares except 86NE & NW, 87, 07NW and 08SE.

Moschatel family
Adoxaceae

Moschatel
Adoxa moschatellina

Widespread in woodlands and hedgerows on the more nutrient-rich soils of the S and E.

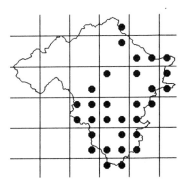

Valerian family
Valerianaceae

Common cornsalad
Valerianella locusta

A rare sp. of shallow soil over rocks kept open by summer drought. Boulder by lay-by on road (formerly railway) below Twyn y Garth, Llandeilo Graban 1043; Knucklas Castle 2574 PJP & CP.

Keeled-fruited cornsalad
Valerianella carinata

Noted from by the R. Wye near Llyswen c13NW GP (1961 NMW) and from dry, calcareous mudstone cliff in wood E of Knucklas 2673 ! and PJP.

Narrow-fruited cornsalad
Valerianella dentata

"One small piece seen in a tillage field" reports AL in his "Notes on a flora of Aberedw" c04NE made between 1886 and 1890 (*Trans Woolhope Club* 1890-92, 189). No other records.

Common valerian
Valeriana officinalis

Common in marshes, damp woodlands, ditches, on streamsides, pool and lake margins and in damp pastures. In all squares except 86NE & NW, 97, 25NW, 26SW and 36NW.

Marsh valerian
Valeriana dioica

Frequent in wet flushes in woodlands and in peaty flushes and marshes, especially on the more base-rich rock types. The sites are often grazed, with well-developed stands of brown mosses.

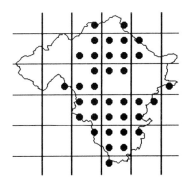

*Red valerian

Centranthus ruber

Occasionally naturalized on walls and waste ground in towns and villages. Rhayader 9668; near Wye bridge, Llanelwedd 0451; Llandrindod 0561; Gladestry churchyard 2355 DRD, IHS & CC; walls, New Radnor 2160; near gas works, Knighton 2972; Presteigne c3164 (MsFlR).

Teasel family
Dipsacaceae

Teasel *Dipsacus fullonum*

Not common, but occurring sporadically on disturbed roadside verges, river banks and waste ground.

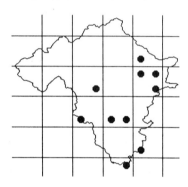

Small teasel *Dipsacus pilosus*

A scarce sp. of open habitats in woodland. First recorded in the gorge woodland at Aberedw 0747 by HNR in 1880 (*J Bot* 1881, 172) and still present; near Llowes c14SE FA Rogers and WMR in 1898 (*J Bot* **XXXVII**, 21) and in woodland on the bank of the Wye at Cabalva 2446, JE Messenger and DPH.

Devil's-bit scabious

Succisa pratensis

Common in damp pastures, heathland, marshes, damp roadsides and churchyards, especially in the uplands. In all squares except 86NW, 87SW, 04SE, 26SW & NW and 27SE.

Field scabious

Knautia arvensis

Not seen during this survey but reported as "rather scarce" in MsFlR with records from Rhayader c96NE; Disserth c0358 and Howey, c0558. WMR notes it at Aberedw cO4NE and Boughrood c13NW in 1898 (*J Bot* **XXXVII**, 21) and MP in Glasbury cutting 1839 in 1969. Perhaps a victim of acidification or changed roadside verge maintenance practices.

Bellflower family
Campanulaceae

Spreading bellflower

Campanula patula

Rare on dry, sunny roadside banks and possibly not seen for over 10 years. The cause of this decline is not known but as a biennial, flowering late in the summer it is likely that regular mechanical cutting of verges could have reduced its abundance. Roadside bank, Glasbury 1639 ACP and MR Pugh; bank near Rhulen 14 or 15 M Taylor (1961, *Nat. Wales* **8**, 28); near Llowes c14 WMR (1899, *J Bot* **XXXVII**, 17-25); bank at Llanstephan c14 MH Sykes and JAW (1949, *Proc Swansea Sci and Fld Nats Soc* Vol **II**, pt 10, 326); one plant, roadside near Clyro c24 HNR (1880, *J Bot*, 1881, 172); one plant on roadside between Stanner and Dolyhir 25 JW Gough and NY Sandwith (1945 *Rep Bot Exc Club* 1945-47, 61); corner of wood, Knill 2861, PJP (*Nat Wales* **16**, 211); bushy banks by roadsides between Stanner Rocks and Presteigne 26 and /or 36 WMR (1899, *J Bot* **XXXVII**, 22); near Little Cwmgilla, Knighton 2672 PR (1965). A search of the Teme Valley may identify further sites since it is known nearby in Shropshire (Sinker et al. 1985).

Clustered bellflower

Campanula glomerata

A probable garden escape under scrub by the top of limestone quarries, Dolyhir 2458 ACP.

*Adria bellflower

Campanula portenschlagiana

Naturalized on walls Llandrindod Wells 0560 and Presteigne 3164.

*Trailing bellflower

Campanula poscharskyana

Naturalized on rocky bank of R. Wye above Boughrood 1240 GM Kay (*BSBI News* **19**, 14).

Giant bellflower

Campanula latifolium

An uncommon sp. of shady, moist woodlands and streamsides, mostly on the ORS and more calcareous shales of the SE, with an outlier by the Clywedog Brook, Crossgates 0865 CAC. 12 sites.

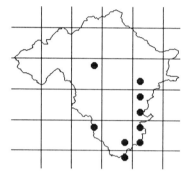

Nettle-leaved bellflower

Campanula trachelium

A scarce sp. of woodland edges, waysides and streambanks mostly in the E of the county in the Lugg catchment. A record from Ysfa 9964 F Lancaster on a forestry track is probably an introduction with road stone. 12 sites.

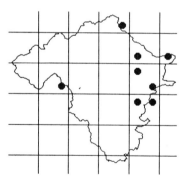

*Creeping bellflower

Campanula rapunculoides

Naturalized in Downton farmyard, New Radnor c2360 JFH and PCH (1956 *Proc BSBI* **2**, 375).

Harebell
Campanula rotundifolia

Common in freely-drained, acidic grasslands in unimproved pastures, on roadverges and rock outcrops, heathland and sunny hedgebanks. In all squares except 86NW, 87SW and 27SE.

Ivy-leaved bellflower
Wahlenbergia hederacea

Of occasional occurrence by wet, peaty flushes, in damp, acidic grasslands, by streams and springs and in wet glades in woodlands, particularly in the uplands of the west.

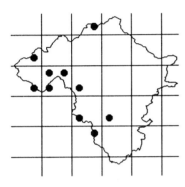

Sheep's-bit *Jasione montana*
On sunny, open, freely-drained soil of banks, rock outcrops, scree and cliffs and occasionally the tops of ant hills in unimproved pasture and woodland glades.

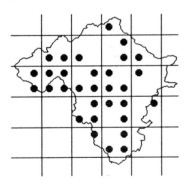

Water lobelia
Lobelia dortmanna

A scarce sp. growing amongst the gravel of upland lake beds. Common in Llyn Cerrigllwydion Uchaf 8469 and Llyn Cerrigllwydion Isaf 8470 in Llansantffraed Cwmdeuddwr. Reported from Llyn Gwyn Gweddw near Rhayader by HJR in 1909 (*Proc Cotteswold Nat Fld Club* (1910) **XVII** 57-61) but not seen in recent surveys.

*Garden lobelia
Lobelia erinus

A rare casual of tips. Roadside tip Knighton 2872.

Daisy family
Compositae

Hemp agrimony
Eupatorium cannabinum

An uncommon sp. of sunny and damp ditch, stream and riverbanks mostly in the east.

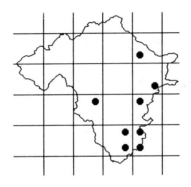

Goldenrod
Solidago virgaurea

Frequent and widespread in well-drained, lightly or ungrazed woodlands, on hedgebanks, cliffs, heath and riverbanks, especially favouring areas of dappled sunlight.

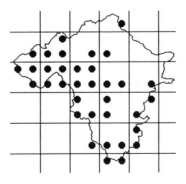

*Canadian goldenrod
Solidago canadensis

This alien appears to have rarely established itself in Radnor. Nantmel tip 0165; rubbish tip, Llandrindod Wells c06SE (MsFlR 1954); bank of R. Wye above Glasbury 13NE CAC.

Daisy *Bellis perennis*
Common in meadows, pastures, mowed grass, upland flushes and river banks. In all squares except 86NW and 87SW.

*Michaelmas-daisy
Aster novi-belgii

A rare alien noted in MsFlR from Builth Road c0253; Cwmbach Llechryd c0254 and on the N side of the Lake, Llandrindod Wells 0660. Seen recently only by the Wye near Penddol, Llanelwedd 0352 ! and PJP and in a roadside hedge SE of Dol-y-Fan Hill, Newbridge on Wye 0261 FMS.

Blue fleabane *Erigeron acer*
First noted in the county at Stanner 2658 by the Swansea Field Naturalists Society (in MsFlR) in 1948, where it still occurs on summer-droughted soil on scree and on rockledges. Also beside tracks and in the old limestone quarries at Dolyhir 2458.

Common cudweed
Filago vulgaris

A scarce sp. of dry, sunny, open conditions. Aberedw 04NE AEW (1927 MsFlR); common on scree, tracksides, quarry spoil and in short turf about Llanelwedd Quarries 0551 and area; Stanner Rocks 2658 (noted in 1899 by WMR, *J Bot,* **XXXVII**, 21 and still there).

Small cudweed
Logfia minima

In similar habitats to the last. Scree below road by the filter beds, Elan Village 9264; dry grassland and disused quarry roads, Rhayader Quarries and S facing bank near Glaslyn, Ysfa 9765; Llandeilo Graban c04SW AEW (1927 in MsFlR); widespread, Llanelwedd Rocks 0551 and area; Stanner Rocks 2658, AH Trow (1923 MsFlR and still present).

Compositae

Heath cudweed
Omalotheca sylvatica

Very rare and not seen recently. Rhayader c96 K Richards (NMW records, no date); Newbridge on Wye c05NW WMR (1899 *J Bot* **XXXVII**, 21); Doldowlod 96 or 06 (1947 MsFlR).

Marsh cudweed
Filaginella uliginosa

A common weed of arable fields, gardens, field gateways, tracksides, muddy lake, pool and river margins. In all squares except 86NE & NW, 87, 96SW & NW, 97NW, 16SE, 18SW and 36SW.

Mountain everlasting
Antennaria dioica

A plant of short turf and the edges of rock ledges in the uplands. Much decreased if not even extinct. The Rev A Wentworth Powell in WJ and JO Bufton's Illustrated Guide to Llandrindod Wells (F Hodgson London 1906) says "the hillsides in places are white with wild everlasting". The MsFlR notes it from near Rhayader 96, Claerwen 86; Camllo Hills 06 and 07, round Abbeycwmhir 07, hills above Llanbadarn Fynydd c17 or 18, Forest Inn 15, Radnor Forest 16, and the Kerry Hills above Felindre 18. There are no more recent records. The cause of its decline is unknown. In Brecknock it is now frequent only on the limestone. Acidification due to atmospheric pollutants may be partly the cause, together with afforestation, ploughing, fertilizers and heavier grazing pressures.

*Pearly everlasting
Anaphalis margaritacea

A rare casual on dry banks. Reported in FPW from 06, the source of this record cannot now be traced. Otherwise only known from a hillside on Upper Goytre farm, Beguildy 27NW AJ Hant (1947 NMW).

*Elacampane *Inula helenium*
An alien, no longer widely grown in gardens, it has only recently been seen growing on a lane verge near Llanstephan 1141 ACP and in a field near Norton 36NW SF Bond.

Ploughman's-spikenard
Inula conyza

A scarce sp. of dry, open, sunny woodland clearings and scrub edges on lime-rich soil. Steep ORS bank by road W of Cwmbach, Glasbury 1639; openings in wooded cliff on bank of Wye, Wyecliff, Clyro 2242 ! & ACP; limestone scree and old quarries in and near Yatt Wood, Dolyhir 2458; scree, Stanner Rocks 2658. Noted in MsFlR from, in addition, Aberedw 04NE, Walton c2559 and Presteigne 36.

Common fleabane
Pulicaria dysenterica

In view of the abundance of this sp. in damp pastures and flushes in southern Wales (see FPW) its scarcity in Radnor has come as a surprise. Not apparently ever seen by Wade and Webb (MsFlR) seven localities have been traced. All recent records are from damp sites on heavy clay soils. The plant is grazed by sheep but left comparatively untouched by cattle and ponies. Builth Road c0253 WMR (1899, *J Bot* **XXXVII**, 21); one clump by the Bog, Newbridge on Wye 0157; roadside, Disserth 0358 J Blashill; roadside W of Tyncoed, Disserth 0255; edge of damp pasture, Castle Crab, Disserth 0454 IHS & CC; pasture, Cwm Gwanon, Clyro 1944 MEM; pasture, Lower Caer Faelog, Llanbister 1074 DRD & CC.

Trifid bur-marigold
Bidens tripartita

A scarce sp. of mud and silt banks by pools and rivers. Abercamlo Farm, Gwystre 0764 LMP; by R. Wye near Glasbury 13 DC Hutt; oxbow lakes, Fforddfawr, Glasbury 1840 MEM & MP; bank of R. Wye, Bronydd 2345 ACP; by pool, Hendregenny, Pilleth 2569 ACP.

Nodding bur-marigold
Bidens cernua

Only known from the edge of a pond near Cloggie, Llangunllo 2271. First noted by PR in 1964 and still there in 1985 (PJP & CP).

*Common sunflower
Helianthus annuus

A casual on waste ground, now carpark, Llandrindod Wells 0561 and on Clyro tip 2143.

Sneezewort
Achillea ptarmica

Common in unimproved meadows, pastures, on streambanks and road verges on damp heavy soils. In all squares except 86NW, 96SW, 04, 16SW, 18SW, 26, 27SE & SW, 36SW and 37SW.

Yarrow
Achillea millefolium

A common plant on dry, sunny banks, meadows, verges, churchyards and lawns. Noted from all squares except 86NW and 87SW.

Scentless mayweed
Tripleurospermum inodorum

A widespread weed of arable fields, disturbed road verges, gateways, farmyards and tips.

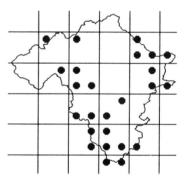

Scented mayweed
Matricaria recutita

Often with the last sp. in similar habitats but less common in the Wye Valley.

Artemisia

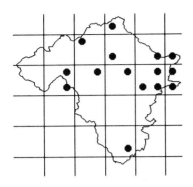

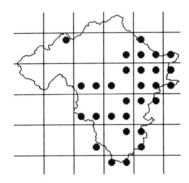

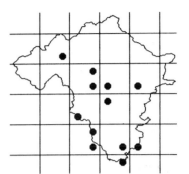

*Wormwood
Artemisia absinthium

Probably formerly widely grown as a herb, it now finds little favour and survives in dry sunny places around farmyards, on walls, shale banks and rarely on silt and shingle by rivers. 14 sites.

*Pineappleweed
Matricaria matricarioides

Spread by vehicles, this fairly recent introduction has colonised most of Radnor, occurring typically on the compressed soil of tracks, gateways and along the edge of roads. Occasionally a field weed. The earliest record so far traced is of HJR from roadsides, Llanelwedd 05 in 1909 *(Proc Cotteswold Club* **XVII** (1), 58). Recorded from all squares except 86, 87, 96SW, 97NW, 16SE and 18SW.

Corn marigold
Chrysanthemum segetum

An attractive weed of arable crops. Considered "very local" in MsFlR its status is probably much the same today. In turnip fields near Llidiardau, St Harmon 9778; Llanstephan 14SW GP (1952); cornfield between Llowes and Glasbury 1840 ACP; arable field, Ffynnon Gynydd 1641 PJP; Bailey Farm, Pilleth 2569 DEG. Noted in MsFlR from Welfield and Cwmbach Llechrhyd 05SW, Llangunllo c27SW, Llaithddu c07NE and Nant-yr-haidd, Ysfa c0064.

*Feverfew
Tanacetum parthenium

Regularly escaping from cultivation in gardens to colonise roadsides, tips and walls, generally close to its place of origin. Recently increased in popularity and widely grown as a cure for migraine.

Colt's-foot   *Tussilago farfara*

Common on heavy and damp disturbed soil of road and railway banks, tips and streamsides. Rare on unstable clay banks by streams in the uplands, its probable natural habitat. In all squares except 86NE & NW, 87, 96NW & SW, 97, 04SE, 16SE, 18SW and 27NW.

Mugwort   *Artemisia vulgaris*

In similar places to, and often associated with, tansy on silt banks by rivers and on roadsides. Widespread but not common and perhaps more a lowland plant.

Butterbur   *Petasites hybridus*

Widespread on streamsides and damp ditchsides in central and S Radnor. Absent from many apparently suitable sites by the Teme and Wye. The apparent lack of female plants may account for its patchy distribution.

Tansy
Tanacetum vulgare

Possibly native beside the main rivers on silt, shingle and rock outcrops. Elsewhere scarce on roadverges, old railway lines, tips and quarry spoil.

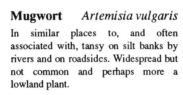

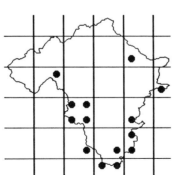

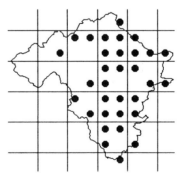

Compositae

*Winter heliotrope
Petasites fragrans

Doubtfully naturalized anywhere except by the brook in Clyro 2143. Occasionally planted in shrubberies and spreading slowly into adjacent woodlands and hedges, these records have been disregarded.

*Plantain-leaved leopard's-bane
Doronicum plantagineum var. *excelsum* N.E. Brown

Established on woodland bank beside lay-by on road S of Knighton near Gwernaffel 2770 coll. S Gooch det. !.

*Leopard's-bane
Doronicum pardalianches

Rarely naturalized in woods, and on grassy roadside banks. Road verge, Disserth 0458 AS Mackintosh; railway bank near Penybont Station 0964; hedgebank, Glasbury 1838; roadside hedge, Beggars Bush 2664 ACP.

Common ragwort
Senecio jacobaea

Widespread on road verges, though removed annually by the highways authority, by gateways, on river banks, railway banks and in pastures on well-drained soil. It does not appear to be a significant agricultural weed and is absent from large areas of the sheepgrazed uplands.

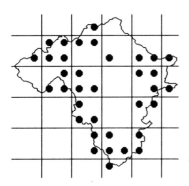

Marsh ragwort
Senecio aquaticus

Common in damp pastures and marshes, by ditches, streams and on pool margins. In all squares except 86, 87, 96NW & SW, 26SW & NE, 27NW, 36 and 37.

*Oxford ragwort
Senecio squalidus

To describe this as a rare and apparently declining sp. speaks eloquently of the scarcity of waste and disturbed "urban" ground in the county and perhaps also of the paucity of opportunity for adventives to be moved into and about the area. Noted in only three places. The old station yard, Builth Road 0253, found in 1983 but not seen in a search in 1986. By the old railway station Glasbury 1839 MP and beside the A438 between Clyro and Llowes 1942 (first noted here by ACP in 1980, a few plants appear each year).

Heath groundsel
Senecio sylvaticus

Widespread on freely drained, acidic, open soil in sunny sites such as road banks, about rock outcrops and scree and in woodland clearings, especially those disturbed by forestry work.

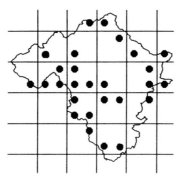

Sticky groundsel
Senecio viscosus

Occasionally with the last sp. but generally favouring more nutrient-rich sites such as tips, road verges and about farmyards.

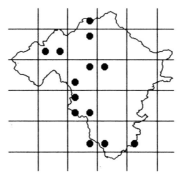

Groundsel
Senecio vulgaris ssp. *vulgaris*

A common weed of cultivated ground in gardens and fields. Common on open disturbed ground by roads, tips, farmyards and on riverbanks. It is recorded from all squares except 86, 87, 96SW, 97SW, 04NE, 15SE, 16SE, 17NW and 18SW. The var. *hibernicus* with rayed flowers is a recent colonist and is still scarce. First noted by ACP near Aberedw 0747 in 1973 it is elsewhere known from a few pavement and road edges near Cwmbach 0254, Newbridge on Wye 0158, Howey 0558, Boughrood 1238 and Stanner 2658.

*Pot Marigold
Calendula officinalis

Rarely escaping from cultivation. Roadside tip between Disserth and Newbridge on Wye 0358.

Carline thistle
Carlina vulgaris

A scarce plant of short, sheep-grazed turf on sunny, freely-drained banks, ledges and stable scree. Favours base-rich sites. 17 localities.

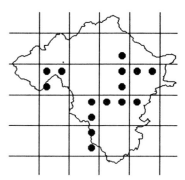

Greater burdock
Arctium lappa

Not seen in this survey and possibly now extinct in the localities mentioned in MsFlR ie Erwood c04, Llanfaredd 05, and Builth 05 together with Clyro 24 where noted by WMR (1898 *J Bot* 37, 21). Sinker et al. 1985 record it from 17SW. The source of this record cannot now be traced.

Cirsium

Lesser burdock

Arctium minus agg.

Common in woodland glades in the lowlands, on road and hedgebanks, waste places, and occasionally as a weed of arable land or grass leys. The sspp. and hybrids have not been seperated in this survey. *A. minus* s.s. and *A. nemorosum* are both reported from the county in FPW. In all squares except 86, 87, 96SE & SW, 97SE & NW, 07SW & NW, 17NE & NW and 18SW.

Musk thistle   *Carduus nutans*

A scarce sp. of disturbed, freely-drained soil persisting only in calcareous sites. Quarries, Llanelwedd 0551 (first reported in 1938, MsFlR and still present); roadverge, Llanbadarn fawr 0863 (one plant 1984, now gone); sunny, thin-soiled bank near Frank's Bridge 1056 CP, PJP & ACP; roadside below Llynheillyn 1658 ACP (abundant in 1972 following roadworks but declined and gone by 1983); reclaimed grassland, Garth Hill, Knighton 2772 PR (1964); limestone spoil and verges, Dolyhir Quarry 2458 ACP (1980 and still present).

Welted thistle

Carduus acanthoides

Of widespread occurrence, but not common, in the S and E of the county on roadside banks, rubbish tips, quarry spoil and river gravels. 14 sites.

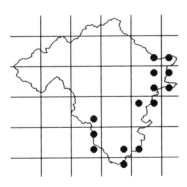

Woolly thistle

Cirsium eriophorum

Known only from roadside banks about the limestone quarries at Dolyhir 2458. First reported by FH Perring, D Bayliss and DE de Vesian in 1956 (*Proc BSBI* **2**, 374). Still present in 1988, though in small quantity but not seen in 1990.

Spear thistle   *Cirsium vulgare*

A common weed of grasslands, hedgerows, verges, riverbanks and waste ground on well-drained soils, in all squares except 86NE & NW, 87SW and 97SW.

Meadow thistle

Cirsium dissectum

Widespread, but not common, in unimproved wet pastures, hay meadows and peaty flushes. Long persistent, it seems rarely to set seed and unlike its close relatives is a poor colonist. Agricultural drainage has reduced its abundance recently. c50 sites.

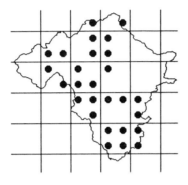

Melancholy thistle

Cirsium helenioides

This thistle reached the southern-most edge of its British range in Radnor. As reported in MsFlR, JAW found it on a bank by the Ithon near Penithon 2 mls N of Llanbadarn Fynydd c0879 in 1938. It was also found on banks near Llanbadarn Fynydd but during the second world war road widening apparently destroyed it entirely. In 1948 however, a "relatively enormous" patch was found on a bank a mile or so S of Camnant bridge. Its grid ref. would have been c084813. No other records were made until 1992 when the author found a small colony on a roadside beside the A483 2Km N. of Llanbadarn Ffynnydd 0879 and a large colony on a verge near the chapel at Pantydwr 9874. The species also appears in a list from Llandrindod Wells c06 (WJ and JO Buftons's *Illustrated guide to Llandrindod Wells*, F Hodgson, London 1906).

Marsh thistle

Cirsium palustre

A common plant of wet soil in poorly-drained meadows and pastures, stream and ditch sides and in peaty flushes found in all squares. White and pink flowered forms are frequent.

Creeping thistle

Cirsium arvense

A widespread and pernicious weed of pastures and gardens, roadsides, riverbanks and wasteground. In all squares except 86NE & NW and 87SW.

Saw-wort   *Serratula tinctoria*

Widespread but not common in damp, herb-rich pastures, woodland glades, especially in alder woods and on river banks. c40 sites. Much decreased through agricultural drainage.

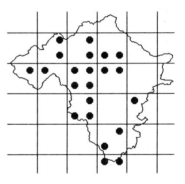

Greater knapweed

Centaurea scabiosa

Reported from 2173 by Sinker et al. 1985, some doubt now surrounds this record. It should be sought in calcareous grassland in the NE of the county.

Knapweed

Centaurea nigra

Common in agriculturally unimproved pastures, roadverges, railway banks, churchyards, hedgerows and riverbanks on well-drained soil in sunny sites. In all squares except 86NW, 87SW and 27NW. The various segregates have not been separated., in this survey. Both *C. debeauxii* ssp. *nemoralis* and *C. nigra* s.s. are reported from the county in FPW.

Compositae

*Perennial cornflower
Centaurea montana

Commonly cultivated, it persists well when planted on grassy banks. Records close to gardens have been disregarded. On a remote railway embankment S of Llangunllo Station 2172.

Cornflower
Centaurea cyanus

A few plants were noted in 1985 on the disturbed soil behind Boughrood churchyard wall 1239 following its reconstruction. A single plant appeared on disturbed soil in pasture adjacent to the withybeds, Presteigne 3164 in 1992 DPH. Perhaps never common as a field weed, MsFlR records it only from turfy walls between New Radnor and Kinnerton 26SW.

Chicory   *Cichorium intybus*

A rare introduction. Roadside, Newbridge on Wye 0158 DA Barnes; field near Disserth c0358 (1948 & 1954 MsFlR); reseeded meadow near Graig Fawr, Bettws Disserth 1257 ACP. In the early 1970's it was sown in agricultural grass mixtures and often persisted for a few years.

Cat's-ear
Hypochoeris radicata

A common flower of unimproved pastures, hay meadows, lawns, churchyards and freely-drained, sunny hedge and stream banks. Extends along the edges of flushes and screes into upland rough grazings. In all squares except 86NW, 87SW, 17SW and 18SW.

Autumn hawkbit
Leontodon autumnalis

In similar sites to the last but extending also into wetter pastures. Common and in all squares except 86NW, 96SW and 18.

Rough hawkbit
Leontodon hispidus

Widespread on sunny, freely-drained soil of banks, unimproved pastures, churchyards, roadside verges and disused quarries, in all squares except 86NW & SE, 87SW and 27NW.

Lesser hawkbit
Leontodon taraxacoides

The least common of the hawkbits but occurring in similar short turf of pastures, road verges etc, it may perhaps favour the stiffer clay soils.

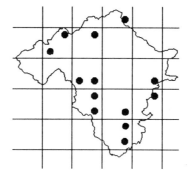

Bristly ox-tongue
Picris echioides

Reported from Radnor in FPW but probably in error.

Hawkweed ox-tongue
Picris hieracioides

A scarce sp. of sunny, open, calcareous soil in the S and E of the county. Noted in MsFlR from Aberedw c04, hedges between Pistyll and Boughrood Station 13NW, Llanstephan 14SW and Llowes 14SE. It has recently only been seen on the verge of the A438 E of Llowes 1942, on a rubbish tip at Clyro 2143 and on a road verge S of Burfa Camp 2860.

Goat's-beard
Tragopogon pratensis ssp. *minor*

A plant of roadside verges and occasionally hayfields, churchyards, hedgebanks and railway banks. Mostly in the lowlands. 18 sites.

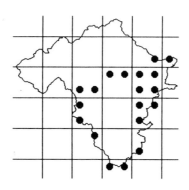

Prickly sow-thistle
Sonchus asper

The commonest sp. of this genus, found on hedgebanks, roadsides, as a weed of fields, gardens and tips, on riverbanks or wherever relatively nutrient-rich soil is disturbed. In all squares except 86NE & NW, 87, 96SW & NW, 97NE, 15SE, 16SE, 17SE & NW and 18SW.

Smooth sow-thistle
Sonchus oleraceus

As widespread as the last and in similar places but less common. A persistent weed, however, once established in a garden.

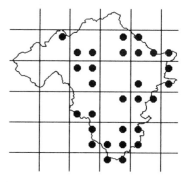

Perennial sow-thistle
Sonchus arvensis

Frequent in the lowlands, in hedgerows, on roadside verges, tips about farmyards and occasionally as a weed of arable fields and gardens.

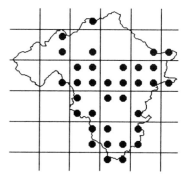

Taraxacum

*Blue sow-thistle

Cicerbita macrophylla ssp. *uralensis*

A rare introduction, persisting on roadside banks in a few sites. Near the church, Beguildy 1979 ACP; W of Clyro Court, Clyro 2042; Gladestry 2355 DRHumphreys and a garden weed at Old Radnor 25 DM Phillips (NMW).

Wall lettuce *Mycelis muralis*

Common on shaded calcareous cliffs, walls, banks, and scree in the S and E of the county. Less common elsewhere.

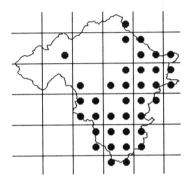

Dandelions *Taraxacum* spp.

A preliminary account is provided below. All identifications have been made by AJ Richards or CC Haworth. Where no collectors name is given the record was made on a recording meeting organised by BSBI in 1985.

The taxonomy follows Howarth, CC, 1988 in ed., *An Annotated List of British and Irish Dandelions*.

Sect. Erythrosperma

T. argutum (Dt.) Raunk.

Dry grassland on rocky slopes. Llanelwedd Rocks 0452; Cefnllys Castle 0861; roadside bank, Dolyhir 2458.

T. brachyglossum Dt.

Dry grassland on rocky slopes. Llanelwedd Rocks 0452; R. Wye, Llyswen c13NW CT & E Vachell (1931, NMW); Boughrood 13 F Houseman ; Stanner Rocks 2658.

T. canulum Hagl.

Rocky outcrops, Llanelwedd Rocks 0452.

T. degelii Hagl.

Rocky grassland. Llanelwedd Rocks 0452.

T. fulviforme Dt.

Dry grassland, Llanelwedd Rocks 0452. Reported also from 97 and 14 in FPW, most of which squares occur in Radnor.

T. lacistophyllum (Dt.) Raunk:

Rocky outcrops. Llanelwedd Rocks 0452 and bank of Bach Howey, Llanstephan 1042.

T. oxoniense Dt.

Rocky grassland, Llanelwedd Rocks 0452 and Stanner Rocks 2658 AJ Wilmott (1932 BM).

T. proximum (Dt.) Raunk.

Rocky grassland, Llanelwedd Rocks 0452.

T. rubicundum (Dt.) Dt.

Rocky grassland, Llanelwedd Rocks 0452; cracks in tarmac of garage forecourt, Ridgebourne, Llandrindod Wells 0560; rock outcrops, Stanner Rocks 2658.

Sect. Spectabilia

T. faeroense (Dt.) Dt.

Wet pasture, Cefnllys 0861; roadside bank and ditch, side of Clyro hill 2046 J Rogerson (*Nat Wales* 16, 211).

Sect. Naevosa

T. euryphyllum (Dt.) M.P.

Grasslands and waysides. Llanelwedd Quarries 0452; Llandrindod Wells 0560; Cefnllys 0861; Dolyhir 2458.

T. maculosa A.J. Rich.

Roadside verge by the Bach Howey Viaduct, Llanstephan 1042.

"T. richardsianum" in.ed.

Roadside banks. Llanelwedd Quarries; by the Bach Howey Viaduct, Llanstephan 1042; Dolyhir 2458.

Sect. Celtica

T. bracteatum Dt.

On grassy verges, near Bach Howey Viaduct, Llanstephan 1042; Burlingjobb 2458 MW Marsden and Stanner Rocks 2658.

T. gelertii Raunk.

Grassland, Llanelwedd Rocks 0452 and Cwmbach, Glasbury 1639 MV Marsden (Hb. PVT

T. nordstedtii Dt.

Grassland, Llanelwedd Rocks 0452; wet field, Cefnllys 0861; marshy field near Clyro 1944 MV Marsden; roadside, Dolyhir 2458; mapped also from 97 and 05 in FPW, parts of which squares occur in Radnor.

T. raunkiaerii Wiinst.

Grassland, Llanelwedd Rocks 0452 and Dolyhir 2458.

T. subbracteatum A.J. Rich.

Common in grasslands and waste places. Llanelwedd Quarries 0452; Cefnllys 0861; roadside and riverside by the Bach Howey Viaduct, Llanstephan 1042; Clyro 1941 MV Marsden; Stanner Rocks 2658; Dolyhir 2458.

T. unguilobum Dt.

Reported from 13 and 14 in FPW, parts of both squares occurring in Radnor.

Sect. Hamata

T. atactum Sahlin & vS.

Roadside verge, near Bach Howey Viaduct, Llanstephan 1042; verge near Clyro 1944 MV Marsden.

Compositae

T. hamatiforme Dt.

Grassland of road verge near Clyro 1942 MV Marsden; Dolyhir 2458 and Stanner Rocks 2658.

T. hamatum Raunk.

Grassland, Llanelwedd Rocks 0452; Cefnllys 0861; roadside near Bach Howey Viaduct, Llanstephan 1042 & 1141; roadside near Clyro 1944 MV Marsden; Dolyhir 2458.

T. hamiferum Dt.

Road and riverside near the Bach Howey, Llanstephan 1141 and 1042 and bank of stream near Cwm Cottage, Clyro 2044 MV Marsden.

"*T. hesperium*" ined.

Roadside, Newbridge on Wye 0158 and on bank of the Bach Howey, Llanstephan 1042.

T. marklundii Palmgr.

Grassland, Cefnllys 0861

T. pseudohamatum Dt.

Grassland and roadsides, Newbridge on Wye 0158; roadside and riverside near Bach Howey Viaduct, Llanstephan 1042; garage forecourt, Ridgebourne, Llandrindod Wells 0560; Stanner Rocks 2658

T. quadrans H.Ollg.

Grassland, Llanelwedd Rocks 0452; road verge near Bach Howey Viaduct, Llanstephan 1042.

Sect. Ruderalia

T. alatum H. Lb.f

Grassland, Cefnllys 0861; lane verge, Llanstephan 1042; lane verge, Clyro 1944 MV Marsden; Stanner Rocks 2658.

T. ancistrolobum Dt.

Roadside and grassland, Dolyhir 2458 and Stanner Rocks 2658.

T. cordatum Palmgr.

Grassland, Cefnllys 0861; road verge near Clyro c14 J Rogerson (*Nat Wales* **16**, 211); Stanner Rocks 2658.

T. croceiflorum Dt.

Grassy places. Roadside, Bach Howey Viaduct and Llanstephan nearby 1042 & 1141; grassland, Stanner Rocks 2658; road verge, Dolyhir 2458.

T. dilaceratum M.P. Chr.

Grassland, Dolyhir 2458 and Stanner Rocks 2658.

T. ekmanii Dt.

Common in grassy places and on roadsides. Newbridge on Wye 0158; roadside near Bach Howey viaduct 1042 & 1141; garage forecourt, Ridgebourne, Llandrindod Wells 0560; Stanner Rocks 2658.

T. exacutum Markl.

Pathside, Newbridge on Wye 0158 and Stanner Rocks 2658.

T. expallidiforme Dt.

Cwm Cottage near Clyro 2024 MV Marsden .

T. fasciatum Dt.

Grassland, Llanelwedd Rocks 0452; roadside near Bach Howey Viaduct, Llanstephan 1042; Dolyhir 2458; Stanner Rocks 2658.

T. hemicyclum Hayl.

Bank of the Bach Howey, Llanstephen 1042; road verge, Dolyhir 2458 and Stanner Rocks 2658.

T. horridifrons Railonsaia

Stanner Rocks 2658.

T. insigne M.P. Chr. & Wiinst.

Grassland, Dolyhir 2458 and Stanner Rocks 2658.

T. laeticolor Dt.

Roadside verge, near Bach Howey Viaduct, Llanstephan 1042 & 1141.

T. laticordatum Markl.

Garage forecourt, Llandrindod Wells 0560; Clyro 2143 MV Marsden; roadside verges at Dolyhir 2458 and Stanner 2658.

T. linguatum M.P. Chr. & Wiinst.

Grassland, Stanner Rocks 2658.

T. lingulatum Markl.

Road verges, near Bach Howey Viaduct, Llanstephan 1042 & 1141 & riverside nearby 1042; Stanner Rocks 2658.

T. necessarium H. Ollg.

Pasture, Cefnllys 0861.

T. oblongatum Dt.

Roadside and riverbank, near Bach Howey Viaduct, Llanstephan 1042.

T. pachymerum Hagl.

Stanner Rocks 2658.

T. pannucium Dt.

Llanelwedd Rocks 0452.

T. pannulatiforme Dt.

Grassland, Cefnllys 0861; riverbank near Bach Howey Viaduct, Llanstephan 1042 & nearby roadside 1141 and Stanner Rocks 2658.

T. pannulatum Dt.

Grassland on road verge, Dolyhir 2458 and Stanner Rocks 2658.

T. pectinatiforme H. Lb.f

Stanner Rocks 2658.

T. piceatum Dt.

Roadside, Llanstephan 1141 and grassland, Stanner Rocks 2658.

T. rhamphodes Hagl.

Clyro 1944 MV Marsden and Stanner Rocks 2658.

T. sagittipotens Dt. & R. Ohisen

Roadside verge, Dolyhir 2458.

Hieracium

T. sellandii Dt.

Bank of Bach Howey, Llanstephan 1042 & nearby roadside 1141 and grassland, Stanner Rocks 2658.

T. sublaeticolor Dt.

Roadside, Dolyhir 2458.

T. sublongisquameum M.P. Chr.

Grassland, Stanner Rocks 2658.

T. undulatum H. Lb.f & Markl.

Stanner Rocks 2658.

T. xanthostigma H. Lb.f

Grassy places, Dolyhir 2458 and Stanner Rocks 2658.

Nipplewort

Lapsana communis ssp. *communis*

A common weed of shady banks, gardens, arable fields, tips, stream banks, rock ledges, scree and woodland rides, wherever moist, nutrient-rich soil is exposed. In all squares except 86, 87 and 96SW.

Marsh hawk's-beard

Crepis paludosa

A northern sp. close to the southernmost edge of its range in Radnor. It favours wet, shaded soil and rocky slopes by streams and in woodlands. Cwm Elan c96 (1918 MsFlR); by Wye near Llanfaredd c05SE (AEW and JAW, 1947 *Trans Rad Soc* **XVII**, 9); Llanbister c17SE JAW (*Rep Bot Exc Club 1929*, 121); damp shaded slope by R. Ithon above Llanbadarn Fynydd 0978 CAC; Water-break-its-neck, Llanfihangel-Nant-Melan 1860 (MsFlR 1948). Only recently seen in the Llanbadarn Fynydd site.

Smooth hawk's-beard

Crepis capillaris

Common on roadside verges, tips, arable fields, streamsides, churchyards and hedgerows. In all squares except 86NE & NW, 87SW, 96SW, 15NW, 16SW, 17NE, 18SW, 25NW and 27NW.

*Beaked hawk's-beard

Crepis vesicaria ssp. *haenseleri*

Recorded from 26 in FPW, the source of this record could not be traced. A colonist of road verges it has not been found by this survey.

Mouse-ear hawkweed

Pilosella officinarum

A common plant of short turf on dry, sunny banks, in fields, on rock outcrops and quarry spoil. The ssp. have not been separated. FPW notes sspp. *micradenia*, *officinarum* and *trichosoma* as being recorded from Radnor. The sp. is in all squares except 86NE & NW, 87SW and 24NW.

Pilosella praealta

An alien hawkweed established on a wall top, Knucklas c2574 JR Palmer (1967 det. CW, *Watsonia* **8**, 309).

*Fox and cubs

Pilosella aurantiaca ssp. *carpathicola*

A widespread introduction, persisting close to habitation amongst short turf on banks and mowed grass, such as on roadside verge, E end Rhayader 9768 DPH; verge near station, Llandrindod Wells 0561 DPH; churchyard and waste ground, Gladestry 2355 ACP; roadside bank, Evenjobb 2662 ! and ACP; tip and roadside bank, Knighton 2872; roadverges, Presteigne 3164, etc.

Hawkweeds

Hieracium spp.

This account is of a very preliminary nature. The Hieracia of Radnor require detailed study.

H. sublepistoides

A plant of roadsides and waste places, collected on railway bank Aberedw 04 WC Barton (1920 NMW) and from Boughrood 13 GC Druce (1931 OXF).

H. cinderella

Another wayside sp. collected at Rhayader 96 CW, det. PDS and CW (1954, Hb. C West). Its British distribution centres on Herefordshire and Brecknock and this sp. is likely to be more widespread than the single record suggest.

H. stenstroemii

A hawkweed of rocky places in SE Wales and adjacent parts of England. The Radnor population lies at the NW edge of its range. By the Wye, Boughrood 13 PDS det. PDS and CW (1953 CGE); Aberedw Rocks 04 HJR (1909 CGE); railway embankment, Aberedw 04 AEW (1927 NMW).

H. vulgatum

On rocky banks near Aberedw 04 AEW (1927 NMW) and Stanner Rocks 2658 ND Simpson det. PDS and CW (1937 Hb. Simpson).

H. submutabile

On railway bank, Aberedw 04 WC Barton (1920 NMW); Boughrood 13 AL det. PDS and CW (1907 Hb. JE Marshall). The distribution of this species is centred on S Wales and further populations might be expected.

H. acuminatum

A hawkweed of rocky places and probably the commonest sp. in Central Britain. Near Aberedw Rocks 04 PDS (1953 CGE); grassy bank between road and railway, Aberedw 04 AL det. PDS and CW (1889 CGE); railway embankment, Builth Road 05 AEW det. PDS and CW (1929 NMW); Lovers' Leap, Llandrindod Wells 0560 WC Barton (1916 NMW); Cwmhir 07 WH Painter (undated in Pugsley 1948 pg. 217); Boughrood 13 CW det. PDS and CW(1952 CGE); Water-break-its-neck, Llanfihangel-Nant-Melan 16 AL det. PDS and CW. (1908 CGE); Stanner Rocks 2658 JEL (1951 NMW); a record of this sp. from 26 by EG Philp is held by BRC (no date).

H. subamplifolium

The only record is from Boughrood c13 GC Druce (no date given in Pugsley 1948 pg. 205).

Compositae

H. diaphanum

Widespread in rocky places. A variable and common sp. in central Britain. Bank of Upper R. Elan 87 or 97 AL det. PDS and CW (1888 CGE); Rhayader 96 CW det. PDS and CW (1954 CGE); Aberedw Rocks 04 PDS det. PDS and CW (1953 CGE); Boughrood c13 AL det. PDS and CW (1907 CGE) Stanner Rocks 2658 AL det. PDS and CW (1902 CGE).

H. diaphanoides

Collected only from Stanner Rocks 2658 JA Hyde det. PDS and CW (1927 NMW), an outpost of a predominantly northern species.

H. schmidtii

A single record from Aberedw 04 AL det. PDS and CW (1888 KEW).

H. lasiophyllum

An attractive large-flowered sp. of rock outcrops. "Rhayader waterfall" (probably rocks by Wye at 9667) AL det. PDS and CW (1891 BM); Aberedw Rocks 04 collected by AL det. PDS and CW (1904 CGE), WR Linton (1907 NMW), HJR (1909 NMW) and AEW (1927 NMW); Craigau Rocks, New Radnor "(probably Creigiau Rocks in Harley Dingle 1963) AL det. PDS and CW (1895 CGE); Stanner Rocks 2658 D Combes det. PDS (NCC records); Burfa Camp, Evenjobb 2861 ACP det. PDS.

H. carneddorum

The only record from rocky outcrops, Burfa Camp, Evenjobb 2861 ACP det. PDS.

H. vagense

The rocky banks of the Wye in the gorge below Llandeilo Graban is the locus classicus for this sp.. Erwood 04 in both Brecon and Radnor AL (1887 and 1907 NMW); Boughrood 13 HJR (1907 CGE) and upstream in 14 PDS (1954 CGE); Llanstephan 14 CW (1954 CGE).

H. argenteum

Probably surviving on rocks in the Elan Valley though the sites from which the following records were made may have been flooded by the construction of the Elan Valley Reservoirs. Rocks by the Afon Calettwr near the Elan confluence 87NE FJ Hambury det. PDS and CW (1889 BM); Elan Valley AL (1902 in Pugsley 1948). The record from 14 in FPW is probably an error.

H. subcrocatum

Collected from Afon Calettwr - Elan confluence in the Upper Elan Valley 86NE FJ Hambury det. PDS and CW (1890 CGE) and a road bank near Builth Road 05 (collector unrecorded) det. PDS and CW (1900 CGE), This common sp. of upland Britain might be expected elsewhere in Radnor.

H. strictiforme

Collected by AL in the Elan Valley c96 (1902 NMW).

H. reticulatum

Like the last sp. it was collected by AL in the Elan Valley c96 det. PDS and CW (1902 York Museum). Confirmation that it still exists is desirable, as this is one of only two Welsh records for this predominantly Scottish hawkweed.

H. vagum

A hawkweed of rocky outcrops. Near the River Elan between Cwm Elan and Allt Goch 96 (collector unknown) det. PDS and CW (1892 CGE, site probably now flooded by Elan Valley Reservoirs); Aberedw 04 AL det. PDS and CW (1907 York Museum); Wye banks Aberedw 04 E Flinton det. CEA Andrews (1894 NMW); near Erwood 04 AL (1906 NMW); Boughrood 13 CW det. PDS and CW (1954 CGE).

H. perpropinquum

Widespread on grassy banks, roadsides and woodland glades. Hedgerows near Rhayader 96 (collector unknown) det. PDS and CW (1892 CGE); Llandrindod Wells 06 WC Barton (date unknown in Pugsley 1948 pg. 301); grassy bank by Wye above Boughrood 14 CW det PDS and CW (1954 Hb. C West); in beech and oak wood 3 miles SW of Knighton 26 VS Summerhayes det. PDS and CW (1957 CGE); road bank 1.25 miles east of Knighton 37 collector unknown det. PDS and CW (1910 CGE).

H. umbellatum ssp. umbellatum

Roadside bank 1 ml E of Crossgates c0964 CEA Andrews (1955 Proc BSBI **2**, 34); bank of R. Wye near Boughrood 13 ECW det. CEA Andrews (1938 NMW); Boughrood 13 CW det. PDS and CW (1954 CGE); overgrown track, Brilley Mountain, Michaelchurch on Arrow 2349 ACP. This is the common ssp. in southern Britain.

Ssp. bichlorophyllum

Occurred at Rhayader 96 AL det. PDS and CW (1886 CGE); Boughrood 13 CW det. PDS and CW (1954 Hb. C West); Rhosgoch c14 HNR (date unknown in Pugsley 1948). This ssp. is confined to SW England, Wales and Ireland.

H. uiginskyense

On rocky streambanks. Pont ar Elan and Allt Goch, Elan Valley c86 and 97 AL (1889 CGE); Boughrood 13 AL (1907) and BA Miles (1957) both in CGE. An endemic sp. occurring in mid Wales, central and western Scotland and in two sites in Ireland.

H. substrigosum

Collected only from near Rhayader 96 AL det. PDS and CW (1888 Hb. Melville), this hawkweed of rocky stream banks, almost exclusive to Wales, might be expected elsewhere in Radnor.

H. placerophylloides

A single specimen from by the rocky bank of the Afon Elan near Pont ar Elan 97 and Allt Goch, Elan Valley c86 and/or 96 AL (1889 BIRM).

H. scabrisetum

A hawkweed of rocky places mostly by streams and largely confined to Wales. Riverside, Cwm Elan, Elan Valley 96 AL det. PDS and CW (1896 NMW, site probably flooded by construction of Elan Valley reservoirs); banks of Wye, Aberedw 04 EF Linton (undated specimen NMW); Llanstephan 14 AL det. PDS and CW (1907 BIRM).

Alisma

H. calcaricola

Only known from roadside 2 miles S of Llanbister c1071 CEA Andrews (1955 *Proc BSBI* **2**, 34).

Monocotyledones

Water-plantain family Alismataceae

Water plantain

Alisma plantago-aquatica

In the shallow-water and mud on the edge of slow flowing rivers and streams, by ponds and lakes mostly in the lowlands.

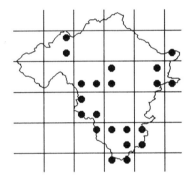

Flowering rush family Butomaceae

Flowering rush

Butomus umbellatus

Reported from Llanbwchllyn 1146 by Lewis Davies in *Radnorshire*, a book in the Cambridge County Geography series published 1920, pg. 58, along with royal fern. That neither plant has ever been reported from this site by numerous visiting botanists again must put this record in doubt. Rare beside R. Wye, Cabalva, Clyro 2445 DRD et al.

Frogbit family Hydrocharitaceae

*Canadian pondweed

Elodea canadensis

A relatively recent addition to Radnor's flora, it occurs in a number of pools on the larger rivers and in ponds and ox-bows in the lowlands. In view of the recent spread and now widespread occurrence of similar looking spp., a reappraisal of this sp. and its relatives is required. 9 sites.

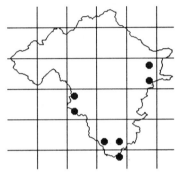

*Curly water-thyme

Lagarosiphon major

Probably introduced along with white water-lily to the quarry pool in the E quarry, Llanelwedd 0551. Noted in 1980, the water now seems too turbid to support submerged aquatic plants.

Arrowgrass family Juncaginaceae

Marsh arrowgrass

Triglochin palustris

Frequent in unimproved pastures and on upland commons in the wettest of basic peaty flushes. 35 sites.

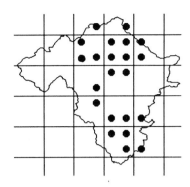

Pondweed family Potamogetonaceae

Broad-leaved pondweed

Potamogeton natans

Widespread in pools, lakes and slow flowing ditches and streams with shallow to deep water.

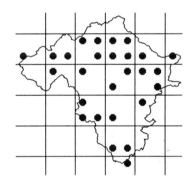

Bog pondweed

Potamogeton polygonifolius

Common in peat-bottomed streams, in pools, ditches, ponds and lakes in the uplands and in peaty or organic-bottomed sites elsewhere, even if very shallow.

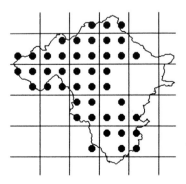

Potamogetonaceae

[Shining pondweed

Potamogeton lucens]

No Radnor records. It was reported by BA Seddon from Llanbwchllyn 1146 in 1964 (*Nat. Wales* 9, 5) but the specimen was later redetermined by him as *P. praelongus* (1967 *Proc BSBI* 6, 397). Unfortunately a number of authors of reports on Llanbwchllyn have overlooked this redetermination.

Red pondweed

Potamogeton alpinus

A record of this sp. appears in Sinker et al. (1985) from tetrad F in 27, near Llangunllo. The source of this record could not be traced and its presence in the county requires confirmation.

Long-stalked pondweed

Potamogeton praelongus

A scarce sp. growing submerged in the larger lakes. Llanbwchllyn 1146 (noted by AL in 1886 BM and by BA Seddon in 1964 in *Proc BSBI* 6, 397, it was not found in 1977 (Thurley, SS MSc thesis, Univ of Wales) but was refound by DRD & IDS in 1990); reported as subfossil seeds from the peat 2.5 m below surface of Rhosgoch Common 1948 Bartley, DD (1960a); Llynheilyn 1658 JAW (1938 MsFlR), and refound by NY Sandwith 1945 (BM) and G Taylor 1948 (NMW). Could not be refound by BA Seddon in 1964 (*Proc BSBI* 6, 397) and there are no other recent records.

Perfoliate pondweed

Potamogeton perfoliatus

Very rare. Reported from the R. Ithon near Disserth 0459 CAC and from Llynheilyn 1658 (1945, record on card index BM, recorder unknown).

Flat-stalked pondweed

Potamogeton friesii

A record of this sp. appears in Sinker et al. (1985) from tetrad K in 27 near Llangunllo. The source of this record cannot be traced and its presence in the county requires confirmation.

Blunt-leaved pondweed

Potamogeton obtusifolius

Known only from Llanbwchllyn 1246 DRD & IDS in 1990, a pool NW of Gogia on the Begwns, Llowes 1643 and a pond near Upper Weston, Llangunllo 2070 PR (1964 conf. JE Dandy, BM).

Small pondweed

Potamogeton berchtoldii

The commonest of the fine leaved sp., but still very scarce in nutrient-rich lakes and pools. Collected by a number of people from Llandrindod Lake 0660 viz C Bailey 1890, A Wedgewood 1905 (MBH), SJ Chamberlain 1912 (UCNW), ES Todd 1939 (BM), all det. or conf. JE Dandy and G Taylor and described as "often abundant" there in MsFlR. It has not been seen there for many years.and no submerged aquatics survive following the deliberate dumping of superphosphate in the lake by the local authority in the early 1950's (Llandrindod Wells Lake, Powys. Harrison et al. unpub. rep. UWIST, Cardiff 1981). Pool in old quarry, Llanbadarn Fynydd 0481, CP & PJP; small deep pool near Glan Gwy Farm, Glasbury 1538 ACP (det. JE Dandy); Llanbwchllyn 1146 SS Thurley; hill stream, Painscastle 1448 ACP (*Nat Wales* 16, 211); pool W of Gogia, The Begwns, Llowes 1643 ACP (det. NTH Holmes); Llynheilyn 1658 ACP et al. (conf. NTH Holmes. Also noted here in MsFlR).

Curled pondweed

Potamogeton crispus

Frequently seen growing on the bed of slow-flowing reaches of the larger, more nutrient-rich rivers such as the Ithon and lower Wye and in ponds and deep ditches in the lowlands. 17 sites.

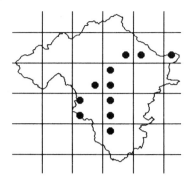

Fennel pondweed

Potamogeton pectinatus

Only known from Llanbwchllyn 1246 DRD & IDS in 1990 and the pond by Hindwell Farm, Walton 2560 JFH and PCH (1956 *Proc BSBI* 2, 380).

Horned pondweed family Zannichelliaceae

Horned pondweed

Zannichellia palustris

Favouring nutrient-rich ponds and lakes it is known from two sites. Llynheilyn 1658 PJP, ACP and IDS (conf. NTH Holmes and noted there for the first time in 1985) and Hindwell Farm pond, Walton 2560 JFH and PCH (1956 *Proc BSBI* 2, 380).

Lily family Liliaceae

Bog asphodel

Narthecium ossifragum

A common sp. in wet upland peaty flushes, on moorland and in rough pastures and around acidic pools and flushes associated with lowland peatlands.

Hyacinthoides

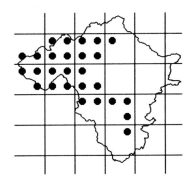

Meadow saffron
Colchichum autumnale

A rare sp. of unimproved pastures and haymeadows. Pasture near Coedgwgan Hall, Crossgates 0865 D Stephens; hayfield by lane NE of Boughrood 1339; field by the station, Boughrood 13 (AL in *Trans Woolhope Club* 1890-92, 192) but not seen for many years; steep pasture, Cwm Gwanon, Clyro 1944 MEM; field at top of hill, Llanstephan 14SW GP (1952); pasture and rough grassland, Dolyhir 2458.

Radnor lily
Gagea bohemica

Occurs in shallow, summer-droughted pockets of soil on S facing dolerite cliffs at Stanner Rocks 2658, its only British station. Originally unwittingly collected by RFO Kemp in 1965 in moss samples which were grown on for some time in Cardiff before a withered specimen was noticed and incorrectly identified as the Snowdon lily - *Lloydia serotina*. The discovery of a plant on Stanner Rocks in full bloom amongst many hundreds of vegetative specimens by the author in January 1975 confirmed the suspicion that this was a species of *Gagea* previously unknown in Britain. (see Rix and Woods 1981 for details).

The plant had presumably escaped detection in undoubtably the most heavily botanized locality in Radnor by its relative flower-shyness, its small size, the earliness of its flowers (Jan-April) and the annual loss of its leaves, usually before May and generally long before any higher plant botanist would consider visiting the site. Slater, FM (1990) reviews its biology. The common name of Radnor lily is proposed for the plant, instead of the recently concocted name of early star of Bethlehem.

*Pyrenean lily
Lilium pyrenaicum

Rarely naturalized in grassland. Wasteground, Frank's Bridge c15NW RCLH (1956 *Proc BSBI* 2, 379); by railway near Corner Farm, Llangunllo 2174 PR (1964); woodland edge Gwernaffel Dingle, Knighton 2770.

Bluebell
Hyacinthoides non-scripta

Common in woodlands, under bracken in the open and in unimproved upland meadows and pastures. Absent from some oakwoods through its sensitivity to heavy grazing and eliminated from former rough grazings by ploughing and reseeding in recent years. Recorded from all squares except 86NW and 87SW.

*Spanish bluebell
Hyacinthoides hispanica

Reported as an escape from cultivation only on a road verge near Gelli-cadwgan, Llanelwedd 0551 and at Clyro 24NW JAW (1958 record in NMW). These records may refer to the hybrid between this and the last sp. (see Paye, KW, *BSBI News* 47, 9) and require confirmation.

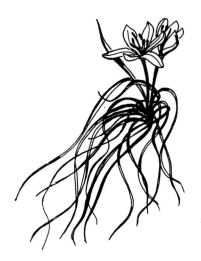

Radnor lily

Chives
Allium schoenoprasum

A large population exists on rocks and shingle in the winter flood zone of the R. Wye from Penddol Rocks, Llanelwedd 0352 to Maesllwch shingle bank, Glasbury 1840. First reported by members of the Woolhope Club in their *Transactions* for 1874, p. 7 from Penddol, and then by AL in 1886 (*Bot Rec Club* 1884-1886, 145-146) from near Llanstephan 14SW.

Ramsons
Allium ursinum

In damp woodlands on the more lime-rich soils and extending onto shaded riverbanks and roadsides. Frequent in the S and E. Possibly introduced in the N of the county where it seems to be associated with habitations.

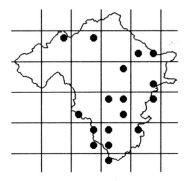

Wild onion — Allium vineale

A rare sp. of open, summer-droughted grassland. ORS block scree by R. Wye, Wyecliff, Clyro 2242 ACP & !; Stanner Rocks 2658 in abundance (known since 1853. AL *Trans Woolhope Club* 1853, 91).

Lily-of-the-valley
Convallaria majalis

Recorded from Cwm Elan c96 by MA Tooke (1879 *Sci Gossip* p. 111) and from 1 mile N of Cwm Elan "opposite Dol Folau" 9165 in the Elan Valley by the Woolhope Club in 1896 (*Trans Woolhope Club*, 1897, 179). Parts of this valley were subsequently flooded by the construction of reservoirs, though woodland survives in the Dol Folau area. One colony on inaccessible, wooded cliffs, Cerrig Gwalch, Rhayader 9370 IJL Tillotson et al.. Aberedw 04NE by JG Williams (1938 WFP), but no other recent records.

Iridaceae

Solomon's-seal
Polygonatum multiflorum

Doubtfully native, it grows on the Wye bank near Builth Road 0153 ACP, CP and PJP and was noted at Clyro c2143 by Rev JE Beckerlegge in 1946 (WFP). These records need to be confirmed since confusion can occur with the widely cultivated *P. x hybridum*.

Herb Paris *Paris quadrifolia*

An uncommon sp. of moist, shaded, somewhat base-rich woodlands. Most frequent on the boulder clay of mid Radnor, a large population also exists on peat in the willow carr around Rhosgoch Common 1948 RK. 10 sites.

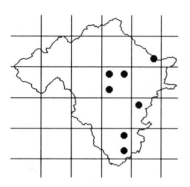

Daffodil family
Amaryllidaceae

Snowdrop *Galanthus nivalis*
Widespread but not common and possibly always an escape from cultivation. River banks, hedgerows, woodlands, field edges, churchyards. 10 sites omitting churchyards.

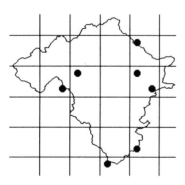

Wild daffodil
Narcissus pseudonarcissus ssp. *pseudonarcissus*

MsFlR reports Dr AH Trow as stating the wild daffodil to be "plentiful in the Mid Wye area". The authors had not, however, encountered it, not visiting the area in the spring. Apart from probably planted populations in churchyards, it is now very rare. Reported by NMW staff from a field near St Harmon 97SE, it elsewhere occurs in an orchard at Newbridge on Wye 0158 and a small field at Dolyhir 2458. In this latter site, cultivars have been planted in recent years. It also occurs in Daffodil Wood, Whitton 2767, DPH.

Yam family
Dioscoreaceae

Black bryony
Tamus communis

Frequent in hedgerows and open woodlands in the warmer and more base-rich valleys of the S and E. Absent in the uplands.

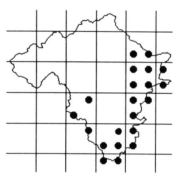

Iris family
Iridaceae

Stinking Iris
Iris foetidissima

Rarely naturalized and uncommon even in gardens. Wooded bank of the Wye near The Vicarage, Boughroood 1338; roadside hedge Beggar's Bush 2664 and hedge, Upper Thorn Cottage 2763. Both latter sites are near Discoed and noted by ACP.

Yellow Iris
Iris pseudacorus

Widespread in clayey flushes in unimproved meadows, by slow flowing streams and rivers, lake and pond margins.

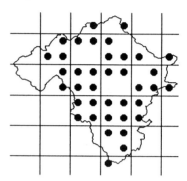

Rush family
Juncaceae

Hard rush *Juncus inflexus*

Not common and widespread only on the flushed boulder clay of central Radnor, where it is found in damp unimproved pastures and by flushes, streams, ditches and ponds. 27 sites.

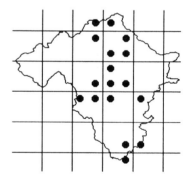

Soft rush *Juncus effusus*
The commonest rush and an aggressive colonizer of poorly-drained grassland where it germinates readily in poached areas. Elsewhere by rivers, lakes and ponds and in peaty flushes. Recorded in all squares.

Juncus

Compact rush

Juncus conglomeratus

Widespread but less common than the last, occurring in unimproved pastures, damp open scrub, heathland and in the wetter areas of the extensive common grazings.

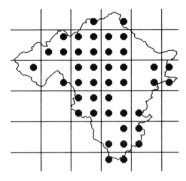

Heath rush

Juncus squarrosus

Common to abundant in the heavily sheep grazed, peaty turf of the uplands. Being unpalatable to grazing stock except in spring, grows at the expense of heather.

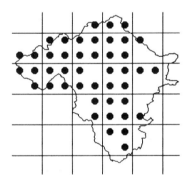

*Slender rush Juncus tenuis

This recent introduction has established itself on grassy paths and by roads where its ability to withstand trampling is remarkable. First noted on a roadside between Walton and Stanner c2659 by JAW in 1938 (MsFlR), it now occurs on a roadside, Aberceuthon, Llansantffraed Cwmdeuddwr 9566; on rocks by the Wye in the park, Rhayader 9668; roadside near Rhydfelin, Newbridge on Wye 0255 MEM; path by the R. Wye, Newbridge on Wye 0158; grassy path through a playing field, Llandrindod Wells 0560 and roadside E of Bryn, St Harmon 0171.

Toad rush

Juncus bufonius agg.

Common in muddy ditches, on damp tracks, by ponds, pools and streams and as a weed of arable land, especially on the heavier soil. The segregates have not been recorded in this survey. Recorded in all squares except 86NW, 87, 04NE, 16SE, 18SW, 26SW and 36NW.

Blunt-flowered rush

Juncus subnodulosus

Noted in MsFlR from a "bog near Llandrindod Wells Lake" c06SE and by HN Ridley in 1880 (*J. Bot.* **19**, 178) from Rhosgoch Common 1948 and a marsh between there and Clyro 14 or 24.

Bulbous rush

Juncus bulbosus agg.

Common in peaty flushes, by streams and ditches, on damp, open soil of tracks and submerged in pools in the uplands. *J. kochii*, the common sp. in the uplands, has not always been separated by recorders, so only records of the aggregate are given. A fine-leaved form is frequently found submerged in peaty pools. The agg. is found in all squares except 04SE, 13, 16NE, 26SE, SW & NW, 27SW & NW, 36 and 37.

Sharp-flowered rush

Juncus acutiflorus

A common plant of wet flushes in the uplands and in unimproved pastures, fens and on lake margins. Many populations fail to set seed and may be hybrids with the next species.

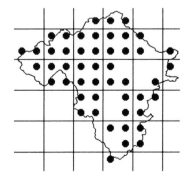

Jointed rush

Juncus articulatus

Often in similar habitats to the last but extending onto more mineral soil such as on damp tracks and in gateways, by ponds, ditches, streams and rivers. Recorded in all squares except 14NW, 16SE, 26SW and 27NW.

Field wood-rush

Luzula campestris

Common in unimproved pastures and meadows, in short turf on dry, neutral to acidic roadside banks, churchyards, and lawns. In all squares except 86SE.

Heath wood-rush

Luzula multiflora

Widespread in the uplands in damp peaty grasslands and moorlands about the edges of flushes and on stream banks.

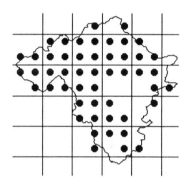

Great wood-rush

Luzula sylvatica

Widespread on woodland and river banks and shady cliffs in sites subject to little or no grazing. Occasionally found also on the more lightly grazed moorlands.

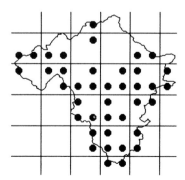

Hairy wood-rush

Luzula pilosa

Frequent on shady woodland and hedge banks throughout.

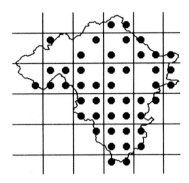

Southern wood-rush

Luzula forsteri

Confined to freely-drained, somewhat shaded but warm, sheltered banks in the Wye Valley between Aberedw and Clyro. Noted by AL in "the wooded glen" at Aberedw 04NE (1884 *Trans Woolhope Club* pg. 193); R. Wye bank above Boughrood 13NW RM Harley and AB Emden (1966 *Proc BSBI* **7**, 198 and still present); frequent on road verges of former railway about the Bach Howey viaduct, Llanstephan and Llandeilo Graban 1042 MEM and 1043; on lane bank below church, Llanstephan 1241; riverside cliff ledges, Wyecliff, Clyro 2242 ACP & !.

Grass family Gramineae

Wood fescue

Festuca altissima

Known only from the precipitous wooded cliffs of the Bach Howey gorge below Trewern Hill, Llanstephan 1143 where it occurs in some quantity. First noted there by the author and MP in 1987.

Giant fescue

Festuca gigantea

Absent from the NW uplands, it is frequent in damp woodlands, hedgerows and on streambanks in the valleys on the more nutrient-rich soils.

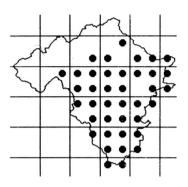

Meadow fescue

Festuca pratensis

A widespread but uncommon sp. of unimproved pastures, meadows, river banks and road verges on the damper clay soils.

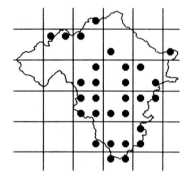

Tall fescue

Festuca arundinacea

Possibly native on the banks of the larger rivers and in a few damp pastures. It appears to have been a recent colonist or perhaps a deliberately sown plant on trunk road verges.

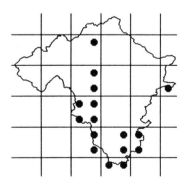

Red fescue

Festuca rubra agg.

The fine-leaved fescues grouped into this and the next aggregate require a critical review. Red fescue is a common sp. of well-drained, sunny banks, hedgerows, cliffs, scree, lawns, churchyards and, with its rhizomes, at times becomes a weed of flowerbeds. Recorded from all squares.

Sheep's-fescue

Festuca ovina agg.

The members of this aggregate, including the fine-leaved fescue *Festuca tenuifolia*, have not been separated. Sheep's-fescue is common on dry, sunny banks by tracks, hedges, roads, on cliff and scree, in old quarries and as a major component of the heavily grazed grasslands on the upland commons. Recorded from all squares except 27NW, 36 and 37.

Perennial rye-grass

Lolium perenne

Common in meadows, pastures, road verges, river banks and tips on the more nutrient-rich soils. The agriculturalists have developed many strains of this grass and it is by far the most widely sown sp. to establish grass leys for hay and silage. It is also sown on disturbed road verges, on playing fields and lawns. Whilst a native population might once have been present in unimproved meadows and pastures and on river banks, it is no longer possible to distinguish it from the product of seeds spread from introduced populations. Recorded from all squares except 86NE & NW, 87SW, 96SW and 16SW & NW.

*Italian rye-grass

Lolium multiflorum

Regularly sown with the last sp. to form grass leys. Though fast growing, it is less persistent and in consequence, is much less frequently discovered naturalized, but has been noted on road verges and tips in 96SE, 97SW, 05SW, 13NW, 24SW, 27SW and 36SW.

Brachypodium

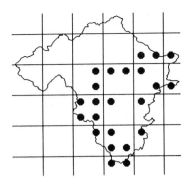
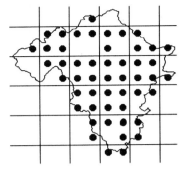
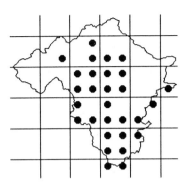

especially where extensive rock outcrops occur. Rare in woods away from rivers.

Lesser hairy-brome

Bromus benekenii

Reported only once from wooded rocks at Aberedw 04NE by AL in 1886 (BM).

Upright brome

Bromus erectus

Rare in grassland below Stanner Rocks 2658. First noticed there in 1988. No other Radnor records.

*Rye brome

Bromus secalinus

Only known from a cornfield at Gaer near Llansantffraed-in-Elvel 0854 HNR (1880 BM). No more recent records.

Meadow brome

Bromus commutatus

Known only from a field border at Aberedw c 04NE AL (1886 *Bot Loc Rec Club 1884-1886*, 146). No other more recent records from Mid Wales.

Soft brome

Bromus hordeaceus ssp. *hordeaceus*

Common and widespread on roadside verges, waste ground and in hay meadows. In some upland hay-meadows, especially in fields heavily grazed by sheep in the spring and cut late, this grass may dominate large areas, shedding its seed before the hay is cut.

*Slender soft-brome

Bromus lepidus

Reported E of Abbeycwmhir in 07 by DE de Vesian in 1956 (*Proc BSBI* **2**, (1957), 381). The source of the 26 record in the *Atlas Brit Fl* has not been traced.

False brome

Brachypodium sylvaticum

Common in scrub and woodlands where it favours the more lime-rich, freely-draining soils of the S and E of Radnor. Occasional on cliffs and in hedgerows. As a meadow plant it is confined to the limestone at Dolyhir 2458.

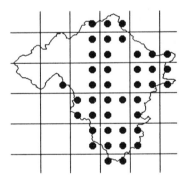

Tor-grass

Brachypodium pinnatum

Appears on a BRC card compiled at Stanner Rocks 2658 by RSR Fitter in 1955. There are no other records.

Bearded couch

Elymus caninus

Uncommon but widespread on the wooded banks of the major rivers,

Common couch or Squitch

Elymus repens

Common on cultivated ground and a pernicious weed in gardens, extending into nearby hedgerows, roadsides and on tips.

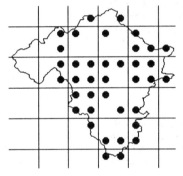

*Bread wheat

Triticum aestivum

Rarely cultivated, except in the lower Wye Valley, it occurs regularly on tips and about farmyards, probably from spilt stock feed. Not mapped.

*Barley

Hordeum spp.

As with the bread wheat, barley is cultivated only in the S & E of the county but appears on tips, roadsides and about farmyards where grain for domestic animals is spilt or discarded. The 2 and 6-rowed barley were not separated in the early years of the survey. All recent records are of 2-rowed barley *H. distichon*. Not mapped.

Gramineae

Wall barley
Hordeum murinum

Very rare. A single plant on the Wye Bridge Llanelwedd 0451 in 1992 and noted in 06 VG in 1952.

Cultivated oat *Avena sativa*
Very little is cultivated but, as with wheat and barley, spilt and discarded animal feed ensures a regular crop of oats on tips, disturbed roadsides and about farmyards. Not mapped.

Downy oat-grass
Avenula pubescens

A rare sp. of long undisturbed grassland. Pasture Rhiw'r-ogo-Fach near Crossway 0757; roadside bank and old hay meadows, Llanbadarn Fynydd 0978; meadow, Painscastle 1645; Glascwm churchyard 1553 PJP & CP; pasture, Far Hall, Dolau 1468; grassland below Stanner Rocks 2658.

False oat-grass
Arrhenatherum elatius

Common on road verges and railway banks, churchyards and waste ground, in hedgerows and field margins in sunny, freely-drained sites, especially in the lowlands. Recorded in all squares except 86NE & NW and 87SW, the sspp. *elatius* and *bulbosum* both appear to be common and have not been recorded separately.

Yellow oat-grass
Trisetum flavescens

Widespread, but not common, in short turf on well-drained, somewhat basic soil of churchyards, ancient monuments, road verges, unimproved hay meadows and pastures.

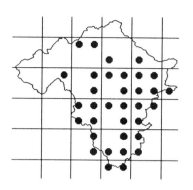

Tufted hair-grass
Deschampsia cespitosa ssp. *cespitosa*

Common in damp woodlands and pastures, on stream and river banks and in flushes in the uplands. Recorded from all squares except 86NE.

Wavy hair-grass
Deschampsia flexuosa

A common grass of moorland and freely-drained acidic woodland and hedge banks. Recorded in all squares except 15SW.

Early hair-grass
Aira praecox

Common on open freely-drained soil of sunny banks, rock outcrops, scree, quarry spoil and tracksides. Recorded in all squares except 87SW, 05NW, 06NE, 08SE, 13NE, 16NE, 18SE, 24NW, 26SE, 27NW and 36SW.

Silver hair-grass
Aira caryophyllea

In similar sites to the last but less frequent.

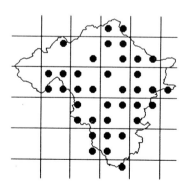

Sweet vernal-grass
Anthoxanthum odoratum

Common in meadows, pastures, roadside verges, churchyards, around flushes and on the edge of heathland in the upland and in open woodland lawns on damp to fairly dry sites. Reported from all squares.

Yorkshire-fog
Holcus lanatus

A common weed of cultivated ground and in pastures, haymeadows, hedgebanks, marshes and upland flushes. Reported from all squares except 86NW and 87SW.

Creeping soft-grass
Holcus mollis

A common grass of well-drained, acidic woodland soils, hedgebanks, unimproved pastures and heathland. Particularly common under bracken in the hills. Recorded from all squares except 86NW & NE and 87SW.

Brown bent & Heath bent
Agrostis canina and *A. vinealis*

The records of these two spp. have been amalgamated as they were not separated in the early years of the survey. The former is common in damp pastures, marshes and on pool margins, whilst the latter favours slightly better drained sites such as around rock outcrops in flushes and on banks at higher elevations. In combination they have been recorded from all squares except 13NE, 15SW, 18SE, 24SW, 26SW & NW, 27SW & NW and 36NW.

Common bent
Agrostis capillaris

The common grass of heavily-grazed, upland pastures and commons on freely-drained soils and widespread in lawns, roadverges, hedgerows and open woodlands. Reported from all squares.

Black bent
Agrostis gigantea

A scarce or overlooked sp. of disturbed ground. Noted on waste ground by a brook in Llandrindod Wells 06SE by AEW (*Nat Wales* NS **1**, 4 in supplement) and on waste ground behind foundry, Presteigne 3164.

Creeping bent

Agrostis stolonifera

Common on river and stream banks, by ditches, on field margins, roadside verges and hedgerows and in damp grassland. Recorded from all squares except 86NW & NE and 87SW.

Timothy

Phleum pratense ssp. *pratense* and ssp. *bertolonii*

The sspp. have not been separated, though both appear to occur frequently, often together and with intermediate forms. Common in meadows, pastures and roadverges, especially on the heavier clay soils. Occasionally sown and probably frequently introduced accidentally by feeding hay containing timothy to stock. Recorded from all squares except 86, 87SW, 97NW, 15SE and 24NW.

Meadow foxtail

Alopecurus pratensis

Common in unimproved meadows, pastures, on roadside verges, in churchyards, railway banks and waste ground on the lower ground. Scarce in the uplands. Recorded from all squares except 86, 87, 96SW, 97SW & NW, 07NW, 13NE, 14SW, 27NW and 36NW.

Marsh foxtail

Alopecurus geniculatus

Common along the edges of rivers, streams, lakes and pools, in wet ditches and especially in the seasonally-wet ruts of tracks and about gateways. Recorded from all squares except 86SW & NW, 87SW, 96SW & SE, 97NW, 13SW, 25NE, 26SW and 27SE.

Reed canary-grass

Phalaris arundinacea

Widespread on heavy, ill-drained soil by streams, rivers, lakes and ponds, in ditches, marshes, on roadside verges and in woodland glades. Recorded from all squares except 86, 87SW, 96SW, 97SW, 16SE, 18SW and 26SW.

Alopecurus

*Canary-grass

Phalaris canariensis

One of the few alien plants germinating from spilt and waste bird seed that turns up regularly on tips. Nantmel rubbish tip 0165; Clyro rubbish tip 2143.

Wood millet

Milium effusum

A rare grass of woodland banks on the more nutrient-rich soils. Skynlas Dingle wood, Glasbury 1639 LMP; Bach Howey gorge, Llanstephan 1143 ! & MP; banks of the R. Aran, Llanbister 1571 G Pell, 1573 E Matheson, 1575 M Mosey; Bettws Dingle Wood, Clyro 2246 DEG & JSB; Cabalva Dingle Wood 2346 DEG & JSB; Worsell Wood, Stanner 2657; Gwernaffel Dingle Wood, Knighton 2771 IHS et al.; Stocking Farm Wood, Presteigne 3166 NCC woodland survey.

Common reed

Phragmites australis

A rare sp. of bogs and lake margins. Edge of Gwynllyn, Llansantffraid Cwmdeuddwr 9469; basin mire, The Bog, Newbridge on Wye 0157; marsh, Colwyn Brook, Carneddau Farm, Llansantffraid-in-Elvel 0755 NCC survey; marsh near Llwynbarried Hall, Nantmel 0165 FE & DRD; shore of Llanbwchllyn, Painscastle 1146; willow carr, Rhosgoch Common 1948 and by Presteigne Football Ground, 3264 DPH.

Heath-grass

Danthonia decumbens

A widespread sp. of heathy grassland, moorland edges, short unimproved dry or wet-flushed turf. Recorded from all squares except 97NW, 04, 14NW, 24NW, 26NW & NE, 27SW & SE and 36SW.

Purple moor-grass, Feg or Disco grass

Molinia caerulea

Abundant on the wet, acidic peatlands covering the rounded ridges and valley floors of the western uplands. Its tussocks make walking difficult and earn it a recently coined local name of "disco grass" - presumably because the walker staggering along looks not unlike a dancer in a disco. In the E it occurs in marshes and damp hollows.

Agricultural drainage has recently reduced the area of this grass, particularly in parishes such as St Harmon. Sitka spruce grows particularly well in situations favoured by *Molinia* and forestry plantations have further reduced its area. The grass is still cut late in the year in the Elan Valley to provide "rhos hay". It is also burned in the spring to stimulate palatable shoots for livestock. The ssp. have not been separated though both occur in Radnor (WFP)

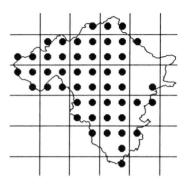

Mat-grass

Nardus stricta

Unpalatable to stock, it grows extensively on the heavily-grazed, acidic, peaty upland commons, replacing dwarf shrubs as they are eliminated by fire and grazing. It is a colonizer of ground following the erosion of blanket bogs and is frequent in peaty flushes. Elsewhere it occurs on acidic road verges and in unimproved pastures. Recorded from all squares except 04SE, 13NE, 24SW, 25NW & NE, 26SW, 27SW & SE and 36.

*Maize   *Zea mays*

A rare casual of waste ground. Clyro tip 2143.

Arum family
Araceae

*Sweet flag   *Acorus calamus*

A rare introduction, long established by the lake near the house, Stanage Park 3371 ACP.

Lords-and-ladies, Adder's horns

Arum maculatum

Common in woodlands, hedgebanks, riverbanks and occasionally road verges on the more nutrient-rich soil types. Forms with white spadix are rare; noted from Gilwern Brook, Dolyhir 2356 ACP. Forms with spotted and unspotted leaves are common.

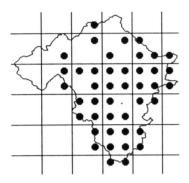

Duckweed family
Lemnaceae

Ivy-leaved duckweed

Lemna trisulca

A scarce sp. of the more nutrient-rich pools. Llanbwchllyn 1146 AL (1886 BM) and still present and pond near Tiled House, Knighton 2770 PR (1965).

Common duckweed

Lemna minor

Common on the surface of pools, lakes, marshes and slow-flowing ditches in the lowlands.

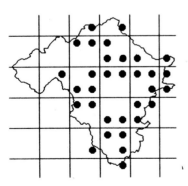

Bur-reed family
Sparganiaceae

Branched bur-reed

Sparganium erectum

Frequent on the margins of pools, lakes, rivers and in marshes and ditches in the lowlands. Rarer in the uplands. The sspp. have not been distingished in this survey. FPW notes ssp. *erectum* is widespread, whilst ssp. *microcarpum* is only recorded from 06. Ssp. *neglectum*, apparently recorded from Radnor, could not be localized.

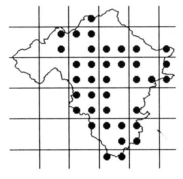

Unbranched bur-reed

Sparganium emersum

An uncommon plant of muddy pond and reservoir margins. Shore of Garreg Ddu Resr., near Cwm Coel, Elan Valley 9063; pool by R. Wye, Llanelwedd 0451; pool on common W of Penygraig, Llandeilo Graban 1045; Llanbwchllyn 1146 BA Seddon (1964 NMW records); pool NW of Gogia, Begwns, Llowes 1643 ! & JE Messenger; pool, Graig, Llanfihangel Rhydithon 1767 PJP & CP; pond, Warren House Farm, Knighton 2570 ACP; ponds near Llangunllo 2271 ACP.

Bulrush family
Typhaceae

Lesser bulrush

Typha angustifolia

A rare sp. known only from the reedswamp at Llanbwchllyn 1146 BA Seddon (1964 *Nat Wales* **9**, 69 and noted in NCC survey 1984) and in the marsh at the SW end of Rhosgoch Common 1948 BSBI exc..

Bulrush   *Typha latifolia*

Widespread but not common and probably introduced to many ponds. Occurs also in oxbows, marshes and on lake margins.

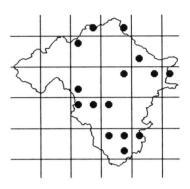

Sedge family
Cyperaceae

Wood club-rush

Scirpus sylvaticus

Frequent in marshy hollows in unimproved pastures, on river banks and in clearings in wet woodland on the clay soils of the Ithon Valley and by the Wye below Newbridge on Wye 0158 to the English border. Notably rare in many apparently suitable areas eg the Teme valley. 44 sites.

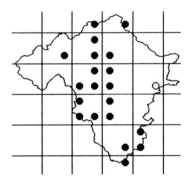

Bristle club-rush

Scirpus setaceus

An inconspicuous plant of wet soil in open, short turf of tracksides, grazed and stony flushes and streamsides.

Eleocharis

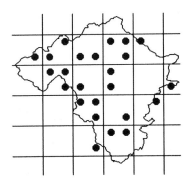

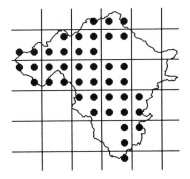

Floating club-rush

Scirpus fluitans

An uncommon plant of muddy pool edges and peatland pools. Gwynllyn, Llansantffraed Cwmdeuddwr 9468 (1969 NCC records); peaty pool, The Bog, Newbridge on Wye 0157; pools, Rhosgoch Common 1948 (NCC records); St Michael's Pool, Bleddfa 1869 ACP; pools Maelienydd Common 1270 and 1371 ! & FMS and Warren Bank 1470 PJP & CP, both in Llanddewi Ystradenni. Noted in MsFlR from moors near Llandrindod c06; pools, Llandegley Rhos 16SW and near Penybont c16SW.

Deergrass

Scirpus cespitosus ssp. *germanicus*

Common on heavily grazed and regularly burned blanket bog, where it may become the dominant sp.. Scattered on the more lowland peatlands and heathlands.

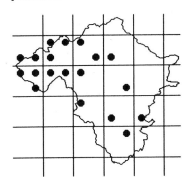

Common cottongrass

Eriophorum angustifolium

Common in the wetter parts of peatlands and in peaty pools.

Broad-leaved cottongrass

Eriophorum latifolium

A scarce sedge of calcareous, peaty flushes. Pentrosfa Bog, Llandrindod Wells 0559; peaty meadow, Portway, Rhosgoch 1948 NCC records, though first reported from Rhosgoch in 1843 (T Westcombe in letter to the *Phytologist* 1, 1844, 781); glade in alder carr, Lloyney, Clyro 2044 NCC records.

Hare's-tail cottongrass

Eriophorum vaginatum

Frequent on blanket bogs on the more level upland ridges and occasionally on lowland peatlands such as Cors y Llyn, Newbridge on Wye 0155 and Rhosgoch Common 1948.

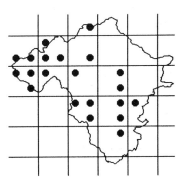

Few-flowered spike-rush

Eleocharis quinqueflora

An uncommon sp. of peaty, calcareous flushes, generally kept open by grazing. Margins of Gwynllyn, Llansantffraed Cwmdeuddwr c9469 AL (1886 *Bot Rec Club* 1884-1886, 146); Pentrosfa, Llandrindod Wells 0559; Pen Common, Llanbedr 1447 N Penford; The Begwyns 1643 ACP et al.; S of Ireland, Llanbedr Hill 1647; Portway Meadow by Rhosgoch Common 1948 NCC survey (noted on Rhosgoch Common by HNR in 1880, BM); peaty pool, Maelienydd 1270; by stream near New House, Llanbister 1575 ACP det. SG Harrison.

Needle spike-rush

Eleocharis acicularis

A rare sp. not seen recently. The origin of the 14 record in FPW has not been traced. JAW noted it on Llandegley Rhos 16SW in 1934 (MsFlR).

Common spike-rush

Eleocharis palustris

A common sp. of marshes and lake, pool and river margins.

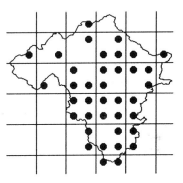

Many-stalked spike-rush

Eleocharis multicaulis

An uncommon spike-rush of wet peatlands. Flush S of Glogfawr by Garreg Ddu Resr., Elan Valley 9165; bog NE of Pont ar Elan, Elan Valley 97SW WE Warren (1959 *Proc BSBI* 5, 38 and still present); basin mire, The Bog, Newbridge on Wye 0157; muddy stream, Rhosgoch Common 1947 NCC survey and peaty flush in valley W of Stanky Hill, Beguildy 1576.

Cyperaceae

Great fen-sedge
Cladium mariscus

Probably extinct. BA Seddon in *Nat Wales* **9**, 69 reports finding a stand only a few feet in extent fronting the reedswamp on the N side of Llanbwchllyn 1146 in 1964. The Water Authority have raised the level of the lake on more than one occasion in recent years. This sp. was never reported again in many subsequent surveys. Subfossil remains are reported by Bartley, DD, (1960) in the basal peat deposits of Rhosgoch Common 1947.

White beak-sedge
Rhynchospora alba

Generally an uncommon plant of wet peaty flushes, but frequent in the Claerwen Valley in 8763, 8862 and 9061. Also by R. Elan above Aberglanhirin, Upper Elan Valley 8872; by Nant y Blwnbren, Pen-y-Garreg, Elan Valley c9167 RCH & VG (1956 *Proc BSBI* **2**, 380); W of Treheslog, Llansantffraed Cwmdeuddwr 9368 JH; N of Pont ar Elan, Upper Elan Valley 9071; basin mire, Cors y Llyn, Newbridge on Wye 0155; moorland near Bryngwyn 1850 J Sankey-Barker.

Greater tussock-sedge
Carex paniculata

The large tussocks are a feature of wet alder woods, valley mires, marshes and the more extensive flushes in the uplands.

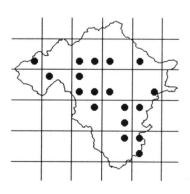

Lesser tussock-sedge
Carex diandra

Reported twice - from Llandrindod Wells c06SE by ES Todd (*Rep Bot Exch Club* 1919, 687) and Rhosgoch Common 1948 coll. DDB, det. AEW (1956 NCC records). A search should be made to assess the current status of the plant.

False fox-sedge
Carex otrubae

Scarce and known only from a damp hollow below ORS roadside bank and in wood W of Cwmbach, Glasbury 1639.

Spiked sedge   *Carex spicata*
Known only from Stanner Rocks 2658 where it was first found in 1974 by RW David (BRC records).

Prickly sedge
Carex muricata ssp. *lamprocarpa*

Widespread but not common on sunny and dry grassy banks, often by roads.

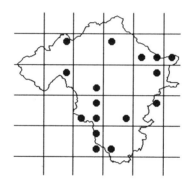

Grey sedge
Carex divulsa ssp. *divulsa*

Collected by DO Jones on Clyro Hill c14NE or 24NW pre 1950 (confirmed RW David, BRC records). Growing in similar habitats to the last species it may still occur in the county.

Sand sedge   *Carex arenaria*
Reported only from ballast by the side of the railway NE of Dolau Station 1568 ITE railway survey 1978 (Int. rep. NCC).

Broom sedge   *Carex disticha*
Scarce Newmead Farm, Llanelwedd 0553 SM Gooch; in calcareous, peaty flushes near Pentrosfa, Llandrindod Wells 0559 and from near Upper Hall Farm, Heyope 2274 JM Roper.

Remote sedge   *Carex remota*
Common on damp soil and muddy flushes in woodlands and on stream banks in somewhat shaded places. Reported from all squares except 86NE & NW, 87, 96NW & SW, 97NE & NW, 17NW, 16SE, 18SW and 26SW.

Oval sedge   *Carex ovalis*
Widespread on open, damp, heavy soil in meadows, on streamsides, road verges and in rough grazings. Reported from all squares except 86NW, 87SW, 04SE, 14SW, 24SW, 26SW and 27SW.

Star sedge   *Carex echinata*
Common on the damp, peaty soil of upland heaths, flushes, rough grazings and on peatlands in the lowlands. Reported from all squares except 04, 06SE, 13, 16SE, 25NE, 26SW, 27SE, 36 and 37.

Dioecious sedge
Carex dioica

A rare sedge of wet flushes. Below Craig y bwlch, Claerwen Valley 9061 D Reed and in a spring head flush in the bottom of a pasture below Upper Hall Farm, Heyope 2274 JM Roper.

White sedge   *Carex curta*
Confined to the wettest parts of acidic peatlands, particularly on floating lawns of vegetation.

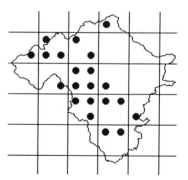

Hairy sedge   *Carex hirta*
Of scattered and unpredictable occurrence in unimproved pastures, roadside verges, riverbanks and waste ground, often on the heavier soils.

Carex

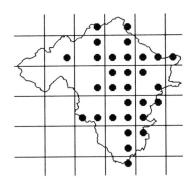 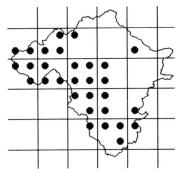 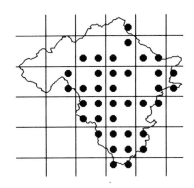

Lesser pond-sedge
Carex acutiformis

Widespread, but not common, typically forming extensive stands in wet hollows such as partially filled river oxbows, alder woodland, wet unimproved pastures, the flood zone of lakes and the peripheral drainage zones of peatlands such as Rhosgoch Common 1948. 13 sites.

Bladder sedge
Carex vesicaria

Scarce or overlooked in the more nutrient-rich wetlands such as riverside marshes, by ponds and in damp hollows in pastures, mostly in the lowlands.

Thin-spiked wood-sedge
Carex strigosa

A rare or overlooked sedge of base-rich flushes in woodland. Cwm Gwanon Dingle, Clyro 1944 MEM and Cilkenny Dingle, Llowes 1741 DRD.

Glaucous sedge *Carex flacca*

Frequent in base-rich flushes in woodland glades, unimproved pastures, road and railway banks, ditches and marshes.

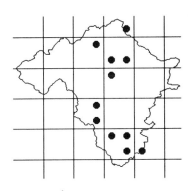 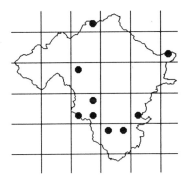 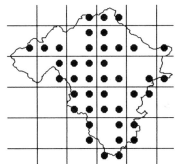

Greater pond-sedge
Carex riparia

A rare sedge of marshy hollows, reported only from near Llananno c07SE IM Vaughan det. BA Seddon (1964 *Proc BSBI* **5**, 239) and in the flood zone of Llanbwchllyn 1146 BA Seddon (1964 NMW records and noted in 1984 by members of the Herefordshire Botanical Society).

Pendulous sedge
Carex pendula

A scarce sp. of damp soil in woodlands. Possibly native in some sites but certainly introduced and naturalized in others. Rocky gorge to N of Rock Park, Llandrindod Wells 0561; common in gorge of the Bach Howey, Llanstephan 1143; wet flushes in woodland, Cilkenny Dingle, Llowes 1741 MEM; by ornamental lake, Stanage Park, Stanage 3371.

Wood sedge *Carex sylvatica*

Common on heavy, moist soil in woodland glades and clearings in the more nutrient-rich sites.

Carnation sedge
Carex panicea

In similar sites to the last but extending into less base-rich sites such as peaty flushes in the NW uplands. Recorded from all squares except 04SE, 26SW & NW, 27SE and 36SW.

Bottle sedge *Carex rostrata*

Commonly growing out of acidic water in slow-flowing flushes in peatbogs and spreading out from the margins into lakes and pools.

Cyperaceae

Smooth-stalked sedge
Carex laevigata

Widespread but rarely common in peaty flushes in wet woodlands, on clay banks by streams and occasionally around the edges of flushes on moorland.

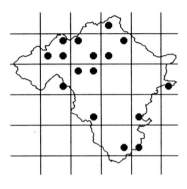

Green-ribbed sedge
Carex binervis

Common on moorland, heath and in acidic upland grasslands.

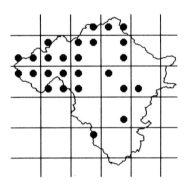

Tawny sedge   *Carex hostiana*
Widespread, but rarely common, in nutrient-rich, frequently peaty flushes in unimproved pastures and on rough grazings in the hills. Probably decreased due to agricultural drainage.

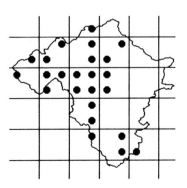

Long-stalked yellow-sedge
Carex lepidocarpa

A rare sp. of base-rich flushes. Tall forms of *C. demissa* are frequent and may cause confusion. Carregwiber, Llandrindod Wells 0859 CP & PJP conf. AO Chater; hay meadow, Coch-y-dwst Farm, Llanyre 0563 LMP; Rhosgoch Common 1948 DDB (a BRC card dated 1955 to 1957); Llanfihangel-Nant-Melan 15 FM Day det. E Nelmes (1936 *Rep Bot Exc Club* **11**, 285); Stanner Rocks 2658 HNR det. E Nelmes (1880 BM). Also recorded from 17 in WFP.

Common yellow sedge
Carex demissa

Common in wet, stony, peaty and clay flushes in grassland and moorland and by ditches and streams throughout in the more acidic sites. In all squares except 04NE, 13, 16NE, 26SE, SW & NW, 27SE and 36.

Pale sedge   *Carex pallescens*
Widespread, but not common, in damp, unimproved pastures and meadows and rarely on roadside verges. 32 sites and much reduced due to the use of artificial fertilizers and reseeding of fields.

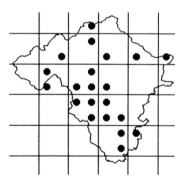

Spring sedge
Carex caryophyllea

Frequent on well-drained, sunny banks on road verges, below hedges, on rock outcrops and in the short turf of agriculturally unimproved pastures.

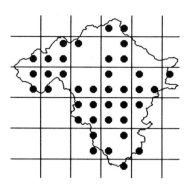

Pill sedge   *Carex pilulifera*
Frequent in the uplands in heather moor, grass heath and acidic grassland.

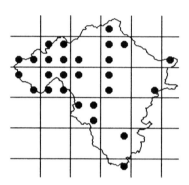

Common sedge   *Carex nigra*
Widespread and common in flushed peatlands, marshes, by lakes, ponds and rivers.

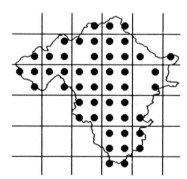

Slender tufted-sedge
Carex acuta

Frequent only along the muddy banks of the larger streams and rivers. Rare in reedswamp elsewhere.

Platanthera

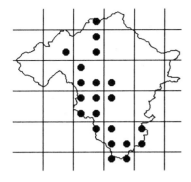
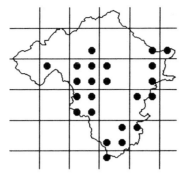

Flea sedge *Carex pulicaris*
Widespread in the short turf of base-rich, peaty and stony flushes, mostly on the upland commons.

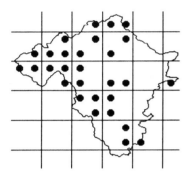

Orchid family
Orchidaceae

Marsh helleborine
Epipactis palustris

Noted in a calcareous flush at the edge of an alder carr near Gwernfythen, Clyro 2044 by ACP in 1968. The habitat remains but this orchid has not been seen there for over a decade. This sp. is listed for Radnor in Watson's *Topographical Botany* of 1883. The source of this record is not known.

Broad-leaved helleborine
Epipactis helleborine

Widespread and easily overlooked in woodland, especially around the edges of glades and along the margins, in scrub and by hedgerows. Rarely occurring in old hay meadows some distance from scrub as at Llanfaredd 0650. Favours moist, but not wet, neutral to basic clay soils.

Bird's-nest orchid
Neottia nidus-avis

Noted once in the wood above the Edw at Aberedw c04NE by HNR (1880 *J Bot* 1881, 173). There are no more recent records. Suitable habitat persists in the area and this sp. may yet be refound.

Common twayblade
Listera ovata

Scarce, but widespread and recently seen in no more than a dozen sites. Wide in its habitat preferences, which range from unimproved hay meadows and churchyards, through dry, base-rich woodlands to those on heavy clay soil and flushed peat. All share a freedom from grazing stock during its growing period. Considered a 'local' species in MsFIR.

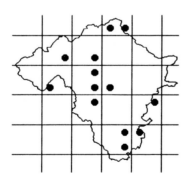

Lesser butterfly-orchid
Platanthera bifolia

Rare in damp, peaty, unimproved pastures and mires. Rhos-yr-Hafod, Pen-y-garreg, Elan Valley 9167 NCC survey; Rhos Dyfnant 0260 and Gweunydd Dyfnant 0360 between Newbridge on Wye and Llanyre, NCC survey; near Cefnbychan Farm, Llansantffraed-in-Elvel 0855 ACP and at the edge of the birch carr, Rhosgoch Common 1948 NCC survey.

Greater butterfly-orchid
Platanthera chlorantha

More common than the last, but still rare and known from fourteen sites. The largest populations occur in unimproved haymeadows and pastures in the NW uplands. Noted elsewhere from Cefnbychan, Llansantffraed-in-Elvel 0755 NCC survey; on a roadside bank NE of Llanfihangel Rhydithon 1767 IDS; churchyard, Crug y Byddar 1581 CC & DRD; old pasture, Gladestry 2355 ACP & IDS; dry hill pasture, Upper Thorn, Evenjobb 2763 M Brown and wet pasture S of Stanage Park 3371.

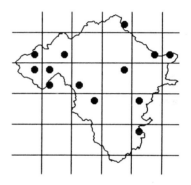

Fragrant orchid
Gymnadenia conopsea

Very rare. Peaty bank at edge of hayfield between Moelfryn and Pen-y-garreg Resr; 8967 JH; old hay meadow, Henfron farm, Elan Valley 9064 MEM and ! ; pasture on SE shore of Pen-y-garreg Resr; Elan Valley 9067 CC et al.; meadow near Disserth c0358 WH Painter (1899 NMW and BM); a single spike on dry bank in Portway Meadow, Rhosgoch 1948 NCC survey. Further work is required to determine which ssp. these records refer to.

Frog orchid
Coeloglossum viride

A single record by a Mrs Debenham from near Presteigne c36 in *Rep Bot Exc Club* 1923, 215 is the only evidence of this sp. ever occurring in the county.

Orchidaceae

Early marsh-orchid

Dactylorhiza incarnata

A scarce orchid of peaty, somewhat calcareous, flushes in unimproved pastures. Cefnbychan, Llansantffraed-in-Elvel 0855 ACP; wet meadows, Portway, Rhosgoch 1948 NCC survey (noted at Rhosgoch by AL in 1881, BM); many colour variants noted near Llanbwchllyn 1146 M Taylor (1961 *Nat Wales* **8**, 28), though only a few plants found here in recent surveys. Peaty flush in pasture SW of Wern-newydd, Painscastle 1744 NCC survey; wet pasture, Bettws Disserth 1257 ACP (*Nat Wales* **8**, 28. RH Roberts determined a specimen from this field as ssp. *pulchella*, but the site was totally destroyed by agricultural drainage c1975).

Northern marsh-orchid

Dactylorhiza majalis ssp. *purpurella*

A scarce plant of peaty flushes. Wet pasture Crossway, Howey 0558 DRD & FE (building work commenced on this site in 1987); near Esgair-draenllwyn, Llanbadarn Ffynydd 0882 IHS & CC; marshy field NE of Pen Ithon Hall, Llanbadarn Ffynydd 0881 CP & PJP; peaty flushes, Lanes Farm, Rhosgoch 1846 & 1847 HLJ Drewett; wet flush in bottom of dingle, Moelfre City, Llanbister 1377 ACP (det. RH Roberts); Burfa Bog, Evenjobb 2761.

Southern marsh-orchid

Dactylorhiza majalis ssp. *praetermissa*

The only certain record is from a building plot in Newbridge on Wye 0157 FMS from where 30 plants were transplanted to The Bog 0157 in 1981. Recent searches have failed to locate the sp. in either site. Problems in determining the identity of this and the above members of this genus and the undoubted existence of hybrids have led to the rejection of some records pending further study.

It is, however, clear that they now occur in very few sites, populations suffering from drainage, fertilizers, ploughing, overgrazing and building development in recent years. They can no longer be described as "locally frequent up to c800 ft" as noted in MsFlR.

Heath spotted-orchid

Dactylorchis maculata

Frequent in damp and peaty unimproved upland pastures, in peat bogs and on moorland and rarer in heathy lowland pastures.

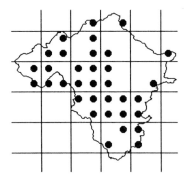

Common spotted-orchid

Dactylorchis fuchsii

Frequent, in similar sites to the last and often hybridizing with it, but extending more into the lowlands in damp clay pastures, wet road banks and flushes in open woodlands.

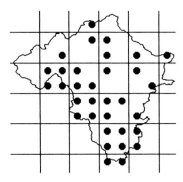

Green-winged orchid

Orchis morio

Very rare and possibly now confined to a dwindling population in a single locality. "Very frequent at Llanelwedd" c05 JAW (1924-26 *Proc of Swansea Sci & Fld Nat Soc* **1**, 6, but no subsequent records); "locally fairly frequent, Builth Road towards Llandrindod by the direct road in pastures and Builth Road to Builth" c05 (pre 1955 MsFlR). A search of this area recently has failed to locate any sites. On side of old railway cutting (and formerly in field above), Glasbury 1839 MP. This is the only known surviving population. Llanstephan c14SW GP (1957); Beguildy c17NE (1931 MsFlR).

Early-purple orchid

Orchis mascula

Widespread on the more base-rich, clay soils. Mostly in woodlands, particularly the dingle woods of the lower Wye Valley and on the boulder clay of the Ithon Valley and rarely on roadside verges and in hedge bottoms. 40 sites. White flowered form reported by PJP & ACP from Llanelwedd 0552, and by T Price from Penybont 1063.

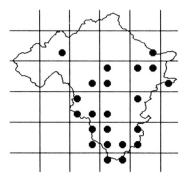

Bog orchid

Hammarbya paludosa

A small population occurs in a peaty flush near Pont ar Elan, Upper Elan Valley 9071. First noted in 1988.

Non-Vascular Plants

Introduction

A complete treatment of the non-vascular plants of Radnorshire is not at this stage possible. A few groups have been studied in detail. The stoneworts, mosses, liverworts, hornworts and lichens have been examined in sufficient detail to provide an account which, though far from complete, probably reflects their true status in Radnor. The accounts of the remaining green algae, lichenicolous fungi and rust and smut fungi are far less complete. Other groups still await study with little more than preliminary lists available for a small selection of sites.

11. Algae and Cyanobacteria

History of Recording

The first detailed survey of algae other than stoneworts in any site appears to be that of **F.E. Round** (1956). He had commenced a study of the phytoplankton of the three Elan Valley Reservoirs in April 1952, and extended this study to the then newly-constructed Claerwen Reservoir as it filled during the course of 1952. Thirty-five species of algae and cyanobacteria were identified. The majority (14 spp.) were desmids. *Staurastrum paradoxum* (a taxon which Brook, A.J. (1959) in a paper entitled 'The published figures of the desmid *Staurastrum paradoxum*" in *Rev. Algol.* NS **4**, 239-255, has been shown to consist of 6 distinct taxa, including *S. anatinum* Cooke and Wills, a species also recorded by Round from all 3 reservoirs) dominated the plankton in late summer. Other desmids collected included *Xanthidium antilopaeum* (Bréb.) Kütz., *X. armatum* (Bréb.) Rabenh., *Closterium kützingii* Bréb. and 3 species of *Micrasterias*.

Other algae recorded included 3 species of planktonic diatoms (*Tabellaria flocculosa* (Roth) Kütz., *T. fenestrata* (Lyngb.) Kütz. and a *Fragilaria* species), a dinoflagellate (*Peridinium willei* Huitf.-Kaas), 2 genera of green algae (*Ulothrix* sp. and *Hormidium* sp.) and 2 species of yellow-green algae (*Dinobryon sertularia* Ehrenb. and *Malomonas longiseta* Lemn.).

The occasional desmid cell in the plankton of what were thought to be benthic species

But he the man of science and of taste,
Sees wealth far richer in the worthless waste
Where bits of lichen and a sprig of moss
Will all the raptures of his mind engross.

John Clare   Shadows of Taste 1882

prompted a search of the sediments in Craig Goch, Caban Coch and Dol-y-Mynach reservoirs (Round 1957). A rich algal flora was discovered with an abundant growth of both desmids (47 spp.) and diatoms (75 spp.). Notably abundant species included the diatoms *Nitzschia dissipata* (Kütz). Grun., *Navicula pupula* var. *capitata* Skvortz. & Meyer and *Neidium iridis* (Ehrenb.) Cleve and the desmids *Pleurotaenium trabecula* (Ehrenb.) Näg., *Closterium didymotocum* Ralfs and *Euastrum ansatum* Ralfs.

The planktonic algae of the Rivers Wye, Elan and Ithon were studied by Furet, J.E. (1979). His results are summarized in Edwards, R.W. & Brooker, M.P. (1982). During 1975-76 he collected water samples at approximately 2 weekly intervals from 10 sites on the Wye and at one site on the Elan and Ithon. In the upper Wye ie. above the confluence with the R. Irfon, the diatom *Meridion circulare* (Grev.) Ag. was prevalent in the winter, whilst in the summer the diatoms *Rhoicosphenia curvata* (Kütz.) Grun. and *Navicula cryptocephala* Kütz. and the green algae *Ankistrodesmus falcatus* (Corda) Ralfs and *Scenedesmus* spp. dominated the plankton. There were no seasonal blooms of algae.

The middle reaches of the R. Wye between Builth Wells and Erwood were richer in calcium salts. As in the upper Wye *M. circulare* dominated in the winter, but in the summer blooms of algae occurred of, for example, the diatoms *Cyclotella meneghiniana* Kütz. and *Thalassiosira fluvialilis* Hust. and green algae *Scenedesmus* spp. and *Carteria* spp.. Below Hay on Wye larger summer algal blooms can occur with species such as *Monoraphidium tortile* (W. & G.S. West) Kom-Leg and *Oocystis crassa* Wittrock increasing their proportional abundance.

The R. Elan was poor in planktonic algae. Fed mostly in summer by water taken from the lower layer of Caban Coch Reservoir its flora perhaps better reflected that of the reservoirs. The high concentrations of *Tabellaria flocculosa* and *T. fenestrata* noted by Furet were probably from this source. In contrast to the paucity of algae in the R. Elan the R. Ithon was found to be very rich. It supported diatoms such as *Diatoma vulgare* Bory, *Cocconeis placentula* Ehrenb. and *Nitszchia* spp. more typical of the lower Wye.

Also in the Ithon catchment and supporting a rich algal flora is Llandrindod Lake. Its plankton has been studied by A. J. Brook (see Brook, 1982) who found that throughout the year there was an abundant phytoplankton population. Blue-green algae (Cyanobacteria) dominate the plankton with *Lyngbya limnetica* Lemm. being the commonest species. Other blue-green algae noted included *Oscillatoria redekii* Van Goor, *Anabaena flos-aquae* (Lyngb.) Bréb. and *Microcystis aeruginosa* Kütz. emend Elekin. The following green algal genera of the order Chlorococcales were always present:- *Ankistrodesmus, Coelastrum, Crucigenia, Lagerheimia, Micractinium, Oocystis, Pediastrum, Tetradesmus, Tetraedron* and *Scenedesmus* as were the Euglenophyta *Euglena, Phacus, Lepocinclis* and *Trachelomonas*.

Most notable was the presence of 7 species of desmid. In nutrient-rich water Brook (1982) notes that seldom are more than 2 or 3 species found. In Llandrindod Lake he found *Staurodesmus cuspidatus* (Bréb.) Teiling, *Staurastrum tetracerum* (Kütz.) Ralfs, *S. pseudotetracerum* (Nordst.) W. & G.S. West, *S. irregulare* W. & G.S. West, *S. bibrachiatum* Reinsch, *Closterium acutum* var. *linea* (Perty) W. & G.S. West and *C. limneticum* var. *limneticum* Ruzicka. This latter species appeared to be new to the British plankton flora, as was *S. bibrachiatum*.

Antoine and Benson-Evans (1985) made an extensive benthic algal survey of the Wye and its tributaries, the Lugg, Ithon and Elan, between June 1979 and August 1981. They identified 212 algal taxa from natural rocks, soft sediments and perspex slides placed in the water. Six of their 9 sample sites lay in Radnorshire or along its border with Brecknock. Pennate diatoms dominated the samples from rocks in the Wye, Ithon and Elan. Whilst the total species list in this paper provides useful information as to the substrata from which a particular species was recorded, it provides little detailed geographical information on algae in the Wye, lumping most records from English and Welsh sites together.

To provide a clearer understanding of the distribution of algae in the Wye the Countryside Council for Wales in 1991 commissioned the Department of Botany of the Natural History Museum to survey this river. Their report entitled *Monitoring Algae in the River Wye* by D.M. John, L.R. Johnson and P. Sims records the presence of over 250 species of algae, over three quarters of which had never previously been recorded from the Wye.

Fourteen sites were sampled, two of which, one at the confluence with the Afon Marteg and one at Glasbury bridge were entirely in Radnorshire vice-county, whilst those sites at Llanwrthwl bridge, Builth Wells, Llanstephan and Boughrood straddled the boundary with Brecknock. Natural surfaces eg. pebbles and logs were sampled monthly from May to September 1991 whilst artificial surfaces ie. petri dishes and plastic bags placed in the river were sampled monthly between June and September. Further smples of diatoms were collected in February 1992 from 6 sites. Planktonic species were not recorded. The sampling sites were selected to coincide with water sampling points of the National Rivers Authority.

At the Marteg confluence the red alga *Lemanea fluviatilis* (L.) Ag. was frequent with tufts of filamentous green algae such as species of *Hormidium, Microspora, Mougeotia, Spirogyra, Stigeoclonium* and *Ulothrix*. Forty-seven species of diatom were also recorded, including several species of *Eunotia*, a genus largely confined to the upper reaches of the river. In May the normally brackish-water diatom *Fragilaria pinnata* Ehrenb. was noted in some abundance. It occurred in lesser abundance in sampling stations lower down the Wye at sites close to road bridges. Its presence here may be linked to the widespread use of salt on the adjacent trunk road.

Lemanea was also abundant at Llanwrthwl bridge where it occurred with the large thalloid green alga *Enteromorpha flexuosa* (Wulfen ex Roth) J. Ag.. Filamentous species were represented by similar genera to those noted at the Marteg confluence, though in addition the filamentous blue-green algae *Tolypothrix distorta* Kütz and *Lyngbya putealis* Mont. were also recorded. Fifty-seven species of diatom were also reported from this site. *Surirella brebissonii* Krammer & Lange-Bertalot, *Eunotia* and *Achnanthes* species were the dominant diatoms, this dominance also being noted at Builth Wells Bridge.

The river becomes mesotrophic in these reaches and at Builth the distinctive water-net alga *Hydrodictyon reticulatum* (L.) Lagerh. was noted for the first time. In the dry, warm summer of 1991, from August until the first winter floods, large growths of this alga formed a prominant feature in the slower-flowing margins of the middle and lower Wye. This may be a recent phenomenon which potentially could effect many other organisms in the river. The large green algae *Cladophora glomerata* (L.) Kütz. also reaches its upstream limit at Builth Wells.

At Boughrood and at Glasbury pebbles and rocks supported growths of *Cladophora glomerata, Oedogonium* and *Spirogyra* species. *Rhizoclonium hieroglyphicum* Kütz. was reported, in addition, from the extensive pebble beds at Glasbury. The diatom flora was also rich in species with 60 taxa recorded from Boughrood and 51 taxa from Glasbury bridge. In all, this study reported 104 species of diatoms and 43 other species of algae from sites in, or adjacent to the Radnor border.

For a number of years, Dr Alan Brook of the University of Buckingham has brought students to Llysdinam Field Centre, Newbridge on Wye. His study of Llandrindod Lake is noted above. Latterely with David Williamson from Leicester they have studied the desmid flora of a number of mid-Wales sites. An exceptionally rich flora has been discovered in a few acidic wetland sites close to the Radnor border in Brecknock and similar sites seem likely to exist in Radnor. The richest site so far discovered in Radnor is Aberithon Turbary, near Newbridge on Wye, with 30 species of desmids living in the pools and associated with the floating lawns of vegetation.

The above represents a brief survey of our current knowledge of the distribution of algae other than stoneworts in Radnor. Many interesting substrata await study. Priority should perhaps be given to the study of the mud-bottomed pools of the upland commons of central Radnor. The drying mud of many of these pools in summer crackles beneath the foot as thousands of plants of the balloon-like alga *Botrydium* sp. burst. Streams unaffected by agricultural drainage work, forestry or nutrient enrichment are now scarce in England and Wales. Radnor's extensive commons provide numerous examples.

The stoneworts have traditionally been studied by higher plant botanists enabling a fuller treatment of this group to be given below.

Stonewort family

Characeae

The county seems rather poor in stonewort spp., almost all plants examined so far being *Nitella flexilis*. The scarcity of standing water rich in lime salts is no doubt responsible for this paucity.

Chara globularis var. *virgata*

Rare and confined to lakes and pools. In pool on Llanbedr Hill, Llanbedr Painscastle 1448 ACP det. JA Moore; Llynheilyn, Llanfihangel-Nant-Melan 1658 ACP det. JA Moore.

Nitella flexilis var. *flexilis*

Widespread in streams, pools, lakes and ditches where the water is clear, the current not too strong and there is a fairly stable mud bottom present. Acidic, peaty sites are avoided.

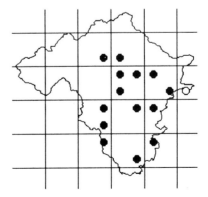

12. HORNWORTS, LIVERWORTS AND MOSSES - THE BRYOPHYTA

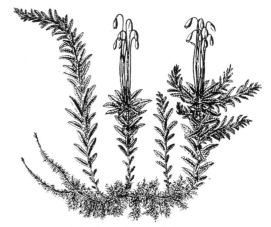

Plagiomnium undulatum

A) History of Recording

No comprehensive flora has ever been produced for Radnor. Old records lie scattered through the literature with specimens held in a number of herbaria. *The Transactions of the Woolhope Naturalists' Field Club*, begun in 1852, are perhaps the richest source of information. The **Revd Augustin Ley** of Sellack, near Ross-on-Wye, provided most of the moss records. He visited Stanner Rocks in 1888 and reported his findings in the *Transactions* for 1886-1889, published in 1892 (p. 218). *Tortella nitida*, *Funaria calcarea* (a species now split into two entities) and *Grimmia trichophylla* var. *subsquarrosa* are mentioned amongst others.

He had either forgotten these records or held the *Transactions* in very low esteem for in his *Notes on a Flora of Aberedw*, published in this journal in 1894 (pp. 184-198) he writes that he believes "that no mosses have been formally recorded for Radnorshire".

Certainly nothing as substantial as the 120 species of moss recorded here as a result of surveys between 1886 and 1890 had ever appeared before. Some species such as *Epipterygium tozeri* have eluded recorders in Radnor since, whilst others such as *Orthotrichum striatum*, *Bartramia halleriana*, *Leptodontium flexifolium* and *Taxiphyllum wissgrillii*, though known now from other sites in the county, have never been recorded since in the Aberedw area. Much suitable habitat remains and many of these species may still be present. Take care if you refer to this list. Many exciting species are named, but the small print reveals that they "remained unseen", were "absent, unnoted", or more ambiguously were "absent from the hills".

In July 1899 the **Revd W.H. Painter** of Stirchley Rectory, Shifnal, Salop stayed for a short time in Llandrindod Wells. He reports his visit under the title of *Mosses near Llandrindod Wells* in *Science Gossip* 1901, **8** (NS), 7-8. Noting the increasing fame of the town's medicinal waters he made a number of excursions locally to Lover's Leap, Alpine Bridge and the then primitively suspended Shaky Bridge. In excursions further afield he visited Abbeycwmhir and the Elan Valley. Other interesting observations concerned the good state of cultivation of the district and consequent lack of "bog-land". Nevertheless he records 60 species of moss and 2 liverworts, adding 7 species not reported by the Revd Ley from Aberedw. Most notable were *Dicranum bonjeani* from Llanyre, *Hedwigia ciliata* from Lover's Leap, Llandrindod Wells and *Hylocomium brevirostre* from an unspecified part of Llandrindod.

Unfortunately neither of these clerics appeared to have taken much notice of the liverworts, so by 1905 when the **Revd S.M. Macvicar** compiled the *Census Catalogue of British Hepatics*, only one species was reported from Radnor. This and its companion volume, the *Census Catalogue of British Mosses*, listing 89 Radnor mosses, compiled by **W. Ingham** in 1907, detail the occurrence of species in the British Isles by vice-counties. Sadly they provide no information as to where or when a record was made and the origins of some of these early records are still unknown. A few records appear in the notebooks of, or were published in, the *Reports of the Moss Exchange Club*, founded in 1896. Later records are more easy to trace, such as the useful notes which appear in the margin of a copy of the 1st edition of the *Moss Census Catalogue* held by the present moss recorder of the **British Bryological Society** (BBS). An official records book was subsequently kept by successive moss and liverwort recorders and in most cases supporting specimens were deposited in the herbarium of the BBS, a society founded in 1923.

The origins of the records of the following species which appear in the 1st edition of the *Census Catalogue* (1907) remain untraced and the species never appear to have been recorded

Plate 5

Llanbwchllyn from Llandeilo Hill. (*RWT*)

Llanbwchllyn. The largest naturally nutrient-rich lake in Radnorshire in August with amphibious bistort and white water-lily in flower. (*D.C. Boyce-CCW*)

Flower-rich agriculturally unimproved hay meadow in the Elan Valley with greater butterfly-orchid, wood bitter-vetch and eyebright amongst many other species. (*CCW*)

Sticky catchfly.
(*R.J. Haycock - CCW*)

Burfa bog - A Radnorshire Wildlife Trust reserve managed to conserve species-rich wet pasture. (*The author - CCW*)

Mud-bottomed pools on upland commons in Radnorshire are a key habitat for the scarce aquatic fern called pillwort. (*The author - CCW*)

in the county subsequently:- *Anoectangium aestivum*, *Grimmia orbicularis*, *Schistidium alpicola* var. *alpicola* and *Tetraplodon mnioides*.

The next significant contribution to the Radnor bryophyte flora came in April 1918 when **Mr. Harry Bendorf** of Manchester visited Aberedw. Within a 2 mile radius he collected 43 species, which were identified by his friend **Mr. William Henry Pearson**, who in 1919 published the records in the *Journal of Botany* 57, 193-195 under the title *Notes on Radnorshire Hepatics*.

Notable amongst his finds were *Lophocolea fragrans*, *Marchesinia mackaii*, *Marsupella funckii*, *Preissia quadrata*, *Plagiochila punctata*, *Blepharostoma trichophyllum* and *Frullania fragilifolia*.

From 1919 until 1966 no significant lists of bryophytes appear to have been published. Individual members of the BBS visited the county and a steady stream of records of species new to the county were published in the *Transactions* of the society and latterly in the *Journal of Bryology* and *Bulletin* of the society. Revisions of the *Census Catalogue of Liverworts* were published in 1930 and 1965 and of mosses in 1926 and 1963.

In 1965 the BBS held its Annual General and Field Meeting at Llandrindod Wells. The results, written up by **Dr. A.J.E. Smith**, were published in its *Proceedings* in 1966 (*TBBS* 5, 207-209). Sites visited included the Elan Valley, Aberedw Rocks, the Bach Howey gorge near Llanstephan, Water-break-its-neck and Stanner Rocks. Twenty six species were reported from the county for the first time. Most notable were *Solenostoma caespiticium*, *Weissia rutilans*, *Bryum riparium* and *Tortula papillosa*. These records, along with all earlier ones, are summarized in *Distribution of Bryophytes in the British Isles* by **Corley and Hill**, 1981, updating the earlier census catalogues mentioned above.

In 1988 **M.O. Hill** published *A Bryophyte Flora of North Wales* in *J. Bryol.* 15, 377-491. That part of Radnor lying in grid squares SO08 and SO18 is included in this account but was not treated systematically. The only localized record is of *Barbula spadicea* from the gorge of the R. Teme in 18, whilst only *Bryum bornholmense* and *Calliergon giganteum* appear in the vice-county summary. This is, however, the only recent account of bryophytes from an area contiguous with Radnor.

The history of recent recording for this flora has already been recounted above. It is, however, worth restating the important part played by **John Port** and **Alan Orange**. Without their field records and help in many ways this account would probably never have been put together.

To date 115 species of liverworts and hornworts and 365 species of mosses are known from Radnor.

B. Biogeography of the Hornworts, Liverworts and Mosses.

No comprehensive account of the major world distribution patterns shown by bryophytes is available. Those species which show a markedly Oceanic or Atlantic distribution have been the subject of a masterly paper by Ratcliffe (1968) on which this account draws heavily. Averis (1990) in his studies of western Scottish woodlands has also considered the biogeographical affinities of the species he encountered and his treatment has been followed here. The British distributions of the remaining species have been examined and those showing a restricted distribution have been assigned to one of five groups. In all, ten phytogeographical groups have been recognised and 131 species have been assigned to a group. They are listed below the account of each group and a distribution map is provided which indicates the number of species in that group found in each 5km square or quadrant of the Ordnance Survey's national grid. The larger the dot size the more species are recorded in the quadrant.

Northern Atlantic Group

Species in this group are strongly represented in Scandinavia and Western Scotland, but become scarce as you travel south and are absent from the Mediterranean region. They favour regions with cool summers and generally at least 220 wet days per year. Poorly represented in Radnor, 2 of the 3 species (12% of the British total) are confined to the uplands. The best example of a habitat for this group is provided by the E-facing cwm above Gwynllyn, Llansantffraed Cwmdeuddwr. *Rhabdoweisia crenulata* is found in shady crevices in cliffs, whilst *Bryum riparium*, a moss only known outside Britain from Atlantic Norway, is found beside the rocky stream. *Grimmia retracta*, a British endemic, is the exception, having been found in the last century on rocks by the R. Wye at Erwood in what must be a sunny and warm site.

Bryophyte Biogeography

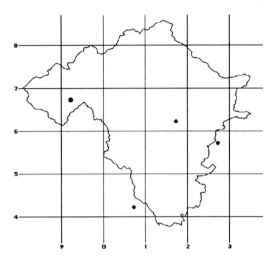

Northern Atlantic Group Species

Bryum riparium
Grimmia retracta
Rhabdoweisia crenulata

Southern Atlantic Group

This group, best represented in south-west Ireland and south-west Wales, becomes scarce in north-west Scotland and is absent from Scandinavia. The 7 species noted from Radnor (26% of the British total) are mostly confined to the low ground, favouring sheltered and warm ravines such as at Aberedw, in the lower Bach Howey Valley and the lower Wye Valley where *Jubula hutchinsiae*, *Marchesinia mackaii*, *Orthotrichum sprucei*, *O. rivulare* and *Porella pinnata* occur.

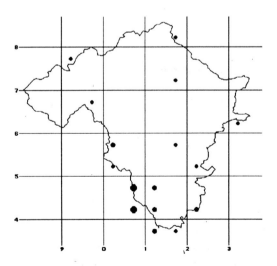

Southern Atlantic Group Species

Cololejeunea minutissima	Fissidens celticus
Jubula hutchinsiae	Orthotrichum rivulare
Marchesinia mackaii	O. sprucei
Porella pinnata	

Widespread Atlantic Group

Species in this group encompass the area occupied by the above two groups and favour humid sites. Many are distributed from Iberia to Scandinavia. Some species such as *Saccogyna viticulosa* and *Lophocolea fragrans* extend into the Mediterranean. There are 9 species noted from Radnor (34% of the British total). Two species which Ratcliffe (1968) placed in the Sub-Atlantic Group (*Breutelia chrysochoma* and *Plagiochila spinulosa*) have been transferred to this group by Averis (1990). The former is widespread in damp, flushed grassland and on rock ledges whilst the latter is found in sheltered ravines and woodland. The species of this group are found most abundantly in the humid valleys of the Elan and Claerwen in NW Radnor.

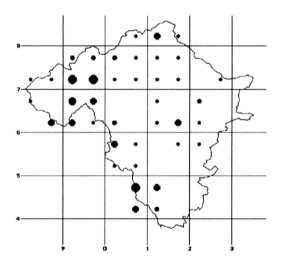

Widespread Atlantic Group Species

Lejeunea lamacerina	Breutelia chrysochoma
Lophocolea fragrans	Fissidens curnowii
Plagiochila punctata	Isothecium holtii
P. spinulosa	Rhynchostegium lusitanicum
Saccogyna viticulosa	

Sub-Atlantic Group

The species in this group, whilst showing a western tendency in Europe, extend further east than any of the species in the above groups. Many occur widely in Britain and it is no suprise that this group is very well represented in Radnor with 25 species (58% of the British total) recorded. They occupy many and varied habitats throughout the county. Twelve species grow as epiphytes, though a number of these also regularly grow on rocks. A few, such as *Campylopus atrovirens, Hyocomium armoricum* and *Scapania gracilis*, have a distinctly western distribution in Radnor. Both *Scleropodium caespitans* and *S. tourettii* show affinities with the Mediterranean-Atlantic group and in Radnor are found most commonly in warm and sunny sites in the lower Wye Valley.

Mediterranean-Atlantic Group

The species in this group extend from their headquarters in the Mediteranean north along the western seaboard of Europe. Many show a preference for sunny, base-rich sites. Of the 11 species (26% of the British total) noted from Radnor, 7 have been recorded from Stanner Rocks. *Bartramia stricta, Tortula canescens, Tortella nitida, Riccia beyrichiana* and *R. nigrella* have not been reported from anywhere else in Radnor.

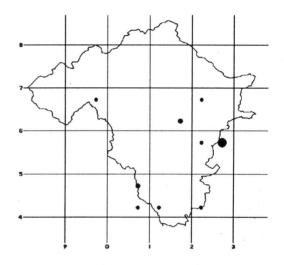

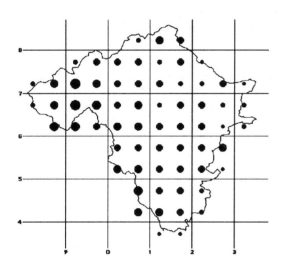

Sub-Atlantic Group Species

Calypogeia arguta	Heterocladium heteropterum
Lejeunea patens	Hookera lucens
L. ulicina	Hyocomium armoricum
Metzgeria temperata	Hypnum cupressiforme var. resupinatum
Odontoschisma sphagni	
Porella arboris-vitae	Leptodontium flexifolium
Scapania gracilis	Orthotrichum pulchellum
Campylopus atrovirens	Pterogonium gracile
C. paradoxus	Ptychomitrium polyphyllum
C. subulatus	Scleropodium caespitans
Fontinalis squamosa	S. tourettii
Funaria attenuata	Ulota phyllantha
F. obtusa	Zygodon conoideus

Mediterranean-Atlantic Group Species

Riccia beyrichiana	Fissidens algarvicus
R. nigrella	F. rivularis
R. subbifurca	Philonotis rigida
Targionia hypophylla	Tortella nitida
Bartramia stricta	Tortula canescens
Epipterygium tozeri	

The following groups are characterized by their distribution within Britain.

Western British Group

A group of 35 species was identified by Ratcliffe (1968) as showing a western bias in Britain. When their world distribution was considered they did not show any other preference for oceanic areas. Nineteen species (54% of the British total) have been reported from Radnor. Only 8, however, show a distinctly western distribution in Radnor. Most of these are confined to damp, long-undisturbed woodland such as those found in the Elan Valley eg. *Dicranodontium denudatum, Sphagnum quinquefarium* and *Bazzania trilobata*. The remaining species occur in a range of habitats from rock outcrops to soil.

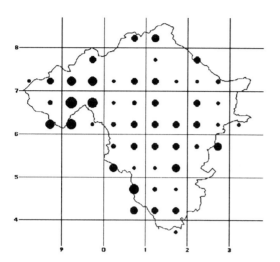

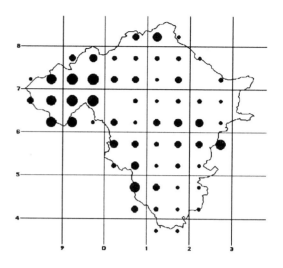

Western British Group Species

Bazzania trilobata	Scapania compacta
Diplophyllum obtusifolium	S. umbrosa
Frullania fragilifolia	Bartramia halleriana
Jamesoniella autumnalis	Dicranodontium denudatum
Marsupella funckii	Grimmia hartmanii
Metzgeria conjugata	G. montana
Mylia taylorii	Oxystegus tenuirostris
Nowellia curvifolia	Sphagnum quinquefarium
Riccardia chamaedrifolia	Thuidium delicatulum
R. palmata	

Western and Northern Group

These tend to be mostly upland species since almost all the British upland is found in the north and west. They differ from the last group in extending into eastern Scotland. With 44 species this is the largest group. About a third of them are confined to the uplands in Radnor with *Nardia compressa, Polytrichum alpinum, Oligotrichum hercynicum, Grimmia donniana,* and *Orthothecium intricatum* providing the most extreme examples. Others such as *Bryum alpinum, Amphidium mougeotii* and *Blindia acuta* favour rock crevices or stony flushes, which are commoner in the uplands.

Western and Northern Group Species

Blepharostoma trichophyllum	Drepanocladus uncinatus
Barbilophozia attenuata	Encalypta ciliata
B. barbata	Fissidens osmundoides
B. floerkei	Grimmia donniana
Jungermannia atrovirens	Hygrohypnum ochraceum
J. exsertifolia ssp. cordifolia	Isopterygium pulchellum
Lophozia sudetica	Oligotrichum hercynicum
Nardia compressa	Orthothecium intricatum
Scapania subalpina	Plagiobryum zieri
Tritomaria quinquedentata	Plagiopus oederi
Andreaea rothii	Pohlia cruda
A. rupestris	P. elongata
Amphidium lapponicum	Polytrichum alpinum
A. mougeotii	Racomitrium affine
Anomobryum filiforme	R. aquaticum
Blindia acuta	Rhabdoweisia crispata
Bryum alpinum	R. fugax
Calliergon sarmentosum	Schistidium alpicola
Campylopus fragilis	Sphagnum girgensohnii
Cynodontium bruntonii	S. russowii
Dicranella subulata	S. warnstorfii
Diphyscium foliosum	Splachnum sphaericum

South-Eastern Group

The 4 species in this group are commonest in south-east England. For two species Radnor lies on, or close to, the north-western edge of their range i.e. *Cinclidotus mucronatus* and *Weissia longifolia* var. *angustifolia*.

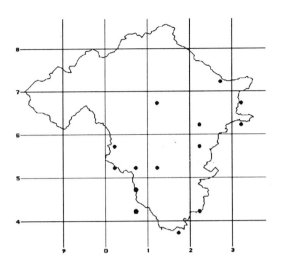

South-Eastern Group Species

Cinclidotus mucronatus　　Tortula latifolia
Eurhynchium schleicheri　　Weissia longifolia
　　　　　　　　　　　　　　var. angustifolia

Eastern Group

This is a small group of 5 species which are commonest in eastern Britain. *Lophocolea heterophylla, Aulacomnium androgynum* and *Tortula papillosa* are commonest in eastern Radnor.

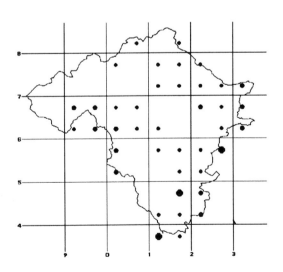

Eastern Group Species

Lophocolea heterophylla　　Dicranum montanum
Ptilidium pulcherrimum　　Tortula papillosa
Aulacomnium androgynum

Southern Group

The 7 species of this group are commonest in southern Britain. All are scarce in Radnor and are mostly confined to the low ground. *Pylaisia polyantha* is a good example.

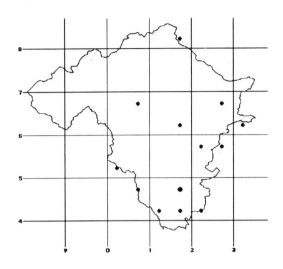

Southern Group Species

Riccia fluitans　　Plagiothecium latebricola
Bryum radiculosum　　Pylaisia polyantha
Fissidens crassipes　　Taxiphyllum wissgrillii
F. exilis

C. Conservation of Mosses, Liverworts and Hornworts

Britain has one of the most diverse bryophyte floras in the world. Of all the bryophyte species recorded from Europe, 70% can be found in Britain, compared to less than 20% of the flowering plants. In Atlantic bryophyte species it is one of the richest areas of the world (Ratcliffe 1965). With such a diverse flora we have a special responsibility to conserve it.

i) World Rarities

Longton (1992) produced a list of 76 British moss species which he considered to be rare on a world scale. Five are recorded from Radnor ie *Fissidens algarvicus, F. celticus, Grimmia retracta, Hyophila stanfordensis* and *Isothecium holtii*. Interestingly only *F. celticus* appears on Stewart's Red Data list of British and European rarities (see below). Britain, however, supports internationally important populations of these species and significant populations should be conserved. In Radnor sympathetic management of the R. Wye and its banks, Stanner Rocks and Water-break-its-neck would conserve important populations of these species.

ii) National and European Rarities

In 1991 Nick Stewart, funded by the Nature Conservancy Council, produced a report entitled *Conservation requirements of Threatened Lower Plants* (NCC CSD Contract Report No. 12272). The most clearly threatened species on a British and international scale were identified. Of the 90 species of mosses and liverworts placed by Stewart on his British Red Data list 5 are known from Radnor. To be included on this list a species should be well-recorded, occur, post-1960, in less than 15 10km grid squares of the national grid, or if more widespread than this, must have declined markedly in extent in recent years, or if only slightly more widespread, be known to occur in only a small quantity in most of its localities.

Radnor's Red Data List species are shown in table 12.1. Their site details are given in the species accounts below. This table summarizes the statutory protection given them by the Countryside Council for Wales through the notification of Sites of Special Scientific Interest (SSSI) and in a few cases the additional protection of National Nature Reserve (NNR) status. At Stanner Rocks the small *Bartramia stricta* population has been monitored for a number of years and encroaching scrub managed. The distributions of *Riccia nigrella* and *Grimmia ovalis* have recently been studied by the author and A. Orange respectively. The spread of grass and trees in the absence of grazing animals will need to be kept in check. Elsewhere recent searches of once-known localities for *Grimmia ovalis*, *Jungermannia caespiticia* and *Myrinia pulvinata* have proved unsuccessful, though apparently suitable habitat persists.

Table 12.1 British Red Data List Bryophytes found in Radnor

Species	SSSI	NNR	Unprotected	Total
Jungermannia caespiticia	1	0	1	2
Riccia nigrella	1	1	0	1
Bartramia stricta	1	1	0	1
Grimmia ovalis	2	1	0	2
Myrinia pulvinata	1	0	0	1

In 1990 R. Schumacker produced a list of threatened bryophytes in European Economic Community (EEC) countries. To this list Stewart (1991) has added British and Irish endemic species. Disregarding those species which appear on the British Red Data List a total of 36 additional liverworts and 44 mosses should perhaps be considered for special protection. Those that occur in Radnor are listed in Table 12.2.

Table 12.2 Species threatened in the EEC

Species	SSSI	NNR	Unprotected	Total
Cryptothallus mirabilis	1	0	0	1
Jungermannia paroica	1	0	1	2
Plagiochila britannica	1	0	0	1
Porella pinnata	1	0	0	2
Riccia beyrichiana	1	1	0	1
Campylopus atrovirens	4	0	46+	50+
Discelium nudum	0	0	2	2
Fissidens celticus	0	0	5	5
Grimmia retracta	1	0	0	1
Orthotrichum sprucei	1	0	0	4
Splachnum ampullaceum	1	1	0	1
Splachnum sphaericum	1	0	0	2

Threats to the species on this list are also difficult to quantify. All except 2 species are offered some statutory protection. Unprotected are *Fissidens celticus* and *Discelium nudum*. Since both occur on bare soil in frequently naturally short-lived habitats such protection may not be appropriate. Amongst the species of protected sites, *Splachnum ampullaceum* is feared extinct due to changes in farming practice. *Orthotrichum sprucei* is threatened by the felling of riverside trees to improve agricultural drainage. *Grimmia retracta* does not appear to have been seen for nearly 90 years in its R. Wye locality.

Whilst protective designations are a step towards conserving species a knowledge of their detailed distribution within a site and their management requirements are also essential. At present this information is almost entirely lacking. Even with this knowledge, unless a site is managed by a conservation body, the voluntary cooperation of owners and occupiers is needed should positive management be required. Too few of these notable species are found in nature reserves in Radnor for their future to be completely secure.

Bryophyte Conservation

iii) Nationally Threatened Species

The final group which requires consideration at a national level are those species listed by Stewart (1991) whose status requires monitoring. Included here are those species confined to clearly threatened habitats or those which have shown a marked decline in abundance in recent years and may soon fall into one of the groups above. Of his 25 species possibly 8 may occur in Radnor. Uncertainty over identification surrounds *Funaria muhlenbergii*, whilst *Pogonatum nanum* may be extinct. The remaining 6 are shown in Table 12.3. A good degree of statutory protection is afforded them, though as with other groups, inadequate information exists to be certain the correct management of sites is being undertaken.

Table 12.3 Species whose status nationally requires monitoring

Species	SSSI	NNR	Unprotected	Total
Cladopodiella fluitans	2	1	1	3
Reboulia hemisphaerica	6	1	5	11
Riccia fluitans	1	1	0	1
Targionia hypophylla	5	1	4	9
Leucodon sciuroides	3	1	22	25
Rhodobryum roseum	2	1	3	5

iv) Local Rarities

To maintain the local diversity of bryophytes some consideration should be given to the conservation of locally rare species. Adopting the definition of a local rarity as one which occurs in 5 or fewer sites in Radnor, 121 additional species need to be considered (28% of the remaining total). This is a very high proportion of the flora. Undoubtedly a number of species will prove with further survey to be more widespread than the present records suggest. Many, however, due to their narrow habitat requirements, appear likely to be genuinely rare. Most of the Atlantic species and many calcicoles probably fall in this group. Locally rare species can be identified from the species accounts below since individual site details are provided for all species with 6 of fewer localities in Radnor. It is hoped this information might form the basis for a conservation strategy.

v) Local Extinctions

Without the security of properly managed nature reserves or active conservation in the wider countryside more species will have to be added to the list of possible Radnor extinctions. To date it numbers 18.

Table 12.4 Species not seen in Radnor for at least 50 years

Species	Year last seen	Place
Lophocolea fragrans	1919	Aberedw
Preissia quadrata	1890	Radnor Forest
	1919	Aberedw
Barbula ferruginascens	1899	R. Wye, Erwood
	1926	R. Wye, Builth
Barbula revoluta	c.1890	Aberedw
Barbula spadicea	c.1899	R. Wye, Erwood
Bryum caespiticium	1891	Aberedw
	1903	R. Wye, Erwood
	1905	Stanner Rocks
Campylium polygamum	1906	Rhosgoch Common
Dicranella cerviculata	1908	Rhosgoch Common
Epipterygium tozeri	1891	R. Wye, Aberedw
Eurhynchium schleicheri	1886	Aberdew
Funaria muhlenbergii &/or Funaria calcarea	1888	Dolyhir and/or Stanner Rocks
Grimmia laevigata	c.1904	Stanner Rocks
Grimmia montana	1894	Stanner Rocks
Plagiomnium ellipticum	1886	Llanbwchllyn
Pogonatum nanum	c. 1910	Aberedw
Polytrichum longisetum	1908	Rhosgoch Common

Table 12.4 lists those species not seen in the county for at least 50 years. It is hoped that most still survive, having been overlooked. All have recently been sought and it is of some consolation that in almost all cases a suitable-looking habitat exists close to where it is believed they were originally recorded. Only one other species, *Splachnum ampullaceum*, seems to have become extinct in recent years.

D. Arrangement and Nomenclature.

In the accounts of species below the arrangement and nomenclature follows Corley and Hill (1981) to which reference should be made for details of authorities. Rarely, a subsequent nomenclatural change has been adopted. In these cases an authority has been quoted. The same criteria as those adopted for the higher plants have been used to determine whether detailed species distribution information or a map is provided in the species accounts below. The layout, mapping policy and level of information provided is similar to that of the higher plants and reference should be made to the introductory sections above for details.

Reference is made in some species accounts to capsules being present. No systematic study has been made of the occurrence of sporophyte stages, but where they have been noted in passing they have been mentioned. Occasionally a species is referred to as probably having been overlooked. The author is conscious of being unfamiliar with a number of species, particularly the smaller leafy liverworts and ruderal mosses. They, in particular, may, on further study, prove to be more widespread than the present records suggest. Lastly this account, perhaps more so than the higher plants, should be treated as but a very preliminary account of the distribution of Radnor's hornworts, liverworts and mosses. It has been compiled by a very small team of workers and though they have been industrious, much of Radnor still awaits exploration.

Map 12.1 Map of the number of bryophyte taxa recorded in each quadrant.

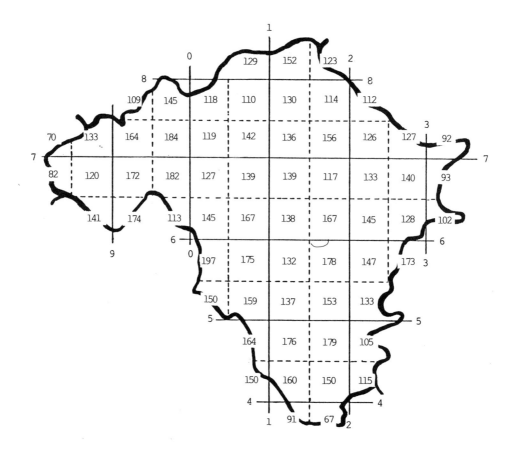

HORNWORTS ANTHOCEROTAE

Anthoceros punctatus

Known only from the damp soil of an arable field near Ditchyeld Bridge, Burfa 2760 PJP and in a ditch, Burfa Camp, Burfa 2761, PJP.

Phaeoceros laevis ssp. *laevis*

Rare on damp soil, often by streams. In forestry on tracksides and ditch sides, SE of Caerhyddwen, Ysfa 9864 and 9865; by the R. Edw near Rhulen 1250; by waterfall in Harley Dingle, Llanfihangel-Nant-Melan 1864; Brilley Mountain, Michaelchurch on Arrow 2651 EA (1916, NMW); arable field, New Radnor 2261 PJP.

LIVERWORTS HEPATICAE

Targionia hypophylla

Scarce except in a limited area of the Wye Valley between Aberedw 0746 and Llandeilo Graban 0943 where it occurs in some abundance on warm, south-facing, though often shaded soil in crevices in basic mudstones. Elsewhere at Water-break-its-neck, Llanfihangel-Nant-Melan 1860 CI and NY Sandwith (1945 NMW) and JAP 1961; Stanner Rocks 2658; rocks near Monaughty Camp 2368 PJP. Capsules frequent.

Reboulia hemisphaerica

In a number of widely scattered sites on damp, base-rich soil on somewhat shady rock ledges. Capsules frequent.

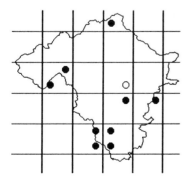

Conocephalum conicum

A common sp. of damp soil on shady banks, especially by water. Capsules occasional.

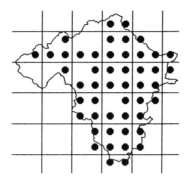

Lunularia cruciata

Frequent on soil by streams and rivers in the lowlands and occasionally in woods, on hedgebanks and in gardens.

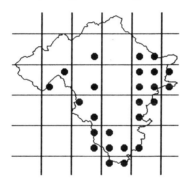

Preissia quadrata

Only two old records have been traced. WHP reports H Bendorf as seeing this sp. at Aberedw 04 (1919, *J Bot* **57**, 194). AL records it from the head of Llandegley Dingle, Radnor Forest c16 (1890, BM).

Marchantia polymorpha

An uncommon sp. recorded from reedswamp by ponds and in bogs, eg Rhosgoch Common 1948, on soil of stream banks and on damp soil around buildings. Gemma cups frequent, gametophores less common, but both male and female present.

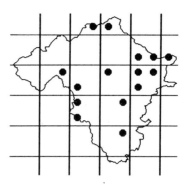

Riccia fluitans

Known only from pools in the reedswamp on Rhosgoch Common 1948 where it was first seen in 1987.

Riccia glauca

A scarce sp. of damp soil in open habitats. Stream bank near Yr Allt, Llandeilo Graban 0844; soil on rocks by stream, Llandeilo Hill, Llandeilo Graban 1146; track by Bach Howey, Llanstephan 1042 JAP et al. (1966, *TBBS* **5**, 180); Stanner Rocks 2658 HHK (1910, NMW); arable field, New Radnor 2261 PJP; arable field near Burfa 2760 PJP.

Riccia subbifurca Warnst. ex Crozals

Rare on thin, summer-droughted soil over rock. Aberedw 04 H Bendorf, det. JAP (MANCH. Originally det. WHP as *R. crozalsii*); rock ledges, Yatt Wood, Dolyhir 2458; Stanner Rocks 2658 JW Fitzgerald et al. (1954, BBSUK).

Riccia beyrichiana

On thin, summer-droughted soil overlying dolerite rocks at base of cliff, Stanner Rocks 2658. Known here for over 50 years (HHK, 1929, NMW).

Marchantia polymorpha

Riccardia

Riccia sorocarpa

The commonest sp. of the genus and frequent on moist, open soil on rock outcrops, by streams and tracks and on ant hills.

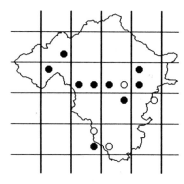

Riccia nigrella

This nationally rare sp. has been known from thin soil overlying dolerite below the cliffs at Stanner Rocks 2658 for many years.

Metzgeria fruticulosa

Occasionally encountered on elder trunks in woods and hedges. Rare on dogwood and hazel by the limestone quarries at Dolyhir 2457. 10 sites.

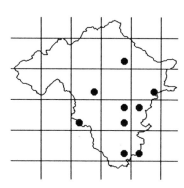

Metzgeria temperata

Rare on trees and shrubs in damp woodland. On beech in gorge, Nant Gwyllt, Elan Valley 9162; on hazel by R. Wye, Nannerth, St Harmon 9571; on hazel in gorge woodland, Cwmbach Llechrhyd 0354; on hazel in damp woodland, Llanboidy, Llanfihangel-Nant-Melan 1557; on oak, Ddol Wood, Llanbister 1372 ! and PJP.

Metzgeria furcata

Common on the bark of a wide range of tree spp. in woods and hedgerows and occasionally on shaded rocks. Capsules frequent. Recorded from all squares except 86NE & NW, 87, 97NW, 16NE and 25SW.

Metzgeria conjugata

Of scattered occurrence on shady rocks and very occasionally spreading to nearby tree trunks. Capsules frequent.

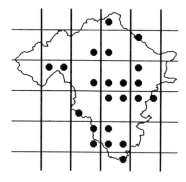

Aneura pinguis

Frequent on wet clay, soil and peat of flushes, stream and roadside banks throughout the county. Occasionally with capsules.

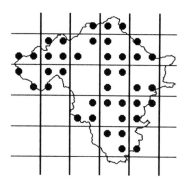

Cryptothallus mirabilis

Rare or overlooked. Accidentally discovered by the author, when a wet flush in an oak-rusty-willow wood on a steep slope slipped away underneath him to reveal this liverwort, buried in the peaty soil. Limited searching in similar habitats elsewhere has failed to locate further specimens. Cwm Coel, Elan Valley 9063.

Riccardia multifida

Frequent on soil in wet flushes throughout the county. Capsules occasional in spring.

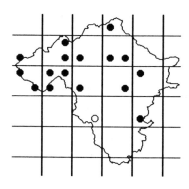

Riccardia chamaedrifolia

Uncommon on damp soil in ditches and probably much overlooked.

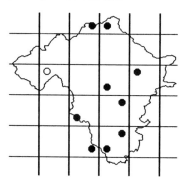

Riccardia palmata

Rare on rotting log in damp oakwood, Cwm Coel, Elan Valley 8963 conf. D Long.

Riccardia latifrons

Known only from rotting logs in wet oak-alder woodland, Cwm Coel, Elan Valley 8963 and 9063 BBS exc. in a site close to the previous species.

Pellia epiphylla

Common on moist clay and soil in shaded habitats such as woodland, stream and hedge banks. Capsules common. Recorded from all squares except 87, 04NE and 13NW.

Pellia neesiana

Widespread around the edges of wet flushes in upland rough grazings. Capsules occasional.

Barbilophozia

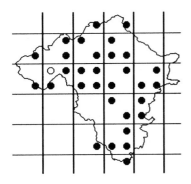
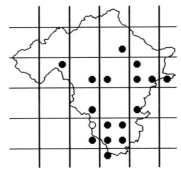
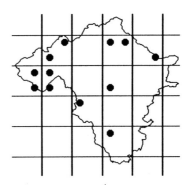

Pellia endiviifolia

Frequent on damp, shaded soil with a high base status, on stream banks and wet rock outcrops. Capsules common.

Fossombronia wondraczekii

More widespread than the last and appearing to favour damp soil on stream and reservoir banks and around flushes.

Barbilophozia barbata

Similarly distributed to the last sp. but less frequent and apparently preferring grassy places by streams as well as rock ledges.

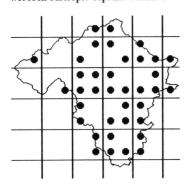
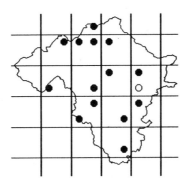
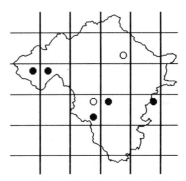

Blasia pusilla

A scarce sp. of damp soil. On bank of R. Edw, Cregrina 1252; bank of R. Ithon, Dolau Jenkin, Penybont 1065, AO; in grazed grassland by stream, Pound Alehouse, Llanbister 1374; Brilley Mountain, Michaelchurch on Arrow 2651 EA (*Rep MEC* 1917,195); disturbed soil, Park Bank Wood, Stanage 3371 AO and !. Both forms of gemmae present. Capsules not seen.

Fossombronia pusilla var. *pusilla*

The var. *pusilla* occurs occasionally on soil around rock outcrops and ditch and roadside banks in the S of the county. It is rare elsewhere. As with all spp. in the genus it was only determined when capsules were present.

Barbilophozia floerkei

Frequent on the acidic soil of peaty banks and crevices in rock outcrops in the uplands.

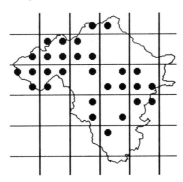

Barbilophozia attenuata

Uncommon on peaty soil of shady rock ledges and on decaying logs in damp woodlands.

Lophozia ventricosa

Common throughout on shady banks, heaths, woodlands on acidic soil and decaying logs. The vars. have not been separated, though both var. *ventricosa* and var. *silvicola* are reported from Radnor.

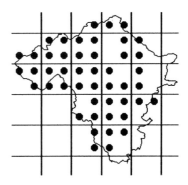

Jungermannia

Lophozia sudetica

Rare or overlooked on damp soil. Nant Caletwr, Craig Goch, Elan Valley 8869 JAP (BBS records); Craig-ddu, Treheslog, Llansantffraed Cwmdeuddwr 9469 JAP (BBS records); Aberedw Rocks 04 (1965, BBS exc.); Water-break-its-neck, Llanfihangel-Nant-Melan 1860 JAP (1961, NCC files); Coxhead Bank Common, Llanbister 1570 (1965, BBS exc.); on the slope below rocks, Dolyhir 2458 CI and NY Sandwith (1945, BBSUK).

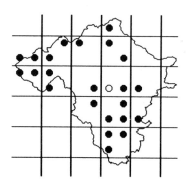

Lophozia excisa var. *excisa*

Uncommon amongst mosses and on soil on shady, acidic banks. Nant Caletwr, Elan Valley 8868 JAP; rocks above Middle Hall, Aberedw 0749; Aberedw Rocks 04 JAP (1965 *TBBS*, 5,183); wood, Crossway, Howey 0658 (1965 BBS exc.); Carregwiber Bank, Llandrindod Wells 0859; roadside bank near Bach Howey viaduct, Llandeilo Graban 1042; Stanner Rocks 2658 PJP; Hanter Hill 2557; Knucklas Castle 2474.

Lophozia incisa

Rare on damp, acidic, shady banks. On the edge of blanket bog, Grafia 8371 and on stream bank 8572, both in the Gwngu Valley, Llansantffraed Cwmdeuddwr; Craig-ddu, Treheslog, Llansantffraed Cwmdeuddwr 9369 JAP (1965, BBS records); peaty ledges, Cerrig Gwalch, N of Rhayader 9370; Marteg Valley below Gilfach Farm, St.Harmon 9671.

Lophozia bicrenata

Very rare or overlooked. On bank amongst heather, 457m(1500ft) Harley Dingle, Radnor Forest 16SE EWJ and EFW (1957,BBSUK); on hard earth of path amongst heather, Rhiw Lawr, Radnor Forest 2166 EWJ (1945,NMW). This area has since been planted with conifers.

Gymnocolea inflata var. *inflata*

Frequent on peat in shallow pools subject to seasonal flooding in the uplands.

Tritomaria exsectiformis

Rare on decaying log in damp oakwood. Cwm Coel, Elan Valley 9063.

Tritomaria quinquedentata

Widely scattered, but not common on damp, acidic soil on rock ledges and in unimproved acidic grassland by streams.

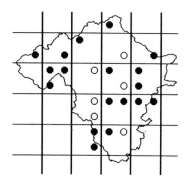

Jamesoniella autumnalis

A scarce Atlantic liverwort at the edge of its range in Radnor. On blocks of gritstone and decaying tree stump in damp oakwood at head of Caban Coch Resr., Llansantffraed Cwmdeuddwr 9061.

Mylia taylorii

Rare and noted only from blanket bog, Ffos Trosol, Claerwen Valley 8271 and 8371; damp rock ledges on the acidic cliffs of Cerrig-gwalch, N of Rhayader 9370 and on a decaying log lying over a stream in a wood near Dolau Jenkin, Penybont 1066.

Mylia anomala

An uncommon sp. of *Sphagnum* lawns in blanket, basin and raised bogs.

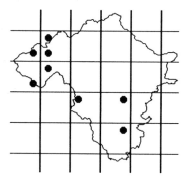

Jungermannia atrovirens

Scarce or overlooked on damp rocks and gravel by streams in somewhat base-rich places.

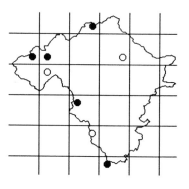

Jungermannia pumila

In similar habitats to the last but confined to the western uplands.

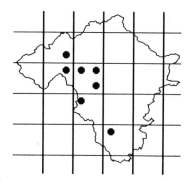

Plagiochila

Jungermannia exsertifolia ssp. *cordifolia*

Reported by NTH Holmes from the Afon Elan at Pont ar Elan, Llansantffraed Cwmdeuddwr 9071 (NCC files) but with no confirmatory specimen.

Jungermannia gracillima

Common on damp, shaded, acidic soil, gravel and clay in woodland and on stream and hedge banks throughout.

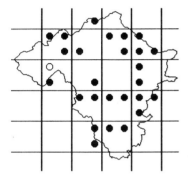

Jungermannia caespiticia

This nationally rare sp. was recorded new to Wales from a gravelly slope at the N end of Garreg Ddu Resr., 96 by JAP (1965, *Nat Wales* **10**, 9) and subsequently also from a narrow earthy ledge on a slate rock outcrop by the Nant Calettwr, W of Craig Goch Resr. 8868 in the Elan Valley.

Jungermannia hyalina

Of scattered occurrence on damp, shaded rocks and soil by streams and rivers.

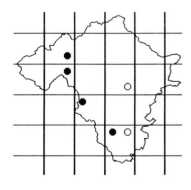

Jungermannia paroica

Rare on wet rocks in oakwood by the R. Wye, Nannerth, N of Rhayader 9571 F Rose (1958, *TBBS* **4**, 152) and on soil in crevice of rocky stream bank, by the Afon Dulas, Sychnant, St Harmon 9877 AO.

Jungermannia obovata

Scarce on wet soil and rocks. Upper end of Garreg Ddu Resr., Elan Valley 9166 (1965 BBS exc.); Water-break-its-neck, Llanfihangel-Nant-Melan 1860 JAP (1961 NCC records); wet bank, Coxhead Bank Common, Llanbister 1570 JAP (1965, *TBBS* **5**, 184); Cwm Mawr, Ednol 2064 PJP.

Nardia compressa

Frequent on rock in fast flowing streams in the western uplands.

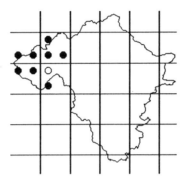

Nardia scalaris

Common on clay, soil and gravel in damp, shady places on cliffs, in woods, and on track, hedge and stream banks throughout. Capsules occasional. In all squares except 87SW, 05SW & SE, 13NW, 15NW, 25NE, 27SE, 36 and 37.

Marsupella emarginata

The var. *emarginata* is common on damp, acidic rocks in flushes and streams in the uplands.

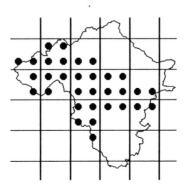

The var. *aquatica* is rare and found on rocks in upland streams. Rocky stream-side by Garreg Ddu Resr., Elan valley 9066 ARP (1966, *TBBS* **5**, 184); near St. Harmon 97 ARP (NMW recording meeting); R. Elan, Pont ar Elan, Elan Valley 9071.

Marsupella funckii

Rare or overlooked on shallow, acidic gravelly soil. Near Pen-y-garreg Resr. dam, Elan Valley c96NW (1965, BBS exc); Aberedw 04 H Bendorf (1919, reported by WHP, *J Bot*, **57**, 194); on hard earth of path amongst heather, Rhiw Lawr, Radnor Forest 26NW EWJ (1945, NMW), this area has since been planted with conifers; on thin soil over sloping rocks, Stanner Rocks 2658 EWJ (1945, NMW).

Plagiochila porelloides

Common on neutral to basic clay, soil, boulder tops and tree bases on shaded banks, cliffs, stream sides and in woodlands and hedgerows. Capsules rare. In all squares except 86NE & NW, 87, 07NE & NW, 08 and 36SW.

Plagiochila asplenioides

Somewhat less common than the last and perhaps preferring more basic sites, but otherwise in similar habitats. Capsules rare.

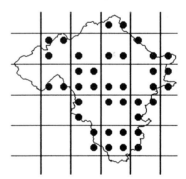

Plagiochila britannica

A recently described sp. favouring base-rich rock outcrops. It may prove to be more widespread than the present record suggests. On cliff in woodland above Pontshoni, Aberedw 0746 AO.

Scapania

Plagiochila spinulosa

Rather scarce on rock ledges and occasionally tree trunks in possibly long undisturbed, humid and sheltered valleys, eg Aberedw Wood 0847, the Bach Howey Gorge, Llanstephan 1243 and Water-break-its-neck 1860.

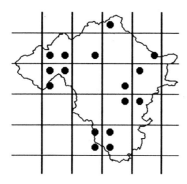

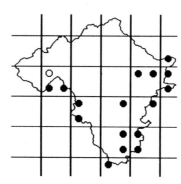

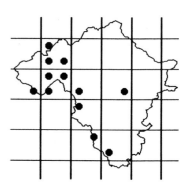

Plagiochila punctata

Rare in long undisturbed, damp woodlands. On oak trunks and rocks, Cwm Coel, Elan Valley 8964; old oak in wood between road and Caban Coch Resr., Elan Valley 9163; wet rocks on wooded cliff, Cerrig-gwalch, N of Rhayader 9370; on mudstone cliffs, Aberedw Rocks 0746; Harley Dingle, Radnor Forest 16SE ARP.

Lophocolea bidentata and *L. cuspidata*

Smith (1990) treats the above two taxa as, at best, varieties. The former, becoming *L. bidentata* var. *rivularis* (Raddi) Warnst, is common on the ground amongst herbs and grasses in moist, shady places and is recorded from all squares except 86NW, 87, 96SW, 15NW, 16NE, 17NE and 26SE. Capsules are rarely produced. The latter becomes *L. bidentata* (L.) Dum. var. *bidentata* and is found commonly on decaying logs and occasionally living trees and shale rocks in shady places. It is reported from all squares except 86NE & NW and 87. Capsules are common.

Lophocolea heterophylla

An uncommon sp. except in SE Radnor. On tree stumps, bark-free logs and occasionally on living bark on the edge of sap runs or in runnels below decaying branches. Capsules occasionally noted.

Lophocolea fragrans

There is a single, old record from Aberedw 04 of H Bendorf (1919, reported by WHP in *J Bot*, **57**, 194). Sheltered, rocky woodland remains in the area and the plant may yet be rediscovered.

Chiloscyphus polyanthos

Infrequently fertile and consequently not always reliably separated from the next sp.. This appears to be the commoner sp., occurring frequently in flushes, spring heads and along the edge of small streams.

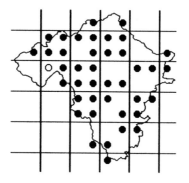

Chiloscyphus pallescens

Apparently rather rare (but see above) and confined to marshy ground. Rhosgoch Common c1948 DDB (1954, NMW); bank of R. Arrow, Glascwm 1754 AO.

Saccogyna viticulosa

Frequent only in the W where it occurs on moist, shaded rocks in sheltered sites such as gorge woodlands. Capsules not seen.

Diplophyllum albicans

Abundant on shady, acidic soils and rocks in all squares in the county except 13NW, 15NW and 36. Capsules occasional.

Diplophyllum obtusifolium

Rare, but probably overlooked as it grows on acidic soil, often with the last sp.. Trackside near Water-break-its-neck, Llanfihangel-Nant-Melan 1860 JAP (1961 NMW record); The Wimble, Radnor Forest 2062 W Watson (*Rep BBS*, 1937, 44); near Heyop 2374 PJP.

Scapania scandica

A rare or overlooked sp. of soil on rock outcrops or by paths. Soil on rock ledge, Nant Calettwr, Elan Valley 8868 JAP (*J Bryol* **7**, 506); near Pentre, Rhulen 1450 PJP; rock outcrops SW of Worsell Wood, Stanner 2557 PJP.

Scapania umbrosa

Known only from decaying logs in damp valleys. By the Afon Dulas above Tylwch 9779 and in a dingle woodland, Glasnant, Bryngwyn 1751 PJP.

Scapania nemorea Grolle

Occasional on shaded rocks and logs in damp woodlands and in gorges.

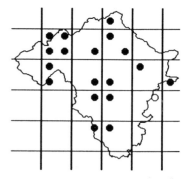

Cephaloziella

Scapania irrigua

Widespread but not common amongst grass and sedges on stream banks and along the edges of flushes, especially on heavily grazed common land.

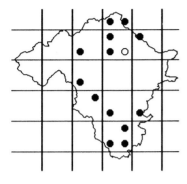

Scapania compacta

Frequent on the shallow, acidic soil of sunny rock ledges on cliffs throughout the county. Capsules occasional.

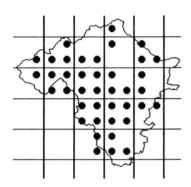

Scapania subalpina

Only recorded from sand and silt-filled hollows in rocks in the flood zone of the R. Wye and its tributaries. R. Wye at Rhayader 9668; by R. Elan, Pont ar Elan 9071; R. Marteg, Nannerth, St Harmon 9671 and from the R. Wye at Erwood Bridge 0843 (1965 BBS recording meeting); Llanelwedd 05 HHK (1923, NMW) and Boughrood 1338.

Scapania undulata

Common on rocks in fast-flowing streams and wet stony flushes. Capsules occasional.

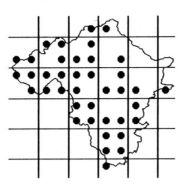

Scapania aspera

A single record from calcareous mudstone cliffs in the bottom of Harley Dingle, Radnor Forest 1962 EWJ and EFW (1957, BBSUK).

Scapania gracilis

On block scree, cliffs and occasionally trees in acidic, moist and sheltered woods and valleys. Capsules very rare.

Odontoschisma sphagni

An uncommon sp. of *Sphagnum* lawns in some of the least disturbed blanket, basin and raised bogs.

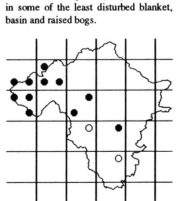

Odontoschisma denudatum

Rare on decaying logs in damp ravines. Valley at the SW end of Garreg Ddu Resr., Elan Valley 8964 JAP (1965, BBSUK) and nearby in the Nant Gwyllt Valley 9162. Also in ravine S of Rhodoldog, Llansantffraed Cwmdeuddwr 9467.

Cephaloziella hampeana

A scarce or overlooked sp. of damp soil, ranging from the bank of the Wye, Erwood Bridge 0843 to the top of the Radnor Forest 16.

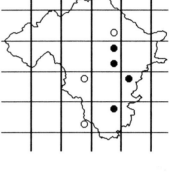

Cephaloziella divaricata

Like the last, rare or overlooked on soil, peat or on decaying wood.

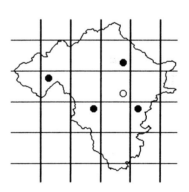

Calypogeia

Cephalozia bicuspidata

The ssp. *bicuspidata* is frequent in damp, acidic habitats such as bogs, decaying logs and peaty soil. The map below is of this ssp. which frequently forms capsules. The ssp. *lammersiana* is rare or overlooked. Nant Caletwr, Craig Goch, Elan Valley 8868 JAP (BBS exc); wet crevice in boulder in stream below Craig-ddu, Llansantffraed Cwmdeuddwr 9469 ARP (1965 BBSUK).

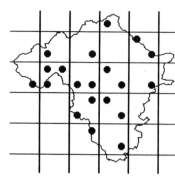

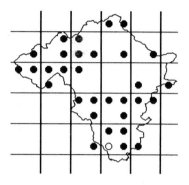

Cephalozia connivens

An uncommon sp. of damp, humus-rich sites such as decaying logs and *Sphagnum* bogs.

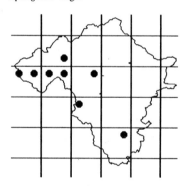

Cephalozia lunulifolia

In the absence of herbarium specimens some doubt surrounds the existence of this liverwort in the county. Reported by WHP as having been collected by H Bendorf at Aberedw 04 (1919, *J Bot* **57**, 194) and by DDB on Rhosgoch Common 1948 (1954, NCC records).

Nowellia curvifolia

Widespread on damp, decaying logs in sheltered sites. Capsules occasional.

Cladopodiella fluitans

Known only from *Sphagnum* lawns in mires. Gors Goch, Elan Valley 8963; Cors y Llyn, Newbridge on Wye 0155 from where it has been known since 1908 (HHK, NMW) and a mire E of Ty'n-y-berth, Llananno 0773. This latter site was damaged recently by attempted drainage.

Kurzia pauciflora

Uncommon in *Sphagnum* lawns in bogs. Blanket bog, Hirnant Claerwen 8466; Gors Goch 8963; by the Elan, Aberglanhirin 8872, all in Llansantffraed Cwmdeuddwr; basin mire, Cors y Llyn, Newbridge on Wye 0155; Camnant Brook near Castle Bank, Llansantfraed-in-Elvel 0857 (1965, BBS exc.); Rhosgoch Common 1948.

Lepidozia reptans

Widespread and common in the W on peaty soil, stumps and tree bases in moist and shady woodlands and gorges.

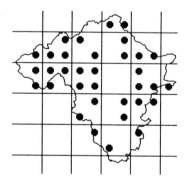

[Lepidozia cupressina]

The record of this sp. from Radnor in Corley and Hill (1981) is believed to be in error. The site was in Brecknock.

Bazzania trilobata

A scarce and possibly relict sp., surviving in a few of the least disturbed woodlands, where it grows on shady, humus covered boulders and decaying logs. 10 sites.

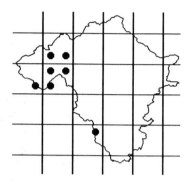

Calypogeia muelleriana

Widespread on damp clay and soil banks in shady places such as woods.

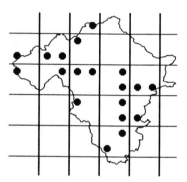

Calypogeia fissa

Common on clay and soil of acidic, moist, shady banks.

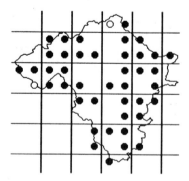

Calypogeia arguta

In similar sites to the last, though tolerating greater shade.

Plate 7

Cors y Llyn National Nature Reserve near Newbridge on Wye in spring. A basin mire of international importance with a floating lawn of bogmoss species with heather and cranberry. Scot's pine is a recent colonist. (*P. Wakely - CCW*)

Hare's-tail cottongrass in early summer around a mawn pool in the hills west of Pant y dwr. (*The author - CCW*)

Plate 8

Dyffryn Wood near Rhayader, a RSPB reserve. A fine example of a western sessile oak woodland with a rich moss and liverwort flora. *(The author - CCW)*

River Wye from Erwood Bridge looking upstream. Wild chives are a feature of rocky crevices in the flood zone of the river from Builth Wells to Boughrood. *(The author - CCW)*

Porella

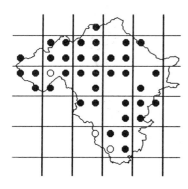

Ptilidium pulcherrimum

Rare on shaded rocks and scarce on acidic, leaning or horizontal tree trunks in damp woodland. Rocks by R. Elan, Elan Village 9365; horizontal oak in wet woodland, Caerhyddwen, Ysfa 9865; on alder, Tyncoed Wood, Cwmbach Llechrhyd 0354; oak in wood, S of Wernfawr, Painscastle 1845 ! and PJP.

Radula complanata

Uncommon, though widespread on the bark of trees and shrubs such as hazel, ash and elder in moist places and occasionally amongst mosses and on shallow soil over somewhat basic rocks in gorges. Capsules frequent.

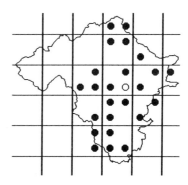

Porella cordeana var. *cordeana*

Occasional on the sheltered, downstream facing sides of boulders and tree roots in fast-flowing streams and rivers. Possibly avoids the most acidic streams, hence its absence from the NW of the county.

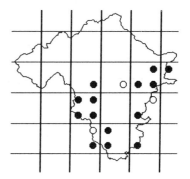

Blepharostoma trichophyllum

Known only from near Pen-y-garreg dam, Elan Valley c96NW (1965, BBS exc.) and from Aberedw 04 H Bendorf (1919, reported by WHP in *J Bot* **57**, 194).

Trichocolea tomentella

A scarce sp. of wet flushes in alder and willow carrs. Apparently absent from many suitable looking sites which are grazed.

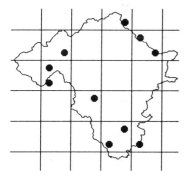

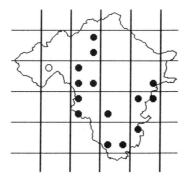

Porella arboris-vitae var. *arboris-vitae*

Scarce on basic rock outcrops such as the volcanic rocks of Graigfawr, Hundred House 1358 and on calcareous mudstones in SE Radnor.

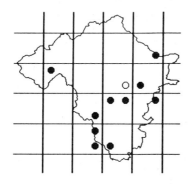

Porella pinnata

This western sp. is known from two sites in the R. Wye, where it occurs on the sheltered downstream faces of large boulders in areas under water for most of the year. Llandeilo Graban 0843 AO; below old railway bridge, Boughrood 1338.

Frullania tamarisci

Forming conspicuous colonies on trunks of trees, notably the larger and older specimens in damp woodlands, but also frequent on sheltered, often slightly basic shale and volcanic rock outcrops.

Ptilidium ciliare

Common in grazed, acidic turf and on shallow peaty soil on boulders and cliffs amongst other bryophytes in open but humid upland sites.

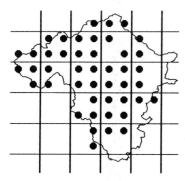

Porella platyphylla

Frequent in base-rich habitats with the last species but more widespread and extending onto trees with basic bark such as elm, ash and field maple. Undoubtably rarer of late due to the loss of large wych elms from Dutch elm disease.

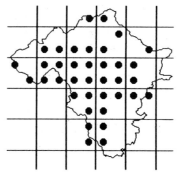

Sphagnum

Frullania fragilifolia

A scarce sp. of sunny, often somewhat basic rocks such as the dolerite at Stanner Rocks 2658 and the mudstones by the R. Teme, Cilfaesty Hill 1283 and occasionally on trees, particularly oak in the Elan Valley.

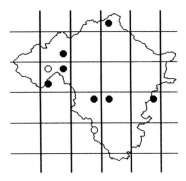

Frullania dilatata

Common on the bark of deciduous trees, colonizing quite small branches, especially in the W. Occasionally present on rock, though less frequent than *F. tamarisci*. Capsules frequent. Recorded from all squares except 86NW and 87SW.

Jubula hutchinsiae

A rare sp., known only from wet rocks, deep in the gorge of the Bach Howey below Trewern Hill, Llanstephan 1143 ! and MP.

Marchesinia mackaii

Known for over 60 years from shallow soil in crevices and overgrowing the adjacent calcareous mudstone rocks in the wooded gorge at Aberedw 0847.

Lejeunea cavifolia

Frequent on damp, shaded rocks by streams and occasionally seen elsewhere on tree trunks and rocks in damp woodlands and gorges. Capsules frequent.

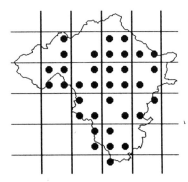

Lejeunea lamacerina

In similar situations to the last but much rarer.

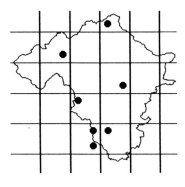

Lejeunea patens

Rare on damp, somewhat base-rich rocks in shady places. Cerrig Gwalch, N of Rhayader 9370; rocky gorge near Graig Fawr, Llandeilo Graban 0944; Teme gorge below The Ring, Felindre 1283; Rock Dingle, Bleddfa 2067 ! and AO. Capsules occasional.

Lejeunea ulicina

A rare sp. of tree trunks in damp woodlands. Despite careful searches the plant remains unaccountably rare compared to its frequency in Brecknock. Alder by stream, Cwm Coel, Elan Valley 8963; on old oak on the shore of Caban Coch Resr., Elan Valley 9163; rusty-willow in wet woodland by the R. Marteg near Nannerth, St Harmon 9671; on hazel in dingle, Cwmbach Llechrhyd 0354; on willow S of Wern-fawr, Painscastle 1845 ! and PJP; on alder in wet woodland, Fforest Colwyn, Hundred House 1153.

Cololejeunea minutissima

Known only from the trunk of a rusty willow on the edge of the basin mire at The Bog, Newbridge on Wye 0157. Capsules abundant.

MOSSES
MUSCI

Sphagnum papillosum

Frequent on heaths and acidic peat bogs, especially in the W, but nowhere abundant. Scarce in the central uplands. Capsules rare.

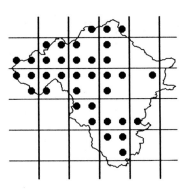

Sphagnum palustre

Common in acidic flushes and bogs on open hills, stream-sides and in woodlands, always where there is water movement.

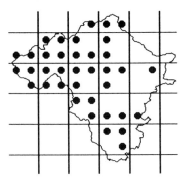

Sphagnum squarrosum

Generally scarce but present on most wooded basin mires near Newbridge on Wye 05 and Llandrindod Wells 06, on Rhosgoch Common 1948 and rarely in soft rush flushes and in damp, peaty, acidic woodlands. Capsules occasional.

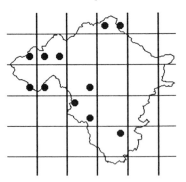

Sphagnum teres

A rather scarce sp. of shaded, usually wooded, peaty flushes in the NW of the county. It frequently occurs with the last sp. and in some sites intergrades with it.

Sphagnum

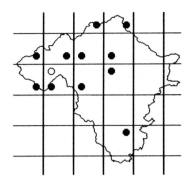
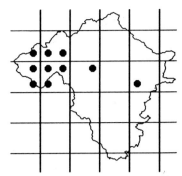

Sphagnum compactum

Rather scarce in flushed, peaty, grazed turf. Valley E of Hillgate, Upper Elan Valley 8773; Craig-ddu, Treheslog, Llansantffraed Cwmdeuddwr 9369 JAP (1965, BBS records); by head of Craig Goch Resr., Elan Valley 9071; Coxhead Bank Common 1570 (BBS exc. 1965); Rhosfallog Common 1374 and Moelfre Hill 1175, Llanbister.

Sphagnum fimbriatum

Of scattered occurrence in soft rush flushes and on most of the least disturbed mire sites such as Cors y Llyn, Newbridge on Wye 0155 and Rhosgoch Common 1948. Also found in willow carr in upland valleys. Capsules occasional.

Sphagnum warnstorfii

Rare in nutrient-rich peaty flushes. In woodland, Cwm Coel, Elan Valley 9063 and beside spring-head flush in the upper part of the Cwmnant Valley near Rhydoldog, Llansantffraed Cwmdeuddwr 9367. Here at the southernmost edge of its range in Britain.

Sphagnum capillifolium

Scarce on blanket bogs, basin mires and in a few damp, peaty upland woods. Capsules occasional.

Sphagnum auriculatum

The var. *auriculatum* is a common moss of stony and peaty flushes and upland stream banks, readily colonizing wet roadside ditches in the hills.

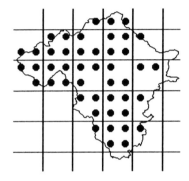

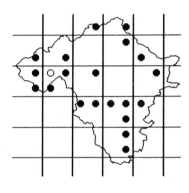

The var. *inundatum* is widespread but not common in somewhat base-rich flushes in unimproved grassland and rarely on the floating lawns of basin mires.

Sphagnum girghensohnii

Scarce on damp, N facing woodland banks. Roadside bank near Pen-y-garreg Resr. dam, Elan Valley 9067; woodland bank, Nannerth, Rhayader 9472 ! and FR and nearby on bank of R. Wye 9572 and in old railway cutting above Pont Marteg near Nannerth 9571.

Sphagnum russowii

Recorded from Llandrindod Wells c06 by SW (*Rep BBS* 1944-45, 277), there are no other records. It occurs close to Radnor in the Elan Valley and should be sought on damp, shaded woodland banks.

Sphagnum quinquefarium

Frequent only in a few moist sessile oakwoods and on damp, acidic, shaded banks in the NW uplands. Capsules occasional.

Sphagnum subnitens

One of the commonest bogmosses occurring on shaded, peaty banks in woodlands and around peaty flushes in the uplands. Capsules frequent.

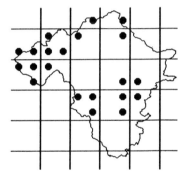

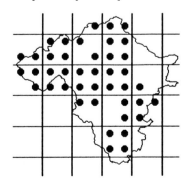

Sphagnum subsecundum

Rare in somewhat base-rich peaty flushes, below Blaencoel, Cwm Coel, Elan Valley 8963 and near Pont ar Elan 9071.

Sphagnum contortum

Rare in peaty flushes. By the R. Elan above Aberglanhirin 8872 and NW of Y Foel 9165 in the Elan Valley. Perhaps overlooked in flushes elsewhere.

Sphagnum cuspidatum

Frequent in peat-bottomed mawn pools on the hills and in lowland mires. Also found on the surface of peat in blanket bogs damaged by burning where the peat has become impermeable to water.

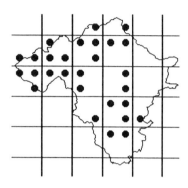

Sphagnum tenellum

A scarce moss of blanket bog and peaty flushes.

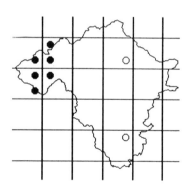

Sphagnum recurvum

The commonest bogmoss, found in soft rush flushes, peat bogs, peaty woodlands and a ready colonist of upland ditches. Capsules occasional. The vars. have not been separated.

Andreaea

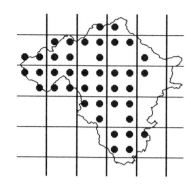

Andreaea spp.

The genus *Andreaea* is the subject of a recent review by Dr. BM Murray. She has determined three gatherings from Radnor as *A. rothii* ssp. *falcatum*. They are from rocks E of Craig-ddu near Rhayader 9469 ARP (1965 Hb. ARP); Glog-fawr, Elan Valley 9166 RD Fitzgerald (1965 NMW); Llandegley Rocks 1361 AEW (1959). The account below was prepared prior to this revision.

Andreaea rupestris

The var. *rupestris* is an uncommon moss of somewhat shaded and frequently damp, acidic rock outcrops. Craig Dyfnant, Claerwen 8664; Dyffryn Wood, Rhayader 9766; Craig-y-foel, Elan Valley 9164; Craig-cefn-lech, St Harmon 9577; on dolerite, Craigfawr, Frank's Bridge 1358 and near Llanwefr Pool, Llandegly 1359 PJP.

Andreaea rothii

The var. *rothii* is frequent on acidic rock outcrops in the NW uplands, favouring sunny, seasonally damp sites. It is rare in the E with an isolated population on the volcanic rocks of Hanter Hill near Stanner 2557.

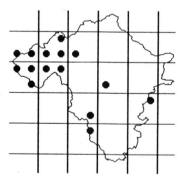

Tetraphis pellucida

Frequent on decaying logs and tree stumps in hedgerows and woodlands and occasionally on the edges of peat hags in blanket bogs. Capsules occasional in the damper sites.

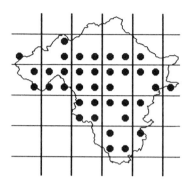

Polytrichum alpinum

The var. *alpinum* is a scarce or overlooked moss of silt or rock debris in open situations in the hills. Silty soil by R. Wye, Nannerth, Rhayader 9571 AO; rocks by road W of Abbeycwmhir 0370; by R. Ithon N of Llanbadarn Fynydd 0983 AO; rock ledge near source of R. Teme 1283 MO Hill. Capsules frequent.

Polytrichum longisetum

A single record from Rhosgoch Common 1948 HHK (1908 NMW). Seen on recently burnt moorland in Brecknock, it should be sought in the Radnor Forest or Elan Valley.

Polytrichum formosum

A common sp. noted in all squares on acidic humus on stumps, woodland, hedge and stream banks and the drier parts of heaths. Capsules common.

Polytrichum commune

Recorded from all squares except 04, 05SW, 13NE, 15SW, 24, 25NE, 26SE and 36, but commonest in the NW where it forms dense stands in the wet-flushed parts of moorland. Also in wet acidic, peaty flushes in woodland. Capsules are common. The vars. have not generally been separated but var. *commune* appears to be the commonest.

Polytrichum piliferum

Common on rock outcrops, dry sandy banks, shingle by rivers and mineral soil on heaths, moorland, rough grazings and road banks. Favouring acidic, well-lit sites, it is commonly with capsules and has been recorded from all squares except 06NE, 08SE and 37SW.

Atrichum

Polytrichum juniperinum

Grows in similar places to the last sp. but is never quite so abundant and perhaps avoids the driest sites. Capsules are common and it has been recorded from all squares except 86NW, 87SE, 97NW, 06NE, 13NE, 24NW and 36SW.

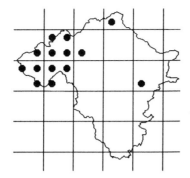

Pogonatum nanum

Noted from the county in the Census *Catalogue of British Mosses* ed. 2 (JD Duncan,1926). A copy of the 1st ed (1907), in the keeping of the BBS moss recorder, has in its margin a note in the hand of JB Duncan that Watson records the species from Aber Edw (sic). It is presumably this record that provides the Radnor entry in the 2nd ed. No specimen or other records have been traced.

Pogonatum aloides

The var. *aloides* is common and is recorded from all squares except 13 and 36NW on shaded, acidic soil and clay on banks in woods, hedgerows, road and streamsides.

Pogonatum urnigerum

Widespread but only common in the uplands in soil-filled crevices in rocks, on gravel and mineral soil beside flushes, streams and tracks, always in damp, acidic situations. Capsules common.

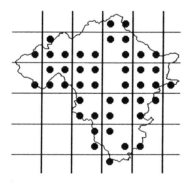

Oligotrichum hercynicum

Frequent in the uplands of the NW where it colonises damp clay and gravel on banks, particularly beside streams. In the E it is only known from the Wimble on the Radnor Forest 2062 ARP and WD Foster. Never seen with capsules.

Atrichum crispum

Forms hummocks on silt in the flood zone of rivers and reservoirs in the NW uplands. Hirnant Claerwen 8366 ! and AO; by the R. Elan, Aber Glanhirin 8872; R. Claerwen near Llanerch-y-cawr 9061; exposed mud at head of Craig Goch Resr., Elan Valley 9071; by the Marcheini Fawr, W of St Harmon 9573 ARP; bank of R. Wye below Ddol Farm, Rhayader 9767. Male plants only.

Atrichum undulatum

The var. *undulatum* occurs commonly on partially shaded soil in all but the most acidic woodlands, hedgerows, stream and roadside banks. Capsules common. Recorded from all squares except 86NW.

Diphyscium foliosum

Almost confined to the western uplands on damp, nearly vertical peaty banks in woodlands and amongst rock outcrops. Its distinctive grain-of-wheat-like capsules are rarely produced.

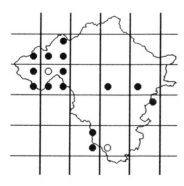

Archidium alternifolium

An inconspicuous moss of open, grazed turf or mown roadside banks which has probably been overlooked. Most frequently found on the heavily grazed turf of the commons in the SE of the county.

Pleuridium acuminatum

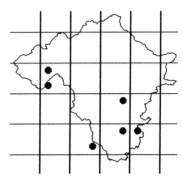

A common sp. of soil in open habitats such as stream and roadside banks, the sides of yellow hill ant mounds and soil in crevices of rock outcrops. Capsules common and only fertile material determined.

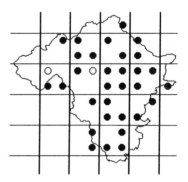

Pleuridium subulatum

Much scarcer than the last sp., from which it has been separated only when fertile and probably favouring similar, but more base-rich sites.

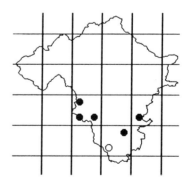

Seligeria

Pseudephemerum nitidum

Widespread on damp, open soil. Perhaps most common on stream banks, but also in fields, woodlands and hedgebanks. Capsules common.

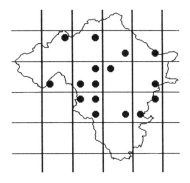

Ditrichum cylindricum

Rare or overlooked on bare soil in the E of Radnor. By the track to Waterbreak-its-neck, Llanfihangel-Nant-Melan 1859 (1965 BBS exc.); arable field, Little Hill, Hundred House 1353 PJP; potato field, Foesidoes, Radnor Forest 2265 EWJ (*Rep BBS*, 1944-45); soil of grass ley, Brunant, Llowes 2042 PJP; arable fields, Walton 2560 PJP; near Norton 2868 PJP.

Ditrichum flexicaule

A rare calcicolous moss only noted on mud encrusted N facing volcanic rocks in disused quarry at Llanelwedd 0452 and in Yatt Wood, Dolyhir 2458, where it occurs in quantity on limestone scree at the edge of the wood.

Ditrichum heteromallum

A common, though inconspicuous sp. of shaded soil on banks by tracks, in woods and beside streams. Capsules common.

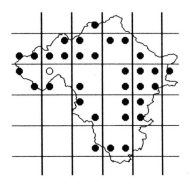

Ditrichum lineare

This nationally uncommon sp., here at the southernmost extremity of its range in Britain was recorded twice by BBS members on excursions in Radnor in 1965. On a steep bank at the N end of Garreg Ddu Resr., Elan Valley 9167 and beside a stream flowing into Gwynllyn, Llansantffraed Cwmdeuddwr 9469.

Brachydontium trichodes

A scarce moss of shaded rock outcrops. On vertical rock face Cefn-hir Dingle, Gladestry 2055 PJP; on the sheltered underside of a mudstone rock outcrop to the NW of Whinyard Rocks, Radnor Forest 2063; mudstone outcrop in wood E of Knucklas 2673 ! and PJP.

Seligeria pusilla

Rare on moist, shaded, basic rocks. Aberedw Rocks c04NE MFV Corley (1969 BBS records) and in the Radnor Forest c16 by SW (1944-45, *Rep BBS*, 1946, 279).

Seligeria recurvata

Rather scarce on moist, shaded, calcareous mudstone rocks in mid and E Radnor. Capsules common.

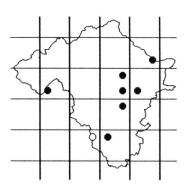

Blindia acuta

Scarce on wet, acidic rocks in stony flushes. Hirnant Claerwen 8367; near Claerwen Dam 8763; Esgair-cywion, Elan Valley 8870; Y Foel, Elan Valley 9165; The Ring, Upper Teme Valley 1283.

Ceratodon purpureus

The var. *purpureus* occurs in all squares except 86NW and 87SW. It is common in urban and rural areas throughout the county on slightly acidic to basic soil and occasionally on decaying wood in sunny situations. Capsules common.

Rhabdoweisia fugax

Scarce on acidic soil on shaded rock ledges in the W of the county. Craig-y-bwch, Elan Valley 9062; Craig-y-foel, Elan Valley 9164; Glanllyn Wood, Llansantffread Cwmdeuddwr 9469; Dyffryn Wood, Rhayader 9766; Aberedw Rocks 0746 BBS exc. (1965, *TBBS* **5**, 192); Dol-y-fan Hill, Newbridge on Wye 0160.

Rhabdoweisia crispata

Rare on soil in crevices in rock outcrops around Gwynllyn ie Treheslog Rocks 9468; cliffs above Glanllyn 9469 and on Craig-ddu 9369, all in Llansantffraed Cwmdeuddwr.

Rhabdoweisia crenulata

Rare in soil-filled crevices in rock outcrops. Craig-ddu, Treheslog, Llansantffraed Cwmdeuddwr 9369 JAP (1965 BBS exc.) and Harley Dingle, Radnor Forest 16 EWJ and EFW (1957 *TBBS* **3**, 474).

Cynodontium bruntonii

Common in the W of the county in well-lit, but sheltered, crevices in acidic rock outcrops. Scarce in the E. Capsules common.

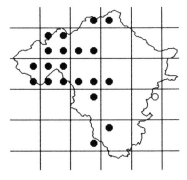

Dichodontium pellucidum

Common on the more base-rich streamside rocks in damp, shady places. Rarely with capsules, it has not been consistently separated from the next sp., all sterile material being referred here.

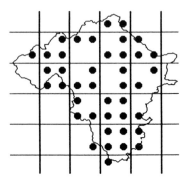

Dicranum

Dichodontium flavescens

There are no recent records, having probably been overlooked as the last sp. when sterile. EM Lobley records it from the R. Ithon (no other details given) in 1954, whilst EWJ reports it occurred in abundance on wet rocks in a ravine to the E of Harley Dingle, Radnor Forest 2062 in 1945 (specimens from both localities in NMW).

Dicranella palustris

Frequent on wet soil and gravel in flushes, on stream banks and in ditches, particularly in the uplands.

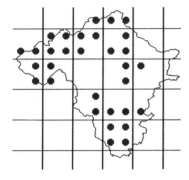

Dicranella schreberana

A nondescript and possibly under-recorded moss of damp soil on banks, particularly of streams. It probably avoids the most acidic sites.

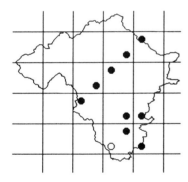

Dicranella subulata

Scarce or overlooked. Noted on a stream bank near Gwynllyn, Llansantffraed Cwmdeuddwr 9469 and on Coxhead Bank Common, Llanbister 1570 by BBS members (1965 in *TBBS* **5**, 207-209) and by EWJ on a steep, moist, N facing stony bank at 530m, Cwmygerwyn, Radnor Forest 2062 (1945, NMW records).

Dicranella rufescens

Of scattered occurrence on shaded, damp soil, particularly of stream banks. Capsules common.

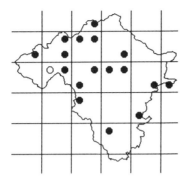

Dicranella varia

Rather scarce on damp, basic soil and rock. Noted on stream banks, shaded rock outcrops and furnace slag on a disused railway

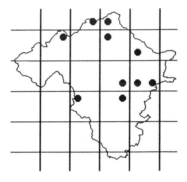

Dicranella staphylina

A tiny moss of bare soil in fields and beside paths. Much overlooked, it is probably more common on the less acidic soils of E Radnor. Reported from 0157 by AO; 2042 by AO and PJP and from 1642, 1749, 2261 and 3169 by PJP.

Dicranella cerviculata

Last reported by HHK and EA from Rhosgoch Common 1948 in 1908 (NMW), it may now be extinct. A moss recorded regularly from peat cuttings in Mid Wales about the turn of the century, it has disappeared along with the peat cutters.

Dicranella heteromalla

A common sp. of freely-drained, acidic soil and tree stumps on banks, in woods, hedgerows, ditches, etc. and noted from all squares.

Dicranoweissia cirrata

Common on the acidic bark of tree trunks of most types of deciduous tree. Occassional on old hardwood posts and rails and rare on acidic rock outcrops. In all squares except 86NE & NW and 96SW and commonly with capsules.

Dicranum bonjeanii

Rather scarce in grazed, slightly basic, damp grassland, particularly on the commons of E and S Radnor. Rare in woods and marshes on basic soils.

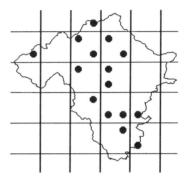

Dicranum scoparium

Common on acidic, humus-rich soil in woodlands, extending onto logs, tree trunks and branches. Frequent in acidic grassland, heather moors, roadsides and streambanks. In all squares and occasionally with capsules.

Dicranum majus

Often with the last sp. in damp, acidic woodlands but less common in disturbed and marginal habitats such as grassland and road verges. Capsules rare.

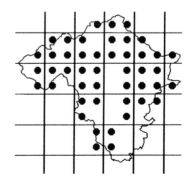

Campylopus

Dicranum fuscesens

Frequent on acidic boulders and tree trunks in the oakwoods of the Elan Valley 96. Elsewhere it is found in the least disturbed woodlands. Capsules rare.

Campylopus pyriformis

No attempt has been made to map the distribution of the vars.. Most records are probably of the var. *pyriformis*. It is frequent on peaty soil and decaying logs but is less common than the next sp.. Capsules occasional.

Campylopus introflexus

An introduced sp., first recorded in Britain in 1941 and now widespread. First noted in Radnor in 1952 (RD Fitzgerald BBSUK), it now occurs commonly on open peaty soil, tree stumps and decaying wood.

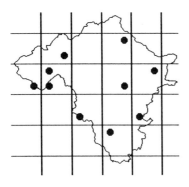
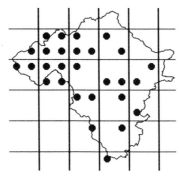
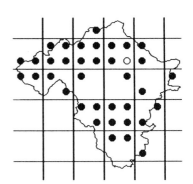

Dicranum montanum

This moss, at present extending its range in Britain, is known only from the trunk of an old pollard oak and nearby birch in damp woodland S of Caerfagu, Llanfihangel Helygen 0464.

Dicranodontium denudatum

The var. *denudatum* is of rare occurrence on decaying logs in damp, shaded woodland in the Elan Valley. Cwm Coel 8963; Nant Gwyllt 9162; near Pen-y-garreg dam 96NW (1965 BBS records); gorge S of Rhydoldog, N of Elan Village 9467.

Campylopus subulatus

Rare on sand and gravel in crevices of rocks, mostly in the flood zone of the R. Wye. By R. Wye in park above Rhayader Bridge 9668; on bend of R. Wye above Newbridge on Wye 0158; by R. Wye above Builth Wells 0252 HHK (1923 NMW); by R. Wye, Erwood Bridge 0843 CHB (*Rep BBS*, 1926, 211); Stanner Rocks 2658 EWJ (1945 NMW records).

Campylopus fragilis

Rather scarce on humus-rich soil on banks and acidic rock outcrops.

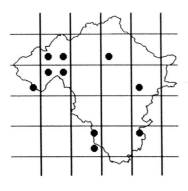

Campylopus paradoxus

Common on peaty banks, decaying wood and shallow soils over acidic rocks throughout the county. Capsules occasional.

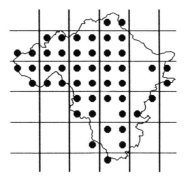

Campylopus atrovirens

The var. *atrovirens* is frequent on seasonally wet, acidic shale rocks in the W of the county.

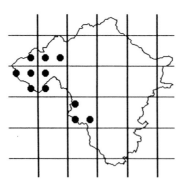

Campylopus brevipilus

Recorded from the county in a BBS supplement of 1935 to the *Census Catalogue of British Mosses* ed. 2 (Duncan 1926). The origin of this record has not been traced.

Leucobryum glaucum

A distinctive, cushion-forming moss of peat bogs and the acidic, humus-rich soil of oakwoods. Ciloerwynt, Claerwen Valley 8862; Cwmcoel 8964; blanket bog by Elan above Aberglanhirin 8873; Craig-y-bwch, Claerwen Valley 9061; Coed-y-cefn, Llansantffraed Cwmdeuddwr 9567; Dyffryn Wood, Rhayader 9766; Penybont Common 1264; Timber Hill, Cold Oak 2963 (1968 NCC records, not refound in 1985).

Leucobryum juniperoideum

Rare in oakwoods. Cwm Coel, Elan Valley c9063 JAP (1965 *J Bryol* 7,4) and on ledge in gorge of the Afon Marteg below Gilfach, St Harmon 9671.

Fissidens

Fissidens viridulus

This and the next sp. are small mosses of damp and shaded, somewhat basic rocks and soil which have probably been overlooked, particularly in the S of Radnor. Various opinions have been expressed as to their taxonomic status. Further work is required to determine their true distribution in the county.

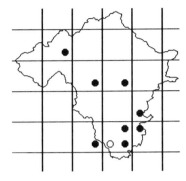

Fissidens pusillus

Perhaps in some cases overlooked as the last sp, the var. *pusillus* is rare on damp rocks by streams. Afon Dulas above Sychnant, Pant-y-dwr 9777 AO and by R. Arrow, Michaelchurch-on-Arrow 2551 AO. The var. *tenuifolius* is reported from 16 by AJE Smith (*TBBS* 6, 66), the source of this record has not been traced.

Fissidens incurvus

Scarce or overlooked on soil banks. Aberedw c04 MFV Corley (1969 BBS records); NW of Llanstephan c14SW (1965 BBS exc.); below Water-break-its-neck, Llanfihangel-Nant-Melan 1860 (1957 *TBBS* 3, 476).

Fissidens bryoides

Abundant on damp, shaded soil banks by streams, in woods, on roadsides, etc.. Avoiding very acidic sites it has nevertheless been recorded from all squares except 86NW, 87SW & SE, 97SW & SE and 27SW. Capsules common.

Fissidens curnovii

Rather scarce on damp soil in shaded crevices in acidic rockfaces, mainly in NW Radnor on rocks close to streams and rivers.

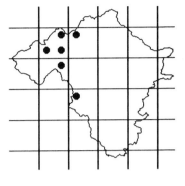

Fissidens rivularis

A nationally scarce sp. of rocks in streams and rivers. Preferring rocks submerged for most of the year it may have beem overlooked. R. Wye above Rhayader Bridge 9668; Bach Howey near confluence with R. Wye 1042 CC Townsend (BBSUK); Clyro Brook above Clyro 2144.

Fissidens crassipes

Rare on rocks in basic streams and rivers. Rocks in R. Wye by Hay Bridge 2242 EC Wallace (1955 BBSUK); rocks in Gilwern Brook near Stanner 2658 PJP; Rock Bridge, R. Lugg 2965 NTH Holmes.

Fissidens rufulus

A nationally scarce moss of frequently submerged rocks in rivers, found in a number of sites in the R. Wye, Ithon and Dulas.

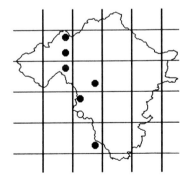

Fissidens algarvicus

A rare moss of sporadic occurrence on shaded clay or rocky banks, it has been recorded only from the gorge below Water-break-its-neck, Llanfihangel-Nant-Melan 1860 and nearby Harley Dingle 16 EWJ and EWF (1957 *TBBS* 3, 476) and noted by JAP in the former site in 1961 (BBS records).

Fissidens exilis

A small and inconspicuous sp. of clay banks seen only near the gorge of the Bach Howey, Llanstephan 14 by DF Chamberlain (1965 BBSUK).

Fissidens celticus

Scarce or overlooked on shaded soil banks, especially by streams. Gigrin Wood, Rhayader 9867; streambank N of Neuadd-ddu, St Harmon 9276; by R. Edw below Llanbadarn-y-garreg 1048; near Llanerch, Llanfihangel-Nant-Melan 1558 PJP; by R. Arrow below Michaelchurch-on-Arrow 2551 AO.

Fissidens osmundoides

Rather scarce beside flushes and on moist rocks. Seen recently only in the W of the county. Peaty flushes by the A. Claerwen in 8862 & 8962; Cwm Coel, Elan Valley 8964; Aberhenllan, Elan Valley 8972; Cerriggwynion, Rhayader 9765; rocks near Cwm-berwyn, Llansantffraed-in-Elvel 0754; Camnant Brook, Gilwern Hill 0857 (1965 BBS exc.); roadside, Lower Cwmhir, Abbeycwmhir 0370 ! and PJP; Llandegley Rocks 16SW (*Rep BBS* 1944-45, 281); Harley Dingle, Radnor Forest 1963 EWJ (1945 NMW records).

Fissidens taxifolius ssp. taxifolius

Common on shaded and moist soil and clay banks and recorded from all squares except 86NE & NW, 87, 96SW and 97SW. Capsules common.

Tortula

Fissidens cristatus

Scattered throughout the county on base-rich rocks. Less common than the next sp. and rarely with capsules.

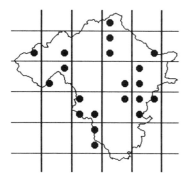

Fissidens adianthoides

Frequent throughout the county on flushed, slightly calcareous rock outcrops, in nutrient-rich mires and flushed grasslands. It is especially frequent in the grazed turf on commons in mid Radnor. Occasionally with capsules.

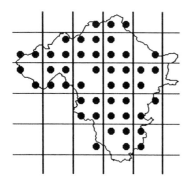

Encalypta vulgaris

Scarce on sunny, soil-capped calcareous rock outcrops. To the E of Middle Hall, Aberedw 0749; railway cutting, Cil-y-byddar, Llanfihangel Rhydithon 1670; railway cutting near Llangunllo Station 2172; cliff W of Knucklas Viaduct 2474.

Encalypta ciliata

In similar habitats to the last sp. and equally scarce. Aberedw Rocks 0746 HHK (1923 NMW); on old wall of limestone quarry and on floor of old disused quarry, Dolyhir 2458; on old limestone wall, Lower Hanter, Burlingjobb 2557; shale rocks in valley NW of Whinyard Rocks, Radnor Forest 2063. As with the last sp., only determined when with capsules.

Encalypta streptocarpa

Common on lime mortar in walls throughout the county. Occasional in seminatural habitats such as on the limestone and Precambrian rocks of Yatt Wood, Dolyhir 2458 where it occurs on the woodland floor and on the dolorite of Baxter's Bank, near Nantmel 0566. Capsules never found.

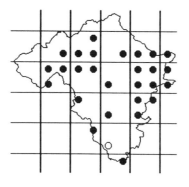

Tortula ruralis ssp. ruralis

Of widespread occurrence in a diverse range of habitats. Frequent on tarmacadam of little used paths and roadside edges, on asbestos cement roofs and once on the wooden boards of a disused farm cart. In natural habitats it was noted on an ash tree trunk near Llanbwchllyn 1146 and on basic volcanic rock outcrops at Baxter's Bank near Nantmel 0566, Alpine Bridge, Llanbadarnfawr 0963 and Craigfawr, Hundred House 1358.

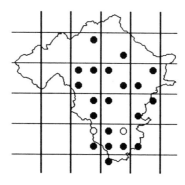

Tortula intermedia

Widespread on man-made and natural lime-rich habitats. Common on asbestos cement and stone slab roofs and more widespread on natural rock outcrops than the last sp., with records from sunny calcareous shale as well as igneous rocks and limestone. Recorded once as an epiphyte from a field maple near Alpine Bridge, Llanbadarnfawr 0963. Capsules rare.

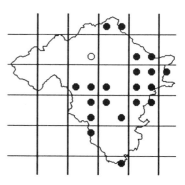

Tortula laevipila

Found on sunlit, somewhat basic or nutrient-rich bark of mature or overmature trees and is widespread but not common. It favours ash and wych elm, though mature specimens of the latter are now scarce due to Dutch elm disease. Also recorded from oak, elder, field maple and sycamore, it would probably be more common on oak but for the effects of acidic atmospheric pollutants. Noted once on asbestos cement roof of a chicken house. Capsules are common. The vars. have not been separated, but a sample survey suggests most records are probably of var. *laevipila*. Var. *laevipiliformis* has only been recorded once by HND from Rhayader c96NE (*Rep MEC* 1906, 216).

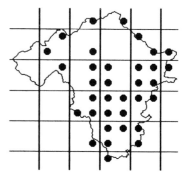

Tortula canescens

A nationally scarce moss of acidic soil subject to summer drought. It was recorded from Stanner Rocks 2658 on more than one occasion about the turn of the century, eg. CHB (1899 and 1903 NMW), but does not seem to have been reported since. Suitable habitat probably remains and it may yet be refound.

Pottia

Tortula muralis

The var. *muralis* is common and abundant on walls constructed from many types of rock, brick and cement and on most basic rock outcrops throughout the county. Capsules are abundant and the moss has been recorded from all squares except 87SW, 97SE, 18SW and 26SW.

Tortula subulata

Frequent on somewhat calcareous soil on rock outcrops and stream banks. Also present on decaying mortar on walls. Difficulty has been experienced in satisfactorily separating the var. *subinermis* from var. *subulata*. The latter var. is almost certainly the commoner, with the former perhaps present on tree bases in the flood zone of the R. Wye. Capsules common.

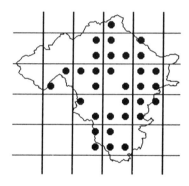

Tortula papillosa

Rare, on an elder branch at a plantation edge, Boughrood 1339; on recently felled ash tree near Yatt Farm, Old Radnor 2458 and on an elm trunk at the foot of Stanner Rocks 2658 BBS exc. (1965, *TBBS* **5**, 207-209). Not seen recently at Stanner and most elms now dead.

Tortula latifolia

On tree bases in the flood zone of mainly lowland, slow flowing streams and rivers and rarely on shady roots elsewhere. At Dolyhir 2458, close to the limestone quarries and in areas affected by wind-blown dust, it grows on trunks and branches well above the flood zone of streams. Capsules never seen.

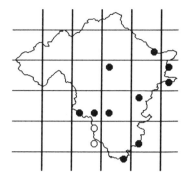

Aloina aloides

The var. *aloides* is a scarce moss of calcareous soil on rock outcrops, notably on calcareous mudstones of NE Radnor and basic volcanic rocks and limestone. Occasionally noted on lime mortar of old walls. Only determined when with capsules.

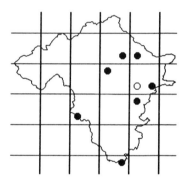

Pottia crinita

A mainly coastal sp. noted once on shallow soil on a stony bank of an old railway NW of Builth Road 0253 AO.

Pottia truncata

Common on all soils except the most acidic, in open habitats in the lowlands. Rarer in the uplands. Capsules common.

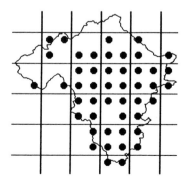

Phascum cuspidatum

The var. *cuspidatum* is uncommon or overlooked on bare soil on river banks, road verges and in arable fields. Capsules common.

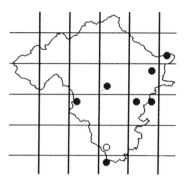

Acaulon muticum

A tiny, possibly overlooked moss of bare soil. On ant hills on the bank of an old railway line, Newbridge on Wye 0157 AO; fields near Foesidoes, Radnor Forest 2265 EWJ (*Rep BBS* 1944-45, 281).

Hyophila stanfordensis

This probable recent colonist of soil on river banks was first recorded in the county by HLK Whitehouse in 1972 on the bank of the Wye E of Llowes 2042 (BBSUK). Otherwise only known from the bank of the R. Llynfi near Glasbury 1738 where it was first seen in 1984.

Barbula convoluta

Common wherever soil accumulates in cracks in paths, walls and on banks, but avoiding the more acidic sites. The vars. have not been separated though both *convoluta* and *commutata* appear to be widespread. Capsules common.

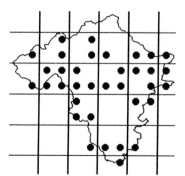

Barbula

Barbula unguiculata

Common in similar habitats to the last sp., it is particularly frequent along the edge of roadways where it grows on the tarmacadam. Capsules frequent.

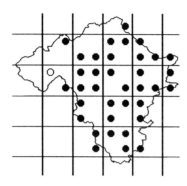

Barbula hornschuchiana

This inconspicuous moss which rarely produces capsules has probably been much overlooked. On bare soil of disturbed ground, mostly in the W of Radnor. Railway bank below Aberedw Castle 0747 AL (*Trans Woolhope Club* 1891,195); stony ground of old railway, Aberithon, Newbridge on Wye 0157 AO; soil on rock outcrop in quarry, Llanelwedd 0452. Also reported by AO from 05NE and 06SE.

Barbula revoluta

There is but a single record made between 1886 and 1890 by AL on a bridge wall and reported in *Notes on a Flora of Aberedw* (*Trans Woolhope Club* 1890-92, 195). It seems very likely that it must occur on lime mortar somewhere in the county.

Barbula acuta

Reported from Radnor in the *Census Catalogue of British Mosses* 2nd ed (Duncan 1926), the origin of this record has not been traced and no other records are known.

Barbula fallax

A somewhat scarce or overlooked sp., avoiding acidic soils and growing on open soil and gravel by roads, paths and associated with lime-rich materials, such as lime mortar.

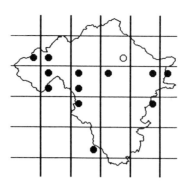

Barbula reflexa

This lime-loving sp. is only known from the limestone quarries at Dolyhir 2458 where it occurs on soil on walls and by the side of tracks.

Barbula spadicea

A rather scarce moss of rocks and alluvium by rivers. Collected by CHB and WHP from by the R. Wye, Llandeilo Graban c0483 (NMW), the collection is undated, but they are known to have visited the site in 1899. Recent searches have failed to locate this moss.

Barbula rigidula

A scarce or overlooked moss of damp, calcareous rocks and walls.

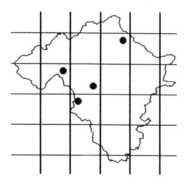

Barbula tophacea

Rare or overlooked. On mortar of old railway bridge near Newbridge on Wye 0157 AO. With capsules.

Barbula cylindrica

Frequent throughout Radnor on rocks and silt by streams and rivers and on roadside banks and the less acidic rock outcrops.

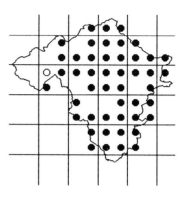

Barbula recurvirostra

Frequent only on calcareous substrates such as lime mortar, lime-rich rock outcrops and roadsides made from calcareous materials. Capsules common.

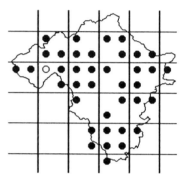

Barbula ferruginascens

Recorded from by the R. Wye at Llandeilo Graban c0843 CHB and WHP (1899 NMW) and above Builth Wells 0352 by HHK (1926 NMW). There are no more recent records.

Gymnostomum aeruginosum

A rather scarce moss of damp, calcareous shale rock outcrops, mainly of Wenlock and Ludlow age.

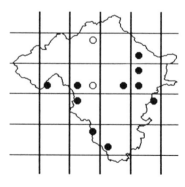

Trichostomum

Anoectangium aestivum

Reported from the county in *Census Catalogue of British Mosses* ed. 1 (Ingham 1907). The origin of this record cannot now been traced, nor have any other records been located.

Eucladium verticillatum

Rare on damp rocks receiving lime-rich water and frequently becoming encrusted with lime salts. By R. Edw, Aberedw 0847 AO; on ORS in Cilkenny Dingle, Llowes 1741; in disused limestone quarry, Burlingjobb 2458.

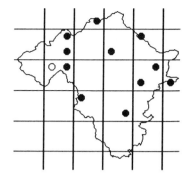

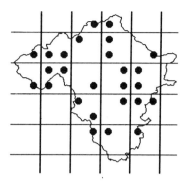

Weissia controversa

The var. *controversa* is found commonly on sandy, freely-drained soil on sunlit banks and rock outcrops. Only recorded when with capsules.

Trichostomum crispulum

Uncommon on calcareous shale rocks such as those at Aberedw 0746 and in the Bach Howey gorge at Llanstephan 1143.

Tortella nitida

Only known from the sunny, S facing rocks at Stanner 2658 where it was noted by AL (*Trans Woolhope Club* 1888, 218) and AJE Smith (1963 *TBBS* 4, 497).

Leptodontium flexifolium

Scarce on acidic soil. Peaty bank, Carreg-Bica, Elan Valley 9266 AO; wooded gorge, Aberedw 04NE AL (*Trans Woolhope Club* 1891, 195); on soil shallowly covering a small shale rock in grassland on common SE of the Begwns, Llowes 1643; Harley Dingle, Radnor Forest c16SE (BBS exc.); on shallow detritus on low rock in steeply sloping, acidic woodland, Timber Hill, Rowley, Presteigne 2863 ! and PJP.

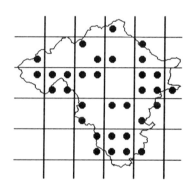

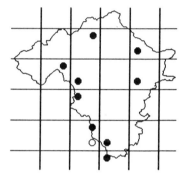

Trichostomum brachydontium

An uncommon moss of shaded soil in rock crevices on cliffs and river banks.

Cinclidotus fontinaloides

Frequent on boulders and occasionally on logs in the most base-rich rivers and streams, eg. the Wye below Builth Wells and the Edw. Capsules common.

Weissia rutilans

Recorded once from a soil-covered ledge on Aberedw Rocks 0746 JAP (1965 BBSUK).

Weissia microstoma

The var. *microstoma* occurs in similar habitats to *W. controversa* but is more scarce and is confined to the Wye Valley between Llanelwedd Rocks and Llanstephan in 04NE, 05SW and 14SW and on The Smatcher, New Radnor 2060 PJP.

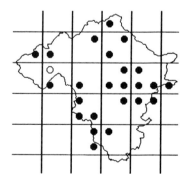

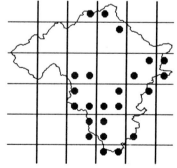

Weissia longifolia

The var. *angustifolia* was recorded in New Radnor 2160 by SW (1947 *TBBS* 1, 30).

Oxystegus tenuirostris

The var. *tenuirostris* is of scattered occurrence on soil, especially of shaded river and stream banks.

Tortella tortuosa

Of widespread occurrence on lime-rich soil in rock crevices, both dry and arround the edge of wet flushes. Frequent in limestone grassland at Dolyhir 2458.

Cinclidotus mucronatus

On the NW edge of its range in Radnor, it is found rarely on sunny rocks in the flood zone of the larger rivers. R. Wye, Erwood Bridge 0843 HND and CHB (1903 NMW); R. Ithon above Disserth 0459 E Matheson det. AO.

Grimmia

Schistidium alpicola

The var. *alpicola* is reported from Radnor by Ingham (1907). The source of this record has not been traced and no other records have been discovered. The var. *rivulare* is frequent on rocks in the flood zones of rivers and the larger streams. Capsules common.

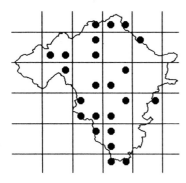

Schistidium apocarpum

The var. *apocarpum* is common on calcareous rocks, mortar and cement and has been reported from all squares except 86NW, 87SW and 37SW. Capsules are common. The var. *confertum* is rare or overlooked. It was reported by BBS members from a small rock outcrop W of Abbeycwmhir 0470 in 1965.

Grimmia laevigata

A moss of sunny, acidic rocks recorded only from Stanner Rocks 2658 by EA and not seen since her collection around 1904 (*Rep MEC* 1904, 160).

[Grimmia montana]

Possibly reported in error from Radnor. Like the last species this moss has only been recorded from Stanner Rocks 2658 i.e. CHB in 1894 (*Braithwaite Moss Fl II*, 258) and HHK in 1910 and 1926 (NMW). All specimens in NMW checked by AO from this site proved to be *G. trichlophylla*. Recent surveys by AO at Stanner have failed to find it.

Grimmia donniana

The var. *donniana* grows in cracks in sunny, acidic, shale rocks in the NW uplands of the county. Capsules common.

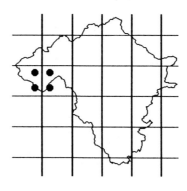

Grimmia ovalis

Scarce on sunny, somewhat basic rocks. Rocks in the flood zone of the R. Wye, Erwood Bridge 0843, HND and CHB (1903 NMW); Stanner Rocks 2658 HHK (1910 and 1926 NMW), and BBS members (1965 *TBBS* 5, 207-209) and still present. (AO, 1993).

Grimmia pulvinata

The var. *pulvinata* is common on sunny, somewhat calcareous rocks and occurring also on concrete, has been recorded from all squares except 87, 96SE, 06SW, 08SE, 17NW & NE. Capsules common.

Grimmia orbicularis

Reported from the county by Ingham (1913), the origin of this record is uncertain. A specimen in NMW collected by CHB in 1899 is labelled "Nash Scar, Presteign, Radnor". Nash Scar is in Herefordshire. If this specimen is the origin of the record for Radnor, then this moss should be deleted from the county list.

Grimmia trichophylla

Fairly frequent on sunny, acidic to somewhat basic rock outcrops throughout the county. No attempt has been made to map the vars. separately. Var. *trichophylla* is widespread, whilst var. *subsquarrosa* is noted from dolerite rocks at Graigfawr near Frank's Bridge 1358 and at Stanner 2658. The var. *stirtonii* is also known from Stanner 2658 and from the Ithon Valley, Llandrindod Wells 06 SW (*Rep BBS* 1944-45, 281).

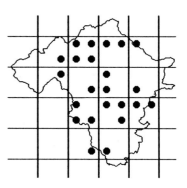

Grimmia hartmanii

A rare moss of rock on the bank of the R. Wye near Erwood Bridge 0843 (1965 BBS exc.) and on shaded dolerite rocks at Stanner 2658.

Grimmia retracta

Recorded from rocks in the R. Wye, Erwood Bridge 0843 by AL (1886 NMW) and by HND and CHB (1903 NMW). Recent searches have failed to refind it

Grimmia decipiens

Collected by CHB from Stanner Rocks 2658 (1897 and 1903 NMW) and still present (AO, 1993) on sunny, slightly basic volcanic rocks. There is also a specimen from Llandrindod Wells 06 of WHP (1900 ABS) and a record from S facing igneous rocks near Llanelwedd Quarries 0552 det. AJE Smith.

Racomitrium aciculare

Common on damp, acidic rocks, especially beside and in the flood zone of streams and rivers. In all squares except 13NE, 15SW, 16NW, 17SE & SW, 24SW, 26SW, 27NW & SE, 36SE and 37. Capsules common.

Racomitrium aquaticum

Frequent on damp, near vertical, shale rocks in usually shady places in the NW of the county. Capsules occasional.

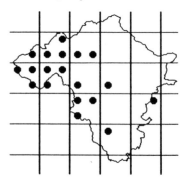

Funaria

Racomitrium fasciculare

Common on acidic rocks, especially on horizontal surfaces in the areas of highest rainfall in the W. Capsules occasional.

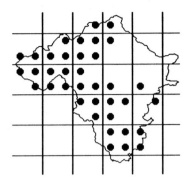

Racomitrium lanuginosum

Frequent on peat and acidic rocks in the uplands of the W and N and in scattered places in the hills of mid Radnor. Capsules occasional.

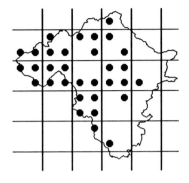

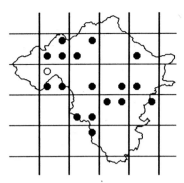

Discelium nudum

A rare moss of damp, bare, acidic, clay soil. Mud on shore of Caban Coch Resr., Elan Valley 9162 JAP and on clay of landslip beside the trunk road near Esgair Draenllwyn Bridge, Llanbadarn Fynydd 0882. Capsules present in the latter site.

Racomitrium heterostichum

The recent separation of *R. affine* leads to a degree of uncertainty as to exactly which taxa are represented in the records shown on the map of this sp.. A search for *R. affine* suggests it is scarce and almost invariably occurs with *R. heterostichum*. This latter sp. is common in the N and W of the county on acidic rocks, stone walls and screes and is found frequently with capsules.

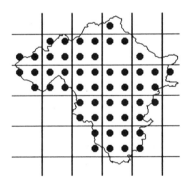

Racomitrium affine

Not reliably separated from the last sp. during most of the survey, it is probably more widespread than the records suggest. On exposed, acidic shale rock outcrops near Craig Goch Resr. dam, Elan Valley 8968; wall beside Garreg Ddu Resr., Elan Valley 9166 SG Harrison (1965 NMW det. T Blockeel); rocks in oakwood, near Glanllyn, Llansantffraed Cwmdeuddwr 9469 SG Harrison (1965 NMW det. T Blockeel) and Aberedw Rocks 0746 SG Harrison (1965 NMW det. T Blockeel).

Racomitrium canescens group

MO Hill in *Bull BBS* **43**, 21-25 drew attention to AA Frisvoll's revision of this group, splitting the above sp. into 4 entities, 2 of which have so far been reported from Radnor. The early records made in this survey cannot be reinterpreted in the light of this revision. *R. ericoides* (Brid.) Brid. appears to be the commoner taxon, found on shingle beside rivers, on rock ledges and by tracks in acidic places. It occurs in 05SE, 08SE, 14NE & NW, 15NE and 25NW. *R. elongatum* Frisvoll is found in similar places and sometimes with the last taxon but may be less common. It occurs in 25NE and NW. The map below is of the group.

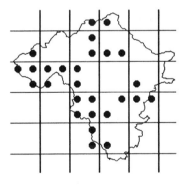

Ptychomitrium polyphyllum

Widespread on hard, acidic rocks throughout the county. Perhaps mostly an upland moss, it is absent from many apparently suitable places, yet appears, for example, on old furnace slag used to edge flower beds in the lowlands. Capsules common.

Funaria hygrometrica

Common on the sites of wood fires, on rubbish tips, about gardens and on the sides of roads and tracks. Capsules abundant.

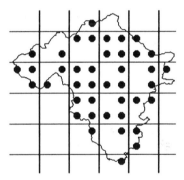

Funaria muhlenbergii and *F. pulchella*

AL records finding *F. calcarea* (now split into the above taxa), on an excursion to Dolyhir c2458 and Stanner 2658 (*Trans Woolhope Club* 1888, 218). A search, in particular, of the limestone at Dolyhir has failed to locate it and in the absence of a specimen the true identity of AL's plant cannot be determined.

Pohlia

Funaria attenuata

Rather scarce on damp soil on banks. Flushed stream bank near Aberhenllan, Elan Valley 8972; rock outcrops, Glaslyn, Nantmel 9765; near Rhayader c96 CHB (*Rep BBS* 1930, 259); rocks in R. Wye near Builth Wells c05 HHK (1923 NMW); sandy bank of R. Wye, Llandeilo Graban c04 CHB (1903 NMW). Capsules common.

Funaria fascicularis

Scarce on damp soil. Wet flush near Rhyd-myheryn, St Harmon 9877; Aberedw Rocks 0746 (1965 BBS exc.); roadside bank near disused Erwood Station 0943; NW of Llanstephan c14SW (1965 BBS exc.); Stanner Rocks 2658 PJP. Capsules common.

Funaria obtusa

On damp soil on banks, often arround springs or flushes and occasionally on stream banks. Rather scarce and only recorded when with capsules.

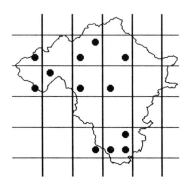

Physcomitrium pyriforme

Rare on damp soil. Near Llanstephan c14SW (1965 BBS exc.); Rhosgoch Common 1948 JAP (undated record from c1960's in NCC's files); Coxhead Bank Common, near Llanbister Road Station 1570 (1965 BBS exc.); The Boglands, Burfa 2761 PJP.

Ephemerum serratum

The var. *serratum* is a minute moss of damp soil which is rare or overlooked. On track below Trewern Hill, Llanstephan c1143 DF Chamberlain (1965 *TBBS* 5, 196); on unshaded soil by the R. Arrow above Kesty, Colva 1754 AO; ditch side, Dolau 1466.

The var. *minutissimum* is a rare moss of soil. Trewern Hill, Llanstephan c1143 DF Chamberlain (1965 BBS records); near Fron Farm, Llanfihangel-Nant-Melan 1959 EWJ and EFW (1957 *TBBS* 3, 481).

Tetraplodon mnioides

Reported from Radnor in Ingham, (1907). The origin of this record has not been traced, nor have any other records been located. Its presence in the county now must be in doubt.

Splachnum sphaericum

Rare on sheep dung lying on moist peat in eroded blanket bogs. Llyn Cerrigllwydion Uchaf, Llansantffraed Cwmdeuddwr 8469 and Gors-goch 8963 in the same parish. Capsules occasional.

Splachnum ampullaceum

Rare on dung in wetland. Reported from Rhosgoch Common 1948 by HHK (1908 NMW) and HJ Birks (1968 NCC records).

Schistostega pennata

The cave moss is rather scarce and confined to the N and W of Radnor. Small cave, near Craig Goch Reservoir dam, Llansantffraed Cwmdeuddwr 8968; roadside cave, Glan-llyn, Llansantffraed Cwmdeuddwr 9469 (1965 BBS exc.); rabbit burrow, Craig Cefn-llech, St Harmon 9577; rabbit burrow, The Banks, Llansantffraed-in-Elvel 0658.

Orthodontium lineare

An alien moss which has rapidly colonized Britain this century. First noted in Radnor in 1954 on a fence post by the R. Ithon (Hill, MO, 1989), it is now widespread on decaying logs, tree bases and peaty banks. Capsules abundant.

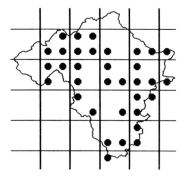

Pohlia elongata ssp. elongata

Rare, though possibly overlooked unless capsules are present. Shaded rock cleft by stream, Aberhenllan, Elan Valley 8972; crevices in Treheslog Rocks, Llansantffraed Cwmdeuddwr 9468; in dry rock crevices in Harley Dingle, Radnor Forest at 550m (1800ft) 16SE EWJ (1945 NMW records).

Pohlia cruda

A pretty pale green moss with a metallic sheen found in dry, shaded crevices in banks and shale rock outcrops, mostly in mid Radnor. It possibly avoids the wettest and most acidic of sites.

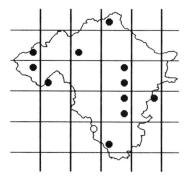

Pohlia nutans

Common and found in all squares except 04SE, 08, 14SE ,16NW, 17NE, 24NW, 26NW, 27 and 36SW on acidic humus, decaying logs and on peat. It tolerates heavy metal rich sites and is frequently the only moss growing below galvanised wire netting fences. Capsules common.

Pohlia annotina agg.

The taxonomy of these soil inhabiting, bulbiferous *Pohlia* spp. was revised by Lewis and Smith in 1978 (*J Bryol* 10, 9-27). Further work is required to determine the true distribution of the species.

Pohlia bulbifera

A scarce moss on the mud of drawn-down reservoirs. Garreg Ddu Resr. 96NW RJ Fisk (*Bull BBS* 50, 24) and at the head of Craig Goch Resr. 9071, both in the Elan Valley.

Bryum

Pohlia proligera

Common on soil exposed in stream banks and ditches, in crevices on rock outcrops and by paths. Often present as a few stems.

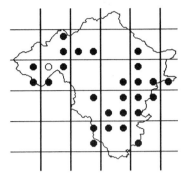

Pohlia camptotrachela

Probably more common on damp soil than the two records suggest. On exposed silt by Garreg Ddu Resr., Elan Valley 96NW RJ Fisk *(Bull BBS* **50**, 24) and on ditch side N of Park, St Harmon 0172.

Pohlia lutescens

A single record from a wheat field on the bank of the R. Wye opposite Hay on Wye 2242 HLK Whitehouse (1972 *J Bryol* **7**, 511).

Pohlia carnea

Frequent on damp soil of ditch and stream banks, especially in mid and N Radnor. Capsules occasional.

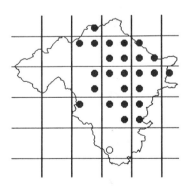

Pohlia wahlenbergii

The var. *wahlenbergii* occurs occasionally on damp soil, often by tracks and on heavily grazed commons, where seasonally wet patches form on shallow soil over rock. Also present on soil covered rock ledges and stream banks.

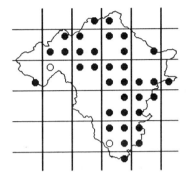

Epipterygium tozeri

Noted from by the R. Wye near Erwood Bridge 0843 by AL (*Trans Woolhope Club* 1891, 197). Not found in recent searches, but present nearby on the Brecknock bank.

Plagiobryum zieri

A scarce moss of open, damp, slightly calcareous shale rocks by streams. By the Gilwern Brook, Cwm-y-bont, Trewern and Gwaithla 1956 PJP; Harley Dingle, Radnor Forest 16SE ARP; valley W of Whinyard Rocks, Radnor Forest 2063.

Anomobryum filiforme

The var. *filiforme* is rare, having been seen recently only from rocks by the R. Wye near Newbridge on Wye 0158 AO. The var. *concinnatum* is scarce on damp rocks. By Afon Elan above Aber Glanhirin 8872; by R. Wye, Llandeilo Graban c04 CHP and WHP (1899 NMW); rocks by R. Wye, Aberedw c04NE CHB (1893 Braithwaite's *British Moss Flora* II, 163); damp volcanic rocks near Cwm-berwyn, Llansantffraed-in-Elvel 0754; by R. Teme below The Ring, Cilfaesty Hill, Felindre 1283; Stanner Rocks 2658 EWJ (1945 NMW records).

Bryum pallens

A distinctive, pink coloured moss of damp, open turf on grazed commons and stream banks. Capsules have not been found and the vars. have not been separated.

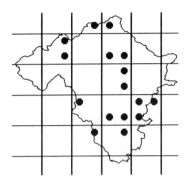

Bryum algovicum

The var. *rutheanum* is a scarce or overlooked moss, noted on the mortar of a railway bridge near the Ridgebourne, Llandrindod Wells 0560 by AO and reported from Coxhead Bank Common, Llanbister 1570 (BBS exc. 1965).

Bryum inclinatum

A lime-loving moss of soil and rocks. Near the R. Wye, Llandeilo Graban c04 CHB and WHP (1899 NMW); old quarry spoil, Llanfair 0661 AO; Stanner Rocks 2658 CHB (1897 NMW).

Bryum capillare

The var. *capillare* is common throughout the county on neutral to basic rock, soil and bark and found in all squares except 87SW. Capsules are common. The var. *rufifolium* is reported from dry rock crevices on Stanner Rocks 2658 by EFW (1955 in Syed, *J Bryol* **1**, 275).

Bryum flaccidum

Of scattered occurrence, usually on elder trunks, but also seen on a decaying tree stump and on an ORS boulder beside a dead wych elm trunk.

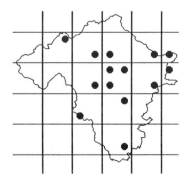

Bryum

Bryum creberrimum

Rare or possibly overlooked. On site of bonfire near Bach-y-graig, Llandrindod Wells 0660 AO. Capsules present.

Bryum pseudotriquetrum

A common moss of wet, stony flushes and stream banks in the uplands. Less common in the lowlands of the S and E. The vars. have not been separated, both being recorded from Radnor. Capsules rare. Recorded from all squares except 86NW, 04NE, 05SW & NW, 06SE, 13NW, 24SW & NW, 27SE & NW, 36 and 37.

Bryum caespiticium

The var. *caespiticium* is scarce or possibly overlooked in dry, calcareous clefts in rocks, walls and on base-rich soil. Aberedw Rocks c0746 AL (*Trans Woolhope Club* 1891, 196). The var. *imbricatum* is similarly scarce. Erwood Bridge 0843 HND (1903 BM). Only a small part of this bridge remains. Stanner Rocks 2658 EA (*Rep of MEC*, 1905, 190).

Bryum alpinum

Commonest in the W on seasonally damp rocks and gravel on cliffs and beside streams and springs. Also noted colonising the mud of the exposed margins of the Elan Valley Resrs. 96.

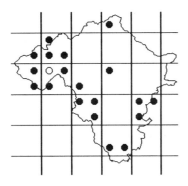

Bryum riparium

Recorded only from the NW uplands, where BBS members in 1965 found it beside the stream above Gwynllyn, Llansantffraed Cwmdeuddwr 9469 and in *TBBS* 5, 207-209 reported it as being not infrequent in shale crevices in and beside streams throughout this area.

Bryum bicolor

A common moss on soil, frequently found where soil accumulates on boulders by streams and on clay banks, but also common on little used tarmac paths and roadsides. Capsules rare. No other members of the *B. bicolor* aggregate have been found in Radnor.

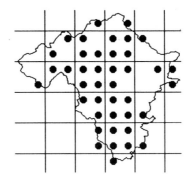

Bryum argenteum

The var. *argenteum* is a common nitrophilous moss in man-made habitats such as pavement edges, roadsides, gardens, gutters, slates on roofs and has been recorded from all squares except 87SW, 04NE, 06NE, 07NE and 18SW & SE. It is less common in seminatural habitats such as in crevices on boulders regularly used by birds. Capsules frequent. The var. *lanatum* is infrequent or overlooked. Coxhead Bank Common, Llanbister 1570 (BBS exc. 1965) and on forestry track, Burfa Hill, Evenjobb 2860 PJP.

Bryum radiculosum

Known only from a bridge over the R. Teme at Felindre 1781 M.O. Hill. Possibly overlooked on mortar in walls.

Bryum ruderale

Favouring somewhat base-rich soils it is not common in Radnor, though probably more widespread than the present records suggest. On lime-rich old railway ballast, Aberithon, Newbridge on Wye 0157 AO; bare soil, Lower Gynydd Common, Glasbury 1640 PJP; arable field, Burlingjobb 2457 PJP; soil in grass ley, Worsell near Stanner 2557 PJP and near New Radnor 2261 PJP. Also recorded from 06SE by AO and 15NE by PJP.

Bryum violaceum

In similar habitats to the last sp., though apparently more rare. Arable field near Ffynnon Gynydd 1641 PJP; on soil of roadside bank near Newchurch 2049; arable field, Burlingjobb 2457 PJP; grassland above Gwernaffel Dingle, Knighton 2770 PJP.

Bryum klinggraeffii

Scarce or overlooked in arable fields. Near Ffynnon Gynydd 1641 PJP.

Bryum sauteri

Scarce or overlooked on soil. Exposed mud of Garreg Ddu Resr., Elan Valley 9064 det. PJP; Rhosgoch Common 1948 JAP (undated card, c1965 in NCC records); roadside bank between Painscastle and Clyro c14 EFW (1957 *TBBS* 4, 617); soil on streamside, Llanfihangel Rhydithon 1666 PJP; soil by R. Wye, Llowes 2042 PJP; arable field, Burlingjobb 2457 PJP.

Bryum tenuisetum

Scarce. On soil of disused railway cutting, Fedw, Tylwch 9879; exposed silt at head of Garreg Ddu Resr., Elan Valley 96NW RJ Fisk; mud beside shallow pool on The Begwns, Painscastle 1544 CC Townsend (1977 *Bull BBS* 34, 30).

Bryum microerythrocarpum

Frequent on soil associated with slightly calcareous mudstone outcrops in the S of the county. Less common or possibly overlooked on somewhat basic soil in the mid and E Radnor.

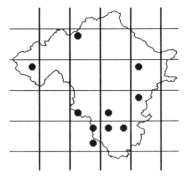

Bryum bornholmense

Reported from an ant hill near the SW margin of Pen-y-garreg Resr., Elan Valley 9067 JAP (1972 *J Bryol* 7, 511).

Plagiomnium

Bryum rubens

Frequent on bare soil in gardens, arable fields, river banks and amongst rocks where the soil is not too acidic.

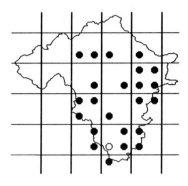

Rhodobryum roseum

A scarce moss of slightly shaded soil on rock outcrops which are at most lightly grazed and somewhat base-rich. Aberedw Rocks 0746; rocks to E of Middle Hall, Aberedw 0749; rocks in ravine SE of Upper Goitre, Llanfaredd 0750; Stanner Rocks 2658; S of Worsell Wood 2557 PJP.

Mnium hornum

A very common moss on the floor of acidic woodlands, on logs and tree bases and in the shade of rocks in the uplands. Recorded from all squares except 86NW. Capsules are produced in abundance in the spring.

Mnium stellare

A rather uncommon moss of shaded, lime-rich soil on banks in woodlands and amongst rock outcrops in the SE of the county.

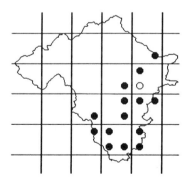

Rhizomnium punctatum

Common beside streams and rivers on soil and rocks and in wet flushes in woodland throughout the county. Capsules common. Recorded from all squares except 04SE, 15NW, 25NW and 36SW.

Rhizomnium pseudopunctatum

A scarce moss of bogs and wet flushes. Flush below Llyn Gwyngu 8472 and near Llyn Cerrigllwydion Uchaf 8469, both in Llansantffraed Cwmdeuddwr; in floating mat of vegetation, The Bog, Newbridge on Wye 0157; flush, Gelli Hill, Bettws Disserth 0958 PJP; soft rush flush by R. Teme, The Ring, Cilfaesty Hill, Felindre 1183.

Plagiomnium cuspidatum

Scarce on shaded, grassy banks by rivers and on fairly base-rich soils amongst rocks. Capsules rare.

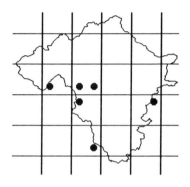

Plagiomnium affine

In similar places to the last sp., but more widespread. Occasionally with capsules. Sterile material may have been confused with *P. rostratum*.

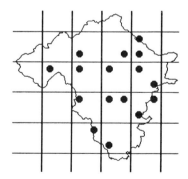

Plagiomnium elatum

A scarce moss of somewhat base-rich flushes in the uplands and around the edges of alder carrs and pools in the lowlands. Capsules rare.

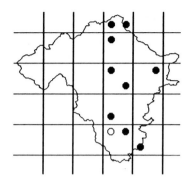

Plagiomnium ellipticum

Reported only from beside Llanbwchllyn 1146 AL (1886 *TBBS* **5**, 641).

Plagiomnium undulatum

Common on damp soil in woodlands, hedgerows and amongst rock outcrops in the more base-rich sites. Reported from all squares except 86NW & NW and 87SW. Capsules rare.

Plagiomnium rostratum

A widespread moss, trailing through turf on hedge banks, in woods, on rock outcrops and stream banks, avoiding acidic sites. Capsules are rare and sterile material may have been confused with *P. affine*.

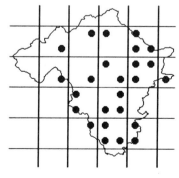

Philonotis

Aulacomnium palustre

The var. *palustre* is common in damp, acidic grasslands, bogs and heaths. Abundant in wet flushes on the heavily grazed, grass dominated mid Radnor commons. Capsules rare, plants with gemmiferous branches are occasionally encountered.

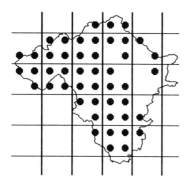

Aulacomnium androgynum

Common in E Radnor, this moss decreases in frequency westwards. It favours well-lit, but sheltered, organically-rich sites such as decaying logs and turf in hedge banks and woods. Occurs on living tree bases, particularly elder and alder. Abundantly gemmiferous, capsules have not been seen.

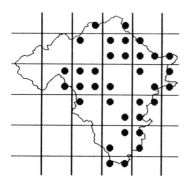

Plagiopus oederi

Known only from base-rich rock clefts, Aberedw Rocks 0746.

Bartramia stricta

This nationally rare moss has been known from Stanner Rocks 2658 for nearly 100 years. It was first reported by CHB (*Trans Woolhope Club* 1897, 305) and still survives in one small area where a few colonies cling to sheltered, S facing crevices in the dolerite rockface. Capsules produced most years.

Bartramia pomiformis

Frequent wherever soil accumulates in sheltered crevices of rock outcrops, hedgebanks and walls. Capsules common.

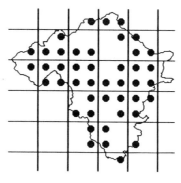

Bartramia halleriana

Rare on sheltered, base-rich shale rock ledges in woodland at Glannau in the Elan Valley 9165. Capsules are produced most years. AL reports this moss from rocks in the wooded gorge of the Edw at Aberedw c0847 (*Trans Woolhope Club* 1891, 196). It may still be present here.

Bartramia ithyphylla

A scarce or overlooked moss of soil-filled rock crevices, recorded from the gorge of the Edw at Aberedw c0847 by AL (*Trans Woolhope Club* 1891, 196); from Aberedw Rocks by HHK (1923 NMW); on rock outcrops in the Edw Valley, Llanbadarn-y-garreg 1048 and from the bank of the Ystol Bach Brook, Radnor Forest 26SW (BBS exc.).

Philonotis rigida

This nationally uncommon moss was collected by ARP in 1965 from the soft, wet shale rock faces of the shaded gorge below Water-break-its-neck, Llanfihangel-Nant-Melan 1860, grown on and identified in Oxford.

Philonotis arnellii

An uncommon moss of freely-drained, sandy and gravelly soil in open habitats. Frequent about the shale rock outcrops in the Aberedw area 04NW and beside the Wye from Newbridge on Wye 0158 to Boughrood 1338 on sandy banks and amongst stable gravel. Also on the exposed shores of the Elan Valley Resrs. 96 and 97.

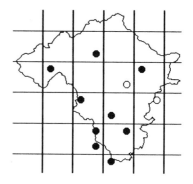

Philonotis caespitosa

This moss of wet places was recorded on Rhosgoch Common 1948 by JAP (1965 BBSUK). It is likely to occur elsewhere and may have been overlooked, resembling small forms of *P. fontana*.

Philonotis fontana

Common and reported from all squares except 07SE, 13, 24, 25NW, 26SE, 27NW, 36NW and 37 beside streams and wet flushes, especially on stony ground and around spring head flushes in the uplands. Capsules common.

Philonotis calcarea

An uncommon moss of calcareous flushes in grazed grasslands. It can form extensive golden carpets in suitable areas. Capsules occasional.

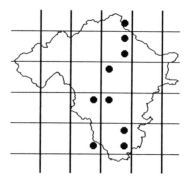

Zygodon

Breutelia chrysocoma

Frequent in the uplands in wet, stony or peaty flushes and occasionally in damp grassland, but always where there is some water movement bringing in additional nutrients. Capsules rare.

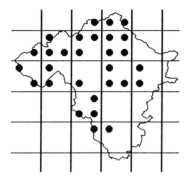

Amphidium lapponicum

This normally montane moss was found in crevices in a shale cliff by the R. Ithon, Disserth 0358 at 160m by AO. Capsules were abundant. It was first reported from the county by Ingham (1907). The origin of this record has not been traced.

Amphidium mougeotii

Frequent on shaded, moist cliff ledges. Capsules not seen.

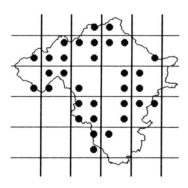

Zygodon viridissimus

The var. *viridissimus* is common on trees, particularly on the trunks of elder, field maple, ash and elm in the S and E of the county. Less frequent elsewhere. Occasional on shaded, slightly calcareous mudstone rocks. Capsules rare. The map below refers to this var.. The var. *stirtonii* is rare on shaded, calcareous mudstone rock outcrops. Aberedw Rocks 0746; cliffs to E of Middle Hall, Aberedw 0749; gorge, Cwm-y-bont, Trewern and Gwaithla 1956; gorge below Water-break-its-neck, Llanfihangel-Nant-Melan 1860 EWJ and EFW (1957 BBSUK); mudstone quarry, Dolau 1367 and rock outcrop, Lloiney, near Knucklas 2475.

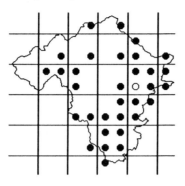

Zygodon baumgartneri

Widespread, particularly on the trunks of old trees of various species eg. ash, oak, wych elm, field maple and elder. Noted twice on calcareous shale rocks at Yr Allt, Erwood 0844 AO and on Rhulen Hill 1349. Capsules rare.

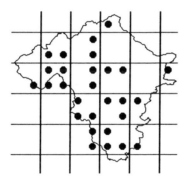

Zygodon conoideus

Rather rare, but most frequent on elder and wych elm trunks in exposed situations in E Radnor. Capsules not found.

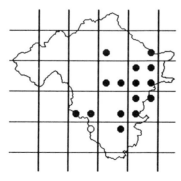

Orthotrichum striatum

Scarce on tree branches. Hazel in woodland, Nannarth Fawr, Rhayader 9472; Aberedw c04 AL (*Trans Woolhope Club* 1891, 196); hazel in Tyncoed Wood, Cwmbach Llechrhyd 0354 and on a branch of a dead wych elm tree, below Stanner Rocks 2658 PJP. Capsules common.

Orthotrichum lyellii

Widespread, though nowhere common, on the S and SW sides of well-lit tree trunks, favouring ash and sycamore but also on wych elm, sessile oak, hazel and horse chestnut. Notably abundant on sessile oak, ash and hazel impregnated with limestone dust arround Dolyhir Quarries 2457. Probably restricted in distribution due to the acidification of bark by atmospheric pollutants.

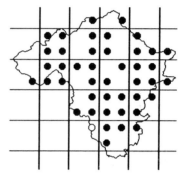

Orthotrichum affine

The most common epiphytic sp. of the genus, recorded in all squares except 86, 87SW, 96SW & NW and 13NE on a wide range of tree spp., especially elder, field maple and wych elm. Capsules common.

Orthotrichum rupestre

A rare moss known only from a disused stone quarry, where it occurs in small quantity on S facing ledges, S of Bryngwyn 1848. Capsules present.

Orthotrichum rivulare

Widespread, but not common in the flood zone of fast-flowing streams and rivers on tree trunks, tree roots and occasionally rocks. The clearance of riverside trees to facilitate land drainage has reduced the habitat available to this and the next species. Capsules common.

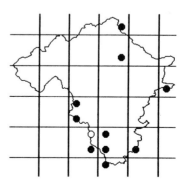

Orthotrichum

Orthotrichum sprucei

Rather scarce on tree trunks and exposed roots in the flood zone of the R. Wye between the Edw confluence 0746 and the English border. Capsules common.

Orthotrichum tenellum

Rare or overlooked. On elder in a windswept hedge near Coch-y-Roosytn, Clyro 2046. Capsules present.

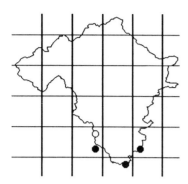

Ulota crispa

The vars. *crispa* and *norvegica* have not been separated in this survey. Both are present and of widespread occurrence on tree and shrub branches in damp woodlands. The sp. is recorded from all squares except 86NW, 87SW, 13NW, 25SW and 36 SW. Capsules common.

Ulota phyllantha

Rare on twigs in damp places. On rusty willow by a stream, Cerriggwynion, Rhayader 9765; on rusty willow in a hedge, W of Cantel Hall Farm, Llanbister 1472 coll. DPH, det.!; near Llandrindod Wells c06 by SW (*Rep BBS* 1944-45, 282).

Hedwigia ciliata

Frequent on sunny, S facing, gently sloping rock surfaces of the hard volcanic rocks such as those near Llanelwedd 0551, Llandrindod 0660, Llandegley 1361 and Stanner 2658. Scattered elsewhere and confined to the harder shales and grits. Noted on stone roof slabs on the parish church in Presteigne 3164. Capsules occasional.

Orthotrichum anomalum

Widespread on concrete in sunny places such as bridge parapets and wall tops. Not apparently known from any natural habitats. Even in Yatt Wood, Dolyhir 2457 where it occurs on ash trunks and branches it only occurs on those covered in limestone dust from the nearby quarries. Capsules abundant.

Orthotrichum diaphanum

Frequent in the valley bottoms on elder trunks, wall tops and boulders, especially in sites such as farmyards enriched with organic materials or bird guano. Capsules common.

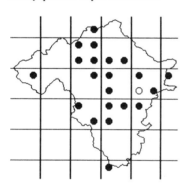

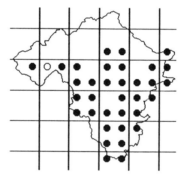

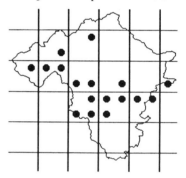

Orthotrichum cupulatum

A few colonies of the var. *cupulatum* grow on the shaded top of a concrete bridge parapet at Llaneon, Frank's Bridge 1255. Capsules present.

Orthotrichum stramineum

The neat colonies of this moss are widespread, if not common, on the trunks and branches of a range of tree spp.. Commonest on elder it also occurs on ash, sycamore, hazel, wych elm, sessile oak and hawthorn in woods and hedgerows. Capsules abundant.

Orthotrichum pulchellum

A scarce moss, mostly on elder branches. Roadside hedge, Pant-y-Dwr 9875; on rusty willow at edge of The Bog, Newbridge on Wye 0157; in spruce plantation near Castle Crab, Disserth 0555; hedge, Great Vaynor, Nantmel 0169; valley woodland, Llanfihangel-Nant-Melan 1857; stream bank, The Hill, Presteigne 2964 PJP; roadside bank W of Heyop 2274. Capsules common.

Fontinalis antipyretica

Common on rocks and tree roots in streams and rivers. The vars. have not always been distinguished. Var. *antipyretica* is the commonest, whilst *gracilis* and *gigantea* occur occasionally. Capsules uncommon.

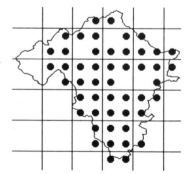

Climacium

Fontinalis squamosa

The var. *squamosa* is frequent on rocks in fast flowing streams in the uplands.

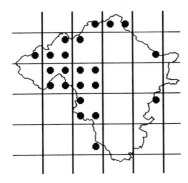

Climacium dendroides

A widespread, though not common moss of three distinct habitats. Rare amongst grass on freely drained, base-rich rock outcrops such as at Llanelwedd 0551 and Stanner 2658. Occasional around the edges of mires and peaty flushes. Widespread in the flood zone of the major rivers where, in shady places, it accumulates about itself coarse sand and gravel. Capsules not seen.

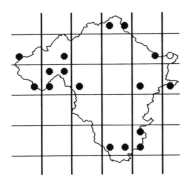

Cryphaea heteromalla

Scarce on the branches of old elders and rarely field maple. In wet woodland, Cwmbach Llechrhyd 0254; roadside hedge near Llyn-gwyn, Nantmel 0065; in woodland near limestone quarries, Dolyhir 2457; field maple, Evenjobb 2862 PJP; roadside bank W of Heyop 2274.

Leucodon sciuroides

The var. *sciuroides* is widespread, but not common, on the sunny sides of exposed, somewhat base-rich tree trunks. Commonest on ash, it is also known from field maple, sycamore, oak and hazel, the latter sp. receiving limestone dust from nearby quarries at Dolyhir 2457. Rare in other habitats such as on dolerite cliffs at Stanner 2658 and on the asbestos cement roof of a chicken house at Llanddewi Hall 1068. Capsules not seen.

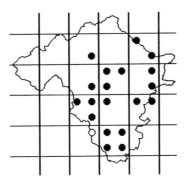

Pterogonium gracile

Uncommon on slightly basic, open but sheltered rock outcrops, such as the volcanic rocks of mid Radnor and basic mudstones. Rare on trees and favouring the base-rich bark of old ash, common oak and field maple. Capsules rare.

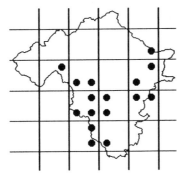

Neckera crispa

Confined to the most calcareous, frequently dry, though sometimes wet, vertical rockfaces. Despite capsules never having been observed, it occurs on some very isolated basic outcrops.

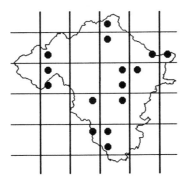

Neckera pumila

Uncommon on base-rich bark. Hazel, Nannerth Fawr, Rhayader 9472; norway maple, Cefndyrys, Llanelwedd 0453; hazel, Tyncoed Lower Wood, Cwmbach Llechrhyd 0354; hazel, Cwm-brith Bank, Cefnllys 0861; field maple, Alpine Bridge, Llanbadarn Fawr 0862; sycamore, Glascwm 1553; hazel, Ddol Wood, Llanbister 1272 ! and PJP.

Neckera complanata

The commonest sp. of the genus on base-rich trees such as ash and field maple and on lime-rich rocks. Capsules not seen.

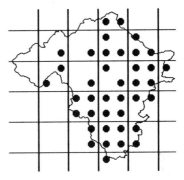

Homalia trichomanoides

Frequent on moist tree bases, especially on ash by streams and rivers and occasionally elsewhere on damp, slightly shaded, basic rocks. Capsules rare.

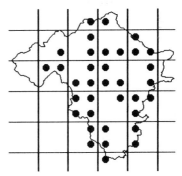

Thamnobryum alopecuroides

Common on shaded rocks and tree bases by waterfalls, streams and rivers, but also present on drier, shaded, base-rich rock outcrops. Capsules rare.

Thuidium

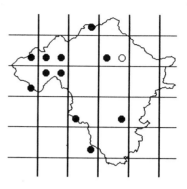

Heterocladium heteropterum

Frequent on rock outcrops in moist and heavily shaded places. The vars. have not generally been separated and are not mapped. Var. *flaccidum* appears to be very scarce and is only reported from Aberedw Rocks 0746. Capsules not seen.

Hookera lucens

A moss of moist soil and decaying wood in shady woodlands. Widespread, but inexplicably absent from many apparently suitable sites in mid Radnor. Capsules occasional.

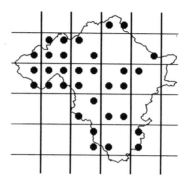

Anomodon viticulosus

Confined to freely drained, base-rich habitats. Frequent on the calcareous mudstone outcrops of the Aberedw area 04 and 14. Elsewhere on base-rich rock outcrops, cliffs and the bark of trees such as old field maple, wych elm, sycamore and ash, often associated with such cliffs. Capsules not seen.

Thuidium philibertii

A rare moss of basic grassland known certainly only from near Burlingjobb 2458 PJP. A specimen, somewhat intermediate in character between this and the last sp. was collected by EA on Stanner Rocks 2658 in 1914 (NMW).

Cratoneuron filicinum

The var. *filicinum* is common in damp, base-rich places such as around flushes, on streambanks and tracksides and in ditches. Capsules rare.

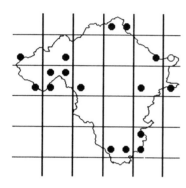

Myrinia pulvinata

An inconspicuous moss of tree trunks in the flood zone of rivers. By R. Wye NE of Clyro 2345 ERB Little and JJ Duckett (1968 BBSUK).

Leskea polycarpa

Frequent on the silt encrusted bases of trees, particularly alders, beside the larger streams and rivers. Also on trees encrusted with limestone dust from the quarries at Dolyhir 2458. Capsules common.

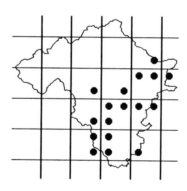

Thuidium tamariscinum

Frequent on the neutral to acidic, humus-rich soil of damp woodlands, stream and hedge banks in all squares. Capsules rare.

Thuidium delicatulum

Uncommon amongst grass on river banks, around the edges of flushes and on rock outcrops in woodland. Capsules not seen.

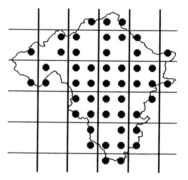

Cratoneuron commutatum

The var. *commutatum* is frequent around base-rich spring heads, often encrusted in lime salts and in associated flushes. Capsules rare.

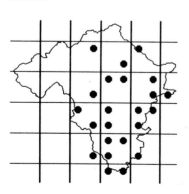

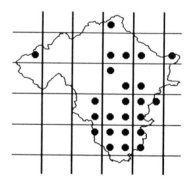

Drepanocladus

The var. *falcatum* is the commonest var. in flushes and bogs, perhaps tolerating slightly less base-rich sites than the last var.. Capsules rare.

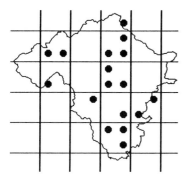

Campylium stellatum

The var. *stellatum* occurs frequently with the last sp. in base- rich flushes and marshes. Capsules rare.

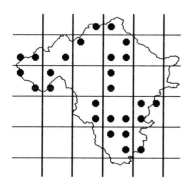

Campylium chrysophyllum

Rare or overlookeed in calcareous turf. Coxhead Bank Common, Llanbister 1570 JAP (1965 BBSUK).

Campylium polygamum

Reported by CHB and EA from turf holes in the bog, Rhosgoch 1948 (1906 NMW). The mire remains but turf has not been cut for many years.

Amblystegium serpens

The var. *serpens* is frequent in shady, humid places on stones, bricks, concrete, soil, logs and bark (especially elder tree bases). Capsules common.

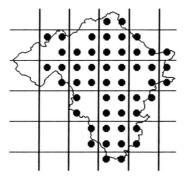

Amblystegium fluviatile

Frequent on rocks submerged for most of the year in streams and rivers.

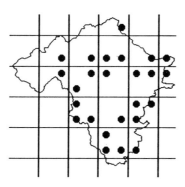

Amblystegium tenax

In similar habitats to the last sp.. Though generally less common, it tolerates greater shade and occurs in smaller streams. Capsules occasional.

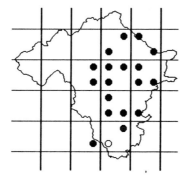

Amblystegium riparium

Uncommon on logs, tree bases, soil and stones in seasonally flooded areas by pools and the larger streams and rivers.

Drepanocladus aduncus

Rare or possibly overlooked in peaty, upland pools. Rhulen Hill 1448 PJP; St Michael's Pool, Bleddfa 1869 PJP.

Drepanocladus sendtneri

Only known from a shallow pool in the bed of a disused railway line, fed by water flowing through base rich-ballast at Builth Road 0253 AO.

Drepanocladus fluitans

The var. *fluitans is* an uncommon moss of shallow, peaty pools which frequently dry out in summer. Capsules not seen.

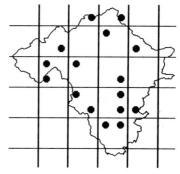

Drepanocladus exannulatus

The vars. *rotae* and *exannulatus* have not always been separated and their distribution is combined on the map below. They both occur in peaty or mawn pools, basin mires and flushes. Capsules occasional.

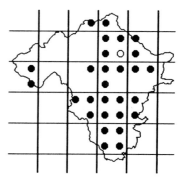

Calliergon

Drepanocladus revolvens

Not common, but widespread in base-rich, peaty flushes in the uplands. Capsules rare.

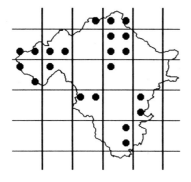

Hygrohypnum luridum

The var. *luridum* is less common than the last sp. on rocks in streams but also occurs occasionally on damp and shaded rocks away from flowing water. Capsules occasional.

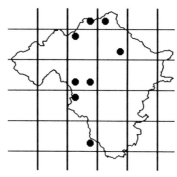

Calliergon cordifolium

In more nutrient-rich sites than the last sp.. By ponds and in wet woodland and rush-dominated flushes. Capsules rare.

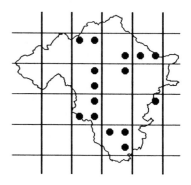

Drepanocladus vernicosus

A rare moss, reported from the Radnor Forest c16 by R Jackett (1929 NMW) and from a flush with *Dicranum bonjeani* at 380m on Glasnant Hill, Bryngwyn 1751 by PJP.

Drepanocladus uncinatus

Widespread on grassy stream banks, wet flushes in the hills, pool edges, rock outcrops and occasionally soil in woodlands in acidic to somewhat base-rich sites. Capsules rare.

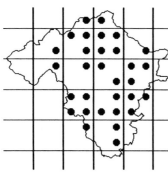

Hygrohypnum ochraceum

Frequent on rocks in fast-flowing streams. Capsules occasional.

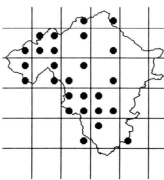

Scorpidium scorpioides

Scarce in base-rich, open peaty and stony flushes in the uplands. Capsules not seen.

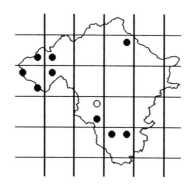

Calliergon stramineum

Frequent amongst *Sphagnum* spp. in blanket bogs, basin mires and peaty flushes. Capsules not seen.

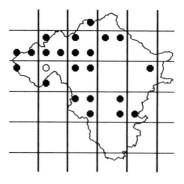

Calliergon giganteum

In similar sites to the last sp. but more frequent in wet birch woodland and willow carr. Capsules occasional.

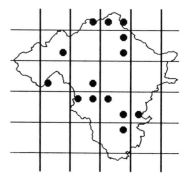

Calliergon sarmentosum

A scarce moss of peaty flushes on moorland. By the Hirnant Claerwen, Llansantffraed Cwmdeuddwr 8466; by the Nant Caletwr, Elan Valley 8769 PJP; Gwastedyn Hill, Rhayader 9766; Pont ar Elan, Elan Valley 9071; bank of the Marcheini Fach, below Llethr Llwyd, St Harmon 9574; Gorddwr Bank, Beguildy c17 or 18 T Laflin (1966 *TBBS* 5, 645).

Calliergon cuspidatum

Common and in all squares in damp terricolous habitats such as grassland, marshes, wet woodland, ditches and stream banks. Capsules occasional.

Brachythecium

Isothecium myurum

Commonest on the more base-rich lower trunks of trees such as ash and elm. Occasional on shaded calcareous rock outcrops. Capsules occasional.

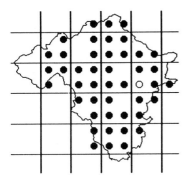

Isothecium myosuroides

The var. *myosuroides* is common on the bases of tree trunks and on acidic boulders in shady places and is found in all squares except 86NW, 87SW and 07NW. Capsules common.

Isothecium holtii

On large boulders, rarely extending onto tree bases, beside and in the flood zone of fast flowing streams and rivers and beside lakes. Mostly in the NW uplands, but extending down the R. Wye to Llandeilo Graban 0843. Capsules rare.

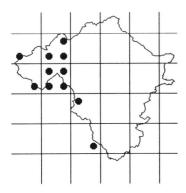

Homalothecium sericeum

Found in all squares except 87SE and SW. It favours the more base-rich habitats such as calcareous shale outcrops, mortared walls or the bark of old elders, ash, elm, field maple and oak trees. Capsules occasional.

Homalothecium lutescens

An uncommon moss of lime-rich soil on rock ledges and in grassland.

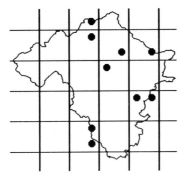

Brachythecium albicans

Frequent on gravelly soil on and beside tracks, in old quarries and on shallow soil overlying rock outcrops, always avoiding the most acidic sites. Capsules not seen.

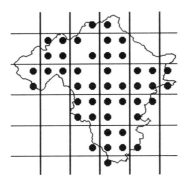

Brachythecium glareosum

In similar places to the last sp. but more scarce. Capsules not seen.

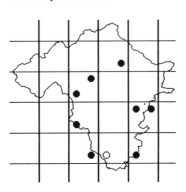

Brachythecium rutabulum

This common moss, found in all squares except 86NW, 87SW and 16SE, occurs on decaying logs, in grassland, on boulders and tree bases, notably on elder and elm in shaded and damp conditions. Capsules common.

Brachythecium rivulare

Common and in all squares except 86NE & NW on stream banks, in wet grassland, woodlands, marshes and flushes. Capsules occasional.

Brachythecium velutinum

On shaded logs, stumps, tree bases and occasionally somewhat basic rocks. Capsules frequent.

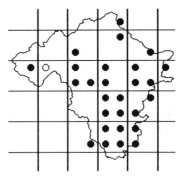

Brachythecium populeum

Commonest on damp, shaded, base-rich rocks in woodlands, lanes and by streams. Capsules occasional.

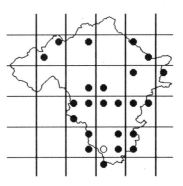

Eurhynchium

Brachythecium plumosum

Common and in all squares except 86NE, 87SW and 36SW on rocks, logs and tree bases by and in streams and wet woodland. Capsules frequent.

Pseudoscleropodium purum

Very common in damp grassland in woods, hedge banks, unimproved pastures and streamsides except in the most acidic sites. Recorded from all squares except 87SE & 17NE. Capsules not seen.

Scleropodium cespitans

Most frequent on silt encrusted, sunny tree trunks beside slow flowing streams and rivers, though noted once on tarmac on a little used path in a cemetery, Llandrindod Wells 0562 AO. Capsules not seen.

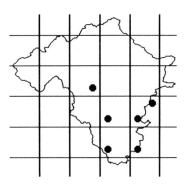

Scleropodium tourettii

Rare on thin soil overlying somewhat calcareous rocks in warm, but shaded places. Aberedw Rocks 0746; boulder by R. Wye, Erwood Bridge 0843 CHB and HND (1903 K); by R. Edw near Llanbadarn-y-Garreg 1048; cliffs NW of Llanstephan House 1142; Craig Pwll du, Llanstephan 1243 HND (1903 ms BM).

Cirrophyllum piliferum

Common in grassland and amongst herbs in moist, somewhat basic, nutrient-rich woodlands, hedgerows and streambanks. Capsules not seen. Recorded from all squares except 86, 87, 97SE, 16SE & NE, 17NW & NE, 26SW and 37.

Cirrophyllum crassinervium

On base-rich rocks on cliffs, scree and beside streams in moist, slightly shaded sites, mainly in the S of the county. Rare on trees, but present on large ash tree receiving limestone dust from the quarries nearby at Dolyhir 2457. Capsules not found.

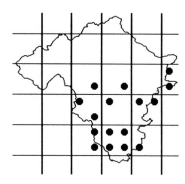

Rhynchostegium riparioides

Common on rocks, large boulders and occasionally logs in streams and rivers except in the NW uplands, where it is replaced by Hyocomium armoricum. It is, in consequence, recorded from all squares except 86NE & NW, 87SE & SW and 96NW & SW. Capsules common.

Rhynchostegium lusitanicum

Only known from rocks in the Nant Pant-y-llyn, Treheslog, Llansantffraed Cwmdeuddwr 9468.

Rhynchostegium murale

Apparently scarce on shaded, somewhat basic rocks and mortared stone walls in a few scattered localities.

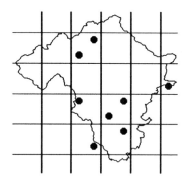

Rhynchostegium confertum

Widespread but not common on tree bases and branches, particularly of elder. Occasional on slightly basic rocks in shady places. Capsules frequent.

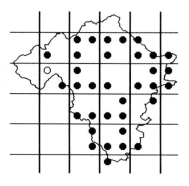

Eurhynchium striatum

Common on soil and tree bases in woodland and on hedgebanks in all except the most acidic sites. In consequence it is reported from all squares except 86NW, 87SE & SW and 97SE. Capsules rare.

Eurhynchium pumilum

A scarce moss of shaded shale outcrops. Aberedw Rocks 0746 (BBS exc. 1965) and cliffs by Craig Pwll Du, Llanstephan 1243 HND (1903 K).

Eurhynchium praelongum

The var. praelongum is common and is reported from all squares except 86NW and 87SW. It occurs on slightly acidic to basic soil in shaded places such as woods, hedgebanks and streams and is also common in grassland. Capsules occasional. The var. stokesii has been separated at times with difficulty. Material probably referable to this var. occurs occasionally on the wooded banks of streams, often in areas where sand and gravel accumulate.

Eurhynchium swartzii

The var. swartzii is found in straggling patches in hedge bottoms, on roadside banks and on soil amongst herbaceous plants in the more base-rich places. Capsules not seen.

Plagiothecium

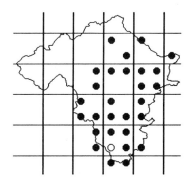

Eurhynchium schleicheri

Known only from shady rocks at Aberedw 04 AL (1886 in the Hb. of HND at K). Possibly overlooked in the lower Wye Valley.

Rhynchostegiella tenella

The var. *tenella* is widespread on damp, calcareous rocks and on the lime mortar of walls in shady places. Capsules common.

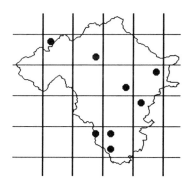

Rhynchostegiella teesdalei

No recent records."In a single shady rivulet" Aberedw 04 AL (*Trans Woolhope Club* 1891, 198); boulders below the falls, Craig Pwll Du, Llanstephan 1243 HND (1903 K); dry cleft by stream near Rhulen 14 or 15 EWJ (1945 NMW records).

Orthothecium intricatum

A scarce moss of sheltered rock crevices. Aberedw Rocks 04 HND (1903 K); rocky valley on N side of Rhulen Hill 1349; Harley Dingle, Radnor Forest 16 EWJ (1945 NMW records); gorge of the R. Teme below The Ring, Cilfaesty Hill, Felindre 1283.

Plagiothecium latebricola

Scarce in damp, peaty places. Stump in marsh by Clewedog Brook, E of Abbeycwmhir 0770 JAP (1965 *TBBS* 5, 207-209); Rhosgoch Common 1948 JAP (undated record, probably mid 1960's, in NCC files).

Plagiothecium denticulatum

The var. *denticulatum* occurs widely on tree bases, logs and boulders in hedgerows and woodlands. Capsules frequent.

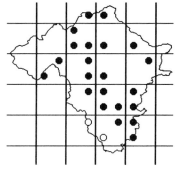

Plagiothecium ruthei

Known only from a wet wooded bank by a disused railway near the Afon Marteg, NW of Rhayader 9571 ARP *(J Bryol* **8**, 176).

Plagiothecium curvifolium

Rare or overlooked in the E of Radnor. On rotten tree stump, Middler Wood, Evenjobb 2861 PJP.

Plagiothecium laetum

A rare moss of tree trunks and bases. Old ash trunk in wood, Llanboidy, Frank's Bridge 1457 AO det. MO Hill; tree bases below Tylcau Hill, Llanbister 1476 PJP; near Whitton 2967 PJP.

Plagiothecium succulentum

The commonest sp. in the genus and reported from all squares except 86NW & NE, 87, 13NE and 26NE. On soil, humus and logs in woodlands, shaded hedge banks and streamsides. Capsules occasional.

Plagiothecium nemorale

In similar habitats to the last sp. but not as common or possibly overlooked. Capsules occasional.

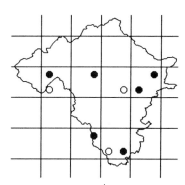

Plagiothecium undulatum

Widespread, but not common in woodlands, except on the floor of damp, shaded, acidic oak woods in the uplands where it may be frequent. Occasionally under heather and bilberry on banks and on blanket bog. Capsules rare.

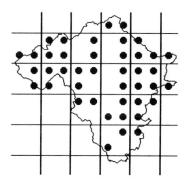

Isopterygium pulchellum

On damp, shaded shale rocks in the gorge below Water-break-its-neck, Llanfihangel-Nant-Melan 1860 JAP (1965 *TBBS* **5**, 204). Possibly overlooked elsewhere.

Isopterygium elegans

Abundant on acidic soil, humus and logs in shaded places. Recorded from all squares except 86NW, 87SW, 07NE, 13NW, 15SW, 17NW and 18SW.

Taxiphyllum wissgrillii

Scarce on somewhat base-rich, shaded mudstone. Wooded gorge, Aberedw 0847 AL (*Trans Woolhope Club* 1891, 198); damp woodland, Cwmbach Llechrhyd 0354; stream bed, Water-break-its-neck, Llanfihangel-Nant-Melan 1860 EWJ (1965 NMW); and seen there recently by ARP and the author.

Hypnum

Pylaisia polyantha

Scarce on basic tree branches. Elder, Cilkenny Dingle, Llowes 1741 AO and ! ; on elder near limestone quarries, Dolyhir 2458; on wych elm in hedge, Rosser's Bridge, Presteigne 3464. Capsules frequent.

Hypnum cupressiforme

The var. *cupressiforme* is common on horizontal tree trunks, logs, rocks, hedgebanks and streamsides and has been recorded from all squares except 86SE, 87SE, 07NE and 08SE. Capsules frequent.

The var. *resupinatum* is frequent on the somewhat basic bark of trees such as elm, field maple, ash and elder. Capsules occasional. In all squares except 86NW, 87SW, 97SE & NW, 17SE, 18SE, 26SE & NE and 27SE & NW.

The var. *lacunosum* is common on the tops of sunlit, basic rocks and scree and occasional in grazed turf in all but the most acidic sites. In all squares except 86NW, 87, 97NE, 06NE, 07NE, 17NE, 18SW and 27SE.

Hypnum mammillatum

Common and reported from all squares except 86NW, 87SW & SE and 16SE on the trunks and branches of trees and shrubs in woods and hedges. Capsules common.

Hypnum jutlandicum

Frequent on acidic soil amongst grass and rock in woods, on stream banks, on heaths and in blanket bog. Capsules rare. In all squares except 04NE, 05SW, 13, 15SW, 17NW, 24NW, 25SW and 36NW.

Hypnum lindbergii

Scattered on stream and river banks, often amongst the sand and silt in the flood zone, on ephemeral pool margins, in heavily grazed turf and on gravel beside tracks. Capsules not seen.

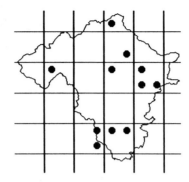

Ctenidium molluscum

The var. *molluscum* is the only var. recorded. It is widespread on soil and rock in base-rich habitats, ranging from wet flushes to dry, sunlit grassland. Capsules very rare. In all squares except 06NW, 15SW, 27SW and 36NW.

Hyocomium armoricum

Common on rocks beside streams in the western uplands. Rare elsewhere. Capsules not seen.

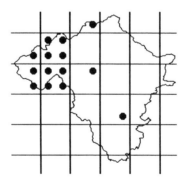

Rhytidiadelphus triquetrus

Common on soil in moist, shady places such as woodlands, scrub and hedgerows in the more nutrient-rich sites. Capsules rare. In all squares except 86, 87, 07SW, 13NW, 24SW, 25SW, 26NE and 36NW.

Rhytidiadelphus squarrosus

Abundant in grassland in fields, gardens, roadsides, woods, hedgerows, etc and recorded from all squares. Capsules rare.

Rhytidiadelphus loreus

Common on peaty soil and decaying logs in acidic oak woods and occasionally amongst heather and bilberry on banks and blanket bogs in the hills. Capsules rare. Recorded from all squares except 07NE, 13, 15SW, 16SW, 17SE, 24SW and 36NW.

Pleurozium schreberi

Common on acidic and usually peaty soils in damp sites such as woodland, hedgerow, heath and mire. Capsules not seen. Reported from all squares except 13NW, 24SW and 36NW.

Hylocomium brevirostre

A rare moss of shady, humus-rich banks, tussocks and boulder tops. In alder carr N of The Banks, Llansantffraed-in-Elvel 0656; wood above Llandrindod Wells Lake 0660 WW Boucher (1934 NMW); wood near Castell Crugerydd, Llanfihangel-Nant-Melan 1558 PJP; Water-break-its-neck, Llanfihangel-Nant-Melan 1860 EWJ and EFW (1957 *Trans BBS* **3**, 489); Fronlace, Llandegley 1663 PJP.

Hylocomium splendens

Common in damp, acidic grassland of woodland edges, heaths, mires, hedgerows and banks, particularly in the uplands. Capsules rare. Recorded from all squares except 13NW, 24NW and 25SW.

13. LICHENS and LICHENICOLOUS FUNGI

Cladonia gracilis

A) History of Lichen Recording

P.W. Carter (1950) notes that the first known published plant record from Radnor was a lichen. In Gough's edition of *Camden's Britannia* ii, 469, published in 1789 there is a record of "Lichen pustulatus on the rocks facing the south under a tower called Keven-lees castle". This lichen, now called *Lasallia pustulata*, is still present on the rocks below Cefnllys Castle 0861. The record was made by the **Rev. Littleton Brown**, a native of Bishop's Castle, Shropshire, where he resided as clergyman until 1731.

In 1871 the **Rev. W.A. Leighton** had privately printed at Shrewsbury *The Lichen Flora of Great Britain, Ireland and the Channel Isles*. It ran to three editions, the last in 1879. An hotel keeper's son, he went to school with Charles Darwin and lived in Shrewsbury, botanizing widely through mid-Wales. Some possible Radnor records made by him, mostly at Stanner and Builth, appear in this book. He, however, ascribes the Stanner records to Herefordshire, whilst the only two records from Builth, a Brecknock town, which are ascribed to a county are said to be from Radnor.

Nevertheless it is interesting that he found the uncommon *Opegrapha ochrocheila* (recorded as *O. atra* f. *ochrocheila*) near Builth, an area in which it still occurs today. Also from the Builth area, though possibly in Brecknock, he recorded *Calicium quercinum* and *Chaenotheca aeruginosa* (as *Calicium trichiale* var. *stemoneum*), lichens never elsewhere reported from Radnor. Leighton also reported the findings of other botanists. Notable where those of the **Rev. T. Salwey** (1791-1877). He was the Rector of Oswestry between 1833 and 1872 and collected lichens in many parts of Wales.

He found *Pycnothelia papillaria* having "never met with it in such perfection as upon the common immediately above the house at Llandrindod Wells" (Salway, 1846, p. 213). This lichen has never been reported again from Radnor. In the same paper he also reports finding *Parmelia stygia* on Llandegley Rocks. Watson (see below) clearly doubted this record and it sent the author searching these rocks in vain. With the nearest known site in the northern Pennines it perhaps did seen unlikely. Yet this distinctive plant has recently turned up on rocks west of Rhayader, adding substance to Salwey's record. The matter has perhaps been resolved following a check of Salwey's specimens in the British Museum. A specimen collected by him from Llandegley Rocks labelled *Parmelia stygia* is clearly *P. disjuncta*.

In 1904 **William H. Wilkinson**, from Birmingham published an account of Radnorshire lichens in the *Journal of Botany* 42, 111-113. His city was at that time constructing the Elan Valley Reservoirs and he visited the area, together with Llandrindod Wells, recording 68 taxa. *Usnea articulata* on trees near Llandrindod Wells was perhaps the most notable record. Very sensitive to sulphur dioxide pollution, it has never been seen since in mid Wales. His record of *Thelocarpon laureri* on trees at Claerwen requires confirmation. This rare species grows normally on fairly recently sawn wood or exposed rock.

No further significant published records have been traced until **W. Watson** produced his *Census Catalogue of British Lichens* in 1953. Published by the Foray Committee of the British

Mycological Society it provides a record of the lichens occuring in each Watsonian vice-county in the British Isles. There are listed 194 taxa for Radnor. This number might have been increased if Watson had not, as appears to be the case, overlooked Wilkinson's records. Watson's records are not localised and contain records of species not otherwise, as far as I have been able to discover, recorded from Radnor. Ignoring those species over which there is some taxonomic confusion or uncertainty this catalogue provides the only evidence of the existence of the following species in the county.

Cladonia mitis	*Lecidella anomaloides*
Cliostomum graniforme	*Leptogium schraderi*
Diploschistes gypsaceus	*Parmelia pulla*
D. muscorum	*Phlyctis agelaea*
Lecania fuscella	*Rhizocarpon umbilicatum*
	Thelidium incavatum

A search of the British Museum and National Museum of Wales herbaria has failed to locate any specimens. It is quite possible some of these records were based on erroneously named specimens. (There is, for example, a specimen of *Parmelia disjuncta* in the British Museum from Radnor that appears once to have borne the name *P. pulla*).

Apart from a few visits to the county by **Arthur Wade**, whilst working for the National Museum of Wales, resulting in the discovery of species such as *Farnoldia jurana*, it was not until the 1970's that any detailed recording began. In 1976 the author moved to near Newbridge on Wye, where a few years previously the University of Wales Institute of Science and Technology had established a field station. During the next decade, through occasional visits by **Brian Coppins**, **Francis Rose**, **Peter James** and **Tony Fletcher** and latterly by **Alan Orange** and much field work by the author, a more comprehensive picture of the lichens of Radnor has been built up than probably ever before. To date 535 taxa have been recorded.

B) Biogeography of Lichens

Relying on Coppins (1976), distribution maps produced by the British Lichen Society, some of which have been published, papers in *The Lichenologist*, monographs such as Coppins (1983) and summarized distributional data in Cannon et al (1985) it is possible to form some opinion as to the British distribution pattern exhibited by many lichens. As noted by Coppins (1976) the grouping of species, which each display a unique distribution pattern, cannot be exact. The classification provided here is a very provisional one, but nevertheless it is hoped that it helps to place the Radnor lichen flora in a national context.

Each geographical element account has appended a map showing the number of species in that element recorded in each 5 x 5 km square of the national grid in Radnor. The more species recorded, the larger the dot size.

Western Group

This group, demanding humid conditions and a relatively mild climate, might be equated with the Atlantic group referred to in the section on bryophytes. Further work is, however, required on world distributions to equate the distributions of lichens with those proposed by Ratcliffe (1968) for the bryophytes. Following Coppins (1976), amongst epiphytic species which he considers to display any sort of an Atlantic distribution, only the sub-Atlantic *Parmelia laevigata* and *P. taylorensis* are noted from Radnor, where they occur in the wettest NW woodlands.

More widespread Western species might be split into two groups. The first contains species largely confined to the western half of the country. Included in this group are *Arthopyrenia ranunculospora, Calicium subquercinum, Catillaria pulverea, Collema fasciculare, C. subflaccidum, Fuscidea lightfootii, Haematomma caesium, Lecidea carrollii, Leptogium cyanescens, Micarea cinerea, Microglaena muscorum, Moelleropsis nebulosa, Opegrapha sorediifera, Pannaria leucophaea, Parmeliella jamesii, P. triptophylla, Pertusaria multipuncta, Phyllopsora rosei, Porina aenea, P. lectissima, P. leptalea, Ramalina calicaris, Sarcogyne privigna, Sticta fuliginosa* and *Tylothallia biformigera*.

The second group, whilst also being common in the west, extends along the south coast of England, typically confined to ancient parks and woodlands or long-undisturbed, damp sites (F. Rose pers. comm.). It consists of *Biatorina atropurpurea, Cladonia digitata, Dimerella lutea, Lecanora jamesii, Lobaria amplissima, Pannaria conoplea, Parmelia reddenda, Polyblastia allobata, Strangospora ochrophora,*

Sticta limbata, Thelopsis rubella and *Usnea florida*.

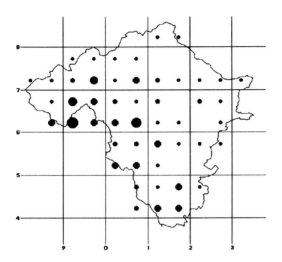

Southern Group

Lichens in this group are generally most abundant south of a line from the Mersey to the Wash. Included here are *Schismatomma cretaceum, Lecanactis premnea* and *Enterographa crassa* which occur on shaded, ancient tree trunks, *Rinodina roboris, Lecanora quercicola* and *Lecidea sublivescens* found on sunny, ancient tree trunks and *Physcia tribacia* which is commonest on sunny but sheltered shale rocks. Most of these species are confined in Radnor to the Wye valley or to sunny and warm sites in the east.

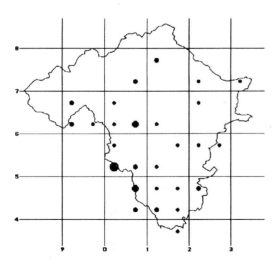

Additional species which might be placed in this group are *Leptogium teretiusculum* and *L. turgidum*. None of the species classified by Coppins (1976) as Extreme Southern have been recorded from Radnor.

South-Eastern Group

I have included here lichens which are commonest south-east of a line from the Wash to the Severn estuary, though many may extend into the lowlands of eastern Scotland. Radnor lies at the north-west edge of their range. Of the epiphytic species *Arthonia impolita, Bacidia incompta, Collema fragrans* and *Opegrapha lyncea* tend to be confined to dry crevices in bark whilst *Anaptychia ciliaris, Caloplaca luteoalba* and *Cyphelium inquinans* tend to prefer sunny, open positions. All the rock-inhabiting lichens favour sunny, lime-rich rocks, whether they be on cliffs or walls ie. *Aspicilia subcircinata, Caloplaca aurantia* and *Candelariella medians*. All probably require long periods of sunshine and perhaps high summer temperatures (July mean over 15 degrees C).

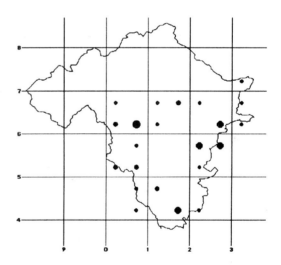

Northern and Eastern Group

This is a fairly distinctive group of lichens which occur most commonly in the eastern Highlands of Scotland and extend via the Pennines into eastern Wales. They tend to be species of the Boreal zone and may have a requirement for, or special tolerance of low winter temperatures since in Britain they are found in some of the frostiest districts. The epiphytes *Bryoria subcana, Buellia schaereri, Catillaria globulosa, Chaenotheca chrysocephala, Cetraria sepincola, Hypocoenomyce friesii, Lecidia hypopta,*

Lopadium disciforme and *Ptychographa xylographoides* may be placed here. Most are lichens of birch or Scot's pine in the Highlands, which are trees with a naturally acidic bark or wood. The introduction of conifers to Radnor and the acidifying effects of atmospheric pollutants on the bark of trees such as oak have created conditions well-suited to these species, which in some cases, might be recent colonists. *Chrysothrix chlorina, Caloplaca subpallida* and *Parmelia stygia* provide examples of rock-inhabiting lichens with a northern and eastern distribution.

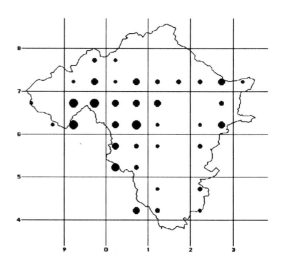

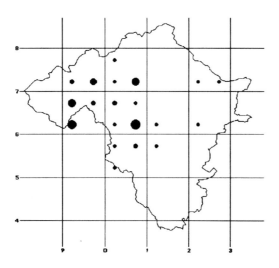

Eastern Group

This group is commonest in eastern Britain and differs from the last group in not showing a northern bias. Most, probably, have a preference for drier conditions, occupying areas with fewer than 140 rain days per annum. In Radnor a high proportion of the epiphytic species are found on sunny or sheltered, bark-free wood ie. *Buellia griseovirens, Chrysothrix chrysophthalma, Hypocoenomyce caradocensis, Chaenotheca ferruginea, C. brunneola, C. trichialis, Lecanora saligna, L. varia* and *Parmeliopsis aleurites*. On bark *Pertusaria coccodes, P. flavida* and *Parmelia elegantula* belong to this element.

Western and Northern

Since most of the upland in the British Isles occurs in the west and north this group tends to be made up of lichens favouring upland conditions eg. on rock *Baeomyces placophyllus, Bryoria bicolor, Fuscidia kochiana, Haematomma ventosum, Micarea subnigrata* and *Sphaerophorus fragilis. Nephroma parile, Ochrolechia tartarea* and *Mycoblastus sanguinarius* also occur on rock but are, in addition, found on trees. On peat and soil *Arthrorhaphis citrinella, Coriscium viride, Micarea leprosula* and *Trapeliopsis glaucolepidia* provide examples of this element.

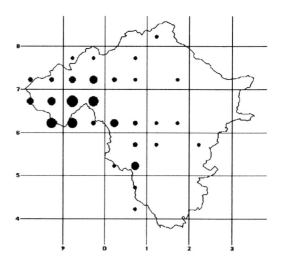

Other species with a western and northern distribution are:- *Acarospora sinopica, Bryophagus gloeocapsa, Cladonia luteoalba, Ephebe lanata, Fuscidea cyathoides, F.*

praeruptorum, Lecanora grumosa, Lecidea chalybeiodes, Massalongia carnosa, Parmelia discordans, Pertusaria dealbescens, Placopsis gelida, Rhizocarpon polycarpum, R. riparium, Umbilicaria deusta, U. polyphylla, and *U. polyrrhiza.*

Maritime Group

The species in this element are normally confined to coastal areas and rarely occur inland. Dry, sunny rock outcrops in Radnor support important populations of maritime species scarcely encountered elsewhere inland in Britain. Further work is required to explain their presence in the county. Lichens in this group include *Anaptychia runcinata, Parmelia britannica* and *Trapeliopsis wallrothii.*

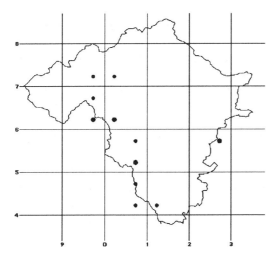

Widespread Group

Around half of the lichen species recorded from Radnor have a widespread distribution in the British Isles. They may not, however, be common eg. *Bactrospora corticola*. Species such as this have most probably been exterminated from large parts of their natural range by the felling of ancient trees and the effects of atmospheric pollution. Other species may be naturally restricted by a preference for a scarce habitat such as heavy-metal rich sites.

C) Atmospheric Pollution and Lichens

If there were pure air to breath it ought to occur in Radnor with its small population and remoteness from centres of industry. Regular air pollution monitoring has never been carried out in the county. Historically levels of at least sulphur dioxide must have been low since sensitive species (Hawksworth, D.L. and Rose, F., 1970) such as *Usnea articulata* were reported from Radnor close to the turn of the century (Wilkinson, 1904). The present widespread abundance of *Usnea* species suggests that sulphur dioxide levels are still fairly low. Most of the county can be placed in zone 9 and 10 (the least polluted zone) on Hawksworth and Rose's scale (op cit.). Yet there is accumulating evidence that sulphur and nitrogen containing compounds are exerting a considerable influence on many lichen communities in Radnor.

The recent policy of emitting potential pollutants from tall chimneys to disperse them widely and so prevent local concentrations from occuring has resulted in them travelling much longer distances. Radnor with its high rainfall and mountain mists is now the recipient of larger quantities of materials than the donor counties in the drier east. Whilst the prevailing winds are westerly, there are sufficient easterly winds to carry significant amounts of oxidized nitrogen and sulphur compounds into Radnor. Their long travel time in the air allows many chemical reactions to take place, some of which lead to the production of acidic hydrogen ions.

Following widespread sampling by the Welsh Water Authority in 1983-84, maps were prepared showing the levels of pollutants deposited in rain (Donald, A.P. and Stoner, J.H., 1988). It should be remembered that the dry deposition of acidic particles and deposition from mist (occult deposition) were not considered by this sampling technique and could be significant sources of additional pollutants in upland areas. Most of Radnor received between 0.2 and 0.4 kilograms of hydrogen ions per hectare per year, with the Elan uplands receiving up to 0.6 kilograms per hectare per year. This compares with levels of up to 0.2 kilograms in Herefordshire and lowland Dyfed.

The rain is probably sufficientlty acidic to reduce the natural base status of such substrates as bark and wood, favouring the spread of acid-loving species such as *Ochrolechia androgyna, Hypogymnia* spp. and *Parmelia saxatilis* at the expense of base-loving species, particularly those associated with the *Lobarion pulmonariae.* Recent work by the Nature Conservancy Council (Looney, J.H.H. and James, P.W., 1988) measuring the growth rate of *Lobaria pulmonaria* throughout mainland Britain have shown that colonies close to the Radnor border in north Brecknock have almost the slowest growth rates of all the sites studied. This was considered to be due to pollutants, though it proved difficult to precisely identify the cause. It seems clear, however, that *Lobaria* species in mid Wales have a very insecure future. They are confined at present to the naturally base-rich bark of ancient trees in mostly very sheltered localities and no longer seem to fruit. With, at the Government's own admission, (*This Common Inheritance. Britain's Environmental Strategy.* Cm 1200, HMSO, London, 1990) no liklehood of a significant reduction in acidifying pollutants into the next century, extinction of the *Lobarion* seems a real possibility in the near future.

Other potential pollutants include ammonium and nitrate ions. Upland Radnor receives some of the highest recorded levels of nitrate and ammonium ions in Wales (Donald and Stoner op cit.). Intensive agriculture with high densities of farm stock, the use of artificial fertilizers and the growing of clover all add nitrogen compounds to the air and contribute to wind-born imported materials. They too can lead to the production of acidic compounds, though close to point sources such as stock-yards eutrophication effects are visible. Trees may develop *Physodion* or *Xanthorion* communities, or with very high concentrations may loose their lichens and become covered in algae.

The only other point source of pollutants significantly modifying the lichen flora of at least its immediate vicinity, are the limestone quarries at Dolyhir. Here it would appear that limestone dust has stimulated the development of the *Xanthorion* community on trees up to a kilometre downwind of the quarry. Species such as *Caloplaca holocarpa*, normally confined to base-rich rocks, can be found commonly on trees in this area.

D) Lichen Conservation

Of all the major plant groups so far studied in Radnor, the lichens are probably most under threat. Though far from completely known, many species show such narrow habitat preferences that, even following a thorough study of the county, they are likely still to be considered rare. At present almost half the species recorded fall into groups considered internationally, nationally or locally rare.

Atmospheric pollution, considered above, is probably the greatest threat, followed by habitat destruction. The loss of ancient trees, the burning, ploughing and reseeding of heathland, the eutrophication of rivers, the conversion of broadleaved woodland to conifer plantations and heavy grazing in the uplands have all led to an impoverishment of the native lichen flora.

Key lichen sites can to some extent be protected from habitat change, if not atmospheric pollution, by designation as Sites of Special Scientific Interest (SSSI) and National Nature Reserves (NNR). SSSI's in Radnor with a notable epiphytic lichen flora include Elenydd (encompassing most of the best Elan and Claerwen Valley woods); Cwm Gwynllyn, near Rhayader; Lakeside Wood, Llandrindod Wells; Aberedw Wood and Cilkenny Dingle, Clyro. Fine epilithic communities occur also within Elenydd and Cwm Gwynllyn SSSIs and in addition in Cerrig Gwalch, Nannerth; the Marcheini Uplands; the River Wye; Craig Fawr, Frank's Bridge; Llanelwedd Rocks; Aberedw to Gwaunceste Hill and in the NNR at Stanner.

i) National Rarities

In 1991, funded by the Nature Conservancy Council, Nick Stewart produced a list of threatened, nationally rare lichens (NCC CSD Contract Report No 12272). To be included in this list a species should be well-recorded and occur in fewer than 15 10km squares of the Ordnance Survey's National Grid in Britain post-1960. If slightly more frequent than this it should have either shown a recent marked decline and be under threat of extinction or only occur in most of its sites in very small quantity. Stewart placed 321 species in this category of which 7 occur in Radnor. They are listed in table 13.1. The most widespread is *Leptogium palmatum*. Its British headquaters appears to lie in mid Wales where it occurs on damp, slightly sloping rock and gravel surfaces receiving some nutrient-rich run off. Occuring on the edges of tracks and farmyards, direct conservation measures present something of a challenge! *Gyalidea subscutellaris*, though only recently discovered in Britain, favours basic, heavy-metal rich ground. In Radnor it is confined to the tip of the Cwm Elan Mine. The remaining species are almost confined to ancient or base-rich tree bark. *Bacidia incompta*, *Caloplaca luteoalba* and *Collema fragrans* have declined due to the death of wych elm through dutch elm disease.

Lichen Conservation

Table 13.1 British Red Data Book lichens recorded from Radnor

Species	SSSI	NNR	Unprotected	Total
Bacidia incompta	1	(1)	1	2
Calopaca luteoalba	0	0	1	0
Catillaria globulosa	1	0	0	1
Collema fragrans	(1)	(1)	1	2
Gyalidea subscutellaris	1	0	0	1
Lecanactis amylacea	0	0	1	1
Leptogium palmatum	3	0	10	13

Where the no. of sites total appears in () the sp. there is thought to be extinct.

Stewart (1991 op cit.) also provides a list of those species at present not sufficiently well known to be included in the Red Data List but which in all probability, when better known, will be included. Only *Bacidia circumspecta* has been found in Radnor. Known from 3 sites on base-rich tree bark, none of which is protected, it is now probably reduced to 2 sites due to the death of wych elm through Dutch elm disease.

He also lists species which on present distribution data would qualify for consideration but which if better known would almost certainly not qualify for inclusion. Thirteen species recorded from Radnor appear in this list. Little attention should perhaps be paid to species such as *Bacidia chloroticula*, which seems poised to take over the vacant niche of tanalized softwood fence posts or *B. caligans*, with its preference for shady concrete. Other lichens such as *Stereocaulon leucophaeopsis* and *Vezdaea retigera* favouring scarce habitats rich in heavy metals may never be common. The remaining species placed by Stewart in this list and found in Radnor are:- *Bacidia delicata, Parmelia disjuncta, Psilolechia clavulifera, Ptychographa xylographoides, Rhizocarpon viridiatrum, Thelocarpon intermediellum, Thrombium epigeum* and *Vezdaea leprosa*.

Another group considered by Stewart (1991 op cit.) were those the status of which required monitoring. These lichens tended to occur in threatened habitats or to have declined markedly in recent years but were too common to include in the Red Data List at present. Twenty-three spp. within this group have been recorded from Radnor. The county, in particular, supports notable populations of *Lasallia pustulata, Lecanora sublivescens, Massalongia carnosa* and *Moelleropsis nebulosa*. A special effort should be made to conserve these species. Table 13.2 lists all the threatened lichens.

Table 13.2 Species whose status nationally requires monitoring

Species	SSSI	NNR	Unprotected	Total
Anaptychia ciliaris var. ciliaris	0	0	5	5
Arthonia impolita	1	0	8	9
Bryoria bicolor	1	0	0	1
Lasallia pustulata	5	1	6	12
Lecanactis premnea	1	0	8	9
Lecanora sublivescens	0	0	3	3
Leptogium cyanescens	1	0	0	1
Lobaria amplissima	0	0	1	1
L. pulmonaria	3	0	5	8
Massalongia carnosa	4	0	3	8
Moelleropsis nebulosa	1	0	8	10
Opegrapha lyncea	0	0	1	1
Pannaria conoplea	3	0	4	7
Parmelia stygia	1	0	0	1
Parmeliopsis aleurites	0	0	3	3
Placopsis gelida	1	0	0	1
Pycnothelia papillaria	0	0	1	1
Ramalina fraxinea	1	0	7	8
Sticta fuliginosa	2	0	0	2
S. limbata	1	0	1	2
Strangospora ochrophora	1	0	1	2
Usnea articulata	0?	0	1?	1?
U. florida	c.10	1	100+	100+

ii) Threatened Lichens of the European Economic Community

In 1989 E. Serusiaux produced a list of lichens which were considered threatened within the European Economic Community (EEC) (*Liste Rouge des Macrolichens dans la Communauté Européene.* Liege). Of the 196 species he listed, 56 have been recorded from Britain. Omitting from this list those species which appear in the Red Data List above, 8 of the 56 species have been recorded from Radnor. They are shown in table 13.3 below. All are scarce and threatened in Radnor except perhaps for *Cladonia luteoalba* and *Cetraria sepincola*, the latter having recently developed a liking for the tops of hardwood crash barrier posts. The remaining species are almost exclusivly epiphytic and favour base-rich bark, a habitat threatened by atmospheric pollution and rendered disjuct by the great scarcity of ancient trees.

Table 13.3 Species threatened in the EEC

Species	SSSI	NNR	Unprotected	Total
Cetraria sepinicola	2	1	4	6
Cladonia luteoalba	2	0	5	7
Lobaria amplissima	0	0	2	2
Pannaria conoplea	4	0	3	7
Parmelia taylorensis	1	0	0	1
Parmeliella jamesii	2	0	0	2
Sticta fuliginosa	2	0	0	2
Sticta limbata	1	0	1	2

iii) Local Rarities

Using the definition of a local rarity as one which occurs in 5 or fewer sites in Radnor, 173 spp. would be included in this category. Futher study will almost certainly reveal additional localities for some species and reduce the number to be considered here. It appears likely that around 60 species may be significantly under-recorded. This would still leave nearly a fifth of Radnor's lichen flora apparently existing as only small populations. The monitoring, management and conservation of this large number of species will not be easy but is nevertheless essential if our lichen flora is not to be further impoverished. In the species accounts below, locality details are provided for all species known from 6 or fewer sites, permitting the identification of local rarities.

iv) Extinctions

It is difficult to prove that a species is extinct. It does seem likely that the following species have been lost. They are mostly easily identified and in some cases are quite habitat specific. A search of likely localities has failed to find them.

Table 13.4 Extinct lichens

Species	Date Last Seen	Possible Cause of Loss
Collema fasciculare	1976	Dutch elm disease
Polyblastia allobata	1986	Dutch elm disease
Pycnothelia papillaria	1879	Loss of heathland
Usnea articulata	1904	Atmospheric pollution?

Almost certainly more species have been lost. With so few pre-1950 records many losses may never be detected. During the course of this study a significant decline has been noted in those species favouring elm bark, due to the wholesale death of mature elms through dutch elm disease. Lichens such as *Acrocordia gemmata, Bacidia incompta, Caloplaca cerina, C. luteoalba, Gyalecta truncigena, Opegrapha vermicellifera* and *Physcia* and *Physconia* spp. have all declined in abundance. A proposed expansion of the limestone quarries at Dolyhir threatens *Acrocordia conoidea, Aspicilia subcircinata* and *Protoblastenia calva*.

E) Arrangement and Nomenclature

In the absence of any generally acceptable phylogenetic scheme for the lichens and lichenicolous fungi, they are presented here in alphabetical order by genus and then by species within genus. The layout of the species accounts generally follows that as detailed for the vascular plants above. It differs only in that, since the lichens were subject to less detailed survey, a number of widespread species unrecorded from a scatter of 5 x 5km. squares would have required a map to be inserted in the text or a fairly lengthy list of 5 x 5 km. squares from which they had not been recorded to be inserted. To save space, at the loss of very little information, their distribution is described at the 10 x 10 km. square level.

The taxonomy follows Cannon, Hawksworth and Sherwood-Pike, *The British Ascomycotina*, Commonwealth Mycological Institute, 1985 in the main. Some minor, more recent taxonomic changes have been followed and a few recently described species have been added. Since the above volume is so excessively priced as not to be generally available the authorities along with the scientific names for these fungi are included here. *

As indicated above, this account must be treated as a preliminary statement concerning the distribution of these plants in Radnor. In particular the lichenicolous fungi have not been systematically studied. Much additional work is, for example, required on the lichens of churchyards and the built environment generally. It is, nevertheless, gratifying to note that a substantial and important lichen flora has been reported from Radnor, to which I am certain, many further records can be added.

---

* Since preparing this account a new British lichen flora has been produced (Purvis et al, 1992). Taxonomic changes are detailed in a footnote below the species account on page 260.

Lichens and Lichenicolous Fungi

Acarospora fuscata (Nyl.) Arnold

Frequent on exposed sunlit rocks, including wall coping, especially those receiving nutrients from bird and livestock droppings.

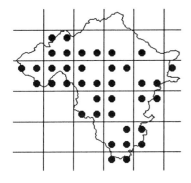

A. sinopica (Wahlenb.) Körber

Rare and only seen on sunlit, iron-rich shale rocks in the Elan and Claerwen valleys. Slate quarry tips near Craig-y-bwch 8961 and 9061; wall, Cwm yr Elan 9064; lead mine tip, Cwm Elan Mine 9065; rocks, Craig-y-foel 9163 and 9264.

A. smaragdula (Wahlenb.) Massal.

Rare, though possibly overlooked on sunlit, acidic rocks. Shale rocks near Caban Coch Resr., Elan Valley 9164 AEW (NMW); on volcanic rocks below Cefnllys Castle 0861 AEW (NMW); shale rocks, Fron Wen, Radnor Forest 1766 AO.

Acrocordia conoidea (Fr.) Körber

Known only from shaded limestone close to active quarry. Yatt Wood, Dolyhir 2458.

A. gemmata (Ach.) Massal.

Rare or overlooked on trees with base-rich bark. Probably much reduced by the loss of wych elm. On old oak by Wye, Nannerth Fawr, Rhayader 9471; wych elm (now dead), Brynwern, Disserth and Trecoed 0156 and old oak, Tyfaenor Park, Abbeycwmhir 0671.

Agonimia tristicula (Nyl.) Zahlbr.

An inconspicuous sp. growing amongst mosses in base-rich places such as old mortared walls and on ancient ash, oak and field maple trees.

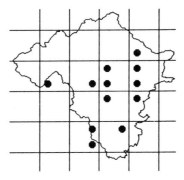

Amygdalaria pelobotryon (Wahlenb.) Norman

Rare or overlooked on wet acidic rocks. S facing shale rocks near Claerwen Dam 8763 and NE facing volcanic rocks near Cwm-berwyn, Llansantffraed-in-Elvel 0754.

Anaptychia ciliaris (L.) Körber

Rare on the base-rich bark of sunlit, isolated trees. Sycamore by road near Yr Allt, Llandeilo Graban 0844; formerly on a now felled ash tree in an arable field below Maesyronnen Chapel, Glasbury 1741; ash at Tynwain, Dolau 1668; ash and nearby old railway sleeper fence post between Burlingjobb and Stanner 2558; field maple, Old Radnor 2459 PWJ; sycamore in Stanage Park 3371. Fertile only at Yr Allt and Old Radnor.

A. runcinata (With.) Laundon

This normally coastal species has been found twice on sunny shale rocks. N of Ashfield, Ysfa 9764 and SE of Wenallt, Abbeycwmhir 0470. It is associated with higher plant spp. of the Thero-sedetum, a common coastal cliff community. No other inland Welsh records are known.

Anisomeridium biforme (Borrer) R. C. Harris

Somewhat uncommon and mostly on the more nutrient-rich bark of ancient ash, wych elm and occasionally oak trees.

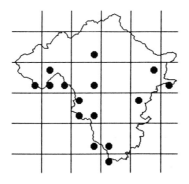

A. juistense (Erichsen) R. C. Harris

Widespread on the bark of elder in hedges and around woodland edges.

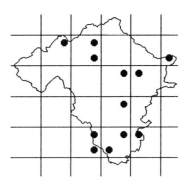

Arthonia didyma Körber

The small spp. of *Arthonia* on smooth bark could not be separated in the field and have, in consequence, probably been much under-recorded. This sp. appears to be rare, having been found only on hazel in woodland N of Llanfihangel Helygen 0464 and 0465.

A. elegans (Ach.) Almq.

Widespread on the smooth bark of hazel and oak branches in moist woodlands.

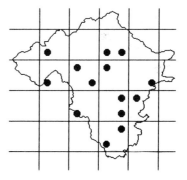

A. glaucomaria (Nyl.) Nyl.

A rare parasite of *Lecanora rupicola*. On volcanic boulders on the Carneddau E of Newmead Farm, Llanelwedd 0654 det. BJC.

Arthonia

A. impolita (Hoffm.) Borrer.

Uncommon in dry, sheltered crevices on the bark of ancient oak trees, usually in open situations.

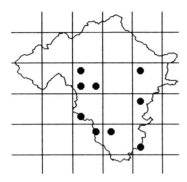

A. punctiformis Ach.

Probably widespread on the smooth bark of oak, hazel and other deciduous tree twigs.

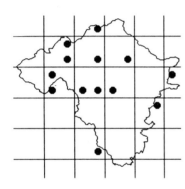

A. radiata (Pers.) Ach.

Common on the smooth bark of trees and shrubs in woods and hedgerows, especially in central and western Radnor.

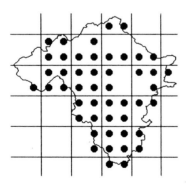

A. spadicea Leighton

A scarce lichen of shaded, ancient tree bark, mostly oak but occasionally ash, in damp woodland and old parkland.

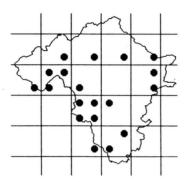

A. tumidula (Ach.) Ach.

Of scattered occurrence on the smooth bark of hazel, oak and ash in damp woodlands, mostly in western Radnor. Further work is required to produce a distribution map since earlier records were confused with *A. elegans*.

A. vinosa Leighton

Scarce and confined to the lower parts of old oak trunks in damp woodland.

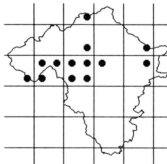

Arthopyrenia

The corticolous spp. of this genus were reviewed by Coppins (1988). The extensive changes made necessitates a revision of the records of all the smaller species. Only a preliminary account can be provided here.

A. cinereopruinosa auct angl. non (Schaerer) Massal.

Most, if not all records of this sp. are probably referable to *A. ranunculospora* Coppins & P. James. It is widespread on the smooth bark plates of mature trees, notably ash, in damp and sheltered, probably ancient woodland.

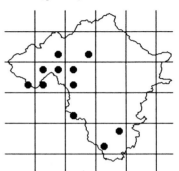

A. lapponina Anzi

Widespread on the smooth, shaded bark of, most commonly, ash, hazel, oak and rowan in woodlands.

A. punctiformis Massal.

Recorded frequently with the last sp. in woodland and displaying a similar distribution. It seems likely, however, that more than one taxon may have been recorded here. *A. fraxini* Massal., *A. nitescens* (Salwey) Mudd, *A. salicis* Massal. and *A. viridescens* Coppins might all be expected in the county (Coppins, 1988).

Arthrorhaphis citrinella (Ach.) Poelt

A conspicuous sp. of acidic soil in well-lit crevices in shale rock outcrops and overgrowing moss cushions, particularly *Andreaea* spp.. Frequently associated with *Leparia neglecta* and *Micarea leprosula*. Common in the western uplands.

Aspicilia

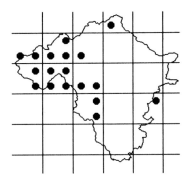

A. grisea Th. Fr.

An uncommon parasite of the lichen *Baeomyces rufus*. Spoil tips of the Cwm Elan Mine, Elan Valley 9065; in Coed yr allt goch, Elan Valley 9067 AO; Fron Wen, Radnor Forest 1866 and Bach Hill, Radnor Forest 2064. Fruits of what appear to be this sp. were found on *Baeomyces placophyllus* on rocks E of Craig-y-bwch in the Claerwen Valley 9061 (spec. in E), believed to be the first record from this species.

Aspicilia caesiocinerea (Nyl. ex Malbr.) Arnold

The grey *Aspicilia* spp. of sunny, nutrient-flushed rocks have not always been determined with certainty. They are common throughout Radnor on shale rock outcrops, especially on the heavily grazed upland sheepwalks where leachates of sheep dung wash over the rocks. This sp. is almost certainly the commonest, though it may have, at times, been confused with *A. gibbosa*.

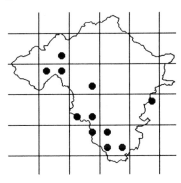

A. calcarea (L.) Mudd.

Rare on natural, basic rock outcrops in well-lit situations. Aberedw Rocks 0746; Yatt Wood, Dolyhir 2458 and Stanner Rocks 2658. Frequent on basic, man-made substrates such as concrete blocks.

A. cinerea (L.) Körber

Uncommon or overlooked on nutrient-flushed, sunny rock outcrops.

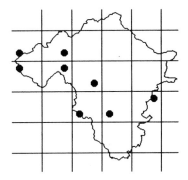

A. contorta (Hoffm.) Krempelh.

Widespread on base-rich rock outcrops, such as calcareous mudstone and dolerite and common on concrete and mortared walls.

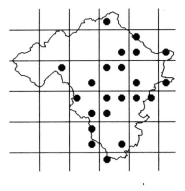

A. gibbosa (Ach.) Körber

Scarce or possibly confused with *A. caesiocinerea* from which it is doubtfully distict. On rocks in the flood zone of streams and rivers. By the Wye, Llanfaredd 0649 and by the Bach Howey, Llanstephan 1143.

A. subcircinata (Nyl.) Coppins

Confined to S facing limestone of a disused quarry, Yatt Farm, Dolyhir 2458.

Bacidia arceutina (Ach.) Arnold

An uncommon epiphyte found only on the well-lit bark of a large field maple at the edge of woodland, Alpine Bridge, Llanbadarnfawr 0962 ! and FR, det. BJC and on oak in wet woodland S of Wern Fawr, Painscastle 1845.

B. bagliettoana (Massal. & de Not.) Jatta

Only found overgrowing the moss *Pterogonium gracile* on a sunny, S facing, basic volcanic rock outcrop, Caer Einon, Llanelwedd 0653 ! and A Fletcher.

B. biatorina (Körber) Vainio

Rare and confined to the bark of ancient, generally pollarded oaks in woodland. Valley woodland, Dderw, Llansantffraed Cwmdeuddwr 9568; bank of R. Wye, Nannerth, Rhayader 9571; gorge of the Edw, Aberedw 0747; wood to N of Llanfihangel Helygen 0464; Lakeside Wood, Llandrindod Wells 0660.

B. caligans (Nyl.) A.L.Sm.

This rare, or more probably overlooked sp. is known only from a concrete slab under a hawthorn bush in Penybont Hall Park, Penybont 1163 det. BJC.

B. carneoglauca (Nyl.) A.L.Sm

A rare sp. of dry, shaded, acidic shale rocks by the Wye below the confluence with the Bach Howey, Llandeilo Graban 1042.

B. circumspecta (Norrlin & Nyl.) Malme

Rare on basic tree bark. Old oak in damp woodland, Dderw, Llansantffraed Cwmdeuddwr 9568; old oak by Wye, Nannerth, Rhayader 9471 and on recently dead wych elm in the gorge of the R. Ithon above Alpine Bridge, Llanbadarnfawr 0962 ! and FR, det. BJC.

Bacidia

B. chloroticula (Nyl.) A.L.Sm.

Reported by AO from a tanalized softwood fence post receiving the runoff from galvanized wire and staples at Cors y llyn, Newbridge on Wye 0155. This sp. is likely to become more widespread if it proves to be well-adapted to this niche.

B. delicata (Larbal. ex Leighton) Coppins

Rare or overlooked on basic, shaded shale rocks. In old railway cutting, now road, S of Erwood Station 0943 AO.

B. epixanthoides (Nyl.) Lettau

A scarce lichen of ancient tree bark. On oak in damp woodland, Tyncoed Wood, Cwmbach Llechrhyd 0354 ! & FR and on recently dead wych elm in gorge of R. Ithon above Alpine Bridge, Llanbadarnfawr 0962 ! & FR.

B. incompta (Borrer ex Hooker) Anzi

Rare in rainwater tracks on the well-lit bark of wych elm. Stanner Rocks 2658 but extinct by 1982 due to the death of elms from Dutch elm disease; Evancoyd Park, Evenjobb 2562.

B. inundata (Fr.) Körber

Probably widespread on acidic boulders in streams and rivers but recorded only in times of low flow.

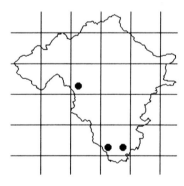

B. laurocerasi (Delise ex Duby) Zahlbr.

A somewhat southern lichen of basic bark, rare in Mid Wales and known only from a field maple in woodland near Shaky Bridge, Cefnllys 0862 det. BJC.

B. naegelii (Hepp) Zahlbr.

Known only from the bark of a field maple near Llanbadarn-y-garreg 1148. A scarce or overlooked sp. in Mid Wales.

B. phacodes Körber

Rare on nutrient-rich bark. Field maple in wood, Bailey Einon, Cefnllys 0861 and wych elm, Stanner Rocks 2658 ! & PWJ, but extinct by 1982 in the latter site.

B. rubella (Hoffm.) Massal.

Uncommon and confined to the nutrient-rich bark of mature and ancient field maple, wych elm, ash, oak, elder and sycamore, usually in well-lit situations.

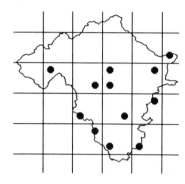

B. sabuletorum (Schreber) Lettau

Frequently overgrowing mosses, soil and rock in base-rich habitats such as mortared walls and concrete and on calcareous rock outcrops.

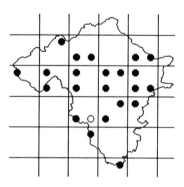

B. saxenii Erichsen

Rare or overlooked. On shaded bricks by old railway, near Pont Marteg, St Harmon 9571 AO.

B. trachona (Ach.) Lettau

A rare sp. of sheltered, slightly basic shale rock outcrops. Aberedw Rocks 0746, BLS field meeting and in a ravine below Pentre Caeau, Llandeilo Graban 0844 AO and !.

B. vezdae Coppins and P. James

An inconspicuous sp. of smooth, shaded bark of hazel and elder in damp woodland. SE of Newmead Farm, Llanelwedd 0553 det. BJC; near Cwmbach Llechrhyd 0354 and Cilkenny Dingle, Llowes 1741 ! & AO.

Bactrospora corticola (Fr.) Almq.

Uncommon and confined to the sheltered crevices in the bark of old oak trees. Henfron, Elan Valley 9064; Noyadd Fach, Elan Valley 9365; woodland by R. Wye N of Rhayader 9668; bank of R. Wye above Erwood Bridge, Llandeilo Graban 0844; woodland near Caerfagau, Llanfihangel Helygen 0465; Tyfaenor Park, Abbeycwmhir 0671; Penybont Hall Park, Penybont 1163 and woodland near Yardro, Harpton and Wolfpits 2258.

Baeomyces placophyllus Ach.

An uncommon sp. of shallow, acidic soil on well-lit rock ledges in the uplands. Most frequent in the NW uplands, it also occurs on Dol-y-fan Hill, Newbridge on Wye 0160 and on the Carneddau, Llansantffraed-in-Elvel 0654.

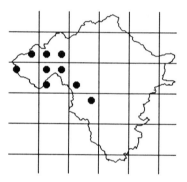

B. roseus Pers.

Widespread and common in the W on acidic soil and clay in open habitats such as roadside banks, cliffs and heathland. Occasionally fertile.

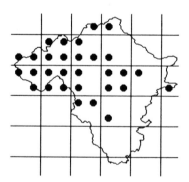

Buellia

B. rufus (Huds.) Rebent.

In similar sites to the last and often with it, though usually more common and extending over small rocks and pebbles. Commonly fertile.

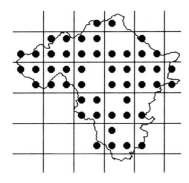

Belonia nidarosiensis (Kindt) P. Jörg and Vezda

A rare sp. of shaded, calcareous rocks. Calcareous mudstone cliffs, Aberedw Rocks 0746; wall of old road bridge, Llananno 0974; shale cliff, Llandeilo Hill 1048; shaded limestone, Yatt Wood, Dolyhir 2458.

Biatorina atropurpurea (Schaerer) Massal.

A rare sp. of tree bark in long undisturbed, damp woodlands. Old oak in wood, Noyadd Fach, Elan Valley 9365; old oak by R. Wye, Nannerth, Rhayader 9472; old oak, Aberedw wood 0746 and on a large ash in wood beside the Lake, Llandrindod Wells 0660.

"*Botrydina vulgaris*" see *Omphalina*

Bryophagus gloeocapsa Nitscke ex Arnold

A scarce or overlooked sp. of decaying mosses and liverworts overhanging banks and rocks. Low bank above shore of Garreg Ddu Resr., near the church, Elan Valley 9063 ! and BJC; in Coed yr allt goch, Elan Valley 9067 AO; peat bank of old turbary near Waun, Pant-y-dwr 9477; N facing rock outcrop, Great Park, Abbeycwmhir 0572 and forestry bank, Ednol Hill, Radnor Forest 2064 ! and AO.

Bryoria bicolor (Ehrh.) Brodo and D. Hawksw.

Known only from a gritstone outcrop in the western uplands near Blaen Rhestr, Llansantffraed Cwmdeuddwr 8469 ! & AO.

B. fuscescens (Gyelnik) Brodo and D. Hawksw.

An uncommon sp., most frequent on the acidic bark of windswept tree trunks such as oak, larch and hawthorn in the uplands. Rare on fence posts and acidic boulders in similar locations.

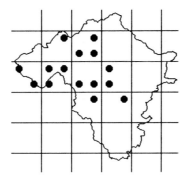

B. subcana (Nyl. ex Stizenb.) Brodo and D. Hawksw.

Scarce or overlooked as *B. fuscescens*. On acid bark of oak and ash in woodland. Nant Gwyllt, Elan Valley 9162 ! and FR; Alpine Bridge, Llanbadarnfawr 0963 ! and FR and Tyfaenor Park, Abbeycwmhir 0672.

Buellia aethalea (Ach.) Th. Fr.

Widespread on sunny, acidic rock outcrops.

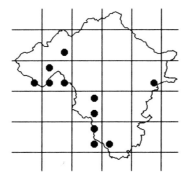

B. griseovirens (Turner and Borrer ex Sm.) Almb.

This sp. is probably far more widespread than the single record suggests. Decorticate oak trunk in open woodland N of Llanfihangel Helygen Church 0464.

B. ocellata (Flotow) Körber

In similar sites to *B. stellulata*, though being less conspicuous has possibly been overlooked in many places. Shale of old lead mine to W of Esgair Dderw, Llansantffraed Cwmdeddur 9469: Aberedw Rocks 0746; wall Llanstephan Churchyard 1142 ! and BJC; Disgwylfa Hill, Newchurch 2251 and Rock Dingle, Bleddfa 2067 ! and AO.

B. pulverea Coppins and P. James

A scarce sp. of acidic wood and bark. On trunk of old Scots pine in woodland E of Nant Gwyllt above the shore of Caban Coch Resr., Elan Valley 9163 and on ancient split oak fence post, Llanelwedd Rocks 0552 ! and BJC. Probably elsewhere.

B. punctata (Hoffm.) Massal.

Frequent on nutrient-rich bark, such as the lower trunks of trees in pastures, but also on twigs and extending onto rocks and walls about farms and settlements.

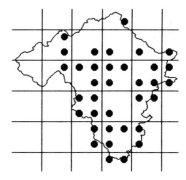

B. schaereri de Not.

A rare sp. on acidic bark. Old larch, Gilfach Farm, St Harmon 9671 and Tyfaenor Park, Abbeycwmhir 0671 ! and FR.

B. stellulata (Taylor) Mudd

Frequent on the harder acidic rocks in sunny positions.

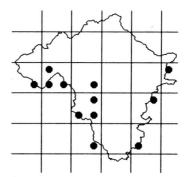

Caloplaca

Calicium glaucellum Ach.

Widespread, but uncommon, on decorticate standing and fallen oak trunks and fence posts where there is some shelter from driving rain.

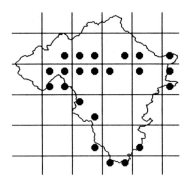

C. salicinum Pers.

Uncommon on the bark of large and ancient oak, rarely ash and elm in open woodlands and parklands.

C. subquercinum Asah.

A scarce sp. in Wales, noted on a number of ancient oak trees in woodland. Wood above the sawmill on the shore of Caban Coch Resr., Elan Valley 9163 det. BJC and on an old oak near Alpine Bridge, Llanbadarnfawr 0963 ! and FR, det. BJC.

C. viride Pers.

Frequent in crevices in the bark of tree trunks in places sheltered from most of the direct effects of rainfall.

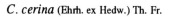

Caloplaca aurantia (Pers.) Steiner

A rare sp. of sunny, base-rich rocks. Volcanic rock outcrops, Caer Einon, Llanelwedd 0652 ! and A Fletcher; on mortared stone wall of porch, Cefnllys Church 0861 AEW; limestone outcrop at top of Yatt Wood, Dolyhir 2458.

C. cerina (Ehrh. ex Hedw.) Th. Fr.

A rare epiphyte noted only on a sunny ash trunk at Llanelwedd 0452 and on a large wych elm at Abbeycwmhir 0571. This latter tree had died by 1989.

C. citrina (Hoffm.) Th. Fr.

Common on calcareous rocks, concrete, mortar, nutrient-enriched bark and wood. Recorded from all 10km squares.

C. crenularia (With.) Laundon

Common on sunny, somewhat calcareous shale and mudstone rock outcrops, stonewalls and monuments.

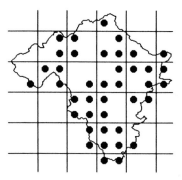

C. flavescens (Huds.) Laundon

Widespread on well-lit, calcareous rocks, including mortared walls and concrete in the lowlands. Rare in the uplands.

C. flavovirescens (Wulfen) Dalla Torre and Sarnth.

Rare on calcareous mudstones and volcanic rock outcrops and extending onto concrete and mortared stones.

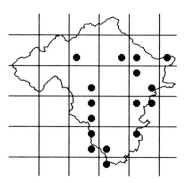

C. holocarpa (Hoffm.) Wade

Common on calcareous rock, mortar, concrete, nutrient-enriched wood and bark, particularly around the limestone quarries at Dolyhir 2458. Recorded from all 10km squares.

C. luteoalba (Turner) Th. Fr.

A rare sp. of sunny, base-rich bark. Wych elm in park, Evancoyd, Evenjobb 2562.

C. obscurella (Lahm) Th. Fr.

Rare or overlooked on basic bark. Old oak in wood Llanboidy, Llanfihangel-Nant-Melan 1457; old ash in park, Penybont Hall 1164; ash near Llandegley 1362; elder, Norton 3067.

C. saxicola (Hoffm.) Nordin

Common on well-lit, calcareous mudstones, especially where incorporated into the walls of farm buildings and occasionally spreading onto wooden boards.

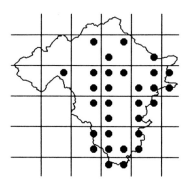

Catillaria

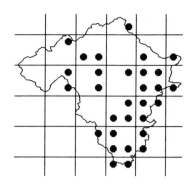
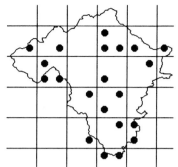
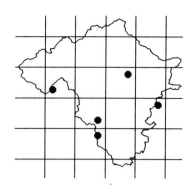

C. subpallida Magnusson

A predominantly E Scottish sp. of rare occurrence on S facing rocks in Wales. In Coed yr allt goch, Elan Valley 9067 AO; Marteg Valley below Wyloer, St Harmon 9571 ! and BJC and on rocks near Llandrindod Wells old church 0660 AO.

Candelariella aurella (Hoffm.) Zahlbr.

Scarce or overlooked on base-rich substrates. On railway ballast at Dolau 1568 AO.

C. coralliza (Nyl.) Magnusson

Widespread, but not common in shallow depressions on sunny, acidic boulders manured by bird droppings.

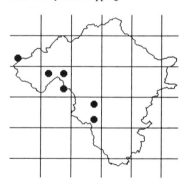

C. medians (Nyl.) A.L.Sm.

A sp. of SE Britain at the edge of its range in Radnor on somewhat calcareous stone of buildings. Stone ledge of shop front, Llandrindod Wells 0561 AO; on stone doorstep, Llowes Church 1941 and on stone walls, Presteigne 3164.

C. reflexa (Nyl.) Lettau

Widespread on the bark of elder and occasionally on other trees with a nutrient-rich bark such as ash in hedgerows and woodland edges.

C. vitellina (Ach.) Müll. Arg.

Common on sunny, nutrient-enriched rocks, including wall tops and monuments and occasionally on tree bark and wood also enriched with nutrients from animal excreta. Recorded from all 10km squares.

C. xanthostigma (Ach.) Lettau

Overlooked in the earlier stages of the survey, this lichen is probably more widespread on well-lit, somewhat nutrient-rich bark than the following records suggest. Sycamore on common below Yr Allt, Llandeilo Graban 0844; ash, Tynwain, Dolau 1668 and sycamore, Beggars' Bush 2564.

Catapyrenium pilosellum O. Breuss

A rare sp. of sunny, shallow, base-rich soil overlying rock. Ledges of old quarry S of Bryngwyn 1848; cliff W of Knucklas Viaduct 2474; rock ledges, Yatt Wood, Dolyhir 2458.

Catillaria atomarioides (Müll. Arg.) Kilias

An inconspicuous lichen of slaty rocks, probably more widespread than the two records suggest. On railway ballast near Tynwain, Dolau 1568 AO and on rock outcrops, Rock Dingle, Bleddfa 2067 AO and !.

C. chalybeia (Borrer) Massal.

The small dark thalli of this sp. are almost certainly more widespread than the records suggest. On sunny, fairly base-rich rock outcrops.

C. globulosa (Flörke) Th. Fr.

As the f. *pallens* it is known only from the bark of an ancient oak in the wood above the sawmill on the shore of Caban Coch Resr., Elan Valley 9162 ! and FR, det. BJC.

C. lenticularis (Ach.) Th. Fr.

Probably overlooked on lime-rich substrates such as mortar and limewashed walls. The only record is from calcareous mudstone cliffs at Aberedw Rocks 04NE, AEW (1959, NMW).

C. pulverea (Borrer) Lettau

A rare sp. of acidic bark in humid sites. On sessile oak in woodland, Cwm Coel, Elan Valley 9063 ! and BJC; on sessile oak by A. Elan, Elan Village 9365 ! and BJC and alder in wet woodland W of Abbeycwmhir 0671.

C. sphaeroides (Dickson) Schuler

A scarce sp. of shady bark in crevices and underhangs of ancient oak and ash trees and rarely on shady and damp, somewhat calcareous shale rocks.

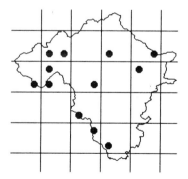

Chaenotheca

Cedidonia xenophana (Körber) Triebel and Rambold

An inconspicuous parasite of Huilia spp.. Vertical rock face, Craig Goch, Elan Valley 8968; shale outcrops below Wyloer, Gilfach Farm, St Harmon 9571 ! and BJC; volcanic rock outcrops, Llandegley Rocks 1362.

Cetraria chlorophylla (Willd.) Vainio

Frequent on the acidic bark of tree and shrub trunks and branches, especially in upland woods and on isolated trees in well lit situations. Also encountered on old wooden fence posts and rarely on acidic shale rock outcrops. Recorded from all 10km squares except 13 and 27.

C. sepincola (Ehrh.) Ach.

An uncommon lichen confined in Wales to central districts. Most abundant on the twigs of birch but also rarely on oak, on, or beside, basin mires around Newbridge on Wye 05NW and Llandrindod Wells 06SE. Also noted on the top of hardwood posts of roadside crash barriers on bare hillsides near Pont ar Elan, Elan Valley 9071 and Bwlch-y-sarnau 0375. It was first reported in the county on rails at Llandrindod by WH Wilkinson about the turn of the century (*J Bot*, **42**, 112).

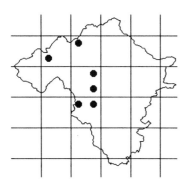

Chaenotheca brunneola (Ach.) Müll. Arg.

A scarce sp. of usually ancient oak trunks in parks and open woodland, favouring sheltered crevices where the bark has been lost.

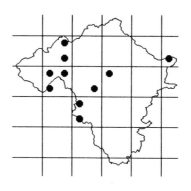

C. carthusiae (Harm.) Lettau

Rare and known only from the underside of a dead, bark-free oak branch in Llanboidy wood, Llanfihangel-Nant-Melan 1457 det. BJC.

C. chrysocephala (Turner ex Ach.) Th. Fr.

A scarce and beautiful sp. occurring in dry cracks in bark and occasionally wood, mostly on the east-facing sides of old oak trees.

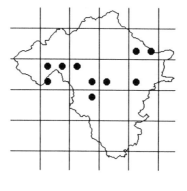

C. ferruginea (Turner ex Ach.) Mig.

Common on dry, acidic bark of oak, larch and occasionally other trees and wooden fence posts in open situations.

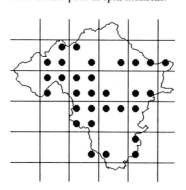

C. trichialis (Ach.) Th. Fr.

A scarce sp. of dry crevices in ancient oak trunks in woodland. Old decorticate oak, Nant Gwyllt, Elan Valley 9162; oak, Treheslog Rocks, Llansantffraed Cwmdeuddwr 9468; Dyffryn Wood, Rhayader 9766; Tyncoed Wood, Cwmbach Llechryd 0354; Alpine Bridge, Llanbadarnfawr 0963; gorge of the Bach Howey below Craig Pwll du, Llanstephan 1143; Dol-wen Wood, Llanddewi Ystradenni 1069.

Chrysothrix candelaris (L.) Laundon

In sheltered, dry crevices in the bark of a wide range of tree spp.; scarce in the wetter uplands, but common in the lowlands in well-lit situations and there rarely extending onto the hardwood boards of barns. Recorded from all 10km squares except 87, 18 & 36.

C. chlorina (Ach.) Laundon

A distinctive chrome yellow sp. of hard, dry, acidic rocks. Only known in small quantity from Cerrig Gwalch, Rhayader 9370.

Chaenotheca chrysocephala x 10

C. chrysophthalma (P. James) P. James and Laundon

Widespread on seasoned hardwood, both on gateposts and on trunks and branches of oak trees in dry, sheltered but well-lit situations. Where abundant it extends onto bark.

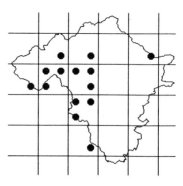

Cladonia

Cladonia arbuscula (Wallr.) Flotow

Common in the acidic grasslands and moorlands of central and western Radnor and extending into upland woods. Rare in the east.

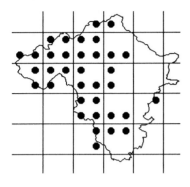

C. caespiticia (Pers.) Flörke

An uncommon plant of stumps, tree bark, peaty banks and moss cushions in moist woodlands. Possibly overlooked when sterile.

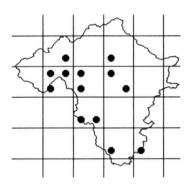

C. cervicornis (Ach.) Flotow

Frequent in open, acidic turf and on shallow peaty soil on rock outcrops, especially in the west.

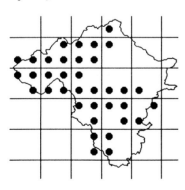

C. cervicornis ssp. *verticillata* (Hoffm.) Ahti

Uncommon on rocks, Gwynllyn, Llansantfraedd Cwmdeuddwr 9469 AO and on the soil of a roadside bank in 07SW.

C. chlorophaea (Flörke ex Sommerf.) Sprengel s. lat.

Common throughout on more or less acidic soil in open habitats, on tree trunks and logs. Recorded from all squares.

C. ciliata Stirton

Both the var. *ciliata* and the var *tenuis* (Flörke) Atiti are widespread on peaty soil amongst block scree, on rock ledges and in open, acidic grassland and heath.

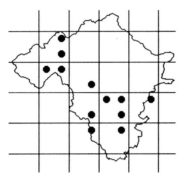

C. coccifera (L.) Willd. s. lat.

Common on peaty banks, stumps, amongst rocks and on wall tops in the western uplands. Rarer in the east.

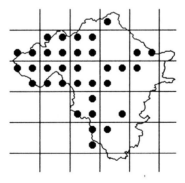

C. coniocraea auct., non (Flörke) Sprengel

Common on logs and the bases of trees throughout. Occasional on organic soils. Recorded from all 10km squares.

C. crispata var. *cetrariiformis* (Delise) Vainio

An uncommon or overlooked sp. of peaty soil in rock crevices and moorland.

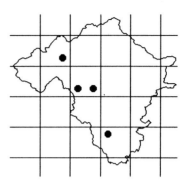

C. digitata (L.) Hoffm.

An uncommon, but widespread lichen of acidic tree bark, particularly oak, and decaying logs and fence posts.

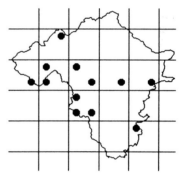

C. fimbriata (L.) Fr.

A ready colonizer of disturbed soil beside roads and tracks, on banks and extending to logs, peat, mossy tree trunks and wall tops. Recorded from all 10km squares.

C. floerkeana (Fr.) Flörke

Common on peaty banks and shallow peaty soil over rock outcrops and decaying logs in the uplands.

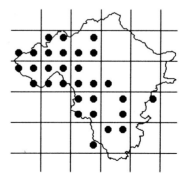

Cladonia

C. foliacea (Huds.) Willd.

A scarce lichen of open, sunny grassland on rocky slopes, especially where there appears to be some dry mineral flushing.

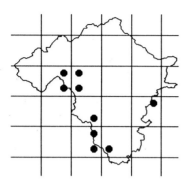

C. furcata (Huds.) Schrader

Frequent on open soil of road and stream banks, rock ledges, scree, peat and peaty banks of heathland and woodland, always in acidic sites. Recorded from all 10km squares except 36 and 37.

C. glauca Flörke

A scarce or overlooked sp. of peaty soil. On rock outcrops in oakwood, Gyrn, Llansantffraed Cwmdeuddwr 9366; rock outcrops Cerriggwynion, Rhayader 9765; old pine logs, Cors y Llyn, Newbridge on Wye 0155 AO; under heather on moorland, Llandeilo Hill, Llanbedr Painscastle 1046.

C. gracilis (L.) Willd.

Widespread on shallow peaty soil on rock outcrops, wall tops and banks in sunlit situations.

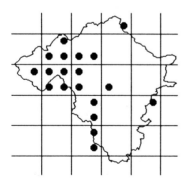

C. humilis (With.) Laundon

Scarce or overlooked on well-drained soil. Old quarry floor, Llanelwedd 0552, AEW (NMW records); roadside bank N of Llanbadarn Fynydd 0879 ! and FR; roadside verge near Bach Howey Viaduct, Llandeilo Graban 1042; roadside bank near Michaelchurch-on-Arrow 2350.

C. luteoalba A. Wilson and Wheldon

Uncommon, on acidic rock ledges and scree. Gyrn, Elan Village 9366; in wood, Glanllyn, Llansantffraed Cwmdeuddwr 9469; Cerriggwynion, Rhayader 9765; Y Gaer, Llansantffraed-in-Elvel 0854; Yr Allt, Doldowlod 0062; Wenallt, Abbeycwmhir 0370; Llandegley Rocks, Llandegley 1362.

C. macilenta Hoffm.

Common on decaying logs, stumps, fence posts and tree bases and on peaty soil in the uplands. Recorded from all 10km squares.

C. ochrochlora Flörke

Scarce, or more probably overlooked as *C. coniocraea*. Decaying stump in woodland, Llandrindod Wells 06SE and on oak stump near Alpine Bridge Bridge, Llanbadarnfawr 0963 ! and FR.

C. parasitica (Hoffm.) Hoffm.

Widespread, but scarce and largely confined to massive oak logs and stumps in ancient woodland and parkland.

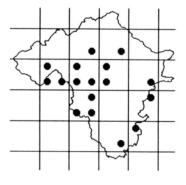

C. phyllophora Ehrh. ex Hoffm.

Reported from Cors y Llyn, Newbridge on Wye 0155 by PDM et al. (1970 unpub. NCC report). As possibly the only record outside the Scottish Highlands, where it occurs on peaty soil, the record is in need of confirmation.

C. pocillum (Ach.) O.- J. Rich.

On soil, mosses or rock, always in calcareous sites such as mortared walls, limestone, base-rich volcanic rocks and mudstones. Separated at times with difficulty from *C. pyxidata* in the field.

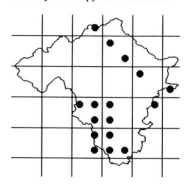

C. polydactyla (Flörke) Sprengel

Widespread on decaying stumps, logs and peaty soil.

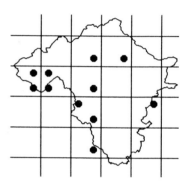

C. portentosa (Dufour) Coem.

Common on peaty soils of moorland, woodlands and roadside banks in the uplands and extending into the lowlands on decaying logs and stumps. Recorded from all 10km squares except 27, 36 and 37.

C. pyxidata (L) Hoffm.

Frequent in the more base-rich sites on shallow soil of rock outcrops, roadside banks and ancient tree bases.

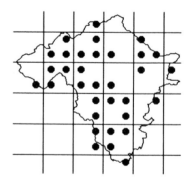

Cliostomum

C. ramulosa (With.) Laundon

Widespread on soil of roadside banks and occasionally on rotting stumps and logs.

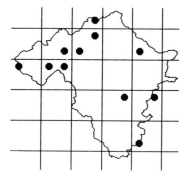

C. rangiformis Hoffm.

On dry, somewhat calcareous sunny slopes about rock outcrops, and scree, in open grassland and roadside banks.

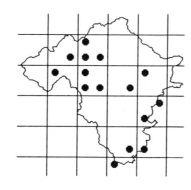

C. squamosa (Scop.) Hoffm.

Common on peaty soil of banks, scree and rock outcrops in woodlands of W Radnor. Less common in woodland elsewhere. The var. *subsquamosa* (Nyl. ex Leighton) Vainio has not been regularly separated during the survey, but was collected from an old oak log near Llwyn Barried Hall, Nantmel 0265.

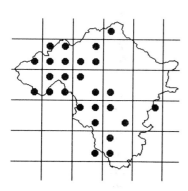

C. subcervicornis (Vainio) Kernst.

Common on exposed to slightly shaded, seasonally damp, acidic shale outcrops in the western uplands.

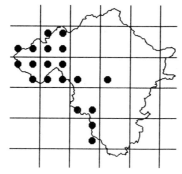

C. subulata (L.) Wigg.

A ready colonizer of disturbed, acidic soil on roadsides and forestry tracksides and occasionally on the peaty soil of heather moorland.

C. uncialis ssp. *biuncialis* (Hoffm.) M. Choisy

Common in the uplands on blanket bog, acidic grassland, moorland and the peaty soil of rock outcrops and block screes.

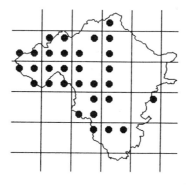

Cliostomum griffithii (Sm.) Coppins

Common in sheltered crevices in the somewhat acidic bark of tree trunks, particularly ash and oak species. Recorded from all 10km squares except 18.

Clypeococcum hypocenomyceae D. Hawksw.

A parasite of *Hypocenomyce scalaris*, common wherever the host occurs in quantity. Not mapped.

Coelocaulon aculeatum (Schreber) Link

Frequent on the acidic, peaty soil of open upland grassland, heath, rock outcrop and scree.

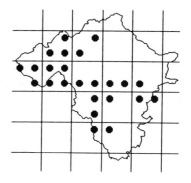

C. muricatum (Ach.) Laundon

Commonly associated with the last sp. and in similar habitats.

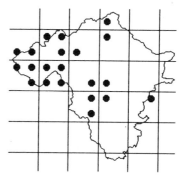

232   Collema

Collema auriforme (With.)
Coppins and Laundon

Of widespread occurrence on damp, basic rock outcrops and mortared walls.

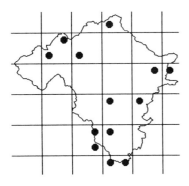

C. crispum (Huds.) Wigg.

In similar habitats to the last sp. but more common.

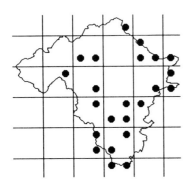

C. fasciculare (L.) Wigg.

Possibly extinct. Known only from a wych elm tree beside the old railway, S of Pont Shoni, Aberedw 0746, the tree has since died and shed its bark.

C. flaccidum (Ach.) Ach.

On mosses and rock on the sheltered sides of large boulders and rock outcrops in the flood zone of the R. Wye and tributaries from Builth Road 0252 to Hay on Wye 2242. Elsewhere rare on basic, moist, rock outcrops and noted once on an ash tree in damp woodland, Cwmbach Llechryd 0354.

C. fragrans (Sm.) Ach.

Very rare on nutrient-rich bark of well-lit trees. Ash, Bailey Einon, Cefnllys 0861 and wych elm at Stanner Rocks 2658. Probably extinct at this last site where the large elms have now died.

C. subflaccidum Degel.

A single record from S facing, basic volcanic rocks at Caer Einon, Llanelwedd 0652 det. BJC.

C. tenax (Swartz) Ach.

Widespread on open, somewhat basic and seasonally damp soil and crumbling mortar of walls. The vars. have not been mapped separately.

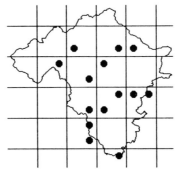

Coniocybe furfuracea (L.) Ach.

Widespread on dry stones, soil and tree roots in sheltered hollows in cliffs, by streams and hedgebanks. Fruits are rare.

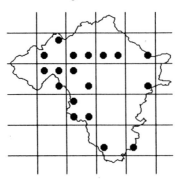

Coriscium viride (Ach.) Vainio

Uncommon on peat in blanket bogs, decaying stumps and peaty soil on rock outcrops. This basidiolichen is now considered to be correctly called *Omphalina hudsoniana* (Jenn.) Bigelow.

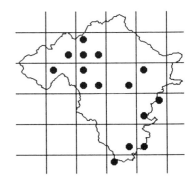
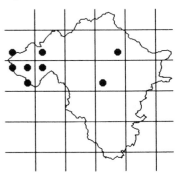

Cyphelium inquinans (Sm.) Trevisan

A widespread but uncommon sp. of old wooden fence posts, rails and gates, bark-free standing oak and rarely on the bark of ancient oak trees, all in well-lit, exposed situations.

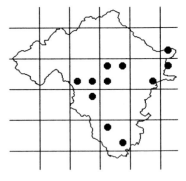

Cystocoleus ebeneus (Dillwyn) Thwaites

Frequent in dry recesses of acidic shale rocks, out of reach of rain and rarely on acidic bark of birch and oak in sheltered valleys in the west.

Dermatocarpon

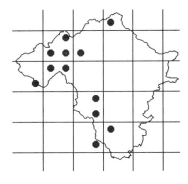

Dermatocarpon luridum (With.) Laundon.

Frequent on rocks and boulders in fast-flowing streams and rivers in the Wye catchment. Rare elsewhere. Recorded as *Lichenoides imbricatum luridum* from " a rivulet in the county of Radnor" by J J Dillenius in *Historia Muscorum*, 224 issued in 1742.

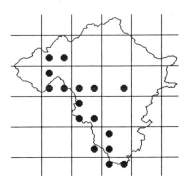

D. meiophyllizum Vainio

The smaller aquatic *Dermatocarpon* spp. are in need of further study. Most are believed to be this sp., but forms somewhat intermediate with the last are encountered. They occur in the Wye and its tributaries, mostly on large rocks in the flood zone.

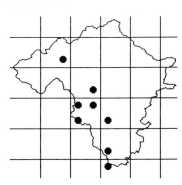

D. miniatum (L.) Mann

Rare on basic rock. Volcanic rock outcrop by R. Ithon N of Cefnllys Castle 0861; limestone outcrop, Yat Wood, Dolyhir 2458; dolerite outcrop, Stanner Rocks 2658; calcareous shale rocks in wood E of Knucklas 2673.

Dimerella diluta (Pers.) Trevisan

Uncommon in shaded and base-rich habitats. Bricks, near Pont Marteg, St Harmon 9571 AO; ash below Cerriggwynion, Ysfa 9765; ash, Tyncoed Wood, Cwmbach Llechryd 0354; ash in wood by lake, Llandrindod Wells 0660; old oak on cliffs N of Llanstephan House 1142; grey willow in wood S of Wern Fawr, Painscastle 1845; wych elm, Glascwm churchyard 1553; ash, Ddol Wood, Llanbister 1272; ash by Presteigne withy beds 3164.

D. lutea (Dickson) Trevisan

Rare on bark and mosses on ancient trees and rarely on rock in woodland. Cwm Coel, Elan Valley 8963 BLS exc.; moss covered rocks by Pen-y-garreg Dam, Elan Valley 9167; ash, old Cwm Elan House gardens, Elan Valley 9064; oak, Llethr Llwyd, Elan Village 9366; on moss covered rock, Glanllyn Wood, Llansantffraed Cwmdeuddwr 9469; oak on Wye bank, Nannerth, Rhayader 9571; oak in Lakeside Wood, Llandrindod Wells 0660; oaks, Tyfaenor Park, Abbeycwmhir 0771 and 0772; oak, Llanboidy wood, Llanfihangel-Nant-Melan 1457; the pycnidial form on ash in a wood near Painscastle 1845 det. BJC.

Diploicia canescens (Dickson) Massal.

Frequent in well-lit, nutrient-rich and basic places such as mudstone outcrops, mortared stone walls and isolated trees in pastures receiving dust and dung from grazing stock.

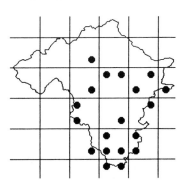

Diploschistes scruposus (Schreber) Norman

Common on well-lit, freely-drained and often sheltered acidic rock outcrops and crevices in grit drystone walls.

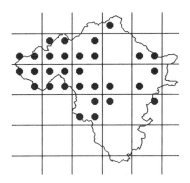

Diplotomma alboatrum (Hoffm. Flotow

Common on sunny mortar of stone walls and occasionally on calcareous rock outcrops. Rare on nutrient-rich tree bark. Probably more common than the mapped records suggest and not separated from *D. epipolium* (Ach.) Arnold.

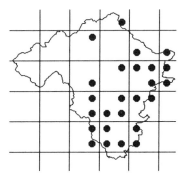

Dirina massiliensis f. *sorediata* (Müll. Arg.) Tehler

On dry and sheltered basic rock outcrops, out of reach of most direct rainfall. Consequently confined to the larger outcrops.

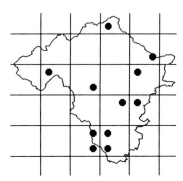

Fuscidea

Endococcus rugulosus Nyl.

Rare or overlooked. On thallus of *Huilia tuberculosa* (Sm.) P. James on coping of Garreg Ddu Resr. dam, Elan Valley 9063 ! and BJC.

Enterographa crassa (DC.) Fée

A rare sp of bark on old trees and rock in shaded places. Oak, Glog Fawr, Elan Valley 9166; oak on bank of R Wye, W of Yr Allt, Llandeilo Graban 0744; oak and ash in gorge of R. Edw, Aberedw 0747; shale in shaded ravine W of Gareg Fawr, Llandeilo Graban 0844; oak in park, Cefndyrys, Llanelwedd 0452; old oak, Bailey Einon, Cefnllys 0861.

Ephebe lanata (L.) Vainio

Widespread but not common on sloping, seasonally wet rocks, on cliffs and beside streams in the western uplands.

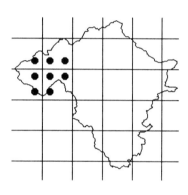

Epigloea soleiformis P. Döbbeler

An inconspicuous lichen of algal crusts overgrowing mosses. On *Andreaea* sp. on damp rocks in ravine N of Hirnant Farm near Craig Goch Resr., Elan Valley 8870.

Evernia prunastri (L.) Ach.

Common on the bark of branches and trunks of almost every species of tree and shrub and recorded from all 10km squares. Not found terricolous or with apothecia in Radnor.

Farnoldia jurana (Schaerer) Hertel

A lichen usually of hard limestone but in Radnor noted only on calcareous mudstone, Aberedw Rocks 04NE, AEW (1959 NMW).

Foraminella ambigua (Wulfen) Friche Meyer

Widespread in central and E Radnor on acidic tree bark and on dry, ancient dead wood on trees and converted to fence and gate posts and rails.

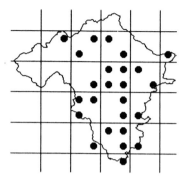

Fuscidea cyathoides (Ach.) V. Wirth and Vezda

Occasional on dry, well-lit, acidic rocks in the uplands.

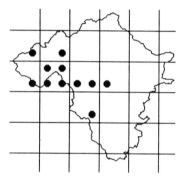

F. kochiana (Hepp) V. Wirth and Vezda

A scarce sp. of upland, acidic shale and gritstone rocks in exposed places. Near Ciloerwynt, Claerwen Valley 8862 and 9061; rocks to W of Craig Goch Resr. dam, Elan Valley 8968; Treheslog Bank, Llansantffraed Cwmdeuddwr 9369; Cerriggwynion, Rhayader 9765.

F. lightfootii (Sm.) Coppins and P. James

An uncommon sp. of smooth bark of trees and shrubs in damp woodlands. Commonest on grey willow but also on ash and rowan, mostly in the west.

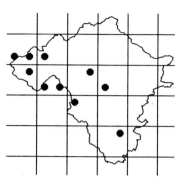

F. praeruptarum (Du Rietz and Magnusson) V. Wirth and Vezda

A scarce or overlooked lichen of sunny to shaded acidic shale rocks. Wall of old lead mine shaft, Cwm Elan Mine, Elan Valley 9065; sheltered shale rock face, Cerrig Gwalch, Rhayader 9370; boulders in grassland S of Wyloer, Gilfach Farm, St Harmon 9571 ! and BJC.

F. recensa (Stirton) Hertel, V. Wirth and Vezda

In similar sites to the last species but extending onto volcanic rocks and monuments. Gyrn, Llansantffraed Cwmdeuddwr 9366; Cerriggwynion, Rhayader 9765; rocks S of Wyloer, Gilfach Farm, St Harmon 9571 ! & BJC; Aberedw Rocks 0746 ! and BJC; Llanelwedd Rocks 0552 ! and BJC; tombstone, Llanstephan Church 1142 ! and BJC.

F. tenebrica (Nyl.) V. Wirth and Vezda

On exposed, acidic shale and grit rock outcrops in the uplands.

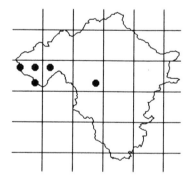

Graphis elegans (Borrer ex Sm.) Ach.

Common on the smooth bark of tree branches and the stems of shrubs in the west. Rarer in the east.

Haematomma

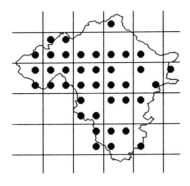

G. scripta (L.) Ach.

In similar habitats to the last and often with it, though perhaps confined to more humid sites.

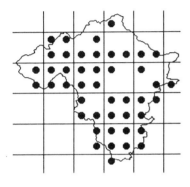

Gyalecta jenensis (Batsch) Zahlbr.

Occasional on damp and shaded, calcareous rock outcrops.

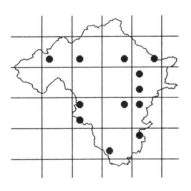

G. truncigena (Ach.) Hepp

A scarce sp. of base-rich bark on ancient trees, probably much reduced in extent by the loss of large elms in recent years. Dying wych elm by Pen-y-garreg Resr. dam, Elan Valley 9167; field maple near Erwood Station 0944; oak in gorge of R. Edw, Aberedw 0747; old oak W of Rhyscog, Aberedw 0948; old ash near Cefndyrys, Llanelwedd 0352; ash, Tyncoed Wood, Cwmbach Llechrhyd 0354; ash near Shaky Bridge, Cefnllys 0861; old oak in gorge of the Bach Howey, Llanstephan 1143; dying wych elm, Stanner Rocks 2658.

Gyalidea subscutellaris (Vezda)Vezda

A nationally rare sp. overgrowing moss cushions on heavy-metal-rich soil. Tips of the disused Cwm Elan Mine, Methan Valley, Llansantffraed Cwmdeuddwr 9065.

Gyalideopsis anastomosans P. James and Vezda

Scarce or overlooked on bark in damp woodlands. On alder by R. Wye above Nannerth Fawr, Llansantffraed Cwmdeuddwr 9472; common on willow, Cors y Llyn, Newbridge on Wye 0155 AO; on elder in damp woodland near Llanfihangel-Nant-Melan 1857.

Haematomma caesium Coppins and P. James

Probably widespread on smooth bark, but much overlooked in this survey. Ash in old garden of Cwm yr Elan House, Elan Valley 9064.

H. elatinum (Ach.) Massal.

Rare on trees in damp woodland. Cwm Coel, Elan Valley 8963, BLS exc; on ash in wood N of Nant Gwyllt, by Caban Coch Resr, Elan Valley 9163 and on alder in Tyncoed Wood, Cwmbach Llechrhyd 0354 ! and FR.

H. ochroleucum (Necker) Laundon

Occasional on sheltered but well-lit, slightly basic to acidic cliff faces. The var. *porphyrium* (Pers.) Laundon is present in the county but has not been separately mapped.

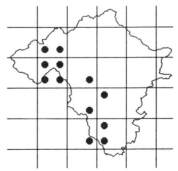

H. ventosum (L.) Massal.

Uncommon on sunlit, exposed hard grit and shale rock outcrops on the summit ridges of the western uplands. Both yellow and grey coloured forms are equally common. Blaen Rhestr 8469; Craig-y-bwch 8962 and 9061; Craig-y-foel 9164; Gym 9366; Treheslog Bank 9369 - all in Llansantffraed Cwmdeuddwr; Cerriggwynion, Rhayader 9765; Wyloer, St Harmon 9571; Dol-y-fan Hill, Llanyre 0161.

Huilia albocaerulescens (Wulfen) Hertel

Frequent on acidic shale and volcanic rock outcrops. Probably most, if not all records should be referred to *H. cinereoatra* (Ach.) Hertel.

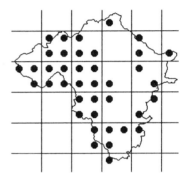

H. crustulata (Ach.) Hertel

Abundant on acidic rock outcrops, stone walls, slates and even small pebbles. Rare on sawn hardwood of fences and old farm carts. Recorded from all 10km squares except 36.

H. hydrophila (Fr.) Hertel

An overlooked sp. recorded by ! and AO on wet rocks by the stream below Craig-y-bwch, Claerwen Valley 8961 and by AO in Coed yr allt goch, Elan Valley 9068.

H. macrocarpa (DC.) Hertel

Common on acidic rock outcrops and stone walls throughout the county. Recorded from all 10km squares except 27, 36 and 37.

H. platycarpoides (Bagl.) Hertel

Recorded from a sandstone wall top in Llanstephan Churchyard 1142 by BJC and !, this lichen has probably been overlooked on acidic shales and sandstones elsewhere.

H. soredizodes (Lamy) Hertel

Noted once from stones by Llynheilyn, Llanfihangel-Nant-Melan 1658 AEW (1959 NMW). Probably overlooked elsewhere.

H. tuberculosa (Sm.) .P James

Common on well-lit, acidic rock outcrops, walls, slate roofs and rarely on sawn timber long exposed to the weather. Recorded from all 10km squares except 18.

Hymenelia lacustris (With.) M. Choisy

Frequent on shale rock outcrops and boulders in fast-flowing streams and rivers.

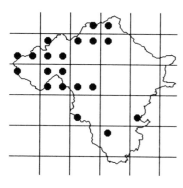

Hyperphyscia adglutinata (Flöerke) Mayrh. and Poelt.

Rare. On Lombardy polar trunk on bank of R. Wye Clyro 2242.

Hypocenomyce caradocensis (Leighton ex Nyl.) P. James and G. Schneider

A scarce but widespread sp. on the bare wood of ancient oak trees, stumps, logs and old fence posts.

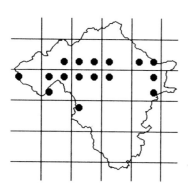

H. friesii (Ach.) P. James and G. Schneider

Reported by AO from the wood of an oak tree near Tyn-y-Graig, Llandrindod Wells 0762.

H. scalaris (Ach. ex Liljeblad) M. Choisy

Common on acidic tree bark and bare wood of stumps, logs, fence posts and of dead standing trees.

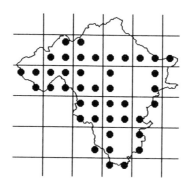

Hypogymia physodes (L.) Nyl.

The commonest foliose lichen on all manner of acidic substrates and one of the earliest colonizers. On twigs, tree trunks, rocks, fences, slate roofs etc. Occasionally with apothecia. Recorded from all squares.

H. tubulosa (Schaerer) Havaas

Common on the acidic bark of branches and trunks of a wide range of tree and shrub spp. and, though never as common, occurs frequently with the last sp. on a wide range of other acidic substrates. Recorded from all 10km squares.

Lasallia pustulata (L.) Mérat

Widespread on acidic, mostly E facing, hard rock outcrops, especially those receiving sporadically slightly nutrient enriched water by run off from soil above. Large populations exist at Cerriggwynion, Ysfa 9765; Cefnllys Castle 0861; Llandegley Rocks 1361 and Graig Fawr, Franks Bridge 1358.

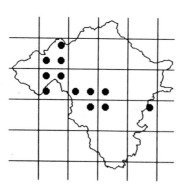

Lecanactis abietina (Ach.) Körber

Of widespread occurrence on the acidic bark of trees, especially in recesses of the lower trunk and extending up the trunk only in sheltered sites. Commonest on oak.

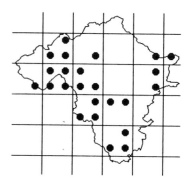

L. amylacea (Ehrh. ex Pers.) Arnold

A nationally rare lichen of ancient woodland sites, known in Radnor from dry, shaded bark on a single ancient oak in a long undisturbed wooded part of the Dulas Valley at Tyncoed, Cwmbach Llechrhyd 0355.

L. dilleniana (Ach.) Körber

Only on the N facing side of a sandstone wall around Llanstephan Churchyard 1142 BJC and !.

L. premnea (Ach.) Arnold

A scarce sp. of mostly shaded bark on the lower trunks of ancient oaks and occasionally ash, sometimes extending onto old ivy trunks associated with ancient trees and once noted on yew wood. In wooded valleys and on small tree groups in parkland.

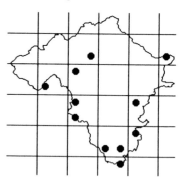

Lecanora

Lecania cyrtella (Ach.) Th. Fr.

A tiny and inconspicuous sp. of wood and bark, notably of elder, ash and sycamore, probably much overlooked.

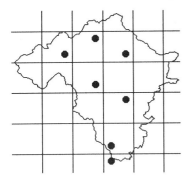

L. erysibe (Ach.) Mudd

As with the last, inconspicuous and much overlooked. In a wide range of habitats. On willow sp. Cwmbach Woods 0354 AP (1969 BLS records); Aberedw Rocks 0746 BLS exc. rocks by R. Wye opposite Llangoed Castle, Llanstephan 1240.

L. nylanderiana Massal.

A single record of this mostly maritime sp. from S facing, slightly calcareous shale rock outcrops by the R. Teme, The Ring, Beguildy 1283 det BJC.

Lecanora aitema (Ach.) Hepp

Not separated from *L. symmicta* (Ach.) Ach. in the early years of the survey. Both sp. often occur together on old sawn timber of posts and gates and on untreated split oak and round softwood posts and have been mapped together. (See below).

L. albescens (Hoffm.) Branth and Rostrup

Common on sunny lime mortar of walls, concrete and on calcareous mudstone and limestone outcrops. More widespread than the mapped records suggest, as not separated from *L. dispersa* (Pers.) Sommerf. in the early part of this survey.

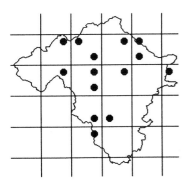

L. atra (Huds.) Ach.

Frequent on well-lit rocks which are neither strongly basic nor acidic. Often more abundant on tombstones and the stone of mortared walls than on native rock outcrops. Recorded from all 10km squares except 87.

L. badia (Pers.) Ach.

Common on sunny, exposed, acidic boulders and rock outcrops receiving nutrient enrichment from the leachate of bird and sheep droppings.

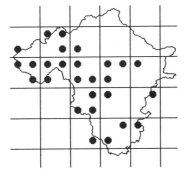

L. campestris (Schaerer) Hue

Common on sunny, dry, usually somewhat basic rocks, mortared stone walls and monuments.

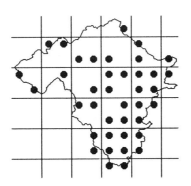

L. carpinea (L.) Vainio

A scarce sp. of smooth, nutrient-rich bark. Sycamore near Llandegley 1362 AEW (1959 NMW); Rhosgoch Common 1948 AEW (1963 NMW); elder, Gwernfythen, Clyro 2044, walnut in Bleddfa churchyard 2068.

L. chlarotera Nyl.

Common on the smooth bark of a wide range of tree and shrub sp. in woods, hedgerows etc. Recorded from all 10km squares.

L. confusa Almb.

An uncommon sp. of smooth bark on trees and shrubs and occasionally sawn timber. Commonest in damp sites in the western uplands.

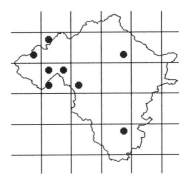

L. conizaeoides Nyl. ex Crombie

Abundant on twigs, branches and trunks of trees and shrub sp., on sawn timber of weather boards and fences etc. Rarer on acidic rock outcrops. Recorded from all 10km squares.

L. crenulata Hook.

Rare or overlooked on calcareous rocks. Mudstone cliff, Aberedw Rocks 0746 BLS exc.; calcareous shale stone of church wall, Heyop 2374.

L. dispersa (L.) Sommerf.

Common on basic rock and one of the commonest early colonizers of mortar, concrete and asbestos cement. Uncommon on trees though frequent on the branches of trees about the limestone quarries at Dolyhir 2458. Recorded from all 10km squares.

L. epanora (Ach.) Ach.

A scarce lichen of iron-rich shale rocks in sheltered sites. Slate quarry waste below Craig-y-bwch, Claerwen Valley 9061; cliffs, Craig-y-foel, Elan Valley 9163, 9164, 9264; cliffs, Treheslog Rocks 9468 and cliffs by the road NW of Glanllyn 9469 all in Llansantffraed Cwmdeuddwr. On rocks in old railway cutting, Gilfach Farm, St Harmon 9571.

L. expallens Ach.

Common on well-lit bark of a wide range of tree sp. on woodland edges and hedgerows and frequent on sawn hardwood timber and on bark-free standing wood. Recorded from all 10km squares.

Lecanora

L. gangaleoides Nyl.

Frequent on acidic shale rock outcrops where overhangs provide some shelter. Rarer on stone walls and monuments.

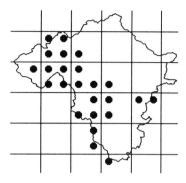

L. grumosa (Pers.) Du Rietz

A scarce or overlooked sp. of sunny, acidic rocks. Gyrn, Llansantffraed Cwmdeuddwr 9366; Cerriggwynion, Rhayader 9765; rocks S of Wyloer, Gilfach Farm, St Harmon 9571 BJC and !; volcanic rocks, Llanelwedd 0552 BJC and !; volcanic rocks, Carreg Wiber, Llandrindod Wells 0859; dolerite cliff, Graig-fawr, Frank's bridge 1358; Old Radnor Hill 2558.

L. intricata (Ach.) Ach.

Frequent on well lit, acidic rock outcrops and wall tops.

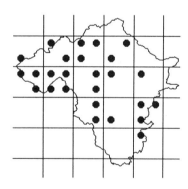

L. intumescens (Rebent.) Rabenh.

Of widespread occurrence on the smooth bark of tree branches and trunks, particularly of rowan and ash in upland woods. Rare in the lowlands.

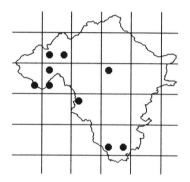

L. jamesii Laundon

Uncommon on the smooth bark of trees and shrubs in damp woodlands, most frequent in the Elan Valley woodlands on rusty willow.

L. muralis (Schreber) Rabenh.

On a wide range of basic and nutrient enriched substrates, eg boulders by streams and pools and concrete fence posts, all used by birds as regular perching sites, asbestos cement, tarmacadam and rock outcrops seasonally flushed by nutrient enriched water.

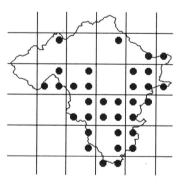

L. polytropa (Hoffm.) Rabenh.

Frequent on well-lit acidic rock outcrops and walls. Rare on sawn timber. Recorded from all 10km squares.

L. pulicaris (Pers.) Ach.

Widespread on smooth, acidic bark of a wide range of tree and shrub species. Also occasionally seen on sawn hardwood and softwood fence posts.

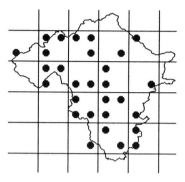

L. quercicola Coppins and P. James

A rare sp. of bark on well-lit, ancient oak trees. By the drive, Cefndyrys, Llanelwedd 0452; roadside oak in wood S of Gaer, Llansantffraed-in-Elvel 0854.

L. rupicola (L.) Zahlbr.

An uncommon lichen of sunny volcanic rock outcrops and sandstone tombstones in churchyards. Much of the native rock seems to be too acidic or friable to support it.

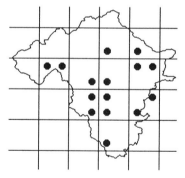

L. saligna (Schrader) Zahlbr.

Scarce or overlooked on sunny and dry wood and bark. On lignum of old field gate, Llanyre 0462; wood of dead wych elm, New House, Llanbister 1172; soft-wood logs still with bark, built into barn, Gwernaffel, Knighton 2670.

Lecidea

L. soralifera (Suza) Räsänen

Frequent on acidic rock outcrops and stone walls in well-lit and usually exposed situations.

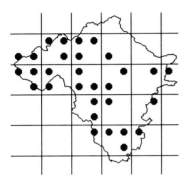

L. subcarnea (Liljeblad) Ach.

Rare, known only from dry underhangs in hard shale cliffs. Cerrig Gwalch, N of Rhayader 9370 det. BJC.

L. symmicta (Ach.) Ach.

Widespread on dry, sunny wooden rails, posts and boards, particularly on fences and gates. Not separated from *L. aitema* (Ach.) Hepp in the early years of the survey, a sp. with which it often occurs. They have been mapped together.

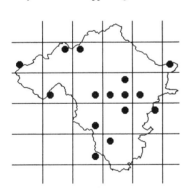

L. varia (Hoffm.) Ach.

A scarce or overlooked sp. of bark and wood. On old railway sleeper fence post by railway line near Cwm-y-geist, Llanbister Road 1771.

Lecidea chalybeioides Nyl.

A nationally rare sp. of seasonally wet, vertical, acidic shale rock outcrops. SW facing outcrops to E of Craig Goch Resr. Dam, Elan Valley 8968.

L. carrollii Coppins and P. James

A scarce sp. of smooth hazel and willow bark in wet woodlands. Tyncoed, Cwmbach Llechrhyd 0354 ! and FR; Bailey Bevan wood, Llanfihangel Helygen 0464; Alpine Bridge, Llanbadarnfawr 0962.

L. erratica Körber

A small, dark and inconspicuous species only noted on shale of old quarry near Llandrindod Wells golf club 0760 AO and on mudstone outcrops in the Bach Howey Gorge, Llanstephan 1243 det. BJC.

L. fuliginosa Taylor

In small crevices on sunny, acidic shale and volcanic rock outcrops. Widespread in the western uplands but not common.

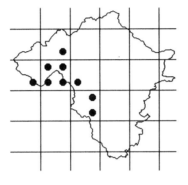

L. furvella Nyl. ex Mudd

The dark, rarely-fruiting crusts of this sp. overgrow the thallus of other crustose lichens, usually on sunny horizontal rock surfaces. Certainly under-recorded. Y Foel, Elan Valley 9165; Llanelwedd Rocks 0552 ! and BJC; wall, Llanstephan church 1142 ! and BJC.

L. fuscoatra (L.) Ach.

Frequent on smooth, well-lit acidic shale and volcanic rocks, often in damp places and by rivers and streams, but also on drier vertical cliffs.

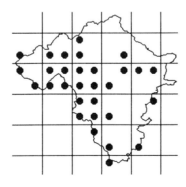

L. hypnorum Lib. s. lat.

An uncommon lichen which overgrows mosses and liverworts on the trunks of ancient trees, mostly oak, in the damp valleys of the NW uplands.

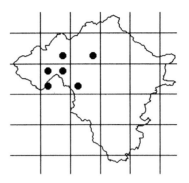

L. hypopta Ach.

A northern sp. with few mid Wales records. On old bark-free oak trunk and on bark of old larch trees, Nant Gwyllt, by Caban Coch Resr., Elan Valley 9162; old conifer wood, Rhydoldog, Llansantffraed Cwmdeuddwr 9468; dead oak by Wye, Nannerth, Rhayader 9571; dead and bark-free oak in the Park, Penybont Hall 1163, det. BJC.

L. insularis Nyl.

A rare sp. overgrowing *Lecanora rupicola* (L.) Zahlbr. On sunlit volcanic rock outcrop, Llanelwedd Rocks 0552 ! and BJC.

L. lactea Flörke ex Schaerer

An uncommon sp. of well-lit, hard, acidic rocks such as the Caban Coch grits and volcanic rock, native or incorporated into walls and monuments.

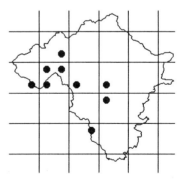

L. lapicida (Ach.) Ach.

In similar habitats to the last but apparently much more scarce. South-facing volcanic rock outcrops, Cefnllys Castle 0861, BLS exc. Stanner Rocks 2658, BLS exc.

Lecidea

L. leucophaea (Flörke ex Rabenh.) Nyl.

A scarce or overlooked species of sunny, acidic rocks. Gyrn, Elan Village 9366; Cefnrhydoldog, Llansantffraed Cwmdeuddwr 9367; Cerriggwynion, Rhayader 9765; Caer Fawr, Llanelwedd 0553 AEW (1959 NMW); Hanter Hill, Burlingjobb 2557.

L. lithophila (Ach.) Ach.

Frequent on acidic shale and volcanic rock outcrops in the uplands.

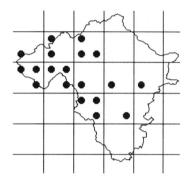

L. monticola (Ach.) Schaerer

Widespread on mortar, concrete, asbestos cement and occasionally on calcareous shale rock outcrops. Probably more common on man-made habitats than the mapped records suggest.

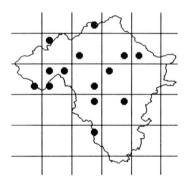

L. orosthea (Ach.) Ach.

Frequent on sunny, but dry and sheltered, acidic rock outcrops and occasionally walls and monuments.

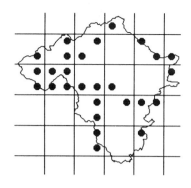

L. phaeops Nyl.

Rare, on damp shale rock outcrop, Gwynllyn, Llansantffraed Cwmdeuddwr 9469 AO.

L. speirea (Ach.) Ach.

Two records only from volcanic rock outcrops. Llanelwedd Rocks 0551, BLS exc. Stanner Rocks 2658 PWJ.

L. sublivescens (Nyl. ex Crombie) P. James

Scarce on old oak trees in open situations. First reported from Cilgynfydd, Builth Road 0354 by AP in 1969 and still present there. Known also from an oak near Builth Road Station 0253 and in the park, Cefndyrys, Llanelwedd 0452.

L. sulphurea (Hoffm.) Wahlenb.

A common sp. of dry, sunny and exposed, slightly basic to acidic rocks on cliffs, walls and monuments.

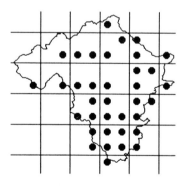

L. valentior Nyl.

Scarce or overlooked on damp acidic rocks in shade. Coed yr allt goch, Elan Valley 9067 AO.

L. vernalis (L.) Ach.

Very rare. On oak trunk in Gwyllyn Wood, Llansantffraed Cwmdeuddwr 9469 AO. The only known Welsh site.

Lecidella elaeochroma (Ach.) M. Choisy

In all 10km squares. Common on smooth bark, especially of ash and elder trees on wood edges and in hedgerows. The form *soralifera* (Erichsen) D. Hawksw. has not been separated.

L. scabra (Taylor) Hertel and Leuckert

Common on somewhat basic and/or nutrient enriched bark and rock such as around the bases of trees in pastures, on wall tops, window sills and other bird perch sites or around farm yards.

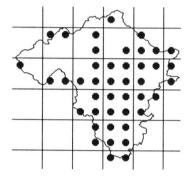

L. stigmatea (Ach.) Hertel and Leuckert

Probably widespread on mortar, asbestos cement sheeting and concrete but under recorded. Also on calcareous mudstone outcrops and on basic volcanic rocks.

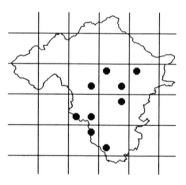

L. subincongrua (Nyl.) Hertel and Leuckert

A maritime sp. with a single mid Wales record from S facing dolerite cliffs, Stanner Rocks 2658 PWJ et al.

Lepraria spp.

The identification of members of this genus has presented problems and much further work is required. Only the more distinctive spp. have been mapped. AO

Leproloma

has begun a survey of the group, the very early findings of which are reported here.

L. incana (L.) Ach. s.l.

In all squares. A number of spp. have probably been recorded here. They occur in dry crevices in bark, rocks, walls and on soil on dry banks. The distinctive apple green thallus of *L. lesdainii* (Hue) R.C. Harris is confined to shaded calcareous sites such as Aberdew Rocks 0746 and the limestone near Burlingjobb 2458. The remaining spp. cannot be certainly identified without a knowledge of their chemistry. AO at NMW provides the following records. *L. incana* s.s. is probably common. On an oak tree by the R. Wye above Pont Marteg, N of Rhayader 9571; on a Scot's pine trunk at Cors y Llyn, Newbridge on Wye 0155 and on dead vegetation on a shale bank, Abbeycwmhir 0370. *L. lobificans* Nyl. is also probably common. On basic rocks, Yr Allt, Llandeilo Graban 0844; Lower Llanelwedd Wood 0452 (coll. AP, 1969, NMW); on shale, Abbeycwmhir 0370; on hawthorn in the Bach Howey valley, Llanstephan 1042 and on a stump in woodland, Worsell Wood, Old Radnor 2657. *Lepraria umbricola* Tonsb. occurs on birch at Cors y Llyn, Newbridge on Wye 0155 and this may be frequent on bark.

L. neglecta Vainio

Common on cushions of mosses such as *Andreaea* spp. on moist, acidic rocks in the western uplands. *L.caesioalba* (B. de Lesd.) Laundon (in ed.) is morphologically indistinguishable and most records here may be this species.

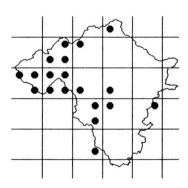

L. nivalis Laundon

A scarce sp. of dry and sheltered calcareous rocks. Aberedw Rocks 0746; gorge of the Bach Howey, Llanstephan 1243; old limestone quarry, Burlingjobb 2458, dolerite cliffs, Stanner Rocks 2658.

Leprocaulon microscopicum (Vill.) Gams ex D. Hawksw.

A widespread, but uncommon sp. on soil in dry and usually sunny crevices in acidic rock faces.

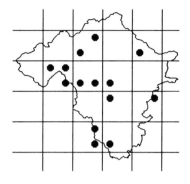

Leproloma spp.

Laundon, (in ed.) J.R. in the *Lichenologist* **21**: 1-22 (1989) revises the spp. in the *Lepraria membranacea* group, placing them in this genus. As the majority of this survey was carried out prior to this revision, further work is clearly required on this group, which are frequent on sheltered, more or less acidic rock underhangs, largely out of reach of rain but may occur in more open situations in sunny and warm places. Occasionally they extend onto the trunks of particularly large oak trees. *Leproloma membranaceum* (Dickson) Vainio may well prove to be the commonest species. *L. vouauxii* (Hue) Laundon is noted by Laundon from Stanner Rocks 2658. *L. diffusum* var. *diffusum* Laundon is recorded by AO from Rock Dingle, Bleddfa 2067. *L. diffusum* var. *chrysodetoides* Laundon was found overgrowing mosses on a cliff in Glannau Wood, Elan Valley 9165 det. BJC.. *L. angardianum* (Övstedal) Laundon. Shaded rocks, Coed yr allt goch, Elan Valley 9067 AO.

Leproplaca chrysodeta (Vainio ex Räsänen) Laundon

An uncommon sp. of dry crevices in calcareous rock faces and on mortar on walls.

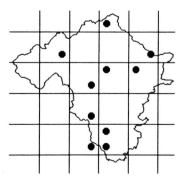

Leptogium cyanescens (Rabenh.) Körber

From a dripping rock outcrop, Gwynllyn Wood, Llansantffraed Cwmdeuddwr 9469, AO and as a single rather depauperate specimen growing on mosses over base-rich volcanic rocks as at Caer Einon, Llanelwedd 0653.

L. gelatinosum (With.) Laundon

A scarce or overlooked sp. of rocks and mortared walls in base-rich areas. Shale outcrops below Yr Allt, Llandeilo Graban 0844 ! & AO; old barn, Abbeycwmhir 0470; by R. Edw, Llanbadarn-y-Garreg 1249; old limestone quarry, Dolyhir 2458; wall, New Radnor 2160.

L. lichenoides (L.) Zahlbr.

Uncommon on mosses growing over base-rich rock outcrops and on tree trunks in moist and shady woodlands and river banks.

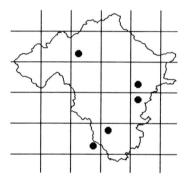

L. palmatum (Huds.) Mont.

Rare on top of low, seasonally damp, somewhat nutrient-rich rock outcrops, often associated with moss cushions. .

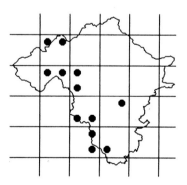

Micarea

L. teretiusculum (Wallr.) Arnold

Widespread, but scarce, on damp, basic rock outcrops and base-rich tree trunks such as ash, walnut and field maple.

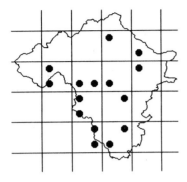

L. turgidum (Ach.) Crombie

A rare lichen known only from calcareous shale rock outcrops at Aberedw 0746, BLS exc..

Leptorhaphis epidermidis (Ach.) Th. Fr.

A single record from downy birch on a basin mire at Cors y Llyn, Newbridge on Wye 0155 FR.

Lobaria amplissima (Scop.) Forss.

A relict species known only from a large roadside oak at Ochr-cefn, Llansantffraed Cwmdeuddwr 9468 and on an ancient pollarded oak in a damp woodland near Nannerth, N of Rhayader 9472. Cephalodia present at both sites.

L. pulmonaria (L.) Hoffm.

A rare sp. of ancient trees. Oak below Llethr Melyn, near shore of Garreg Ddu Resr., Elan Valley 9164; dying wych elm, ash and oak in wood on N side of Pen-y-garreg Resr. dam, Elan Valley 9167; oak, Noyadd Fach, Elan Village 9365; oaks, Nannerth Fawr, N of Rhayader 9471; hazel and ash tree in Tyncoed Wood, Cwmbach Llechrhyd 0354; isolated oak in field, Yr Allt, Doldowlod 0062; oak overhanging gorge, Alpine Bridge Llanbadarnfawr 0963; oak by stream at wood edge, Stanage Park 3371.

Lopadium disciforme (Flotow) Kullh.

The inconspicuous grey crusts of this lichen occur overgrowing mosses on the trunks of mostly mature oak and ash in damp woodlands.

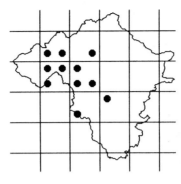

Macentina stigonemoides Orange

This recently described lichen may prove to be widespread on elder in sheltered places. In gorge N of Yr Allt, Llandeilo Graban 0844 and at Water-break-its-neck, Llanfihangel-Nant-Melan 1860 coll. MFVC det. AO.

Massalongia carnosa (Dickson) Körber

A scarce sp. of shaded, seasonally damp acidic rocks. Damp rocks above Claerwen Dam 8763; amongst *Andreaea* sp. on SW facing cliffs below Maen-serth, Glan-llyn, Llansantffraed Cwmdeuddwr 9469; rocks in R. Elan above Pont ar Elan 9071; rocks in old railway cutting E of Tylwch 9779; Penddol Rocks, Llanelwedd 0352; below the Gaer, Llansantffraed-in-Elvel 0854; rocks by road, Yr Allt, Doldowlod 0062. Also found very atypically growing on the lower trunk of a leaning hawthorn, Yr Allt, Llandeilo Graban 0844 conf. BJC.

Micarea adnata Coppins

Rare and only on a damp, decaying log in the gorge of the Wye above Nannerth, N of Rhayader 9571 det. BJC, at the southernmost edge of its range in Britain.

M. bauschiana (Körber) V. Wirth and Vezda

A scarce or overlooked sp. of shaded, acidic rock outcrops near Craig Goch Resr. dam, Elan Valley 8968; rocks N of Ashfield, Ysfa 9764 conf. BJC; on fence post, Cors y Llyn, Newbridge on Wye 0155 AO; in wood SE of Llangunllo 2171.

M. botryoides (Nyl.) Coppins

Probably widespread on mosses, liverworts and decaying vegetation in the uplands along the edges of banks and tussocks. Craig-y-bwch 8962, Bylchau, Abergwngu 8672 and Cwm Coel 9063 in the Elan Valley; Rhyd, Bwlch-y-sarnau 0475; Ty-faenor Park, Abbeycwmhir 0671; Rhulen Hill 1349; Sychwn, Glascwm 1752.

M. cinerea (Schaerer) Hedl.

A scarce sp. of acidic bark in damp woodlands. On oak and alder, Cwm Coel, 9063; on ash by Caban Coch Resr. 9163; oak by Garreg Ddu Resr. 9166, all in the Elan Valley and on oak by the R. Wye above Nannerth, N of Rhayader, 9571.

M. denigrata (Fr.) Hedl.

Probably common but overlooked on decorticate wood and old fence posts. Llanelwedd Rocks 0552 BJC and ! and dead pine trunk, Cors y Llyn, Newbridge on Wye 0155 AO.

M. leprosula (Th.Fr.) Coppins and A. Fletcher

Common on moss cushions growing on acidic shale rock outcrops in the western uplands.

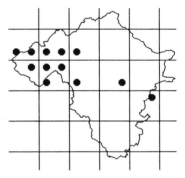

M. lignaria (Ach.) Hedl.

Common on decaying mosses, stumps, peat and heather stems and rocks in damp, acidic sites in the uplands.

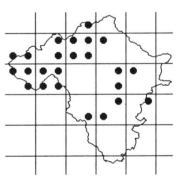

M. lutulata (Nyl.) Coppins

Rare or overlooked on shaded acidic rock underhangs. Coed yr allt goch, Elan Valley 9067 AO.

Mycoblastis

M. melaena (Nyl.) Hedl.

A widespread sp. of old, bark-free oak stumps in woodlands and hedgerows and old sawn timber posts and occasionally peaty turf in the uplands.

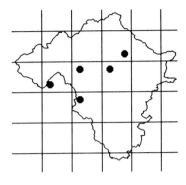

M. sylvicola (Flotow) Vezda and V. Wirth

Widespread on dry and sheltered acidic shale rocks and associated tree roots in underhangs on cliffs and hedgebanks.

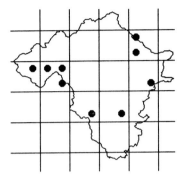

M. sterilis Coppins and P. James

Frequent on acidic bark of branches and trunks of a wide range of tree and shrub sp. and commonest in damp woodlands.

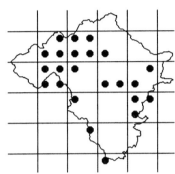

M. nitschkeana (Lahm. ex Rabenh.) Harm.

A scarce or overlooked sp. of branches and wood. A single record from heather stems in the Marteg Valley below Wyloer, Gilfach Farm, St Harmon 9571 BJC and !.

M. peliocarpa (Anzi) Coppins and R. Sant.

A scarce or overlooked sp. recorded from a wide range of acidic habitats in Britain. On oak, Cwm Coel, Elan Valley 9063 BJC and !; on small shale rock in grassland, Begwns, Llowes 1643, conf. BJC.

M. prasina Fr.

Widespread on damp, slowly decaying logs and occasionally on tree trunks in shady woodlands and gorges.

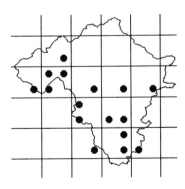

M. subnigrata (Nyl.) Coppins and Kilias

A scarce lichen of sunny, acidic rock outcrops. Rocks S of Wyloer, Gilfach Farm, St Harmon 9571 BJC and !; Llanelwedd Rocks 0552 BJC and !.

Microglaena muscorum (Fr.) Th. Fr.

A scarce sp. overgrowing mosses on ancient trees. Old oak by R. Wye, Nannerth Fawr, N of Rhayader 9471 and old oak in field, Yr Allt, Doldowlod 0062.

Moelleropsis nebulosa (Hoffm.) Gyelnik

An uncommon sp. of well-drained soil, usually around slightly basic, S facing rock outcrops.

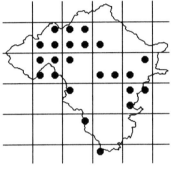

Muellerella pygmaea (Körber) D. Hawksw.

A probably widespread, though under recorded parasite of *Huilia* spp.. SW facing vertical shale rockface, E of Craig Goch Resr., Elan Valley 8968.

Mycoblastus sanguinarius (L.) Norman

Confined to acidic tree bark, wooden fence posts and occasionally rock outcrops in exposed sites in the western uplands.

Mycoporum quercus (Massal.) Müll. Arg.

Frequent on the smooth bark of oak branches in areas of low sulphur dioxide pollution.

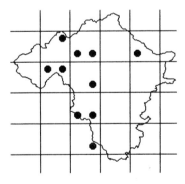

Nephroma laevigatum Ach.

Known only from damp rocks in an oak and ash woodland to the NE of Pen-y-garreg Dam, Elan Valley 9167, where it is fertile, and on damp, somewhat basic rocks in a gulley near Garreg Fawr, Llandeilo Graban 0844.

Ochrolechia

N. parile (Ach.) Ach.

On damp, somewhat basic rocks and rarely on basic tree bark. Ash trunks in old garden, Cwm Elan House, Elan Valley 9064; rocks in wood, Glanllyn, Llansantffraed Cwmdeuddwr 9469; on rock in wood N of Pen-y-garreg Dam, Elan Valley 9167; shale outcrops E of Erwood Station 0944. shaded volcanic rocks, Llanelwedd Rocks 0552 BLS exc..

Normandina pulchella (Borrer) Nyl.

Widespread, but not common, on mosses and liverworts and lichens such as *Parmeliella triptophylla* (Ach.) Müll. Arg. on the bark of usually ancient trees. Noted from oak, ash, field maple, apple cultivars and crack willow. It occurs rarely, also, on mosses over basic mudstones.

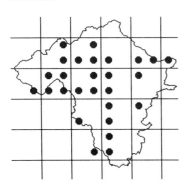

Ochrolechia androgyna (Hoffm.) Arnold

In all 10km squares. Common on acidic bark on a wide range of tree species. Occasionally present on sawn timber of fences and on acidic rock outcrops, especially in the western uplands. Fertile only in the Elan Valley 86 and 96.

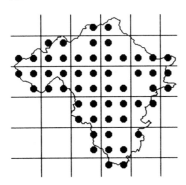

O. parella (L.) Massal.

Common on sunny, somewhat basic to basic rock outcrops, mortared stone walls and monuments. Rare on trees. Noted on ash at Cefndyrys, Llanelwedd 0352, near Michaelchurch 2450 and New Radnor Castle 2160 and on walnut at Bleddfa Church 2068.

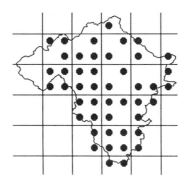

O. subviridis (Höeg) Erichsen

Widespread on the more nutrient-rich bark of ash, elm, sycamore, maple and occasionally other trees on field margins and woodland edges. Rare on slightly basic rock outcrops and sandstone monuments.

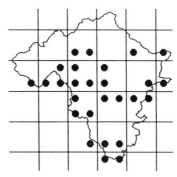

O. tartarea (L.) Massal.

Frequent on acidic rock outcrops, block screes and the acidic bark of trees, notably oak and birch in the western uplands.

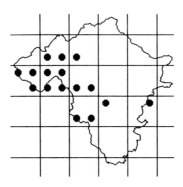

O. turneri (Sm.) Hasselrot

In similar habitats to and often with *O. subviridis* (Höeg) Erichsen.

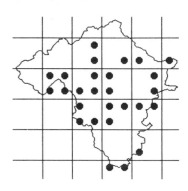

Omphalina ericetorum (Fr.) ex Fr. M. Lange

Further work is required on the species of *Omphalina* forming basidiolichens in Radnor. This sp., occurring widely on damp peaty banks is believed to be the commonest, however AO records *O. velutina* (Quelet) Quelet from soil amongst rocks at 520m on Fron-wen, Radnor Forest 1766 and *O. cupulatoides* P.D. Orton parasitizing *Peltigera* sp. on forest ride, Caerhyddwen, Ysfa 9964 coll. !..

Opegrapha atra Pers.

Widespread on smooth bark of a wide range of trees and shrubs eg wych elm, sycamore, hazel and elder. Often in fairly shaded places.

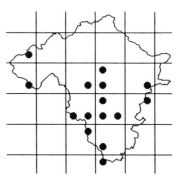

O. gyrocarpa Flotow

Frequent on sheltered and shaded acidic to somewhat basic rock outcrops.

Opegrapha

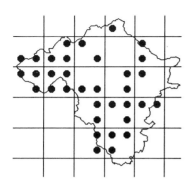

O. herbarum Mont.

A scarce or overlooked lichen of tree trunks in woodland. Ash, Cwmbach Llechrhyd 0354; ash, Alpine Bridge, Llanbadarnfawr 0963; sycamore near Lower Glasnant, Bryngwn 1850; old oak in Dol-wen Wood, Llanddewi Ystradenni 1069; sycamore near Norton 3067.

O. lithyrga Ach.

A scarce or overlooked sp. of shaded, mudstone rocks, By R. Wye, Pont Mareg, St Harmon 9571 AO; Aberedw Rocks 0746 BJC.

O. lyncea (Sm.) Borrer ex Hook.

A rare sp. of sheltered crevices in the bark of ancient trees. On an old oak at edge of Lower Llanelwedd Wood 0452 AP (1969 BLS records). Much of this wood has been cleared in recent years.

O. ochrocheila Nyl.

A distinctive sp. usually on tree bark but on shale rocks in the gorge of the Wye at Nannerth, N of Rhayader 9572. Also on ancient ash at edge of wood, Cefndyrys, Llanelwedd 0352 and elder on Llanelwedd Rocks 0552 BJC and !. Noted in this latter area by Leighton, 1879.

O. rufescens Pers.

Rare or overlooked. A single record from an old ash trunk in Bailey Bevan Wood, Llanfihangel Helygen 0464.

O. saxatilis DC.

A scarce sp. of shaded, slightly basic mudstone rocks. Aberedw Rocks 0746 BJC; rocks by R. Wye above Boughrood 1240 det. BJC.

O. sorediifera P. James

A scarce and distinctive sp. of damp woodland. On hazel on bank of R. Wye, Nannerth, N of Rhayader 9571.

O. varia Pers.

A sp. of nutrient-rich bark, now probably much reduced in abundance due to the death of elm trees, but still persisting on field maple and sycamore.

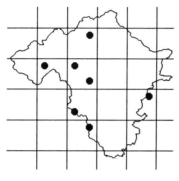

O. vermicellifera (Kunze) Laundon

Frequent on the sheltered and shady bases of the trunks and exposed roots of ancient trees or younger trees with base-rich bark. Favouring wych elm and field maple, the death of many of the former has led to its decline.

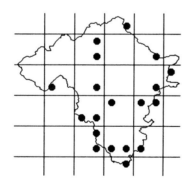

O. vulgata auct., non (Ach.) Ach.

Widespread on shaded, smooth tree bark, particularly of ash.

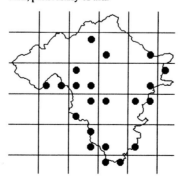

O. zonata Körber

Frequent on sheltered and shaded acidic to somewhat basic rock outcrops. Occasionally spreads onto nearby tree roots and trunk bases.

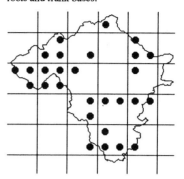

Pannaria conoplea (Ach.) Bory

A rare lichen of mosses overgrowing the bark of ancient trees. Old pollard oak in ravine SE of Henfron, Elan Valley 9064; oak by Pen-y-garreg Dam, Elan Valley 9167; sessile oak in wood S of the Gym, Llansantffraed Cwmdeuddwr 9366; oak by R. Wye, Nannerth, N of Rhayader 9471; on oak, Aberedw Wood 0746; on ash in wood, Tyncoed, Dulas Valley, Cwmbach Llechrhyd 0354; old sessile oak in gorge, Alpine Bridge, Llanbadarnfawr 0962.

P. leucophaea (Vahl) P Jorg.

A single record from south facing basic volcanic rocks. Caer Einon, Llanelwedd 0653.

Parmelia britannica D. Hawksw. and P. James

An uncommon lichen of sunny, S facing acidic volcanic and sedimentary rocks. Cliffs NW of Ashfield, Ysfa 9764 and 9765; Llanelwedd Rocks 0552 BJC and !; Yr Allt, Doldowlodd 0062; Old Radnor Hill 2558.

P. caperata (L.) Ach.

Uncommon on usually well-lit trees, generally in sheltered localities eg on sunlit alders beside the R. Wye.

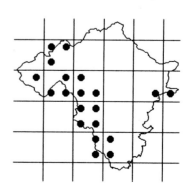

Parmelia

P. conspersa (Erh. ex Ach.) Ach.

Common on sunlit, dry, acidic shale and volcanic rocks. Particularly well developed on rocks receiving seasonal damp flushing. Rare on slates and unused tarmac paths. Occasional on stone walls and monuments.

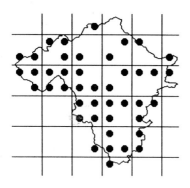

P. discordans Nyl.

Resembling forms of *P. omphalodes* (L.) Ach., many colonies have been chemically tested but only once has *P. discordans* been found. On exposed gritstone cliffs, Craig-cwm-clyd, Llansantffraed Cwmdeuddwr 8962. This sp. is more widespread in Brecknock.

P. disjuncta Erichsen

Known only from hard acidic rocks. Gritstone boulder S of Rhayader Quarries 9765 det. BJC and on acidic, E facing volcanic rock outcrops at Llandegley Rocks 1362. First recorded in the latter site by Salwey (as *P. stygia*) (BM) in 1871.

P. elegantula (Zahlbr.) Szat.

Rare, on wych elm tree branch in Evancoyd Park, Evenjobb 2562.

P. exasperata de Not.

Widespread on the twigs and small branches of a wide range of trees, but commonest on ash. Favouring the upper, well-lit twigs it is probably much under recorded.

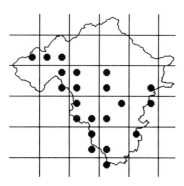

P. exasperatula Nyl.

Rare. On wooden palings and sandstone monument in the Baptist Chapel Churchyard, Newbridge on Wye 0158; on elder in wood, Bailey Einon, Cefnllys 0861; ash tree branches at woodland edge, Llanboidy Wood, Llanfihangel-Nant-Melan 1557 AO and !; tombstone, Dolau Chapel 1367.

P. glabratula (Lamy) Nyl.

In all squares. Common on the bark of a wide variety of tree and shrub sp., particularly on the smooth bark of branches. Occasional on rock outcrops and old sawn timber. Rarely fruiting. Ssp. *fuliginosa* (Fr. ex Duby) Laundon has not been mapped but has been noted as being widespread on sunny rocks, particularly of monuments and wall tops.

P. laciniatula (Flagey ex H. Olivier) Zahlbr.

Rare. On an old sycamore branch on the Cameddau, E of Newmead Farm, Llanelwedd 0553; on base of ash tree at field edge, New Barn, Bronydd 2246; on sycamore branch in New Radnor churchyard 2160.

P. laevigata (Sm.) Ach.

Rare and confined to damp sessile oakwoods in the western uplands. Oak, Nant Gwyllt, Elan Valley 9162; Cwm Coel, Elan Valley 9063; wood below Craig-y-foel, Elan Valley 9163 and on alder, Coed Nannerth-fawr, Rhayader 9472

P. loxodes Nyl.

A scarce sp. of sunlit rock outcrops. SE facing dolerite rocks, Castle Bank, Llansantffraed-in-Elvel 0856 AO; rocks by R. Wye below Boughrood railway bridge 1338.

P. mougeotii Schaerer ex D. Dietr.

Rare on native rock. On quartz boulders by the Afon Gwngu, Llansantffraed Cwmdeuddwr 8472 & 8572. Reported most commonly from slate roofs, where due to their general inaccessibility it may be more widespread than the records suggests. Also on sandstone walls and monuments. Slate roof, Newbridge on Wye 0158; bridge coping stone over R. Ithon, Disserth 0358; slate roof, Dolau, Newbridge on Wye 0260; slate barn roof near Abbeycwmhir 0470; tombstone, Llanfihangel Rhydithon 1566; tombstone Llanfihangel-Nant-Melan 1858.

P. omphalodes (L.) Ach.

Widespread on acidic rock outcrops in the uplands, but rarely in great abundance and frequently absent from apparently suitable habitats.

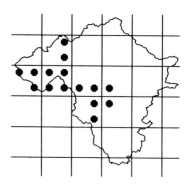

P. pastillifera (Harm.) R. Schubert and Klem.

A scarce sp. of both trees and stonework. On sycamore, Tyfaenor, Abbeycwmhir 0771; on mortared stone walls of Llanfihangel Rhydithon Church 1566; on sycamore trunk in garden, Llanddewi Ystradenni Hall 1068; on wych elm, Evancoyd Park, Evenjobb 2562; on felled elm near Castle Mound, Burfa 2761; sycamore and elder E of Llangunllo 2271; sycamore, Impton, Norton 2967; tombstone, Norton Churchyard 3067.

P. perlata (Huds.) Ach.

An uncommon sp. of generally fairly nutrient-rich, well-lit bark of oak, ash, sycamore and alder, often with *P. caperata*. Rare on rock and noted only from calcareous mudstones, Yr Allt, Llandeilo Graban 0844.

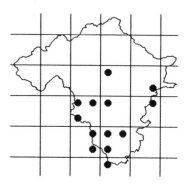

P. reddenda Stirton

Known only from an old ash tree trunk in the orchard of Llanddewi Ystradenni Hall 1068.

Parmelia

P. revoluta Flörke

Frequent on sunny but sheltered tree trunks and branches, especially of ash, alder, hazel and oak. Also on slightly basic mudstone rocks.

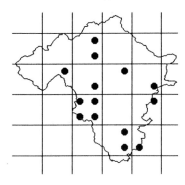

P. saxatilis (L.) Ach.

In all squares. Abundant on tree and shrub stems and branches and on rocks, always in more or less acidic places. Rarely fertile.

P. stygia (L.) Ach.

Reported in Leighton WA (1871) as having been seen by the Rev. T Salwey on Llandegley Rocks 16SW. A specimen collected here by Salwey in the BM, labelled *P. stygia* is *P. disjuncta*. The author and JD Woods, however, in 1986 came across *P. stygia* in a low Silurian shale rock outcrop in grazed, acidic grassland at 550m on the SE slopes of Treheslog Bank, Llansantffraed Cwmdeuddwr 9469. This may possibly be the only extant colony south of the Eastern Scottish Highlands.

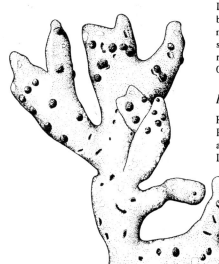

Parmelia stygia x 8

P. subaurifera Nyl.

Common on twigs and branches of trees and shrubs in woods and hedgerows.

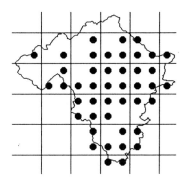

P. subrudecta Nyl.

On the more nutrient-rich bark of trees and shrubs such as wych elm, ash, sycamore, elder and ancient oaks in sheltered but sunny sites.

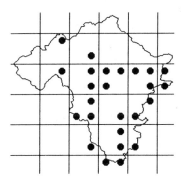

P. sulcata Taylor

In all 10km squares. Abundant on the bark of twigs and trunks of all but the most acidic trees and shrubs. Also on sawn timber and on somewhat basic rock. Rarely fertile but noted in 06NW, 07NE and 25NE.

P. taylorensis Mitchell

Rare. Damp rocks in woodland. By Penygarreg Resr. Dam 9167 ! & FR and above Glanllyn 9469 AO, both in Llansantffraed Cwmdeuddwr .

P. verruculifera Nyl.

Widespread but not common on sunny, acidic sedimentary and volcanic rocks. Particularly abundant on the volcanic rock outcrops at Llanelwedd 05SE and Stanner 2658. Also seen on stone bridge parapets, monuments, slate roofs and on tarmac. Rarely fertile.

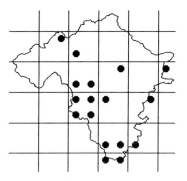

Parmeliella jamesii Ahlner and P. Jörg.

Rare, overgrowing mosses in damp, unpolluted woodland. On old ash tree at the edge of woodland below Craig-y-bwch, Claerwen Valley 8961 ! and AO; on wych elm in old garden of Cwm Elan House, Elan Valley 9064

P. triptophylla (Ach.) Müll. Arg.

A scarce sp. of basic bark on ancient trees and occasionally rocks. Most frequent in the Elan Valley woods in 86 and 96, but scattered elsewhere, especially on old ash trees in sheltered valleys. On rock only on basic mudstone cliffs, Rock Dingle, Bleddfa 2067.

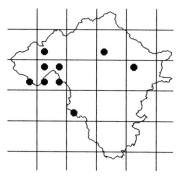

Parmeliopsis aleurites (Ach.) Nyl.

A rare sp. of weathered hardwood. Oak stump, Tyfaenor Park, Abbeycwmhir 0771; old fence rails near E drive, Stanage Park 3371; old gate post to E of Stanage Park 3472.

Peltigera

Peltigera

This survey followed Duncan and James (1970) until 1982 when Orvo Vitikainen's key appeared in the Bulletin of the BLS (**50**, 30-36). This resolved some difficulties but created others. The accounts of *P. collina*, *P. didactyla*, *P. horizontalis* and *P. praetextata* are probably acceptable. Much further work is required on the other spp. and the account here must be considered to be very provisional.

Peltigera collina (Ach.) Schrader

Rare; only noted from a basic volcanic rock outcrop at Caer Einon, Llanelwedd 0652.

P. didactyla (With.) Laundon

An uncommon lichen of soil on roadside banks, cinder heaps, decaying tree stumps and the lime-rich rubble of ruined buildings and walls. Rarely fertile.

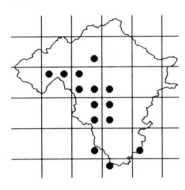

P. horizontalis (Huds.) Baumg.

Widespread on shaded, slightly basic rock outcrops. Rarer on base-rich tree bark, particularly ancient ash and wych elm trees in moist woodlands.

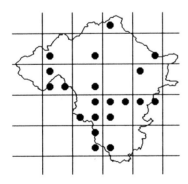

P. lactucifolia (With.) Laundon

Common amongst low herbs and grasses on roadsides, rock outcrops, hedgebanks and on tree bases, boulders, rubble and occasionally on heathland. Frequent on shady lawns and especially in graveyards.

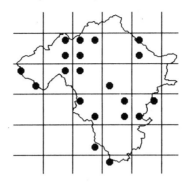

P. leucophlebia (Nyl.) Gyelnik

Rare. On ledges of NE facing, somewhat base-rich shale cliffs, Cerrig Gwalch, Nannerth, N of Rhayader 9370 IDS det. !

P. membranacea (Ach.) Nyl.

Probably more common than the records suggest, occurring in similar habitats to, and often with *P. lactucifolia* (With.) Laundon.

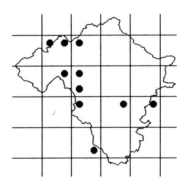

P. praetextata (Flörke ex Sommerf.) Zopf

Frequent on mossy tree trunks, logs, boulders and rock outcrops in moist, somewhat base-rich, shady places. Occasionally fertile.

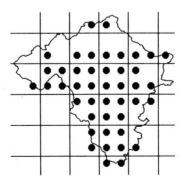

P. rufescens (Weis) Humb.

Widespread, but not common amongst grass, on boulders and on soil by roads and habitations, usually in base-rich areas.

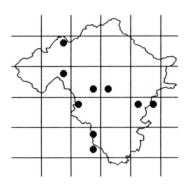

Pertusaria albescens (Huds.) M. Choisy and Werner

The species, including var. *corallina* (Zahlbr.) Laundon is recorded from all 10km squares. The type is much less frequent than the var., but both occur on the bark of well lit, nutrient-rich tree trunks, particularly ash, elm and ancient oaks. The var. *corallina* is also recorded sparingly from basic mudstone outcrops. Not fertile. The var. *albescens* is mapped below.

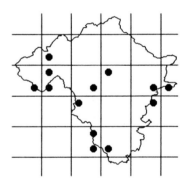

P. amara (Ach.) Nyl.

In all 10km squares except 87. Very common on the bark of a wide range of tree species. Rare on rock but noted from walls and monuments in a number of churchyards.

P. coccodes (Ach.) Nyl.

A rare species known only from the sunny buttress roots of an old oak in a field NW of Bailey Bevan, Llanfihangel Helygen 0464 and a large field maple at Old Radnor 2459 PWJ.

Pertusaria

P. corallina (L.) Arnold

A common species on the harder, well lit, acidic rock outcrops and boulder scree in the uplands. Noted once extending onto bark on the base of a rowan tree in the Upper Elan Valley 87SE.

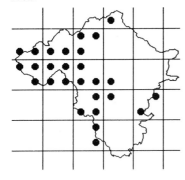

P. coronata (Ach.) Th. Fr.

An ancient woodland species only known from an old oak tree on a woodland edge near Carmel Chapel 0566.

P. dealbescens Erichsen

Uncommon on sunny, acidic to slightly basic rock outcrops. Growing on shales, it is most common on the harder volcanic rock outcrops.

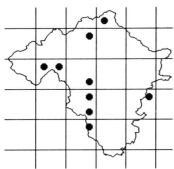

P. excludens Nyl.

On tombstone in Llanstephan Churchyard 1142 BJC and !, this sp. may have been overlooked on sunny, vertical rock faces elsewhere.

P. flavicans Lamy

A scarce species of dry, acidic grit and shale rocks. Craig y Foel 9164 and 9165 and Glog fawr 9166 in the Elan Valley; Treheslog Rocks, Llansantffraed Cwmdeuddwr 9368; Cerrig Gwalch, N of Rhayader 9370; Dol y Fan Hill, Newbridge on Wye 0160.

P. flavida (DC.) Laundon

An uncommon epiphyte on the bark of mature to ancient oak and ash trees in warm, sunny locations.

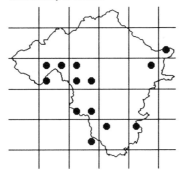

P. hemisphaerica (Flörke) Erichsen

Widespread, but uncommon, on well-lit ancient oak and occaisionally ash trees, generally on the edge of woods or in old parks.

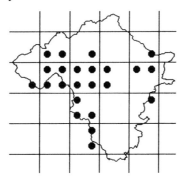

P. hymenea (Ach.) Schaerer

Common on the bark of a wide range of tree spp. in woodlands and hedgerows.

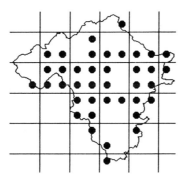

P. lactea (L.) Arnold

Common on acidic shale and volcanic rocks in well-lit and dry sites, but extending into shaded sites such as block scree in woodland.

P. leioplaca DC.

Common in the W on the smooth bark of hazel, young ash and occasionally other trees. In the E confined to sheltered river valleys. Recorded in all 10km squares except 87 and 37.

P. multipuncta (Turner) Nyl.

Frequent in the upland woods of the W where it favours rowan but extends to the smooth bark of other tree spp., particularly hazel. Scarce in the E.

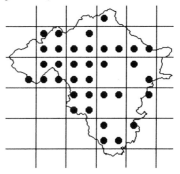

P. pertusa (Weigel) Tuck.

Widespread on the well-lit trunks of a range of tree species. Often with *P. hymenea* but less common. Noted on sandstone rock of Llanfihangel-Rhydithon Church 1566.

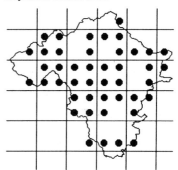

P. pseudocorallina (Liljeblad) Arnold

Widespread, but not common, on sunny shale and volcanic rocks, often in sites occasionally receiving water leached through soil above.

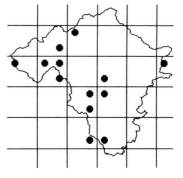

Physcia

P. pupillaris (Nyl.) Th. Fr.

A scarce lichen of smooth bark on trees and shrubs, mainly in ancient woodlands. Ancient oak, Cwm Gwyllt, Elan Valley 9162; Cwm Coel, Elan Valley 9063 BJC and oak, Cwmbach Llechrhyd 0354 AP (1969 BLS records); hazel in wood by lake, Llandrindod Wells 0660; old oak in field, Carmel, Nantmel 0566; oak, Tyfaenor Park, Abbeycwmhir 0671.

Pezizella epithallina (Phill. and Plowr.) Sacc.

A rare fungus living on the thallus of *Peltigera canina* agg. growing on stonework of old railway bridge, Newbridge on Wye 0157. AO.

Phaeographis dendritica (Ach.) Müll. Arg.

Known only from hazel bark in damp woodland. Tyncoed Wood, Cwmbach Llechrhyd 0354 FR and !.

P. smithii (Leighton) B. de Lesd.

Known only on a rowan trunk in Coed Gwarallt, Cregrina 1251 coll. !, det. BJC.

Phaeophyscia orbicularis (Necker) Moberg

Common in a wide range of somewhat base-rich and often nutrient enriched sites such as concrete posts and boulders used as bird perches, asbestos cement and the bark of trees such as ash, sycamore, maple and elder, especially in pastures and around farmyards. In all grid squares except 87 and 08.

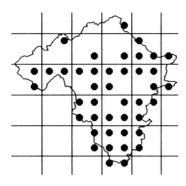

Phaeopyxis varia Coppins, Rambold & Triebel

Rare on thallus of *Trapeliopsis gelatinosa* on soil above shore of Garreg Ddu Resr., Elan Valley 9063 ! & BJC.

Phlyctis argena (Sprengel) Flotow

In all 10km squares. Common on the smooth to somewhat rough bark of a wide range of tree and shrub spp. in woods and hedgerows.

Phragmonaevia fuckelii Rehm

On *Peltigera* sp. on disused railway near Aberithon, Newbridge on Wye 0157 coll. AO det. DL Hawksworth.

Phyllopsora rosei Coppins and P. James

A rare lichen of ancient woodland sites in mid Wales. Old oak near sawmill, Nant Gwyllt, Elan Valley 9162; ash in old garden of Cwm Elan House, Elan Valley 9064; old elders and oak in the Ithon Gorge, Alpine Bridge, Llanbadarnfawr 0962; oak in wet woodland S of Wernfawr, Painscastle 1845.

Physcia adscendens (Fr.) H. Olivir

Widespread on nutrient-rich tree bark, particularly elder, calcareous mudstone outcrops, wall tops and monuments enriched with bird droppings. Sparingly fertile.

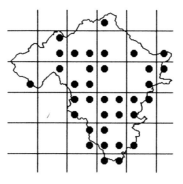

P. aipolia (Ehrh. ex Humb.) Fürnrohr

A frequent sp. on the twigs of elder and ash and occasionally elm, sycamore and maple. Rare on other trees. Abundantly fertile.

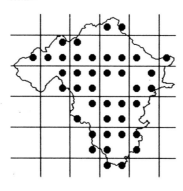

P. caesia (Hoffm.) Fürnrohr

Frequent on basic and/or nutrient-flushed rocks in sunny sites and basic, man-made substrates such as concrete, tarmac and asbestos tiles. Rare on sawn timber near limestone quarry, Dolyhir 2458.

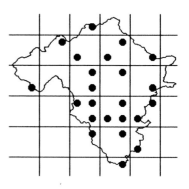

P. dubia (Hoffm.) Lettau

An uncommon sp. of somewhat basic, sunny volcanic and sedimentary rocks on cliffs and rarely walls. Most sites show signs of nutrient enrichment.

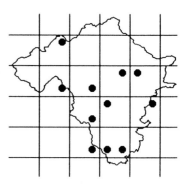

P. tenella (Scop.) DC.

In similar sites to, and often with *P. adscendens* but more common and perhaps demanding less nutrient-rich conditions. Occasionally fertile. In all 10km squares except 08 and 18.

P. tribacia (Ach.) Nyl.

A scarce lichen of sunny, slightly basic rock outcrops. Near Ashfield, Ysfa 9764; Aberedw Rocks 0746; volcanic rocks, Alpine bridge, Llanbadarnfawr 0962; mudstones N of Llanstephan House 1142; cliff, Llanbadarn-y-garreg 1148; cliff, Coed Gwarallt, Cregrina 1251; dolerite cliffs, Stanner Rocks 2658.

Placynthiella

P. wainioi Räsänen

An uncommon lichen of nutrient-enriched rock. Tombstone, Ysfa Church 9964; volcanic rock outcrop, Llanelwedd 0552 and 0652; volcanic rocks, Carreg Wibber, Llandrindod Wells 0859; volcanic rock above spring, Llandegley 1362; tombstone, Norton Church 3067. This sp. is probably only a form of *P. caesia*.

Physconia distorta (With.) Laundon

On the sunny bark of ash, sycamore and black poplar; scarce and mainly in the east.

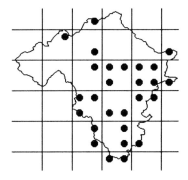

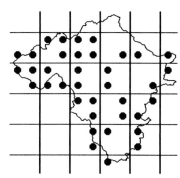

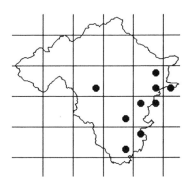

Placopsis gelida (L.) Lindsay

A rare sp. of damp, acidic shale rock outcrops near Craig Goch Resr. Dam, Elan Valley 8968.

P. lambii Hertel and V. Wirth

The thalli of what might be referred to this sp. occur frequently on the blocks of shale spoil, rich in iron, on the tips of the disused Cwm Elan lead mine, Elan Valley 9065 and the old slate quarries below Craig-y-bwch, Claerwen Valley 9061.

Placynthiella icmalea (Ach.) Coppins and P. James

Common on damp acidic substrates such as decaying logs and sawn timber, old turf, peat, occasionally bark and shaded rocks.

Placynthium nigrum (Huds.) Gray

An uncommon sp. of moist, calcareous rocks and mortared walls. Noted only on native rock from calcareous mudstone at Aberedw Rocks 0746; basic volcanic rocks, Llanelwedd 05SE and limestone, Yatt Wood, Dolyhir 2458.

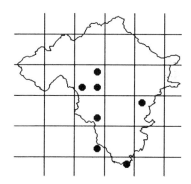

P. enteroxantha (Nyl.) Poelt

A scarce sp. of sunny, nutrient-rich tree bark. Noted from ash, apple and sycamore.

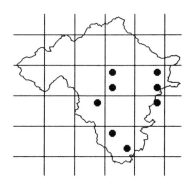

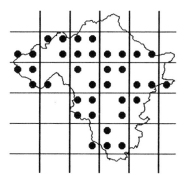

P. grisea (Lam.) Poelt

Widespread on nutrient-rich tree bark, particularly of ash, sycamore, elder and elm and occasionally on basic rock, concrete and asbestos cement roof coverings.

P. uliginosa (Schrader) Coppins and P. James

Frequent on peat and peaty soil, though not as common as the last sp. on decaying wood and bark.

Platismatia glauca (L.) Culb. and C. Culb.

In all squares. Abundant on the acidic bark of a wide range of tree and shrub species, wood and exposed, acidic rocks in the uplands.

Polyblastia allobata (Stizenb.) Zsch.

Known only from the bark of a recently dead wych elm in the gorge of the R. Ithon above Alpine Bridge, Llanbadarnfawr 0963 ! and FR, det. BJC.

Polychidium muscicola (Swatz) Gray

A scarce sp. of damp or seasonally moist rock. Shaded roadside shale rocks, Yr Allt, Doldowlod 0062; Penddol Rocks, Llanelwedd 0352; slightly calcareous mudstone outcrop, Aberedw Rocks 0746 BLS exc.; gritstone, Yatt Wood, Dolyhir 2458.

Porina

Polysporina simplex (Davies) Vezda

A scarce or overlooked sp. of sunny rock outcrops. Shale rocks at S end of Graig Dolfaenog by Garreg Ddu Resr., Elan Valley 9166; shale rocks, S of Wyloer, Gilfach Farm, St Harmon 9571 BJC and ! ; volcanic rock outcrops, Llanelwedd 0551 BLS exc.; old sandstone quarry S of Bryngwyn 1848.

Porina aenea (Wallr.) Zahlbr.

Rare or overlooked. On smooth bark in damp woodland. Hazel, Alpine Bridge, Llanbadarnfawr 0963; hazel, Llanboidy Wood, Llanfihangel-Nant-Melan 1557; rowan in wood, E of Knucklas 2673.

P. ahlesiana (Körber) Zahlbr.

A nationally scarce sp. of shaded rock. On slightly calcareous mudstone cliff in ravine W of Garreg Fawr, Llandeilo Graban 0844.

P. chlorotica (Ach.) Müll. Arg.

An uncommon or unnoticed lichen of shaded, smooth shale rock outcrops and occasionally smooth bark of trees.

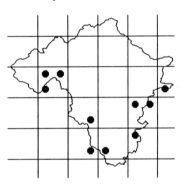

P. guentheri (Flotow) Zahblr.

Rare or overlooked. On damp rock outcrop in wood, Gwynllyn, Llansantffraed Cwmdeuddwr 9469 AO.

P. lectissima (Fr.) Zahlbr.

A scarce sp. of shaded, damp, acidic to basic shale and volcanic rock outcrops.

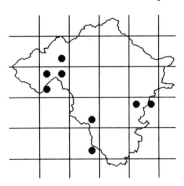

P. leptalea (Durieu and Mont.) A. L. Sm.

A rare or overlooked sp. of hazel bark in damp woodland. Bailey Einon, Cefnllys 0861 and in wood S of Wernfawr, Painscastle 1845.

P. linearis (Leighton) Zahlbr.

A rare sp. noted on a BLS exc. on shaded calcareous mudstone at Aberedw 0746.

Protoblastenia calva (Dickson) Zahlbr.

Known only from the limestone of a disused quarry near Yatt Farm, Old Radnor 2458.

P. rupestris (Scop.) Steiner

A common sp. on basic rock outcrops in well-lit and dry places and abundant on mortar and concrete.

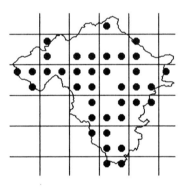

Pseudevernia furfuracea (L.) Zopf.

Occasionally found on acidic tree bark, particularly of oak, birch and conifers in open, windswept sites in the uplands and around bogs. Also on old fence posts and occasionally acidic rocks in exposed sites. Hawksworth, DL and Chapman, DS (1971) in *Lichenologist* 5, 51-58 report only the var. *ceratea* (Ach.) D. Hawksw. from Mid Wales.

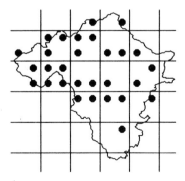

Psilolechia clavulifera (Nyl.) Coppins

A scarce or overlooked sp., noted on exposed tree roots in bank near Garreg Ddu Resr. dam, Elan Valley 9063 BJC and !.

P. leprosa Coppins and Purvis

A copper-loving lichen detected below the copper lightning conductor on Llandrindod Wells Old Church 0660 by AO. Only recently described from Britain, it may be in similar situations elsewhere.

P. lucida (Ach.) M. Choisy

In all squares. The distinctive yellow-green crusts are common on damp, shaded stone and brick work, monuments, acidic rock outcrops and block scree. Rarely fertile.

Ptychographa xylographoides Nyl.

Rare, on old, damp, bark-free conifer trunks in moist woodland. In oakwood with scattered conifers, Nant Gwyllt, by Caban Coch Resr., Elan Valley 9162 det. BJC; in gorge woodland NW of Tyfaenor, Abbeycwmhir 0671. These appear to be its only known localities S of the Scottish Highlands.

Pycnothelia papillaria (Ehrh.) Dufour

Reported from Llandrindod Wells c06 by the Rev. T Salwey in Leighton (1879), this peatland sp. has never been refound.

Pyrenopsis rhodosticta (Taylor) Müll. Arg.

A scarce sp. of seasonally damp rocks. On rocks in the flood zone of the stream flowing into Gwynllyn, Llansantffraed Cwmdeuddwr 9369 ! and AO.

Pyrenula

Members of this genus occur rarely in Radnor on shaded, smooth bark of tree trunks and branches, favouring ash and hazel. Following Duncan, UK and James, PW,(1970), in the early years of this survey two records were made and ascribed to *P. nitida* var. *nitidella* (Flörke) Schaer.:- from hazel and ash at Alpine Bridge, Llanbadarnfawr 0963; ash in the gorge of the Bach Howey, Llanstephan 1143. Due to the rarity of the lichen no specimens were collected so redetermination is not possible. They may probably both be referred to *P. chlorospila* (Nyl.) Arnold, but this requires confirmation.

Ramalina

P. macrospora (Degel.) Coppins and P. James

Rare on the smooth bark of hazel in Aberedw Wood 0847 det. BJC.

P. occidentalis R. C. Harris

Rare on hazel in woodland. Coed yr allt goch, Elan Valley 9067 AO.

Pyrrhospora quernea (Dickson) Körber

Frequent on the sunny bark of, in particular, mature ash and oak and ancient hazel trees. Occasionally fertile.

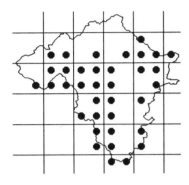

Racodium rupestre Pers.

Scarce in dry, sheltered crevices in acidic rock faces. Less frequent than, but often with *Cystocoleus ebeneus*.

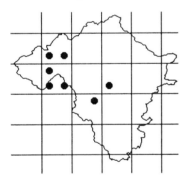

Ramalina calicaris (L.) Fr.

A scarce sp. of isolated, sunny and windswept sycamore and rarely elder trees.

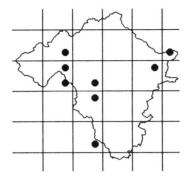

R. farinacea (L.) Ach.

Common on well-lit and exposed tree trunks. Avoiding the most acidic bark it is most frequent on ash and sycamore on field margins and woodland edges. Recorded from all 10km squares.

R. fastigiata (Pers.) Ach.

Widespread, but not common, on exposed and sunny tree trunks, mostly of ash and sycamore in hedgerows and about farms.

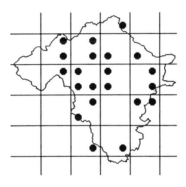

R. fraxinea (L.) Ach.

In similar habitats to the last but less common and frequently poorly developed. Large colonies (over 15cm long) were noted on black poplar near Gwystre 0764.

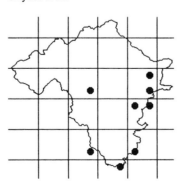

R. pollinaria (Westr.) Ach.

A rare lichen of dry but well-lit rock crevices. Outcrop N of Penygarreg Resr. dam, Elan Valley 9167 and in Glanllyn, Llansantffread Cwmdeu ddwr 9469 AO.

R. subfarinacea (Nyl. ex Crombie) Nyl.

A scarce sp. of sheltered but sunny S and SW facing shale and volcanic rock outcrops.

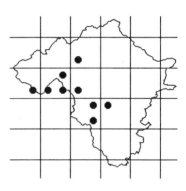

Rhizocarpon concentricum (Davies) Beltr.

Widespread on basic volcanic and mudstone rock outcrops and on mortared stone walls.

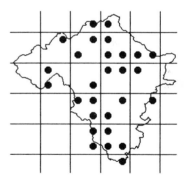

R. distinctum Th. Fr.

A rare or overlooked species of gritstone and volcanic rocks. Near Caban Coch Resr., Elan Valley 9164 AEW (1967 NMW); Llanelwedd Rocks 0552 BJC and !.

R. geminatum Körber

A sp. of damp streamside rocks only noted once in Radnor. Rocks in R. Wye near Erwood 04SE, HHK (1926 NMW).

Rinodina

R. geographicum agg.

Common on sunny, freely-drained, acidic rock outcrops and roofing slates throughout the county. Further work is required to determine the distribution of the spp. within this group.

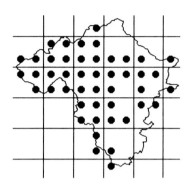

R. hochstetteri (Körber) Vainio

Rare or overlooked on acidic shale rocks. In wood, Gwynllyn, Llansantffared Cwmdeuddwr 9469 AO.

R. lavatum (Fr.) Hazslinsky

Probably more widespread on rocks in streams and rivers than the few records suggest. In A. Marteg near confluence with the R. Wye, Nannerth, N of Rhayader 9571; brook near Cwmbach Llechrhyd 0354; small stream above Llanbwchllyn, Llanbedr 1146; Rhulen Hill 1349.

R. obscuratum (Ach.) Massal.

An early colonizer of bare, acidic rocks and gravel of outcrops and stone walls. The commonest sp. on shale boulders beside recently constructed farm roads. Recorded from all 10km squares except 37.

R. oederi (Web.) Körber

Common on the iron-rich Silurian shales of the western uplands and the volcanic rocks of the Carneddau Hills in sunny, dry places. Extends occasionally onto dry stone walls.

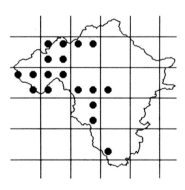

R. polycarpum (Hepp) Th. Fr.

A northern sp. known only in Radnor from sheltered but sunny, hard, acidic shale rocks, Cerrig-gwalch, N of Rhayader 9370 det. BJC.

R. riparium ssp. *lindsayanum* (Räsänen) Thomson

Recorded by AEW from three sites. Mudstone outcrop, Aberedw Rocks 04NE; volcanic rocks, Caer Fawr, Llanelwedd 0553; Llandegley Rocks 16SW (NMW and recorded pre 1970). It may occur elsewhere, but as noted above the *R. geographicum* agg. requires further study.

R. viridiatrum (Wulfen) Körber

Widespread on hard, dry, sunny, acidic rock outcrops.

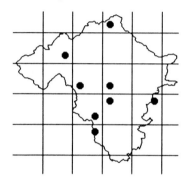

Rinodina exigua (Ach.) Gray

Scarce or overlooked on the dry bark of nutrient-rich trees. Ancient oak near the Vulcan Arms, Ysfa 9763; Aberedw Rocks 0746 BLS exc.; old oak near Llyn Gwyn, Ysfa 0064; wych elm, Abbeycwmhir 0571; oak, Yatt Farm, Dolyhir 2458; ancient oak, Stanage Park, Stanage 3371.

R. griseosoralifera Coppins

A recently described sp. which is scarce or more probably overlooked on nutrient-rich bark. On old elder trunks in sheltered valley E of Newmead Farm, Cwmbach Llechrhyd 0654 det. BJC.

R. lecideina Mayrhofer & Poelt.

Rare or overlooked. On S. facing rock in field near Timber Hill Wood, Lower Rowley, Prestegne 2863 det. BJC.

R. roboris (Dufour ex Nyl.) Arnold

A rare sp. noted only from a single old oak tree trunk at a field edge. New Barn, Clyro 2246.

R. sophodes (Ach.) Massal.

On ash, sycamore, elder and wych elm twigs and young branches, generally in sunny but sheltered sites.

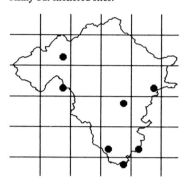

Sarcogyne privigna (Ach.) Massal.

Rare or overlooked. On sunny and dry, slightly basic sandstone rocks of old quarry S of Bryngwyn 1848.

S. regularis Körber

On basic rock outcrops, mortar and concrete in well-lit places. Probably much overlooked, at least in the latter habitats.

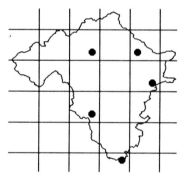

Schaereria cinereorufa (Schaerer) Th. Fr.

Frequent on sunny and exposed acidic shale cliffs in the western uplands; rarer elsewhere. Favours small crevices and extends onto shallow soil.

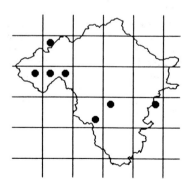

Sphaerophorus

S. tenebrosa (Flotow) Hertel and Poelt

Of local occurrence on exposed, acidic shale and volcanic rock outcrops.

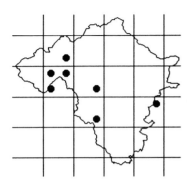

Schismatomma cretaceum (Hue) Laundon

Known only from two shaded, ancient oak trees in long undisturbed woodlands. By the Dulas, Tyncoed Wood, Cwmbach Llechrhyd 0354 and SW end of Tyfaenor Park, Abbeycwmhir 0671.

S. decolorans (Turner and Borrer ex Sm.) Clauz and Vezda

Widespread on the sheltered and shaded bark of mature and ancient tree trunks. Most frequent on oak.

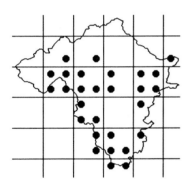

Sclerococcum sphaerale (Ach. ex Ficinus and Schub.) Fr.

A frequent parasite of *Pertusaria corallina*, though much overlooked in this survey. Craig y bwch, Claerwen Valley 8961 AO and !; Gilfach Farm, St Harmon 9671; The Banks, Llansant-ffraed-in-Elvel 0755; volcanic rocks on the Carneddau SE of Newmead Farm, Llanelwedd 0553.

Scoliciosporum chlorococcum (Graewe ex Stenhammar) Vezda

Widespread on a range of woody substrates. Heather, ash and elder twigs are favoured, but it also occurs on wooden poles and gateposts.

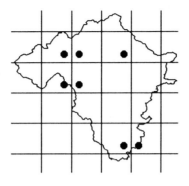

S. umbrinum (Ach.) Arnold

Widespread and probably much overlooked. The dull, scurvy thallus grows on sunny wall tops, rock outcrops and old wood of fence posts, gates and fallen logs, often where nutrient enrichment occurs.

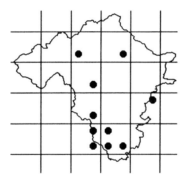

Sphaerophorus fragilis (L.) Pers.

A scarce sp. of hard, exposed, rock outcrops on summit ridges. Y Foel, Elan Valley 9165 and Dol-y-fan Hill, Newbridge on Wye 0161.

S. globosus (Huds.) Vainio

Frequent on acidic rock outcrops and tree trunks in woodlands, block scree and cliffs in the W of the county. Rarely fertile, (in Glannau, Elan Valley 9165 and Nannerth-fawr, Rhayader 9471).

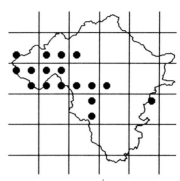

Sphinctrina turbinata (Pers.) de Not.

A parasite of *Pertusaria* sp. growing on a range of substrates. On sycamore, Gwastedyn Hill, Rhayader 9867; on *P. pertusa* on sycamore SE of Newmead Farm, Llanelwedd 0553 and on oak, Alpine Bridge, Llanbadarnfawr 0963.

Staurothele fissa (Taylor) Zwackh

On shale rocks in the flood zone of rivers. R. Wye, Pont Marteg, St Harmon 9571 AO; the Bach Howey near the viaduct, Llandeilo Graban 1042 AO and ! and probably in other streams, but overlooked.

Steinia geophana (Nyl.) B. Stein

A rare or overlooked lichen of freely-drained, acidic soil, which, due to low fertility or possible toxic effects of heavy metals, is kept free of competing higher plants. Old railway track bed N of Rhayader 9668; on S facing roadside (formerly railway) cutting, Llanstephan 1142 det. BJC; on track bed of old railway S of Llanfaredd 0650 and below Yr Allt, Llandeilo Graban 0844. Lead mine spoil may have been used on the track bed of this railway.

Stenocybe pullulata (Ach.) B. Stein

Frequent on alder twigs in damp woodlands by streams and mires in the W of the county.

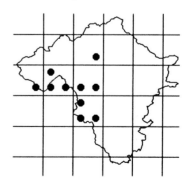

Thelidium

Stereocaulon evolutum Graewe

Widespread, but scarce, on acidic rock outcrops and scree in the uplands. Noted fertile on an old railway sleeper used as fence post near Llangunllo Stn. 2172.

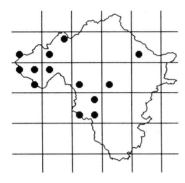

S. leucophaeopsis (Nyl.) P. James and Purvis

A scarce lichen of rocks rich in heavy metals. Well grown, fertile plants occur on old foundry waste probably tipped on the banks of the R. Elan near the Welsh Water Visitor Centre 9264, during the construction of the reservoirs in the 1890's.

S. nanodes Tuck.

Rare on heavy metal-rich rocks on the spoil heaps of the Cwm Elan Mine, Elan Valley 9065.

S. vesuvianum Pers.

In similar habitats to *S. evolutum* Graewe, but less common.

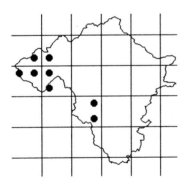

Sticta sylvatica (Huds.) Ach.

Rare. Known only from a damp and shaded, somewhat calcareous shale rock outcrop in a wood N of Penygarreg Resr. dam, Elan Valley 9167 ! & FR.

S. fuliginosa (Hoffm.) Ach.

Only known from the base of an old ash trunk on the edge of a rock outcrop in the wood N of Penygarreg Resr. dam 9167 and on a sycamore by the Nant Methan, Cwm Elan House Garden 9064 in the Elan Valley.

S. limbata (Sm.) Ach.

Rare and known only from Norway maple trees in the old garden of Cwm Elan House, Elan Valley 9064 and rock near Alpine Bridge, Llanbadarnfawr 0962.

Strangospora moriformis (Ach.) Stein

A rare or overlooked sp. of shaded, sawn hard and softwood posts. Near Rhiw-riad, St Harmon 9672; Dolau, Newbridge on Wye 0260; Little Hill, Llandewi Ystradenny 1267 and Norton 3067.

S. ochrophora (Nyl.) R. Anderson

Scarce, in damp woodland. Noted on an ancient ash tree trunk in Tyncoed Wood, Cwmbach Llechrhyd 0354 ! and FR and on an elder in Cilkenny Dingle, Llowes 1841 AO and !.

Thelidium minutulum Körber

A scarce or overlooked sp. of stones by streams. R. Wye, Pont Marteg, St Harmon 9571 AO and Cilkenny Dingle, Llowes 1841 AO and !.

T. pyrenophorum (Ach.) Mudd

Rare or overlooked on somewhat base-rich rock outcrops. Damp rocks by track, Cwm-y-Geist, Llangunllo 1770.

Thelocarpon epibolum Nyl.

A rare or overlooked parasite. On *Baeomyces rufus* in Coed yr allt goch, Elan Valley 9067 AO.

T. intermediellum Nyl.

Rare. On the top of a recently disturbed and exposed mudstone boulder on roadside verge, Llanstephan 1142, det. BJC.

T. laureri (Flotow) Nyl.

Scarce and only known from shale of recently dug quarry, Little Hill, Llandrindod Wells 0760 AO and from the top of a sandstone boulder on a fairly recently constructed track near The Colony, Norton 3167. The earliest record of Wilkinson (1904) "on trees," Claerwen requires confirmation due to this unusual habitat.

T. lichenicola (Fuckel) Poelt and Hafellner

Inconspicuous and possibly overlooked. On acidic soil on top of a recently dug quarry, Little Hill, Llandrindod Wells 0760 AO and in a disused quarry in forestry below Llethr-du, Cascob 1966 AO and !.

Thelopsis rubella Nyl.

A rare sp. only known from the bark of an ancient oak tree on a wooded bank W of Rhyscog, Aberedw 0948.

Thelotrema lepadinum (Ach.) Ach.

Common in a few ancient woodlands such as those on the shore of Caban Coch and Garreg Ddu Resrs. in the Elan Valley 96 where it occurs on the smooth bark of hazel, ash and on old oaks. It is rare elsewhere and largely confined to ancient oak trees in damp woodlands.

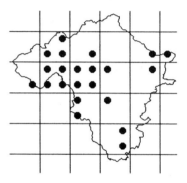

Thrombium epigaeum (Pers.) Wallr.

An inconspicuous lichen of acidic soil. In Coed yr allt goch, Elan Valley 9067 ! and AO; on S facing bank of disused railway cutting below Fedw, Tylwch 9879 and on bank on common W of Yr Allt, Llandeilo Graban 0844 AO and !.

Tomasellia gelatinosa (Chev.) Zahlbr.

A scarce or more probably overlooked sp. of smooth bark of hazel or rarely oak in hedges and woodland.

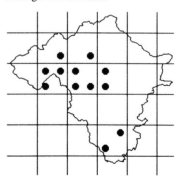

Trapeliopsis

Toninia aromatica (Turner ex Sm.) Massal.

Rare on basic volcanic and limestone rock outcrops. More frequent on lime mortar of old walls.

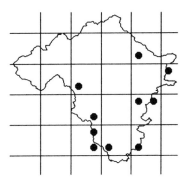

Trapelia coarctata agg.

Duncan, UK and James, PW (1970) were followed in their treatment of this entity as a sp. with forms, in the first half of this survey. The sp. was found in all 10km squares, occurring on well-lit, acidic rocks, particularly in the damper hills of the W of the county. Coppins, BJ and James, PW (1984) were followed in the latter years of the survey, but records are far from complete. *T. coarctata* (Sm) M. Choisy, *T. involuta* (Taylor) Hertel, *T. obtegens* (Th.Fr.) Hertel and *T. placodioides* Coppins and P. James are all widespread on sunny to somewhat shady, seasonally damp, acidic rocks, frequently occurring in various combinations on the same boulder.

T. corticola Coppins and P. James

Widespread, but not common, though overlooked in the early years of the survey. On the acidic bark of trees in damp woodland. Recorded from oak, alder, birch and willow species.

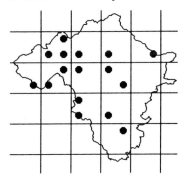

Trapeliopsis flexuosa (Fr.) Coppins and P. James

Common on old wooden posts, rails, gates and on tree stumps. More common than the mapped records suggest.

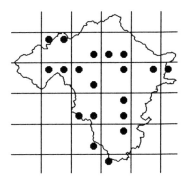

T. gelatinosa (Flörke) Coppins and P. James

Widespread on damp acidic soil and decaying turf on the edges of banks by tracks and streams in the uplands.

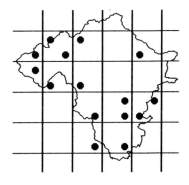

T. glaucolepidea (Nyl.) G. Schneider

A scarce sp. of peaty soil on acidic rock ledges and around the edges of peat haggs in blanket bog in the Elan-Claerwen Uplands. Blaen Rhestr 8469; Gors-goch 8963; Waun Ffaethnant 8371; Craig-y-foel 9164 and Glog-fawr 9266, all in Llansantffraed Cwmdeuddwr.

T. granulosa (Hoffm.) Lumbsch

Abundant on damp organic surfaces such as on decaying wood, peat and occasionally bark. Recorded from all 10km squares.

T. pseudogranulosa Coppins and P. James

In similar habitats to the last, but less common and perhaps preferring greater humidity in more upland sites.

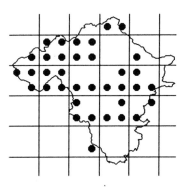

T. wallrothii (Flörke ex Sprengel) Hertel and G. Schneider

The abundance of this normally coastal species in some Radnor sites is notable. It favours peaty soil on S facing ledges on somewhat basic rock faces. It is frequent on Aberedw Rocks 0746; Llanelwedd Rocks 0552 and Stanner Rocks 2658.

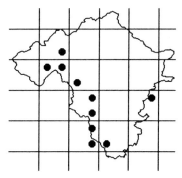

Tremolecia atrata (Ach.) Hertel

Widespread, but rare, on dry and sunny acidic rock outcrops and block scree in the uplands.

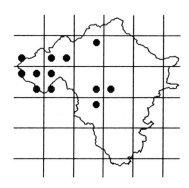

Verrucaria

Tylothallia biformigera (Leighton) P. James and Kilias

A scarce sp. of dry and sheltered underhangs of acidic to somewhat basic, mostly volcanic rock outcrops. Block scree, Craig-y-bwch, Claerwen Valley 9061; Glanllyn Wood, Llansantffraed Cwmdeuddwr 9469; Llanelwedd 0553 AEW (1959 NMW); Caer Einon, Llanelwedd 0653 ! and A Fletcher; Caer Fawr, N end of the Carneddau, Llansantffraed-in-Elvel 0654; Graig Fawr, Frank's Bridge 1358; Llandegley Rocks 1361 AEW (1959 NMW).

Umbilicaria deusta (L.) Baumg.

A rare sp., but common in its only locality on S facing, acidic, seasonally-flushed rock outcrops, S of Wyloer, Gilfach Farm, St Harmon 9570l and 9671 det. BJC.

U. polyphylla (L.) Baumg.

The commonest sp. of the genus but nevertheless generally scarce on dry and sunny acidic rocks in the uplands, particularly the harder grits and volcanic rocks.

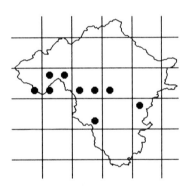

U. polyrrhiza (L.) Fr.

Rare on hard, sunny and dry, acidic rock outcrops. Low gritstone cliff, Craig-cwm-clyd, Claerwen Valley 8962; cliff top, Y Foel, Elan Valley 9165 and on a single boulder, Gwastedyn Hill, Rhayader 9765.

Usnea articulata (L.) Hoffm.

Reported by Wilkinson (1904) as *U. barbata* forma *articulata* from trees, Llandrindod Wells c06, there have been no other records of this very sulphur dioxide sensitive species.

U. filipendula Stirton

Occasional on the acidic bark of conifer, oak and birch trees in upland woods. Becomes abundant in some of the most exposed sites.

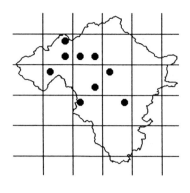

U. florida (L.) Wigg.

Common on the twigs, branches and trunks of a wide range of tree and shrub spp. in the W of the county, favouring damp and sheltered sites.

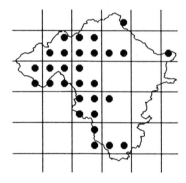

U. inflata Delise

Uncommon or overlooked on the acidic bark of mostly ancient trees in damp woodlands. 7 sites.

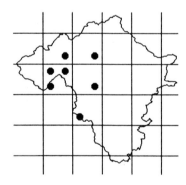

U. rubicunda Stirton

Rare and known from a single old oak trunk in woodland behind the County Hall, Llandrindod Wells 0660 BJC & !.

U. subfloridana Stirton

Common on acidic tree bark of spp. ranging from oak to heather. Rarer on acidic rock outcrops and wooden gates and fence posts. Recorded from all 10km squares except 87 and 36.

Verrucaria spp.

This account provides a very preliminary treatment of this genus. Man-made substrata such as monuments and walls have been but little studied.

V. aethiobola Wahlenb.

Probably widespread on acidic rocks, cliffs and in the flood zone of streams and rivers. Gwynllyn Wood, Llansantfraed Cwmdeuddwr 9469 AO; R. Marteg near the Wye confluence, Gilfach, St Harmon 9571; Afon Dulas, Cwmbach Llechrhyd 0354; R. Wye, Newbridge on Wye 05NW A Fletcher; stream N of Newhouse, Llanbister Road 1670.

V. baldensis Massal.

Probably widespread on limestone memorials and building stone, but with records only from 06NE, 16NW and 26SE. Frequent on native limestone at Dolyhir 2458.

V. bryoctona (Th. Fr.) Orange

Originally recorded as *V. melaenella* Vainio but now redetermined by AO as this sp.. It occurs on soil in a wide range of places. Old railway track bed below Yr Allt, Llandeilo Graban 0843; soil in cracks between paving stones, Builth Road Station platform 0253 and on clay soil of ditch bank, S of Great Vaynor, Nantmel 0169.

V. caerulea DC.

Rare, but probably much overlooked on sunny limestone, in walls and on calcareous mudstone. Bridge parapet, Llannano 0974; S facing calcareous mudstone rocks near Llanfihangel-Nant-Melan 1858 and on Norton Church 3067.

V. dolosa Hepp

Rare or overlooked, on damp shale rocks. Coed yr allt goch, Elan Valley 9067 AO. Reported by AP (as *V. mutabilis*) on roadside rocks near Llanewedd 0452 (1969, BLS records).

V. glaucina Ach.

Probably frequent on lime-rich substrates such as calcareous shales, mortared walls and limestone. Cliffs, Aberedw Rocks 04NE AP (1969,BLS records); mortared wall, Glasbury 13NE; limestone in Yatt Wood, Dolyhir 2458; concrete wall, Heyop 2374 and from 17NE.

Vezdaea

V. hochstetteri Fr.

Widespread on lime mortar and calcareous rocks. Mortar of old garden wall, Cwm Elan House, Elan Valley 9064; shale cliff by road N of Llanbadarn Fynydd 0879; calcareous shale rocks of railway bridge, Cil-y-byddar 1670 and in 27SW.

V. hydrela Ach.

Only noted from calcareous shale rocks on the bank of the Bach Howey below the viaduct, Llandeilo Graban 1042 AO.

V. margacea (Wahlenb.) Wahlenb.

Probably overlooked on rocks by streams. Streambed in Cilkenny Dingle, Llowes 1841 AO.

V. muralis Ach.

Widespread on mortar and calcareous rocks. Calcareous shale in tip of Cwm Elan Mine, Elan Valley 9065; mortar of Llanfihangel Helygen church 0464; mortar of bridge wall in 15SE; limestone, Dolyhir 2458; also recorded in 27SW.

V. nigrescens Pers.

Common on basic rock and particularly concrete in well-lit places.

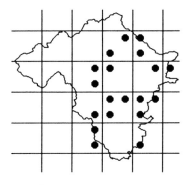

V. praetermissa (Trevisan) Anzi

Probably widespread on streamside rocks, but so far only noted in 05NW, 17SE and 24SW.

V. rheitrophila Zsch.

On stones in stream in Lakeside Wood, Llandrindod Wells 0660 AO.

V. viridula (Schrader) Ach.

Frequent on mortar, concrete and calcareous rocks in sunny sites.

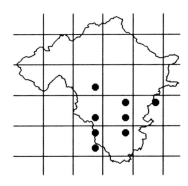

Vezdaea acicularis Coppins

Rare on damp soil and moss protonema on N facing volcanic rock outcrops. Worsell Wood, Stanner 2557.

V. aestivalis (Ohl.) Tsch.- Woess and Poelt

Scarce or overlooked. On mosses on concrete of old railway bridge near Gilfach Farm, St Harmon 9671.

V. leprosa (P.James) Vezda

A rare sp. overgrowing mosses, such as *Barbula convoluta*, in areas possibly subject to the effects of heavy metals. Below corrugated iron roof of old barn, Caerhyddwen, Ysfa 9865; old railway track bed N of Rhayader 9668 and below Yr Allt, Llandeilo Graban 0844.

V. retigera Poelt and P. Döbbeler

A scarce sp. overgrowing moss and soil. On ant hill on S facing slope above head of Garreg Ddu Resr., Elan Valley 9167; on *B. convoluta* Hedw. on cinder heap in yard of old Erwood Station 0843 det. BJC; on *Barbula* sp. on bed of old railway and on soil below old galvanised water tank nearby, Yr Allt, Llandeilo Graban 0844; on *Bryum capillare* Hedw. on mortar of old stone wall by suspension bridge, Llanstephan 1141; soil of roadside verge, Boughrood 1438.

V. rheocarpa Poelt and P. Döbbeler

Rare. Overgrowing mosses, particularly *Barbula* spp., on the track bed of an old railway S of Llanfaredd 0650 det. BJC.

Xanthoria calcicola Oxner

Frequent on stone walls, slates, asbestos cement and on old wooden boards about farmyards. Rarer on trees in pastures receiving dust and dung from grazing livestock.

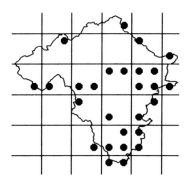

X. candelaria (L.) Th. Fr.

Widespread on nutrient-rich bark of the twigs of trees and shrubs, notably elder, ash, field maple and elm and on rocks and posts enriched with bird droppings.

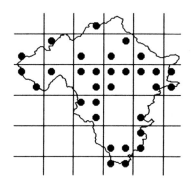

X. elegans (Link.) Th. Fr.

An uncommon but widespread lichen of sunny dry, base-rich stone walls, concrete copings and fence posts. On native rock only on hard, base-rich, sunny volcanic rock outcrop above Alpine Bridge, Llanbadarnfawr 0962.

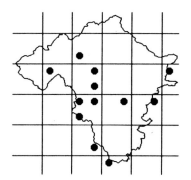

X. parietina (L.) Th. Fr.

Common on nutrient-rich tree bark, particularly ash, elm, sycamore and elder and on dry and sunny walls and posts where enriched by dust and dung near tracks and farmyards. Recorded from all 10km squares.

Xylographa

X. polycarpa (Hoffm.) Rieber

Common in the E of Radnor on twigs and small branches of the more nutrient-rich trees and shrubs, eg. ash, elm, elder, blackthorn, field maple and sycamore. Rarer on sawn timber about farmyards. Abundant on trees near the limestone quarries, Dolyhir 2458 which receive dust.

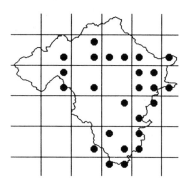

Xylographa vitiligo (Ach.) Laundon

A scarce or overlooked species of old, acidic, bark-free tree trunks in damp sites. Oak log, Glanllyn wood, Llansantffraed Cwmdeuddwr 9469; old oak in ravine SE of Henfron, Elan Valley 9064; old Scot's pine, Cors y Llyn, Newbridge on Wye 0155 AO; old conifer on Carneddau, E of Newmead Farm, Llanelwedd 0553; oak stump on shady bank, Gravel, Llanbister Road 1872.

Addendum to Chapter 13

Since preparing the above account a new lichen flora of Great Britain and Ireland has been published (Purvis et al, 1992). Not all the nomenclatural changes it proposes could be incorporated above. Numbered and listed below in alphabetical order and cross-referenced by number are those species which differ in name between the new lichen flora and the account above. New lichen flora names are shown with their authorities.

1. Anisomeridion juistense = 2
2. Anisomeridion nyssaegenum (Ellis & Everk.) R.C.Harris = 1
3. Arthonia cinnabarina (DC.) Wallr. = 5
4. Arthonia glaucomaria = 6
5. Arthonia tumidula = 3
6. Arthonia varians (Davies) Nyl. = 4
7. Bacidia epixanthoides = 8
8. Biatora epixanthoidiza auct. = 7
9. Biatora sphaeroides (Dickson) Körber = 17
10. Biatora vernalis (L.) Fr. = 47
11. Biatorina atropurpurea = 16
12. Calicium lenticulare Ach. = 13
13. Calicium subquercinum = 12
14. Caloplaca arenaria (Pers.) Müll. Arg. = 15
15. Caloplaca subpallida = 14
16. Catillaria atropurpurea (Schaerer) Massal. = 11
17. Catillaria sphaeroides = 9
18. Chaenotheca carthusiae = 19
19. Chaenotheca chlorella (Ach.) Müll. Arg. = 18
20. Chaenotheca furfuraceae (L.) Tibell = 24
21. Chromatochlamys muscorum (Fr.) Mayrh. & Poelt = 52
22. Claurouxia chalybeioides (Nyl.) D. Hawksw. = 39
23. Clauzadea monticola (Ach.) Hafellner & Bellem. = 41
24. Coniocybe furfuraceae = 20
25. Dimerella diluta = 26
26. Dimerella pineti (Ach.) Vezda = 25
27. Fuscidea lygaea (Ach.) V. Wirth & Vezda = 28
28. Fuscidea tenebrica = 27
29. Haematomma ventosum = 55
30. Huilia = 56
31. Lecanactis lyncea (Sm) Fr. = 54
32. Lecanora atra = 63
33. Lecanora badia = 58
34. Lecanora grumosa = 64
35. Lecanora orosthea (Ach.) Ach. = 42
36. Lecanora sublivescens (Nyl.) A.L. Sm. = 44
37. Lecanora sulphurea (Hoffm.) Ach. = 45
38. Lecidea ahlesii (Körber) Nyl. = 46
39. Lecidea chalybeioides = 22
40. Lecidea leucophaea = 53
41. Lecidea monticola = 23
42. Lecidea orosthea = 35
43. Lecidea speirea = 57
44. Lecidea sublivescens = 36
45. Lecidea sulphurea = 37
46. Lecidea valentior = 38
47. Lecidea vernalis = 10
48. Leproloma angardianum = 49
49. Leproloma cacuminum (Massal.) Laundon = 48
50. Leptogium corniculatum (Hoffm.) Minks = 51
51. Leptogium palmatum = 50
52. Microglaena muscorum = 21
53. Miriquidica leucophaea (Rabenh.) Hertel & Rambold = 40
54. Opegrapha lyncea = 31
55. Ophioparma ventosum (L.) Norman = 29
56. Porpidia Körber = 30
57. Porpidia speirea (Ach.) Krempelh. = 43
58. Protoparmelia badia (Hoffm.) Hafellner = 33
59. Schaereria fuscocinerea (Nyl.) Clauz. & Roux = 60
60. Schaereria tenebrosa = 59
61. Stenocybe pullatula (Ach.) B. Stein = 62
62. Stenocybe pullulata = 61
63. Tephromela atra (Huds.) Hafellner ex Kalb = 32
64. Tephromela grumosa (Pers.) Hafellner & Roux = 34
65. Usnea cornuta Körber = 66
66. Usnea inflata = 65

For additional species accounts see the general addendum at the foot of the index

14. Rust and Smut Fungi

The Uredinales and Ustilaginales.

Many fungi attack higher plants producing symptoms obvious to the field botanist. Sadly these fungi have been largely ignored by local flora writers. With the publication by Croom Helm in 1985 of *Microfungi of Land Plants* by M.B. Ellis and J.P. Ellis, a text became readily available to assist in their identification. A start has been made to describe the rust and smut fungi of Radnor. The account below is very provisional and much further work remains to be done. The majority of the records have been made by the author and A. Orange, with a few additional records gleaned from herbaria at the National Museum of Wales Cardiff and the Royal Botanic Gardens, Edinburgh.

Rusts of grasses and sedges require considerable additional work. Whilst being common, they rarely seem to produce the stages in their life cycle necessary to determine them. No special effort has been made to seek these fungi on garden or crop plants. Where they have been incidentaly observed they are recorded here.

No records have, for example, been made from cereals. Yet smut of wheat must once have been common, since Howse (1949) in his chapter on Radnor folklore refers to a "ceremony of the burning bush" to protect crops from it. A branch of hawthorn fashioned in the shape of a cross was carried from field to field in the early hours of New Year's Day. A fire of straw was kindled and the cross was placed in the flames and then removed to be carried to the next field. So common was the ceremony until the early years of the 19th century that the whole countryside was said to "twinkle like stars" in the darkness of the early morning.

At least 73 species of rust fungi and 9 species of smut fungi are reported from the county. The taxonomy of the rust fungi follows Ellis and Ellis op cit. whilst that of the smut fungi follows Mordue, J.E.M. and Ainsworth, G.C., *Ustilaginales of the British Isles*, Mycological Paper 154, 1984, Commonwealth Mycological Institute, Kew.

Rust Fungi
Uredinales

Coelosporium tussilaginis (Pers.) Berk.

Recorded on butterbur, coltsfoot, groundsel and harebell. Frequent on butterbur. By R. Ithon, Alpine Bridge, Llanbadarn Fawr 0963; Llangunllo 1872; Bleddfa 2168; Whitton c2767 RW Dennis (1969 NMW) and Gwernafel, Knighton 2570. On coltsfoot it is widespread and occurs almost wherever the host is found. On groundsel it is rare and recorded only from Builth Road c0253 AEW (1929 NMW). On harebell it occurs on rock outcrops on Carreg Wiber, near Llandrindod Wells 0859 and at Whitton c2767 RW Dennis (1969 NMW).

Cumminsiella mirabilissima (Peck) Nannf.

On Oregon grape in gardens and hedges. Probably frequent. Hedge, Llansantffraed Cwmdeuddwr 9667; churchyard, Gladestry 2355; roadside, Dolyhir 2458 and Stanner Rocks 2658.

Frommea obtusa (Str.) Arth.

Rare on tormentil and creeping cinquefoil on roadside verges. On creeping cinquefoil near Argoed Mill, Doldowlod 9962 and on tormentil and creeping cinquefoil below Yr Allt, Doldowlod 0061.

Hyalopsora polypodii (Diet.) Morgan

Scarce on brittle bladder-fern. On cliffs in Bach Howey gorge, Llanstephan 1243; dingle N of Llanfihangel-Nant-Melan church 1858; Davy Morgan's Dingle, Radnor Forest 1862.

Melampsora on Willows

Rusts of the genus *Melampsora* occur commonly on willows. The uredinia can frequently colour the young shoots of goat willow, in particular, a bright orange colour. Grey, rusty, purple and crack willows, osier and their hybrids are frequently infected throughout the range of these willows in Radnor. Dr. Stephan Helfer of the Royal Botanic Gardens, Edinburgh has recently reviewed the willow rusts (pers. comm.). Radnor material needs to be re-examined in the light of his findings and in consequence a more detailed account cannot be provided here.

Melampsora euphorbiae Cast.

Noted on sun and petty spurge. Occasional on sun spurge. Roadside verge between Middle Hall and Aberedw 0748; swede field S of Crossway, Howey 0456; swede field near Boughrood 1438; swede field, Stanage 3372. Rare on petty spurge. Roadside verge near the Ridgebourne, Llandrindod Wells 0560.

Puccinia

M. hypericorum Wint.

Scarce on beautiful St. John's-wort and tutsan, it has recently been noted on cultivated species in gardens. On beautiful St. John's-wort in the gorge of the Afon Elan above Aberglanhirin, Elan Valley 8872 and by foresrty track in Cann Wood, Presteigne 3063. On tutsan on lane bank below Llanstephan Church 1241. By 1989 this rust had become established on cultivated species eg. rose-of-Sharon in gardens in Llandrindod Wells 06.

M. larici-populina Kleb.

Widespread on various poplar cultivars. Shaky Bridge, Cefnllys 0861; near Glasbury 1639; Llanfihangel-Nant-Melan 1858.

M. lini (Ehrenb.) Desm.

Widespread on fairy flax. In 87SE, 97SE, 04NE, 05SE, 16NE and 36SW.

M. populnea (Pers.) Karst.

Aecia rare on dog's mercury; woodland edge, Llanstephan 1241 and hedge, Michaelchurch on Arrow 2350. Uredinia and telia probably frequent on aspen; roadside, Rhayader 9767; Llanyre 0461 and near Lower House Farm, Beguildy 1579. Rare on *Populus* x *canescens* in roadside copse, Michaelchurch on Arrow 2350

Melampsoridium betulinum (Fr.) Kleb.

Widespread on seedling birch and young, sappy growths in hedges, developing rather late in the year and uncommon in dry summers. Probably in all 10km squares.

Milesina blechni Syd.

Rare on hard fern. In shady ravine, Craig Gigfran, Elan Village 9264; woodland banks by Gilfach Farm, St Harmon 9671; by the Milo Brook below Gwern Alltcwm, Aberedw 0948; Cilkenni Dingle, Llowes 1841 and Cann Wood, Presteigne 3063.

M. dieteliana (Syd.) Magn.

Rare or overlooked on common polypody. Cliffs E of Esgair Dderw, Llansantffraed Cwmdeuddwr 9469; on fallen tree trunk in wood, Bailey Einon, Cefnllys 0861 and on rocks by R. Wye, Boughrood 1338.

M. kriegeriana (Magn.) Magn.

On broad and narrow buckler-fern and common male-fern. Common on broad buckler-fern in damp woodlands. 87SE, 97NW, 05SE & NW, 06SE, 14SE, 15NW, 24SW, 25NE and 36SW. Rare on narrow buckler-fern. Cors y Llyn, Newbridge on Wye 0154. Rare on common male-fern. Pen y lan Wood, Clyro 2044 and Worsell Wood, Stanner 2557.

M. scolopendrii (Faull) Hend.

Common on hart's-tongue fern in damp woodlands. 04NE, 05SW, 13NE, 14SE, 24SW and 24NW.

M. whitei (Faull) Hirats.

On hard and soft shield-fern. Scarce on soft shield-fern in damp, shady woodland. Dingle below Gwernallt Cwm, Aberedw 0848; in wooded gorge of the Bach Howey near the viaduct, Llandeilo Graban 1042 and on wooded lane bank, Llanstephan 1241. and in ravines near Llowes 1841 and 1941. Rare on hard shield-fern. Bach Howey gorge near the viaduct, Llandeilo Graban 1042.

Miyagia pseudosphaeria (Mont.) Jorst

Scarce on prickly sow-thistle. Disturbed ground on streambank, N of Neuadd-ddu, St Harmon 9176; in farmyard near Crossway 0557; on pavement near The Llannerch, Llandrindod Wells 0561; roadside, Monaughty 2368; ride in conifer plantation, Burfa Camp, Evenjobb 2860.

Phragmidium bulbosum (Str.) Schlecht.

Common on species of blackberry throughout the county.

P. fragariae (DC.) Rabh.

Common on barren strawberry throughout the county.

Phragmidium mucronatum (Pers.) Schlecht.

Rare on dog rose species. Hedgerow NE of Llidiart wen, St Harmon c9771 ARP (NMW); roadside near Wye bridge, Newbridge on Wye 0158; waste ground by old quarry storage area, Llanelwedd 0551 and roadside, Rhyd y Clwydiau, Bwlch y Sarnau 0078.

P. tuberculatum J.Müller

Rare on roses in gardens in Llandrindod Wells. On Japanese rose in 0561 and on hybrid tea in 0562.

P. violaceum (CF Schultz) Wint.

On bramble species, but less frequent than *P. bulbosum*. Roadside verge, Neuadd-ddu, St Harmon 9175; old quarry storage area, Llanelwedd 0551; roadside, Llanstephan 1141; road verge, Monaughty 2368; shrubbery, Womaston, Walton 2660.

Puccinia acetosae Körn

Scarce or overlooked on common sorrel. Roadside verge of A470 N of Neuadd-ddu, St Harmon 9176; roadside S of Llanyre 0461; road verge near Cwrt y Graban, Llanstephan 1142; trackside, Upper Folley, New Radnor 2161; damp grassland, Burfa Bog, Evenjobb 2661; Knighton c2872 RW Dennis (1969 NMW).

P. adoxae DC.

Widespread on moschatel in woodlands and hedgerows. 04NE & SE, 05SW, 15SE, 24SW & NW, 25NE, 26SE and 37SW.

P. aegopodii Rohl.

Rare on ground elder. Roadside banks below Yatt Wood, Dolyhir 2458 and at Discoed 2764.

P. albescens Plowr.

Scarce on moschatel in woodlands and hedgerows. Churchyard, Cefnllys Church 0861 and on wooded river bank, Wyecliff, Clyro 2242.

P. antirrhini Diet. and Holw.

Rare on cultivated snapdragon in gardens. Great House Garage, Newbridge on Wye 0158; Southville, Llandrindod Wells 0560 AO and nearby in North Road 0562.

P. annularis (Str.) Röhl.

Rare on woodsage. Roadside bank S of Erwood Station 0943; roadside near Brynwern Bridge, Newbridge on Wye 0156; in wall near Llanstephan House, Llanstephan 1141 and on scree, Stanner Rocks 2658; Worsell Wood 2557 and Cann Wood, Presteigne 3063.

P. arenariae (Schum.) Wint.

Recorded on red campion, three-nerved sandwort and annual pearlwort. Occasional on red campion; lane bank below Llanstephan Church 1241; Yatt Wood, Dolyhir 2458; woodland, Womaston, Walton 2660; scrub woodland, Burfa Camp, Evenjobb 2860. Scarce on three-nerved sandwort; hedgebank, Cwm Rhocas, Penybont 1064 and in scrub woodland Burfa Camp, Evenjobb 2860. Rare or overlooked on annual pearlwort; old quarry, Aber Main, Dolau 1367.

Puccinia

P. behenis Otth

Rare on red campion. Scrub woodland, Burfa Camp, Evenjobb 2860.

P. betonicae DC.

Rare on betony. Agriculturally unimproved field, Gilfach Farm, St Harmon 9671 coll. DPH.

P. brachypodii Otth

The var. *brachypodii* is probably more widespread on false brome than the few records suggest. Hedgerows and woodland around Llanstephan 1141 and 1241. The var. *poae-nemoralis* (Otth) Cummins & Greene has probably been overlooked on wood meadow-grass. Rock Park, Llandrindod Wells 0560.

P. buxi DC.

Known only from a dwarf form of box (probably 'suffruticosa') used as edging for paths in the now derelict garden of Cwm Elan House in the Elan Valley 9064. Though the ordinary form of box is also present here, it is not infected.

P. calcitrapae DC.

Found on marsh and welted thistle. Scarce or overlooked on marsh thistle. Wet pasture, Abercamlo, Gwystre 0764 and in the dingle N of Llanfihangel-Nant-Melan Church 1858. Rare on welted thistle. Roadside bank between Dolyhir and Old Radnor 2458.

P. calthicola Schröter

Rare on marsh marigold. Willow carr, Rhosgoch Common 1948. A rust on marsh marigold from Burfa Bog, Evenjobb 2761 lacked teliospores and could not be certainly referred to this species.

P. caricina DC.

Aecia on stinging nettle are apparently scarce but have been recorded from 9566, 9571, 0844, 0559, 0861, 2345 and 2062. Much further work is required to unravel its distribution on sedges. Uredinia and/or telia have been found on pendulous sedge in the Bach Howey gorge, Llandeilo Graban 1042, on carnation sedge and black sedge in wet pastures at Abercamlo, Gwystre 0764, on lesser pond sedge at the edge of the raised mire, Rhosgoch Common 1948 and in a streamside bog W of Cwm y Geist, Llanbister 1671, on green-ribbed sedge beside a forest road, Rhyd y Clwydiau, Bwlch y Samau 0078, on prickly sedge on roadside bank, Pant y Dwr 9875, on hairy sedge and slender-tufted sedge in Fforddfawr Mire, Glasbury 1840, on tussock sedge on the edge of Llandrindod Wells Lake 0660 and on Rhosgoch Common 1948, on star sedge at the edge of Rhosgoch Common 1948 and on bottle sedge on Cors y Llyn, Newbridge on Wye 0155.

P. chaerophylli Purton

Rare on sweet cicely. Roadside between Llangunllo and Llanbister Road 17SE or 27SW RW Dennis (1969 NMW).

P. circaeae Pers.

Scarce on enchanter's nightshade in woodland. Bach Howey gorge, Llanstephan 1143; Rhosgoch Common 1948 and in a wood E of Knucklas 2673.

P. cnici Mart.

Probably widespread on spear thistle. On old railway below Yr Allt, Llandeilo Graban 0844; bank of R. Wye, Glasbury 1739 and in grassland N of Llanfihangel-Nant-Melan Church 1858.

P. cnici-oleracei Pers.ex Desm.

Recorded on marsh thistle and yarrow. Probably frequent on marsh thistle. By Afon Claerwen below Craig Cwm Clyd, Llansantffraed Cwmdeuddwr 8862; by Afon Marteg, Gilfach Farm, St Harmon 9571; wet pasture, Abercamlo, Gwystre 0764 and at edge of woodland, Cwm Rhocas, Penybont 1063. Rare on yarrow. Roadside verge, Tircelyn, Aberedw 0745.

Puccinia coronata Corda

Probably the commonest rust fungus on grasses. Noted on common bent, perennial ryegrass, false oatgrass and on giant and wood fescue. A common rust on Yorkshire fog and creeping soft-grass may be this species but telia have yet to be detected. On common bent in garden, Llandrindod Wells 0562. On perennial ryegrass 2km S of Knighton 27SE RW Dennis (1969 NMW). On false oatgrass and giant fescue in lane below Llanstephan Church 1241 and on false oatgrass on roadside verge, Cwmbach Llechrhyd 0253. On wood fescue in the gorge of the Bach Howey, Llanstephan 1143.

P. crepidicola Syd.

Rare or overlooked on smooth hawk's-beard. Roadside verge below Dolyfan Hill, Newbridge on Wye 0161.

P. deschampsiae Arth.

Rare or more probably overlooked on tufted hair-grass. In hazel wood in gorge of the Bach Howey, Llanstephan 1143.

P. galii-verni Ces.

Noted on heath bedstraw and crosswort. Probably widespread on heath bedstraw. Acidic grassland by stream, Neuadd-ddu, St Harmon 9276; roadside bank W of Brynhir, Howey 0658 and on hilltop, Lloyney, Knucklas 2475. Rare on crosswort. Knighton c27SE RW Dennis (1969 NMW).

P. glechomatis DC.

Rare on ground ivy. Scrub woodland, Burfa Camp, Evenjobb 2860.

P. hieracii Mart.

Widespread on mouse-ear and other hawkweeds and hawkbits. On mouse-ear and hawkweeds in all 10km. squares. Rare on autumn hawkbit. Roadside verge, Newbridge on Wye 0158. Rare or overlooked on rough hawkbit. Road verge near Gravel, Llanbister Road 1772 and churchyard, Llanbister 1073. On a hawkbit sp. in Penybont Road, Knighton c27SE RW Dennis (1969 NMW). Rare on cat's-ear. Hedgebank near Cefnllech, Pantydwr 9677 and on road verge Monaughty 2368. A rather frequent rust on common knapweed may be this species. It has only been confirmed where suitable spore stages were present on roadverges S of Llanbister 1072 and S of Burfa Camp, Evenjobb 2860. It is also probably frequent on dandelion. Confirmed S of Llanbister 1072.

P. lagenophorae Cooke

Very common on groundsel throughout the county.

P. lapsanae Fuck.

Very common on nipplewort throughout the county.

P. maculosa (Str.) Rohl.

Rare on wall lettuce. On fallen log in woodland below Gwern Allt Cwm, Aberedw 0948.

P. malvacearum Mont.

Common on common mallow. 96NE, 04SE, 05SE, 13NE, 14NE, 24SW. and 26SW. Rarer on hollyhocks in gardens. Llandrindod Wells 06 and probably elsewhere.

P. menthae Pers.

Uncommon on water mint. By R. Ithon, Abercamlo, Gwystre 0764.

P. obscura Schroet.

Noted on hairy, heath, great and southern wood-rush and probably more common than the records suggest. On hairy wood-rush, Cann Wood, Presteigne 3063. On heath wood-rush in forest ride, Rhyd y Clwydiau 0078. On great wood-rush, 0560, 0861, 1042, 1741, 1842, 2557 and 3063. On southern wood-rush by forest road, Burfa Camp, Evenjobb 2760.

P. pimpinellae (Str.) Röhl.

Scarce on pepper-saxifrage. On old railway bank near Glasbury Station 1839 and on road verge on S side of Old Radnor Hill, Burlingjobb 2558.

P. porri (Sow.) Wint.

Rare on leeks in gardens. Newbridge on Wye 0158 and Glasbury 1838.

P. pulverulenta Grev.

On great willowherb and broad-leaved willowherb. Rare on great willow-herb. By Gilwern Brook near Bridge House, Burlingjobb 2558. Rare on broad-leaved willowherb. Builth Road c0253 AEW (1929 NMW) and on woodland bank between Milebrook and Knighton 3072.

P. punctata Link

On common marsh-bedstraw, crosswort and lady's bedstraw. Occasional or overlooked on common marsh-bedstraw. Roadside verge below Yr Allt, Doldowlod 0061 and in wet pasture, Abercamlo, Gwystre 0764. Rare on crosswort. Roadside verge S of Llanbister 1072. Rare on lady's bedstraw. Grassland below Stanner Rocks 2658.

P. punctiformis (Str.) Röhl.

Common on creeping thistle throughout the county.

P. sessilis Schröter

Scarce or overlooked. Aecia on ramsons in the gorge of the Bach Howey, Llandeilo Graban 1042. Teleutosori on reed canary-grass on the bank of the Wye above Erwood Bridge 0843.

P. tumida Grev.

Occasional on pignut in grassland. Churchyard, Nant Gwyn, St Harmon 9776; Llandeilo Graban churchyard 0944; bank of R. Edw above Aberedw 0947 and in churchyards at Bryngwyn 1849, Llowes 1941 and Gladestry 2355.

P. umbilici Duby

Despite an extensive search it has only been found on navelwort on a S facing volcanic cliff above Alpine Bridge, Llanbadarn Fawr 0963 and on a volcanic rock outcrop to NE of Old Radnor Church 2559 ! and DPH.

P. veronicae Schröter

Uncommon on wood speedwell. Roadside E of Cwmbach Llechrhyd 0353; damp woodland, Lovers Leap, Llandrindod Wells 0560; in grassland, Cefnllys Church 0861 and in woodland in the gorge of the Bach Howey, Llandeilo Graban 1243.

P. violae DC.

Frequent on common dog-violet. Noted in 04SE, 06SE, 07SW, 14SW, 16NE, 17SE, 25NW and 36SW. Rare on hybrid Viola. In garden, Presteigne 3164.

Pucciniastrum vaccinii (Wint.) Jorst.

On bilberry and cowberry. Frequent on bilberry in woodland and on open moorland. 96NE & NW, 97NW, 04NE, 06SE & NE, 07NW & SW, 15NE, 16NE and 36SW. Rare on cowberry. Below Llethr Du, Radnor Forest 1967.

Tachyspora intrusa (Grev.) Arth.

Rare on lady's-mantle. In Crug y Bydder churchyard 1582.

Triphragmium ulmariae (DC.) Link

Frequent on meadowsweet. 96SE & NE, 05SW, 06SE, 14SW & NE, 25SW and 26SE.

Uromyces acetosae Schröter

Scarce or overlooked on common sorrel. Churchyard, Disserth 0358; field by R.Ithon, Cefnllys Castle 0861.

U. dactylidis Otth.

Common on lesser celandine. Probably much overlooked on cock's-foot. On lesser celandine in 96NE, 97SE, 06SE, 15SE, 25NW, 26SE & 27SE. On cock's-foot on roadside verge near Rhyd y Clwydiau, Bwlch y Sarnau 0078.

U. fallens (Arth.) Barth.

Rare on red clover. Edge of forestry track, Cann Wood, Presteigne 3063.

Uromyces

U. ficariae Tul.

Frequent on lesser celandine and probably commoner than the records suggest. Bailey Einon Wood, Cefnllys 0861; Glascwm Churchyard 1553 and old quarry near the church, Old Radnor 2459.

U. muscari (Duby) Graves

Frequent on bluebell throughout Radnor wherever the host is plentiful.

U. pisi-sativi (Pers.) Liro

Rare on bird's-foot-trefoil on roadsides. Near Argoed Mill, Doldowlod 9962; below Yr Allt, Doldowlod 0061 and near Gravel, Llanbister Road 1772.

U. polygoni-aviculariae (Pers.) Karst.

Rare or overlooked on knotgrass. Farmyard, Abercamlo, Gwystre 0764; roadside SE of Llanstephan House 1141 and roadside Cabalva, Clyro 2345.

U. valerianae Fuck.

Occasional on common valerian. Roadside verge, Neuadd-ddu, St Harmon 9176; wet pasture, Abercamlo, Gwystre 0764; roadsides, Rhyd y Clwydiau, Bwlch y Sarnau 0078; Rhosgoch Common 1948; forestry ride, Cann Wood, Presteigne 3063 and roadside between Milebrook and Knighton 3072. It is also expected to occur on marsh valerian, a host it attacks closeby in Brecknock.

U. viciae-fabae (Pers.)Schröter

Noted on bush, tufted and bitter vetch and meadow vetchling. Probably wide-spread on bush vetch. Argoed Mill, Doldowlod 9962; Disserth Churchyard 0358; hedgerow, Henfryn, Llandewi Ystradenni 0768; roadside, Monaughty 2368; wooded bank, Knucklas 2673. Rare on tufted vetch. Edge of forestry track N of New Farm, Bwlch y Sarnau 0077. Probably much overlooked on meadow vetchling on roadverges. Near Ashfield, Ysfa 9764; Argoed Mill, Doldowlod 9962 and Cwmbach Llechrhyd 0353. The var. *orobi* (Schum.) Jorst. is rare on bitter vetch. Wooded road bank, Cwmbach Llechrhyd 0353.

Xenodocus carbonarius Schlecht.

Uncommon on greater burnet. Lane verge between Cwmbach Llechrhyd and Pencerrig 0353 and on railway bank W of Cwm y Geist 1671.

Smut Fungi
Ustilaginales

Entyloma ficariae (Berk.) Fisch. v. Waldh.

Common on lesser celandine throughout Radnor.

Tilletia sphaerococca (Wallr.) Fisch.

Probably widespread but overlooked, stunting common bent grass. Roadside verge, Aber Glanhirin, Elan Valley 8875 and disused playing field near The Rock Park, Llandrindod Wells 0560.

Urocystis ranunculi (Lib.) Moesz

Rare on creeping buttercup. Woodland edge near County Hall, Llandrindod Wells 0660.

U. violae Fisch.v.Waldh.

Rare on common dog-violet. Roadside bank N of Graig, Llanfihangel Rhydithon 1767 and in old quarry near Brook Cottage, Llanbadarn Fynydd 1076.

Ustilago anomala Kunze

Rare on water-pepper, beside pool, Fforddfawr Mire, Glasbury 1840.

U. avenae (Pers.) Rostr.

Occasional on false oat-grass. Road verge, Aberedw 0846; roadside S of The Ridgebourne, Llandrindod Wells 0559; roadside E of Erwood Station 1043 and roadside S of Burfa Camp, Evenjobb 2860.

U. longissima (Sow ex Schl.) Meyen

Rare on sweet-grass sp.. Burfa Bog, Evenjobb 2761.

U. striiformis (Westend.) Niessl

Scarce or overlooked on creeping soft-grass. Farmyard, Downton Farm, New Radnor 2360.

U. succisae P.Magn.

Rare on devil's-bit scabious. Wet pasture, Abercamlo, Gwystre 0764; damp roadside, Rhydyclwydau Brook, Bwlch y Sarnau 0078 and Boughrood Churchyard 1239.

U. tragopogonis-pratensis (Pers.) Roussel

Rare on goatsbeard. Roadside S of The Ridgebourne, Llandrindod Wells 0560 AO.

U. violacea (Pers.) Roussel

On red campion and lesser stitchwort. Occasional on red campion throughout the county. Rare on lesser stitchwort. Roadverge N of Graig, Llanfihangel Rhydithon 1767; hedgebank S of Llanbister 1072 and lane bank near Upper Folley, New Radnor 2161.

Abbreviations

ABS University College of Wales, Aberystwyth Herbarium

agg aggregate

Atlas Brit Fl Atlas of the British Flora (3rd ed.). Perring, F.H. and Walters, S.M. (Eds.)(1983). EP Publishing Ltd. and BSBI, Wakefield

BBS British Bryological Society

BBSUK Herbarium of the British Bryological Society

BIRM Birminghan University Herbarium

BLS British Lichen Society

BM Herbarium of the British Museum (Natural History)

Bot Loc Rec Club Botanical Localities Record Club

Bot Rec Club Botanical Record Club

Braithwaite Moss Fl II British Moss Flora, volume II, Braithwaite, R, 1888.

Brit Pter Soc exc. British Pteridological Society excursion

BRC Biological Records Centre, Monkswood

BSBI Botanical Society of the British Isles

Bull BBS Bulletin of the British Bryological Society

c near to or around

CCW Countryside Council for Wales

CGE Cambridge University Herbarium

coll collected by

conf identification confirmed by

Cotteswold Nat Fld Club Proceedings of the Cotteswold Naturalists Field Club

det determined or identified by

E Royal Botanic Gardens Edinburgh herbarium

exc field excursion

FPW Flowering Plants of Wales. Ellis, R.G. (1983), National Museum of Wales, Cardiff

hb herbarium of

HRNT Herefordshire and Radnorshire Nature Trust Limited

int rep an internal unpublished report

ITE Institute of Terrestrial Ecology

J Bot The Journal of Botany

J Bryol The Journal of Bryology

K Kew herbarium

km kilometre

m metre

MANCH Manchester Museum Herbarium

MBH Marlborough College Natural History Soc. Herbarium

ms manuscript

MsFlR Materials towards a Flora of Radnor. An unpublished manuscript by Wade, A.E. and Webb, J.A. held in the Botany Department of the National Museum of Wales, Cardiff

Nat Wales Nature in Wales

Nat Wales NS Nature in Wales New Series

NCC Nature Conservancy Council

New Phytol The New Phytologist

NMW National Museum of Wales, Cardiff

NNR National Nature Reserve

NW Nat North West Naturalist

ORS Old Red Sandstone

OXF Oxford University Botany School Herbarium

pers comm personal communication

Proc BSBI Proceedings of the Botanical Society of the British Isles

Proc Cotteswold Club Proceedings of the Cotteswold Field Naturalists Club

Proc Swansea Sci & Fld Nats Soc Proceedings of the Swansea Scientific and Field Naturalists' Society

Ray Soc Ray Society

rep report

Rep Bot Exc Club Report of the Botanical Society and Exchange Club of the British Isles

Rep BBS Reports of the British Bryological Society

Rep MEC Reports of the Moss Exchange Club

Resr Reservoir

Sci Gossip NS Science Gossip, New Series

sp species (singular)

spp species (plural)

ss sensu strict - a narrow interpretation of a name

ssp subspecies (singular)

sspp subspecies (plural)

Suppl Supplement

TBBS Transactions of the British Bryological Society

Trans Rad Soc Transactions of the Radnorshire Society

Trans Woolhope Club Transactions of the Woolhope Field Naturalists' Club

UCNW University College of North Wales, Bangor Herbarium

UWIST University of Wales Institute of Science and Technology

var variety

WFP Welsh Flowering Plants (Ed.2) . Hyde, H.A. and Wade, A.E. (1957). National Museum of Wales, Cardiff

! the author

Recorders and Collectors

The following recorders and collectors are abbreviated to their initials in the species accounts above.

ACP	A C Powell	HND	H N Dixon
AEW	A E Wade	HNR	H N Ridley
AL	A Ley	IDS	I D Soane
AO	A Orange	IHS	I H Slater
AP	A Pentecost	JAP	J A Paton
ARP	A R Perry	JAW	J A Webb
BJC	B J Coppins	JFH	J F Hall
BS	B Storer	JH	Sir J Holland
CAC	C A Chadwell	JL	J Langford
CC	C Collingridge	JSB	J S Butler
CHB	C H Binstead	LMP	L M Paul
CP	C Port	MEM	M E Massey
CW	C West	MFVC	M F V Corley
DDB	D D Bartley	MP	M Porter
DRD	D R Drewett	PCH	P C Hall
DEG	D E Gray	PDM	P D Moore
DPH	D P Hargreaves	PDS	P D Sell
EA	E Armitage	PJP	P J Port
EFW	E F Warburg	PWJ	P W James
EWJ	E W Jones	PWR	P W Richards
FE	F Everington	RCLH	R C L Howitt
FMS	F M Slater	RK	R & R Key
FR	F Rose	SW	S Wyard
GP	G Philipps	VG	V Gordon
HHK	H H Knight	WHP	W H Pearson
HJR	H J Riddelsdell	WMR	W M Rogers

Bibliography

This bibliography contains all the major works cited in the Flora except those cited in the chapter on the physical environment which can be found at the end of that chapter and those referring to but one or two species records. The latter are cited in the relevant species accounts. In addition other material relevant to Radnor botany is included here, though a complete bibliography of the Radnor flora lies outside the scope of this work. Material predominantly concerning bryophytes is prefixed with a * whilst that concerning lichens is prefixed with a + . Unpublished works and internal documents are placed in brackets.

Antoine, S.E., & Benson-Evans, K. (1985). Benthic algal flora of the River Wye system, Wales, U.K.. *Nova Hedwigia,* **42**, 31-47.

*****(Averis, A.B.G. (1990a).** *A Survey of the Bryophytes of the Woods of NW Highlands.* Nature Conservancy Council, Edinburgh.)

*****(Averis, A.B.G. (1990b).** *A Survey of the Bryophytes of the Woods of Argyll.* Nature Conservancy Council, Edinburgh.)

*****+Barkman, J.J. (1958).** *Phytosociology and Ecology of Cryptogamic Epiphytes.* Van Gorum, Assen.

Bartley, D.D. (1960a). Rhosgoch Common, Radnorshire: Stratigraphy and pollen analysis. *The New Phytologist* **59**, 238-262.

Bartley, D.D. (1960b). Ecological studies on Rhosgoch Common, Radnorshire. *Journal of Ecology,* **48**, 205-214.

Bufton, J.J. & J.O. (1906). *Illustrated Guide to Llandrindod Wells.* F. Hodgson, London.

+Cannon, P.F., Hawksworth, D.L. and Sherwood-Pike, M.A. (1985). *The British Ascomycotina.* Commonwealth Agricultural Bureaux, Slough.

Carter, P.W. (1950). A history of botanical exploration in Radnorshire. *Transactions of the Radnorshire Society,* **20**, 42-58.

Caseldine, A. (1990). *Environmental Archaeology in Wales.* Dept. of Archaeology, St. David's University College, Lampeter.

Clapham, A.R., Tutin, T.G. and Warburg, E.F. (3rd edn.)(1981). *Excursion Flora of the British Isles.* Cambridge University Press.

Clarke, W.A. (1901). Radnorshire plants. *Journal of Botany,* **39**, 279-280.

(Cooke, R. & Saunders, G.R. (1989). *Woodland Surveys in Dyfed-Powys Region, 1988 Using the National Vegetation Classification.* Internal NCC report.)

(Collingridge, C.E. (1984). *Grassland survey of Radnorshire 1982-1983.* Hereford and Radnor Nature Trust report.)

+Coppins, B.J. (1976). Distribution patterns showed by epiphytic lichens in the British Isles. In Brown, D.H., Hawkesworth, D.L. and Bailey, R.H. (Eds.), *Lichenology: Problems and Progress:* 249-278. London.

+Coppins, B.J. (1983). A taxonomic study of the lichen genus Micarea in Europe. *Bulletin of the British Museum (Natural History), Botany Series,* **11**, 17-214.

+Coppins, B.J. (1988). Notes on the Genus Arthopyrenia in the British Isles. *The Lichenologist,* **20**, 305-325.

+Coppins, B.J. & James, P.W. (1984). New or interesting British Lichens. V. *The Lichenologist,* **16**, 250-263.

(Cragg, B.A. (1977). *A hydrological and vegetational survey of Llangorse Lake and Llanbwchllyn Reservoir in Wales.* MSc. thesis. University of Wales.)

Cragg, B.A., Fry, J.C., Thurley, S.S. and Bacchus, Z. (1981). The aquatic vegetation of Llanbwchllyn Reservoir, Powys. *Nature in Wales* **17**,(4), 217-223.

Davies, L. (1920). *Radnorshire.* Cambridge University Press.

Dony, J.G., Jury, S.L. and Perring, F.H. (1986). *English Names of Wild Flowers,* 2nd ed. Botanical Society of the British Isles, London.

Druce, G.C. (1908). Welsh Records. *Journal of Botany,* **46**, 335-336.

Druce, G.C. (1932). *The Comital Flora of the British Isles.* Buncle, Arbroath.

Bibliography

*Duncan, J.B. (1926). *A Census Catalogue of British Mosses*, 2nd edn. British Bryological Society.

+Duncan, U.K. & James, P.W. (1970). *Introduction to British Lichens*, Buncle, Arbroath.

Edwards, R.M. & Brooker, M.P. (1982). *The Ecology of the River Wye*. Junk, The Hague.

Ellis, M.B. & Ellis, P.J. (1985). *Microfungi on Land Plants*. Croom Helm, London.

Ellis, R.G. (1983). *Flowering Plants of Wales*. National Museum of Wales, Cardiff.

Fojt, W.J. (1986). The vegetation of Gors Goch, an upland mire in Mid Wales. *Nature in Wales NS* **4**, 28-35.

Forestry Commission (1952). *Census of Woodlands 1947-1949*. Census Report No. 1. HMSO, London.

French, C.N. & Moore, P.D. (1986). Deforestation, Cannabis cultivation and schwingmoor formation at Cors Llyn (Llyn Mire), Central Wales. *New Phytologist*, **102**, 469-482.

(Furet, J.E. (1979). *Algal Studies on the River Wye System*. Unpublished Ph D thesis. Univ. of Wales.)

Gibbin, O. (1937). Flora of Radnor. *Transactions of the Radnorshire Society*, **7**, 48-58.

Harrison, J.D., Kirkland, P.H. and Saxon, C. (1981). *Llandrindod Wells Lake, Powys and the Spa Waters of Wales*. Undated int. rep. Powys County Council and University of Wales Institute of Science and Technology, Cardiff.

+Hawksworth, D.L. and Seaward, M.R.D. (1977). *Lichenology in the British Isles* 1568-1975. Richmond Publishing Co. Ltd., Richmond.

*Hill, M.O. (1988). A bryophyte flora of North Wales. *Journal of Bryology*, **15**, 377-491.

*Hill, M.O., Preston, C.D. and Smith, A.J.E. (1991). *Atlas of the Bryophytes of Britain and Ireland:* Vol 1. Liverworts. Harley Books, Colchester.

Howse, W.H. (1949). *Radnorshire*. Hereford. (Reprinted, 1973 for the Radnorshire Soc. by The Scolar Press, Ilkley).

Hyde, H.A. (1931).*Welsh Timber Trees*. (1st edn.) National Museum of Wales, Cardiff.(2nd edn. 1935, 4th edn. 1977)

Hyde, H.A. & Wade, A.E. (1934). *Welsh Flowering Plants* (1st edn.). National Museum of Wales, Cardiff. (2nd edn. 1957).

Hyde, H.A. & Wade, A.E. (1940). *Welsh Ferns*. (1st edn.) National Museum of Wales, Cardiff. (5th edn. 1969, 6th edn. 1978).

*Ingham, W. (ed.) (1917). *A Census Catalogue of British Mosses*, 1st edn. Moss Exchange Club.

*Ingham, W. (ed.)(1913). *A Census Catalogue of British Hepatics*, 2nd edn. Moss Exchange Club.

+James, P.W., Hawkesworth, D.L. & Rose, F. (1977). Lichen Communities in the British Isles in *Lichen Ecology*, Seaward, M.R.D. (ed.), 295-413. Academic Press, London.

Judge, C. (1987). *The Elan Valley Railway*. The Oakwood Press, Oxford.

Kent, D.H. & Allen, D.E. (1984). *British and Irish Herbaria*. BSBI, London.

+Leighton, W.A. (1871). *The Lichen-flora of Great Britain, Ireland and the Channel Islands*. Shrewsbury: privately printed.

Ley, A. (1874). Tabular catalogue of common plants for Breconshire, Radnorshire, Selkirkshire and West Lancashire. *Report of the Botanical Record Club*, **1874 (1875)**, 80-86.

Ley, A. (1886). Radnorshire. *Report of the Botanical Record Club*, **1886**, 144-146.

Ley, A. (1881). Radnorshire. *Report of the Botanical Record Club*, **1881-1882(1883)**, 246-247.

Ley, A. (1891). First contribution towards a Flora of Aberedw, Radnor. *Transactions of the Woolhope Naturalists' Field Club*, **1891**, 180-198.

Ley, A. (1894). Notes on a Flora of Aberedw. *Transactions of the Woolhope Naturalists' Field Club*, **1894**, 184-198.

Linnard, W. (1982). *Welsh Woods and Forests: History and Utilization*. National Museum of Wales, Cardiff.

*Longton, R.E. (1992). Reproduction and rarity in British mosses. *Biological Conservation*, **59**, 89-98.

(Lowther, R.A. (1987). *Macrophyte Assemblages in Welsh Lakes*. Unpub. report of the University Of Wales Institute of Science and Technology, Llysdinam, Newbridge on Wye.)

*Macvicar, S.M. (ed.) (1905). *Census Catalogue of British Hepatics*, 1st edn. Moss Exchange Club.

Bibliography

McAllister, H.A. & Rutherford, A. (1990). *Hedera helix* L. and *H. hibernica* (Kirchner) Bean (Araliacaae) in the British Isles. *Watsonia*, **18**, 7-15.

Matthews, J.R. (1937). Geographical relationships of the British flora. *Journal of Ecology*, **25**, 1-90.

Matthews, J.R. (1955). *Origin and Distribution of the British Flora*. Hutchinson, London.

Merry, D.G. & Slater, F.M. (1978). Plant colonisation under abnormally dry conditions of some reservoir margins in mid-Wales. *Aquatic Botany*, **5**, 149-162.

Merry, D.G., Slater, F.M. & Randerson, P.F. (1981). The riparian and aquatic vegetation of the River Wye. *Journal of Biogeography*, **8**, 313-327.

Moore, H.C. (1898). A visit to the work of the proposed Birmingham water supply from Elan Valley in Wales. *Transactions of the Woolhope Naturalists' Field Club*, **1895-1897**, 178-179.

Moore, J.A. (1986). *Charophytes of Great Britain and Ireland*. Botanical Society of the British Isles, London.

Moore, P.D. (1970). Studies in the vegetational history of mid- Wales II: the late glacial period in Cardiganshire. *New Phytologist*, **69**, 363-375.

Moore, P.D. (1973). The influence of prehistoric cultures upon the initiation and spread of blanket bog in upland Wales. *Nature* (Lond.), **241**, 350-353.

Moore, P.D. (1978). Studies in the vegetational history of mid- Wales.V. Stratigraphy and pollen analysis of Llyn Mire in the Wye Valley. *New Phytologist*, **80**, 281-301

Moore, P.D. and Beckett, P.J. (1971). Vegetation and development of Llyn, a Welsh mire. *Nature* (Lond.), **231**, 363.

Moore, P.D. & Chater, E.H. (1969). The changing vegetation of west-central Wales in the light of human history. *Journal of Ecology*, **57**, 361-379.

Mordue, J.E.M. & Ainsworth, G.C. (1984). *Ustilaginales of the British Isles*. Mycological Paper 154. Commonwealth Mycological Institute, Kew.

Nature Conservancy Council, (1989). *Guidelines for Selection of Biological SSSI's*. Nature Conservancy Council, Peterborough.

Newton, A. (1972). A Welsh bramble foray. *Watsonia*, **9**, 117-130.

*****Painter, W.H. (1908).** Mosses near Llandrindod Wells. *Science Gossip* (New Series), **8**, 7-8.

*****Paton, J.A. (1965).** *Census Catalogue of British Hepatics*, 4th edn. British Bryological Society.

*****Paton, J.A. (1990).** *Riccia subbifurca* Warnst. ex Crozals in the British Isles. *Journal of Bryology*, **16**, 5-8.

*****Pearson, W.H. (1919).** Notes on Radnorshire Hepatics. *Journal of Botany*, **57**, 193-195.

Penford, N., Francis, I.S., Hughes, E.J. & Aitchison, J.W. (1990). *Biological Survey of Common Land, No. 10: Radnor District, Powys*. Nature Conservancy Council, Peterborough.

Perring, F.H. (1957). Report of field meeting, Llandrindod Wells August 4th to 11th, 1956. *Proceedings of the Botanical Society of the British Isles*, **2**, 416-418.

Perring, F.H. & Farrell, L. (1983). *British Red Data Books: 1. Vascular Plants*, (2nd Ed.). Royal Society for Nature Conservation, Lincoln.

Perring, F.H. & Walters, S.M. (Eds.) (1983). *Atlas of the British Flora* (3rd edn.). EP Publishing Ltd. and BSBI, Wakefield. (1st. edn. 1962; 2nd. edn. 1976.)

Perring, F.H. & Sell, P.D. (Eds.) (1968). *Critical Supplement to the Atlas of the British Flora*. BSBI, London.

Plomer, W. (Ed.)(1938). *Kilvert's Diary*. Jonathon Cape, London.

Pugsley, H.W. (1948). A prodromus of the British Hieracia. *Journal of the Linnaean Society of London* (Botany), **54**.

Purvis, O.W., Coppins, B.J., Hawksworth, D.L., James, P.W., Moore, D.M. (Eds.) (1992). *The Lichen Flora of Great Britain and Ireland*. British Lichen Society, London.

*****Ratcliffe, D.A. (1968).** An ecological account of Atlantic bryophytes in the British Isles. *New Phytologist* **67**, 365-439.

Ratcliffe, D.A. (1984). Post-medieval and recent changes in British vegetation: the culmination of human influence. *New Phytologist*, **98**, 73-100.

Redford, K.L. (1940). In Part 35, Radnor, pp 269-289, in *The Land of Britain*, L. Dudley Stamp (Ed.), Geographical Publications Ltd., London.

Bibliography

Riddelsdell, H.J. (1910). The botany of the country round Builth Wells. *Proceedings of the Cotteswold Naturalists' Field Club*, **XVII**, 57-61.

Ridley, H.N. (1881). Notes on Radnorshire plants. *Journal of Botany*, **19**, 170-174.

Ridley, H.N. (1884). Short notes, additions to Topographical Botany, *Journal of Botany*, 1884, 377-378.

Rix, E.M. & Woods, R.G. (1981). *Gagea bohemica* (Zauschner) J.A. & J.H.Schultes in the British Isles and a general review of the *G. bohemica* complex. *Watsonia* **13**, 265-270.

Rodwell, J.S. (Ed.) (1991a). *British Plant Communities. Vol. 1, Woodlands and Scrub.* Cambridge University Press, Cambridge.

Rodwell, J.S. (Ed.) (1991b). *British Plant Communities. Vol. 2, Mires and Heaths.* Cambridge University Press, Cambridge.

Rogers, W.M. (1899). Radnorshire and Breconshire Plants. *Journal of Botany*, **37**, 17-25.

+Salway, T. (1846). A list of the scarcer amongst the lichens which are found in the neighbourhood of Oswestry and Ludlow, with occasional observations upon some of them. *Transactions of the Botanical Society of Edinburgh*, **2**, 203-213.

Sandwith, N.Y. (1947). *Hypericum linariifolium* Vahl in Radnorshire. *Transactions of the Radnorshire Society*, **XVII**, 13.

+Seaward, M.R.D. and Hitch, C.J.B., (1982). *Atlas of the Lichens of the British Isles*, Vol.1. Institute of Terrestrial Ecology, Cambridge.

Simpson, N.D. (1960). *A Bibliographical Index of the British Flora.* Privately published, Bournemouth. (Radnor section pp.314- 315).

Sinclair, J.B. and Fenn, R.W.D. (1991). *The Facility of Locomotion.* Mid-Border Books, Kington.

Sinker, C.A., Packham, J.R., Trueman, I.C., Oswald, P.H., Perring, F.H. and Prestwood, W.V. (1985). *Ecological Flora of Shropshire Region.* Shropshire Trust for Nature Conservation, Shrewsbury.

Slater, F.M. (1980). The rarer plants of Radnorshire. *The Transactions of the Radnorshire Society*, **L**, 73-77.

Slater, F.M. (1987). Llysdinam Field Centre, Newbridge-on-Wye - UWIST in Powys. *The Transactions of the Radnorshire Society*, **XVII**, 92-100.

Slater, F.M. (1990). Biological Flora of the British Isles. No. 168. *Gagea bohemica* (Zauschner) J.A. & I.H. Schultes. *Journal of Ecology*, **78**, 535-546.

Slater, F.M., Helmsley, A. & Wilkinson, D.M. (1991). A new sub-association of the *Piluarietum globuliferae* Tuxen 1955 in upland pools in the mid-Wye catchment of central Wales. *Vegetatio*, **96**, 127-136.

(Slater, F.M. & Moncur, D.G. (Eds.)(1985). *The Flowering Plants of Radnorshire.* Llysdinam Field Centre, Department of Applied Biology, University of Wales Institute of Science and Technology, Newbridge on Wye.)

*Smith, A.J.E. (Ed.)(1978a). *Provisional Atlas of the Bryophytes of the British Isles.* Biological Records Centre, Monks Wood Experimental Station, Abbots Ripton, Huntingdon PE17 2LS.

*Smith, A.J.E. (1978b). *The Moss Flora of Britain and Ireland.* Cambridge University Press.

*Smith, A.J.E. (1990). *The Liverworts of Britain and Ireland.* Cambridge University Press.

Snogerup, B. (1982). *Odontites litoralis* ssp. *litoralis* in the British Isles. *Watsonia* **14**, 35-39.

Stace, C. (1991). *New Flora of the British Isles.* Cambridge University Press.

(Thurley, S.S. (1977). *The vegetation of Llanbwchllyn Lake.* MSc thesis. University of Wales.)

Tooke, M.A. (1879). May flowers in Wales and Shropshire. *Science Gossip*, **XV**, 110.

Turner, D. and Dillwyn, L.W. (1805). *The Botanists Guide through England and Wales.* London.

(Turner, J.C.E., Blackstock, T.H., Jackson, P.K. & Stevens, D.P. (1990). *A Vegetation Survey and Conservation Assessment of some Lowland Grasslands in Radnor.* 2 vols. Nature Conservancy Council, Bangor.)

Wade, A.E. and Webb, J.A. (1945). Radnorshire Plant Records. *North West Naturalist*, **20**, 156-160.

Wade, A.E. and Webb, J.A. (1947). Radnorshire plant records. *Transactions of the Radnorshire Society*, **XVII**, 3-12.

(Wade, A.E. & Webb, J.A. (c.1955). *Materials towards a flora of Radnorshire.* Unpublished manuscript in the library of the Botany Department, National Museum of Wales, Cardiff.)

*Warburg, E.F. (1963). *Census Catalogue of British Mosses*, 3rd edn. British Bryological Society.

Watson, H.C. (1883). *Topographical Botany.* 2nd edn. London.

+Watson, W. (1953). *Census Catalogue of British Lichens.* Cambridge University Press.

Westcombe, T. (1844). List of plants observed in Brecknock and Radnor. The Phytologist. **1**, 781.

Wilkinson, D.M. (1988). A preliminary study of the development of Beilibedws Mire in mid Wales. *Quaternary Newsletter*, **55**, 12-22.

Wilkinson, D.M. (1989). The Flandrian history of Llanbwchllyn Lake in mid Wales. *Quaternary Newsletter*, **57**, 6-10.

+Wilkinson, W.H. (1904). Radnorshire Lichens. *Journal of Botany*, London, **42**, 111-113.

Williams, Rev. J. (1905). *A General History of the County of Radnor.* Davis & Co., Brecon.

*Wilson, A. (1930). *A Census Catalogue of British Hepatics*, 3rd edn. British Bryological Society.

Wiltshire, P.E.J. & Moore, P.D. (1983). Palaeovegetation and Palaeohydrology in Upland Britain, in *Background to Palaeohydrology - A Perspective.* Gregory, K.J. (ed.), 433-451. John Wiley & Sons, Chichester.

Woods, R.G. (1987). The ferns and fern allies of Radnor. *The Transactions of the Radnorshire Society*, **LVII**, 10-21.

Woods, R.G. (1989). The native and naturalized trees and shrubs of Radnorshire. *The Transactions of the Radnorshire Society*, **LIX**, 7-28.

Synonymy of Scientific Vascular Plant Names

A new standard British vascular plant flora (Stace, 1991) was published after the majority of the account for this Radnorshire Flora was produced. A number of scientific plant names have been employed by Stace which differ from those adopted here. Since Stace is likely to become the standard work on the British Flora for some time to come those names used in the "Radnorshire Flora" are cross-referenced by numbers to Stace in the list below. Names employed by Stace are placed in italics.

1. *Anisantha sterilis* = 15
2. *Aphanes inexspectata* = 3
3. Aphanes microcarpa = 2
4. *Asplenium trichomanes - ramosum* = 5
5. Asplenium viride = 4
6. Avenula pubescens = 70
7. Betula pubescens ssp. carpatica = 8
8. *Betula pubescens* ssp. *tortuosa* = 7
9. *Bromopsis benekenii* = 12
10. *Bromopsis erecta* = 13
11. *Bromopsis ramosa* = 14
12. Bromus benekenii = 9
13. Bromus erectus = 10
14. Bromus ramosus = 11
15. Bromus sterilis = 1
16. Bryonia cretica ssp. dioica = 17
17. *Bryonia dioica* = 16
18. Campanula latifolia = 19
19. *Campanula latifolium* = 18
20. Carex demissa = 23
21. Carex lepidocarpa = 22
22. *Carex viridula* ssp. *brachyrrhyncha* = 21
23. *Carex viridula* ssp. *oedocarpa* = 20
24. *Catapodium rigidum* = 49
25. Cerastium fontanum ssp. glabrescens = 26
26. *Cerastium fontanum* ssp. *holosteoides* = 25
27. *Ceratocapnos claviculata* = 35
28. Chaerophyllum temulentum = 29
29. *Chaerophyllum temulum* = 28
30. *Chamerion angustifolium* = 55
31. Chieranthus cheiri = 56
32. Cirsium helenioides = 33
33. *Cirsium heterophyllum* = 32
34. *Claytonia sibirica* = 92
35. Corydalis claviculata = 27
36. Corydalis lutea = 124
37. *Cotoneaster integrifolius* = 38
38. Cotoneaster microphyllus = 37
39. Crepis vesicaria ssp. haenseleri = 40
40. *Crepis vesicaria* ssp.*taraxacifolia* = 39
41. Dactylorchis fuchsii = 43
42. Dactylorchis maculata = 44
43. *Dactylorhiza fuchsii* = 41
44. *Dactylorhiza maculata* = 42
45. Dactylorhiza majalis ssp. praetermissa = 47
46. Dactylorhiza majalis ssp. purpurella = 48
47. *Dactylorhiza praetermissa* = 45
48. *Dactylorhiza purpurella* = 46
49. Dezmazeria rigida = 24
50. Doronicum plantagineum var. excelsum = 51
51. *Doronicum x excelsum* = 50
52. *Eleogiton fluitans* = 142
53. Elymus repens = 54
54. *Elytrigia repens* = 53
55. Epilobium angustifolium = 30
56. *Erysimum cheiri* = 31
57. *Fallopia japonica* = 125
58. *Fallopia sachalinensis* = 126
59. Filaginella uliginosa = 67
60. *Filago minima* = 79
61. Fragaria moschata = 62
62. *Fragaria muricata* = 60

Synonymy

63. Galeobdolon argentatum = 74
64. *Glyceria notata* = 65
65. Glyceria plicata = 64
66. *Gnaphalium sylvaticum* = 101
67. *Gnaphalium uliginosum* = 59
68. *Hedera helix* ssp. *hibernica* = 69
69. Hedera hibernica = 68
70. *Helictotrichon pubescens* = 6
71. Inula conyza = 72
72. *Inula conyzae* = 71
73. *Isolepis setacea* = 143
74. *Lamiastrum galeobdolon* ssp. *argentatum* = 63
75. *Lathyrus linifolius* var. *montanus* = 76
76. Lathyrus montanus = 75
77. *Leontodon saxatilis* = 78
78. Leontodon taraxacoides = 77
79. Logfia minima = 60
80. *Lotus pedunculatus* = 81
81. Lotus uliginosus = 80
82. *Matricaria discoidea* = 83
83. Matricaria maticarioides = 82
84. Melilotus alba = 85
85. *Melilotus albus* = 84
86. Mentha x gentilis = 87
87. *Mentha* x *gracilis* = 86
88. Mentha x piperata = 89
89. *Mentha* x *piperita* = 88
90. Montia fontana ssp. chondrosperma = 91
91. *Montia fontana* ssp. *minor* = 90
92. Montia sibirica = 34
93. Nasturtium microphyllum = 129
94. Nasturtium officinale = 130
95. *Nothofagus nervosa* = 96
96. Nothofagus procera = 95
97. Odontites verna = 98
98. *Odontites vernus* = 97
99. Oenothera erythrosepala = 100
100. *Oenothera glazioviana* = 99
101. Omalotheca sylvatica = 66
102. Oxalis europaea = 103
103. Oxalis stricta = 102
104. *Papaver dubium* ssp. *lecoqii* = 105
105. Papaver lecoqii = 104
106. *Persicaria amphibia* = 116
107. *Persicaria bistorta* = 117
108. *Persicaria hydropiper* = 118
109. *Persicaria lapathifolia* = 119
110. *Persicaria maculosa* = 120
111. *Persicaria wallichii* = 121
112. *Phleum bertolonii* = 113
113. Phleum pratense ssp. bertolonii = 112
114. *Poa humilis* = 115
115. Poa subcaerulea = 114
116. Polygonum amphibium = 106
117. Polygonum bistorta = 107
118. Polygonum hydropiper = 108
119. Polygonum lapathifolium = 109
120. Polygonum persicaria = 110
121. Polygonum polystachyum = 111
122. *Populus candicans* = 123
123. Populus gileadensis = 122
124. *Pseudofumaria lutea* = 36
125. Reynoutria japonica = 57
126. Reynoutria sachalinensis = 58
127. *Rhamnus cathartica* = 128
128. Rhamnus catharticus = 127
129. *Rorippa microphyllum* = 93
130. *Rorippa nasturtium - aquaticum* = 94
131. Rosa afzeliana = 132
132. *Rosa caesia* ssp *glauca* = 131
133. *Rosa caesia* ssp. *caesia* = 134
134. Rosa coriifolia = 133
135. *Rosa mollis* = 136
136. Rosa villosa = 135
137. *Rumex acetosella* ssp. *pyrenaicus* = 139
138. *Rumex acetosella* var. *tenuifolius* = 140
139. Rumex angiocarpus = 137
140. Rumex tenuifolius = 138
141. Scirpus cespitosus ssp. germanicus = 158
142. Scirpus fluitans = 52
143. Scirpus setaceus = 73
144. Sedum reflexum = 145
145. *Sedum rupestre* = 144
146. Senecio vulgaris ssp. vulgaris = 147
147. *Senecio vulgaris* var. *vulgaris* = 146
148. *Silene latifolia* = 149
149. Silene pratensis ssp. pratensis = 148
150. *Stellaria nemorum* ssp. *montana* = 151
151. Stellaria nemorum ssp. glochidiosperma = 150
152. *Thelypteris palustris* = 153
153. Thelypteris thelypteroides = 152
154. Thlaspi alpestre = 155
155. *Thlaspi caerulescens* = 154
156. *Thymus polytrichus* ssp. *arcticus* = 157
157. Thymus praecox = 156
158. *Trichophorum cespitosum* = 141
159. *Triglochin palustre* = 160
160. Triglochin palustris = 159

Gazetteer of place names referred to in the introductory chapters

For each place a 1km grid square reference of the National Grid is provided. For sites which extend beyond a square kilometre the grid square given is in approximately the centre of the site or is the highest point of a hill. For streams or rivers, if a discrete part is referred to in the text, the grid reference given is to that part. If the entire river is referred to, the grid reference is to that part of the stream bearing its name on the 1:50000 OS map.

Aberystwyth	SN 5881
Abbeycwmhir	SO 0571
Aberedw	SO 0847
Aberedw Hill	SO 0850
Aberithon Turbary	SO 0157
Alpine Bridge	SO 0963
Bach Howey	SO 1143
Baxter's Bank	SO 0567
Beacon Hill	SO 1776
Begwns	SO 1544
Beilibedw	SO 1656
Berwyns	SJ 0734
Black Brook	SO 0167
Black Mixen	SO 1964
Blaen Restr	SN 8469
Boughrood	SO 1239
Builth Road	SO 1253
Builth Wells	SO 0451
Burfa	SO 2761
Burfa Boglands	SO 2761
Caban Coch Resr.	SN 9163
Cadair Idris	SH 7113
Caer Einion	SO 0653
Caety Traylow Hill	SO 1956
Carmel	SO 0566
Carneddau Farm	SO 0553
Carneddau Hills	SO 0654
Carreg Bica	SN 9265
Castelltinboeth	SO 0975
Castle Bank	SN 0855
Cefn Cennarth	SN 9676
Cefnllys Castle	SO 0960
Cefn Nantmel	SO 0366
Cerrig Gwalch	SN 9370
Cilkenny Dingle	SO 1741
Claerwen Resr.	SN 8664
Clyro	SO 2143
Clyro Hill	SO 1946
Coed yr Allt Goch	SN 9067
Colwyn Brook	SO 0755
Cors y Llyn	SO 0155
Coxhead Common	SO 1571
Craig Cwm Clyd	SN 8962
Craig Goch Resr.	SN 8969
Craig y bwch	SN 8962
Crickhowell	SO 2118
Cwm Coel	SN 8963
Cwm Elan Mine	SN 9065
Cwm Gilla	SO 2671
Cwm Gwynllyn	SN 9469
Cwmithig	SN 9970
Cwm Ystwyth	SN 8074
Davy Morgan's Dingle	SO 1862
Dolau	SO 1467
Doldowlod	SN 9962
Dol y Fan Hill	SO 0161
Dolyhir	SO 2475
Drysgol	SN 9474
Dyffryn Wood	SN 9766
Erwood Station	SO 0843
Fedw Turbary	SO 0159
Ffordd-fawr Mire	SO 1840
Ffos Trosol	SN 8271
Franksbridge	SO 1155
Fron Barn Dingle	SO 1976
Garreg Ddu Resr.	SN 9164
Gelli-cadwgan	SO 0511
Gilfach Farm	SN 9671
Glanalder	SO 0369
Glannau	SN 9065
Glasbury	SO 1739
Glascwm Hill	SO 1552
Glog fawr	SN 9266
Gogerddan	SN 6383
Gogia	SO 1743
Gorsgoch	SN 8963
Great Creigiau	SO 1963
Gumma	SO 2864
Gwastedyn Hill	SO 9863
Gwaunceste Hill	SO 1555
Gwernyfed	SO 1737
Gwernaffel Dingle	SO 2770
Gwyn Llyn	SN 9469
Gwystre	SO 0665
Hanter Hill	SO 2557
Harley Dingle	SO 1963
Hay-on-Wye	SO 2242
Henllyn Mawr	SO 0845
Hindwell Pool	SO 2560
Hoel y Gaer	SO 1939

Gazetteer

Howey	SO 0558	Newbridge on Wye	SO 0158
Impton	SO 2966	Newchurch	SO 2150
Kington	SO 2956	Newmead Farm	SO 0554
Knighton	SO 2871	New Radnor	SO 2160
Knucklas	SO 2574	Norton	SO 3067
Lakeside Wood	SO 0660	Old Radnor	SO 2459
Lea Hall	SO 2071	Painscastle	SO 1646
Little Creigiau	SO 1963	Pantllwyd	SO 8370
Little Lodge	SO 1838	Pant y Dwr	SN 9874
Llaithdu	SO 0780	Pencerrig Lake	SO 0454
Llanbadarn Fynydd	SO 1077	Pentrosfa Bog	SO 0559
Llanbedr Hill	SO 1348	Pen y Berth	SO 1046
Llanbister	SO 1173	Penybont Common	SO 1264
Llanbister Common	SO 1173	Pen y Garreg	SN 9067
Llanbwchllyn	SO 1146	Plynlimon	SN 7886
Llanddewi Ystredenni	SO 1068	Presteigne	SO 3164
Llandegley	SO 1362	Radnor Forest	SO 1863
Llandegley Rhos	SO 1360	Rhayader	SN 9768
Llandeilo Graban	SO 0944	Rhayader Quarry	SN 9765
Llandeilo Hill	SO 0946	Rhosgoch	SO 1948
Llandrindod Wells	SO 0661	Stanage	SO 3371
Llanelwedd	SO 0451	St Harmon	SN 9872
Llanelwedd Quarries	SO 0552	Stanner	SO 2658
Llanfair Quarries	SO 0661	Sugar Loaf	SN 8342
Llangunllo	SO 2071	Summergil Brook	SO 2360
Llanstephan	SO 1242	The Banks	SO 0655
Llanwrytd Wells	SN 8746	The Bog	SO 0157
Llanyre	SO 0462	The Bower	SO 0656
Llowes	SO 1941	Three Wells	SO 0658
Llyn	SO 0155	Timberhill	SO 2863
Llyn Brianne	SN 7948	Tir-uched	SO 1839
Llyncerrigllwydion Isaf	SN 8470	Trewern Hill	SO 1243
Llyncerrigllwydion Uchaf	SN 8369	Trumau	SN 8667
Llyn Gwyn	SO 1064	Tyncoed Wood	SO 0354
Llyn Heilyn	SO 1658	Tyn-y-Berth	SO 0773
Llysdinam	SO 0058	Upper Weston	SO 2070
Lower Woodhouse	SO 3072	Velindre	SO 1681
Marcheini Uplands	SN 9475	Walton	SO 2559
Marteg Bridge	SN 9571	Water-break-its-neck	SO 1860
Michaelchurch on Arrow	SO 2450	Wenallt	SN 9678
Moity Dingle	SO 1842	Whinyard Rocks	SO 2063
Monaughty	SO 2368	Whitton	SO 2767
Nannerth	SN 9471	Worsell Wood	SO 2557
Nantmel	SO 0366	Wyloer	SN 9572
Nash Quarry	SO 3062	Ysfa	SN 9964

Welsh Place Names in Radnorshire with a Botanical Connotation

PART 1

The following is a list of the more commonly occuring elements in Welsh place names in Radnorshire which may be of interest to botanists. Some elements often appear in slightly different forms. They are all included in the following compilation:

A. Words associated with Woodland

Allt - Wood, often on sloping ground
Berllan - Orchard
Bongam - Twisted trunk
Cellbren - Hollow wood
Celyn/Gelynen/Celynen/Celynnen - Holly
Cerde - Rowan (from 'cerdin')
Coed/goed - Wood
Coety - Woodland dwelling
Colfa/Golfa - place of many trees or boughs
Craifol - Rowan (from 'criafol')
Derw/Dderw/ - Oak
Draenog - Thorny
Drain/Draen - Thorns/briar or bramble
Eirin - Berries or fruits
Fedw/Fedwen/Bedwin/Bedw/Vedw/ - Birch
Fforest - Forest
Gelli/Gelly - grove or copse
Graban - Crabapple trees (from 'crabyn')
Helyg - Willow
Llanerch - Clearing, glade
Llwyn - Bush, grove
Onnen - Ash
Perth/Berth/Berthi/Perthi/Perthy - Bushes
Prysg - Scrub/shrub
Wern - Alder
Ysgawen - Elder

B. Words associated with pasture/agricultural land

Borfa - Grassy area
Cae - Field
Ddol/dolau/dol - Meadow
Ffridd - Hill pasture at edge of enclosed land
Garth - Enclosure, hill
Launt/lawnt - Grassy piece of land near dwelling
Hysfa - piece of rough unimproved land near farmhouse
Maes - Field
Tir - Land

C. Words associated with Wetland

Glas/glais - brook, stream (eg Dulas)
Gors - Bog/marsh
Gwern - place where alders grow
Figyn - Sphagnum bog, quagmire
Nant - Stream
Rhyd - Brook
Rhos - Wet pasture or heathland
Waun/waen/wain - Heathland, moorland

D. Words associated with topographical features or 'location' words

Ar - At edge of or on
Banc - Bank/slope
Blaen - Headwaters or source, upland
Bwlch - Pass
Bryn/Fryn - Hill
Cefn - Ridge
Cil/Gil - At edge of corner of, or in shadow of
Clawdd - Dyke, hedge
Cwm - Valley
Duffnant - Ravine (from 'dyfnant')
Fron/Bron/Vron - Bank
Fynydd/Mynydd - Hill/mountain
Foel - Bare hill
Gwar - Nape
Gwaelod - Bottom
Pant - Depression in the land
Pen - At the end or on the top of
Pentre - Village or homestead
Safn - Opening
Tre/Dre - Homestead
Ty - House (Ty'n y .... - house in the)

E. Adjectives

i)Colours
Coch/Goch/Cochion - Red
Du/Ddu/Duon - Black (used often to mean dark)
Glas/glasson - Blue ("glas" can be used to signify bright green)
Gwyn/wen/gwynion - White
Llwyd/Lwyd - Grey
Melyn - yellow

ii)Others
Bir - Stunted ('byr'), short
Caregle - Stony place
Crin - Withered (often used to describe bracken covered land)
Cwtta - Stunted, short
Dywell/Dywyll/Tywyll - Dark
Faenog - Stony
Fach - Small
Gau - hollow
Hen - Old
Hesglog - sedgy
Hir - Long
Lydan ('llydan') - Wide/broad
Mawr/Fawr - Large
Tew - Thick

F. Other Words

Banal - Broom
Brwyn - Rushes
Cerrig - Stones
Clogau - Crags
Crug/crugyn/cnwch/cnwc - Hillock
Eithen - Gorse
Eithinog - Gorsy
Garn/Carn - Cairn
Graig/Maen/Carreg - Stone
Grug - Heather
Haidd - Barley
Nyth - Nest

Welsh Place Names

Part 2

Radnorshire Place Names

Allt Dderw - Oak wood
Allt Goch - Red wood
Allt Llwyd - Grey wood
Argoed - Edge of wood
Beili Bedw - Birch bailey or enclosure
Banc Du - Dark or black slope
Bancgelli las - Slope of the green grove
Banc gwyn - Light or white bank
Beili Bedw Mawn Pool - Peat pool of the birch enclosure
Berth - Bush
Blaen Bedw - Source of the river of birches
Blaen cerde (Cerdin - cerddinen) - Source of the river of rowans
Bongam bank - Bank of the twisted trunks
Borfa /porfa - Grassy area
Bryn coch - Red hill
Bryn draenog - Thorny hill
Bryn du - Dark or black hill
Bryn Eithinog - Gorsy hill
Bryn Garw - Rough hill
Bryn Glas - Blue hill
Brynmelyn - Yellow hill
Brynwern Bridge - Bridge of the alder hill
Bryn y Garth - Hill of the enclosure
Bryn y Maen - Rock hill
Bwlchyfedwen - Pass of the birch trees
Caebanal - Broom field
Castell foel allt - Castle of the bare hill
Cefn coch - Red ridge
Cefn crin - Rusty or withered ridge
Cefn y Grug - Heather ridge
Cerrig - cochian - Red rocks
Cerrig - gwynian - White rocks
Cil rhos - At the edge of the moor
Cil y Berllan - At edge of orchard
Cnwclas - Green hillock
Coed glasson - Green trees
Coed Mawr - Large woodland
Coed tew isaf - Thick wood (lower)
Coety Bank - Bank of woodland dwelling
Colfa/Golfa - Colfen - Boughs or trees
Craig gelli dywyll - Rock of the dark grove
Craig yr Allt goch - Rock of red hill/wood
Crinfynydd - Withered hill (may probably also be used, to describe a bracken covered hill)
Cringoed Bank - Bank of withered trees
Crofty - perthy - possibly croft of small hedged fields
Crugyn llwyd - Grey hillock
Cwm bedw - Birch valley
Cwm derw - Oak valley
Cwmnantygelli - Valley with grove or woodland stream
Cwmysgawen Common - Elder valley common
Cwrt y graban - 'graban' could be crab - apple trees ('crabyn' or 'craben') ie. court of the crab apple trees
Dolsgallog - Thistle meadow
Drainllwynbir - Place of the small thorn bushes
Drewern - Homestead of alder trees
Esgairdraenllwyn - Ridge of thorny bushes
Fedw - Birch
Fedwllwyd - Grey Birch
Ffridd - Hill pasture between enclosed land and open mountain
Fforest Colwyn/Fforest Fach - Colwyn's Forest/Small forest
Frongoch - Red bank

Gelli dywell - Dark grove
Gelly garn fawr - Wood or grove of the large boulder
Gilwern - Place at the edge of alder trees
Glaslyn - Blue lake
Glasnant - Blue stream
Gorslydan - Broad or wide bog
Graig dolfaenog - Rock of the stony meadow
Graig safn y coed - Rock at woods' opening
Graig yr onnen - Ash rock
Gwaelod y rhos - Bottom of the 'rhos' or moor
Gwar-allt - Wood on the 'nape' of the hill
Gwar-dolau - Meadows on the 'nape' of the land
Gwern-alltcwm - Valley of alder woodland
Gwernduffnant (from Gwerndyfnant) - Ravine of alders
Gwerneirin - Boggy land of alders and berries (Gwern = Alder; eirin = Berries/fruits)
Gwernfach - Place of small alders
Hengoed - Old woodland
Hirllwyn - Long bushes (Possibly extensive bush/scrub cover)
Knucklas - Cnwclas - Blue or green hillock
The Launt - 'lawnt' - Grassy piece of land near house
Llwynbrain - Grove of the crows
Llwyncellbren - Grove of hollow wood
Llwyncwtta - Stunted bush or short wood
Llwynpenderi - Bush at edge of oak
Llyn y Figyn - Lake of the bog moss (mign - figyn Sphagnum bogs or mires, as in 'Migneint')
Llyn y Waun - Lake of the heathland/moorland
Maesyronen - Ash field
Mawn Pool - Peat pool
Mynydd y Perthi- Bushy mountain
Nant Rhyd y Fedw - Stream of the Birch brook
Nant yr Haidd - Stream of the Barley
Nyth grug - possibly heathery nesting place
Pant y Brwyn - Rushy depression in the land
Pant y Caregle - Stony depression in the land
Pen-rhos - End of 'rhos' or moor
Penrhosgoch - Land at end of red coloured 'rhos' or moor
Pentre draen - Thorny 'village' or dwelling place
Pentre wern - Dwelling place amongst alders
Prysgduon - Black or dark scrub
Rhos - crug - Rocky moor
Rhosgoch - Red moor
Rhoshenfryn - Moor on the old hill
Rhosygelynen - Holly moor
Tir celyn - Holly land
Trallwng byr - short piece of wetland in valley bottom
Treallt - Dwelling place on the hillside/in the wood
Trecoed - Wooded homestead/dwelling place
Trehesglog - Sedgy dwelling place
Ty'n y berth - House in bushes
Tynyddole - House in meadows
Ty'n y waun - House in heathland/moorland
Vedwllwyd - Grey birch
Vrongaullwyd - Grey hollow bank
Wainddu - Black or dark moor
Waun wen - Light or white moor
Wenallt - Light woodland
Wern - Alder

Compiled By Elinor Gwynn

INDEX

This index includes all scientific and plant names. Where common names consist of more than one element, they are indexed by their final element followed by their first element e.g. Water Avens is indexed as Avens, Water. Only place names refered to in the introductory chapters are included below.

Abbeycwmhir, 1, 8, 166
Aber-Ceuthon, 3
Aberedw, 2, 12, 26, 27, 33, 67, 166-168, 173, 218
Aberedw Rocks, 167
Aberedw Wood, 218
Aberithon Turbary, 15
Abies alba, 78
Abutilon theophrasti, 111
Acarospora fuscata, 221
   sinopica, 221
   smaragdula, 221
Acaulon muticum, 193
Accumulated temperatures, 18
Acer campestre, 109
   platanoides, 109
   pseudoplatanus, 109
ACERACEAE, 109
Achillea millefolium, 136
   ptarmica, 136
Achnanthes, 164
Acidic pollutants, 24
Aconite, Winter, 87
Aconitum napellus, 87
Acorus calamus, 155
Acrocordia conoidea, 221
   gemmata, 221
Adder's tongue, 74
Adoxa moschatellina, 133
ADOXACEAE, 133
Aegopodium podagraria, 117
Aesculus hippocastanum, 110
Aethusa cynapium, 117
Agonimia tristicula, 221
Agrimonia eupatoria, 100
   procera, 100
Agrimony, 100
Agrimony, Fragrant, 100
   , Hemp, 135
Agrostemma githago, 86
Agrostis canina, 154
   capillaris, 154
   gigantea, 154
   stolonifera, 155
   vinealis, 154
Aira caryophyllea, 154
   praecox, 154
Ajuga reptans, 125
Alchemilla filicaulis ssp. vestita, 101
   glabra, 102
   mollis, 102
   vulgaris, 101
   xanthochlora, 101
Alder, 80
   , Grey, 80
Alisma plantago-aquatica, 145
ALISMATACEAE, 145
Alison, Golden, 93
   , Sweet, 93
Alkanet, Green, 123
Alliaria petiolata, 91
Allium schoenoprasum, 147
   ursinum, 147
Allt y Fran Wood, 4
Alnus glutinosa, 80
   incana, 80
Aloinia aloides, 193
Alopecurus geniculatus, 155
   pratensis, 155
Alpine Bridge, 10, 28, 166
Alyssum saxatile, 93
AMARANTHACEAE, 84
Amaranthus retroflexus, 84
AMARYLLIDACEAE, 148
Amblystegium fluviatile, 207

Amblystegium riparium, 207
   serpens, 207
   tenax, 207
Ammonium ions, 218
Amphidium lapponicum, 203
   mougeotii, 203
Amygdalaria pelobotryon, 221
Anabaena flos-aquae, 164
Anagallis arvensis, 121
   tenella, 120
Anaphalis margaritacea, 136
Anaptychia ciliaris, 221
   runcinata, 221
Andreaea rothii, 186
   rupestris, 186
   spp., 186
Andromeda polifolia, 119
Anemone, Wood, 87
Anemone nemorosa, 87
Aneura pinguis, 176
Angelica, Wild, 118
Angelica sylvestris, 118
ANGIOSPERMAE, 79
Anisomeridium biforme, 221
   juistense, 221
Ankistrodesmus, 164
   falcatus, 163
Anoectangium aestivum, 195
Anomobryum filiforme, 199
Anomodon viticulosus, 206
Antennaria dioica, 136
Anthoceros punctatus, 175
ANTHOCEROTAE, 175
Anthoxanthum odoratum, 154
Anthriscus cerefolium, 116
   sylvestris, 116
Anthyllis vulneraria, 106
Antirrhinum majus, 128
Aphanes arvensis, 102
   microcarpa, 102
Apium inundatum, 118
   nodiflorum, 117
APOCYNACEAE, 121
Apple, Crab, 102
AQUIFOLIACEAE, 110
Aquilegia vulgaris, 89
Arabis, Garden, 92
Arabis caucasica, 92
   hirsuta, 92
   scabra, 92
Arable fields, 40
ARACEAE, 155
ARALIACEAE, 116
Archangel, Variegated Yellow, 126
   , Yellow, 126
Archidium alternifolium, 187
Arctium lappa, 138
   minus agg., 139
Arenaria balearica, 84
   serpyllifolia, 84
   serpyllifolia ssp. leptoclados, 84
Armoracia rusticana, 92
Arrhenatherum elatius, 154
Arrowgrass, Marsh, 145
Artemisia absinthium, 137
   vulgaris, 137
Arthonia didyma, 221
   elegans, 221
   glaucomaria, 221
   impolita, 222
   punctiformis, 222
   radiata, 222
   spadicea, 222
   tumidula, 222
   vinosa, 222

Arthopyrenia cinereopruinosa, 222
   fraxini, 222
   lapponina, 222
   nitescens, 222
   punctiformis, 222
   ranunculospora, 222
   viridescens, 222
Arthrorhaphis citrinella, 222
   grisea, 223
Arum maculatum, 156
Ash, 121
Aspen, 79
Asphodel, Bog, 146
Aspicillia caesiocinerea, 223
   calcarea, 223
   cinerea, 223
   contorta, 223
   gibbosa, 223
   subcircinata, 223
Asplenium adiantum-nigrum, 75
   ruta-muraria, 75
   septentrionale, 75
   trichomanes, 75
   viride, 75
Aster novi-belgii, 135
Athyrium filix-femina, 76
Atmospheric pollution, 24
Atrichum crispum, 187
   undulatum, 187
Atriplex patula, 83
   prostrata, 83
Aubretia, 92
Aubrieta deltoidea, 92
Aulacomnium androgynum, 202
   palustre, 202
Avena sativa, 154
Avens, Water, 100
   , Wood, 100
Avenula pubescens, 154
Bach Howey, 5, 12, 16, 35, 66, 167, 168
Bachell Brook, 15
Bacidia arceutina, 223
   bagliettoana, 223
   biatorina, 223
   caligans, 223
   carneoglauca, 223
   chloroticula, 223
   circumspecta, 223
   delicata, 224
   epixanthoides, 224
   incompta, 224
   inundata, 224
   laurocerasi, 224
   naegelii, 224
   phacodes, 224
   rubella, 224
   sabuletorum, 224
   saxenii, 224
   trachona, 224
   vezdae, 224
Bactrospora corticola, 224
Baeomyces placophyllus, 224
   roseus, 224
   rufus, 225
Bailey Einon, 64
Ballota nigra ssp. foetida, 126
Balm, 126
Balsam, Indian, 110
   , Small, 110
BALSAMINACEAE, 110
Banks, The, 10
Barbarea intermedia, 91
   verna, 91
   vulgaris, 91
Barberry, 89
Barbilophozia attenuata, 177
   barbata, 177
   floerkii, 177
   acuta, 194
   convoluta, 193
   cylindrica, 194
   fallax, 194
   ferruginascens, 194
   hornschuchiana, 194
   recurvirostra, 194
   reflexa, 194
   revoluta, 194
   rigidula, 194

Index

Barbula spadicea, 194
   tophacea, 194
   unguiculata, 194
Barley, 153
   , Wall, 154
Bartramia halleriana, 202
   ithyphylla, 202
   pomiformis, 202
   stricta, 202
Bartsia, Red, 131
Basil, Wild, 126
Baxter's Bank, 10
Bazzania trilobata, 182
Beacon Hill, 1, 12, 17
Beak-sedge, White, 158
Bean, Broad, 105
Bedstraw, Common Marsh, 122
   , Fen, 122
   , Heath, 122
   , Hedge, 122
   , Lady's, 122
Beech, 80
   , Southern, 81
Begwns, 13, 31, 37, 38
Beilibedw Mawn Pool, 15, 55
Bellflower, Clustered, 134
   , Creeping, 134
   , Giant, 134
   , Ivy-leaved, 134
   , Nettle-leaved, 134
   , Spreading, 134
   , Trailing, 134
Bellis perennis, 135
Belonia nidarosiensis, 225
Bent, Black, 154
   , Brown, 154
   , Common, 154
   , Creeping, 155
   , Heath, 154
BERBERIDACEAE, 89
Berberis vulgaris, 89
Betony, 126
Betula pendula, 80
   pubescens, 80
BETULACEAE, 80
Biatorina atropurpurea, 225
Bidens cernua, 136
   tripartita, 136
Bilberry, 119
Bindweed, Black, 82
   , Field, 123
   , Hairy, 123
   , Hedge, 123
   , Large, 123
Birch, Downy, 80
   , Silver, 80
Bird's-foot, 107
Bistort, Amphibious, 82
   , Common, 82
Bitter-cress, Hairy, 92
   , Large, 92
   , Narrow-leaved, 92
   , Wavy, 92
Bitter-vetch, 105
   , Wood, 104
Black Brook, 16
Black Mixen, 17
Blackthorn, 103
Bladder-fern, Brittle, 76
Bladderwort, Lesser, 132
   , Greater, 132
Blaen Restr, 10
Blanket bogs, 16
Blasia pusilla, 177
BLECHNACEAE, 77
Blechnum spicant, 77
Blepharostoma trichophyllum, 183
Blindia acuta, 188
Blinks, 84
Bluebell, 147
   , Spanish, 147
Bog, The, 37
Bog, Orchid, 61
BORAGINACEAE, 123
Boreal zone, 215
Botanical Society of the British Isles, 69
Botrychium lunaria, 74
Botrydina vulgaris, 244

Botrydium sp., 165
Boughrood, 2, 164, 165
Bower, The, 10
Box, 110
Boyce, D.C., 70
Brachydontium trichodes, 188
   pinnatum, 153
   sylvaticum, 153
Brachythecium albicans, 209
   glareosum, 209
   plumosum, 210
   populeum, 209
   rivulare, 209
   rutabulum, 209
   velutinum, 209
Bracken, 74
Bramble, Stone, 97
Brambles, 97
Brassica napus, 94
   nigra, 94
   rapae, 94
Brecknock, 213
Breutelia chrysocoma, 203
Bridewort, 97
Briza media, 152
Brome, False, 153
   , Hairy, 152
   , Lesser Hairy-, 152
   , Meadow, 153
   , Rye, 153
   , Slender Soft-, 153
   , Soft-, 153
   , Sterile, 152
   , Upright, 153
Bromus benekenii, 153
   commutatus, 153
   erectus, 153
   hordeaceus ssp. hordeaceus, 153
   lepidus, 153
   ramosus, 152
   secalinus, 153
   sterilis, 152
Brook A.J., 164
Brooklime, 129
Broom, 103
   , Hairy-fruited, 103
Broomrape, Greater, 132
Bryonia cretica, 114
Bryony, White, 114
   , Black, 148
Bryophagus gloeocarpa, 225
Bryoria bicolor, 225
   fuscescens, 225
   subcana, 225
Bryum algovicum, 199
   alpinum, 200
   argenteum, 200
   bicolor, 200
   bornholmense, 200
Bryum caespiticium, 200
   capillare, 199
   creberrimum, 200
   flaccidum, 199
   inclinatum, 199
   klinggraeffii, 200
   microerythrocarpum, 200
   pallens, 199
   pseudotriquetrum, 200
   radiculosum, 200
   riparium, 200
   rubens, 201
   ruderale, 200
   sauteri, 200
   tenuisetum, 200
   violaceum, 200
Buckler-fern, Broad, 76
   , Narrow, 76
Buckthorn, 110
   , Alder, 110
   , Sea, 111
Buddleja davidii, 128
BUDDLEJACEAE, 128
Buellia aethalea, 225
   griseovirens, 225
   ocellata, 225
   pulverea, 225
   punctata, 225
   schaereri, 225

Buellia stellulata, 225
Bugle, 125
Builth, 213
Builth Road, 2
Builth Wells, 163, 164, 213
Bullace, 103
Bulrush, 156
   , Lesser, 156
Bupleurum obovatum, 117
Bur-marigold, Nodding, 136
   , Trifid, 136
Bur-reed, Branched, 156
   , Unbranched, 156
Burdock, Greater, 138
   , Lesser, 139
Burfa, 16
Burfa Bog, 28, 64
Burnet, Great, 100
   , Salad, 100
Burnet-saxifrage, 117
Bush Butterfly, 128
BUTOMACEAE, 145
Butomus umbellatus, 145
Butterbur, 137
Buttercup, Bulbous, 88
   , Celery-leaved, 88
   , Corn, 88
   , Creeping, 88
   , Goldilocks, 88
   , Meadow, 88
   , Small-flowered, 88
Butterwort, Common, 132
BUXACEAE, 110
Buxus sempervirens, 110
Caban Coch, 10
Caban Coch Resr., 10, 11, 29, 163
Caer Einion, 10
Calendula officinalis, 138
Calicium glaucellum, 226
   salicinum, 216
   subquercinum, 226
   viride, 226
Calliergon cordifolium, 208
   cuspidatum, 208
   giganteum, 208
   sarmentosum, 208
   stramineum, 208
CALLITRICHACEAE, 124
Callitriche brutia, 124
   hamulata, 124
   obtusangula, 124
   platycarpa, 124
   stagnalis, 124
Calluna vulgaris, 119
Caloplaca aurantia, 226
   cerina, 226
   citrina, 226
   crenularia, 226
   flavescens, 226
   flavovirescens, 226
   holocarpa, 226
   luteoalba, 226
   obscurella, 226
   saxicola, 226
   subpallida, 227
Caltha palustris, 87
Calypogeia arguta, 182
   fissa, 182
   muelleriana, 182
Calystegia pulchra, 123
   sepium ssp. sepium, 123
   silvatica, 123
Campanula glomerata, 134
   latifolium, 134
   patula, 134
   poscharskyana, 134
   rapunculoides, 134
   rotundifolia, 134
   trachelium, 134
CAMPANULACEAE, 134
Campion, Bladder, 86
   , Red, 86
   , White, 86
Campylium chrysophyllum, 207
   polygamum, 207
   stellatum, 207
Campylopus atrovirens, 190
   brevipilus, 190

280 Index

Campylopus fragilis, 190
   introflexus, 190
   paradoxus, 190
   pyriformis, 190
   subulatus, 190
Canary-grass, 155
   , Reed, 155
Candelariella aurella, 227
   coralliza, 227
   medians, 227
   reflexa, 227
   vitellina, 227
   xanthostigma, 227
Candytuft, Garden, 93
   , Perennial, 93
CANNABACEAE, 81
Cannabis sativa, 81
CAPRIFOLIACEAE, 133
Capsella bursa-pastoris, 93
Caraway, Whorled, 118
Cardamine amara, 92
   flexuosa, 92
   hirsuta, 92
   impatiens, 92
   pratensis, 92
Carduus acanthoides, 139
   nutans, 139
Carex acuta, 160
   acutiformis, 159
   arenaria, 158
   binervis, 160
   caryophyllea, 160
   curta, 158
   demissa, 160
   diandra, 158
   dioica, 158
   disticha, 158
   divulsa ssp. divulsa, 158
   echinata, 158
   flacca, 159
   hirta, 158
   hostiana, 160
   laevigata, 160
   lepidocarpa, 160
   muricata ssp. lamprocarpa, 158
   nigra, 160
   otrubae, 158
   ovalis, 158
   pallescens, 160
   panicea, 159
   paniculata, 158
   pendula, 159
   pilulifera, 160
   pulicaris, 161
   remota, 158
   riparia, 159
   rostrata, 159
   spicata, 158
   strigosa, 159
   sylvatica, 159
   vesicaria, 159
Carlina vulgaris, 138
Carmel, 10
Carneddau, 8, 10, 67
Carpinus betulus, 80
Carreg Bica, 30
Carrot, Wild, 118
Carter, P.W., 65
Carteria spp., 163
Carum verticillatum, 118
CARYOPHYLLACEAE, 84
Castanea sativa, 80
Castelltinboeth, 11
Castle Bank, 10
Cat's-ear, 140
Catabrosa aquatica, 152
Catapyrenium pilosellum, 227
Catchfly, Small-flowered, 86
   , Sticky, 86
Catillaria atomarioides, 227
   chalybeia, 227
   globulosa, 227
   lenticularis, 227
   pulverea, 227
   sphaeroides, 227
Cattle, 56
Cedar, Western Red-, 78
Cedidonia xenophana, 228

Cefn Cennarth, 27, 64
Cefn Nantmel, 8
Cefnllys, 64
Cefnllys Castle, 32, 213
Celandine, Greater, 90
   , Lesser, 88
CELASTRACEAE, 110
Centaurea cyanus, 140
   montana, 139
   nigra, 139
   scabiosa, 139
Centaurium erythraea, 121
Centaury, Common, 121
Central Wales Railway, 2, 34
Centranthus ruber, 134
Cephalozia bicuspidata, 182
   connivens, 182
   lunulifolia, 182
   divaricata, 181
   hampeana, 181
Cerastium arvense, 85
   diffusum, 85
   fontanum, 85
   glomeratum, 85
   semidecandrum, 85
   tomentosum, 84
Ceratodon purpureus, 188
CERATOPHYLLACEAE, 87
Ceratophyllum demersum, 87
Cerrig Gwalch, 11, 218
Ceterach officinarum, 76
Cetraria chlorophylla, 228
   sepincola, 228
Chaenorhinum minus, 128
Chaenotheca brunneola, 228
   carthusiae, 228
   chrysocephala, 228
   ferruginea, 228
   trichialis, 228
Chaerophyllum temulentum, 116
Chaffweed, 120
Chamaecyparis lawsoniana, 78
Chara globularis var. virgata, 165
CHARACEAE, 165
Chelidonium majus, 90
CHENOPODIACEAE, 83
Chenopodium album, 83
   bonus-henricus, 83
   polyspermum, 83
   rubrum, 83
Cherry, Bird, 103
   , Dwarf, 103
   , Wild, 103
Cherry Laurel, 103
Chervil, Garden, 116
   , Rough, 116
Chestnut, Horse, 110
   , Sweet, 80
Chickweed, Upright, 85
   , Water, 85
Chicory, 140
Chiloscyphus pallescens, 180
   polyanthos, 180
Chives, 147
Chlorococcales, 164
Chrysanthemum segetum, 137
Chrysosplenium alternifolium, 97
   oppositifolium, 97
Chrysothrix candelaris, 228
   chlorina, 228
   chrysophthalma, 228
Cicerbita macrophylla ssp. uralensis, 141
Cichorium intybus, 140
Cicuta virosa, 118
Cilkenny Dingle, 27, 64, 218
Cinclidotus fontinaloides, 195
Cinquefoil, Creeping, 101
   , Marsh, 101
   , Rock, 66, 101
Circaea alpina, 114
   lutetiana, 114
   x intermedia, 114
Cirrophyllum crassinervium, 210
   piliferum, 210
Cirsium arvense, 139
   dissectum, 139
   eriophorum, 139
   helenioides, 139

Cirsium palustre, 139
   vulgare, 139
CISTACEAE, 113
Cladium mariscus, 158
Cladonia arbuscula, 229
   caespiticia, 229
   cervicornis, 229
   cervicornis spp. verticillata, 229
   chlorophaea, 229
   ciliata, 229
   coccifera, 229
   coniocraea, 229
   crispata var. cetrariiformis, 229
   digitata, 229
   fimbriata, 229
   floerkeana, 229
   foliacea, 230
   furcata, 230
   glauca, 230
   gracilis, 230
   humilis, 230
   luteoalba, 230
   macilenta, 230
   ochrochlora, 230
   parasitica, 230
   phyllophora, 230
   pocillum, 230
   polydactyla, 230
   portentosa, 230
   pyxidata, 230
   ramulosa, 231
   rangiformis, 231
   squamosa, 231
   subcervicornis, 231
   subulata, 231
   uncialis ssp. biuncialis, 231
Cladophora glomerata, 165
Cladopodiella fluitans, 182
Claerwen, 168, 213, 218
Claerwen Valley, 5
Clarke, W.A., 68
Cleavers, 122
Clematis vitalba, 87
Climacium dendroides, 205
Clinclidotus mucronatus, 195
Clinopodium vulgare, 126
Cliostomum griffithii, 231
Closterium actum var. linea, 164
   didymotocum, 163
   kuetzingii, 163
   limneticum var. limneticum, 164
Clover, Alsike, 106
   , Hare's-foot, 106
   , Knotted, 106
   , Red, 106
   , Rough, 106
   , Upright, 105
   , White, 105
   , Zigzag, 106
Club-rush, Bristle, 156
   , Floating, 157
   , Wood, 156
Clubmoss, Alpine, 73
   , Fir, 73
   , Stag's-horn, 73
Clypeococcum hypocenomyceae, 231
Clyro, 12, 218
Cocconeis placentula, 164
Cock's-foot, 152
Coed Cymru, 58
Coed yr Allt Goch, 26, 27
Coelastrum, 164
Coelocaulon aculeatum, 231
   muricatum, 231
Coeloglossum viride, 161
Coelosporium tussilaginis, 261
Colchichum autumnale, 147
Collema auriforme, 232
   crispum, 232
   fasciculare, 232
   flaccidum, 232
   fragrans, 232
   subflaccidum, 232
   tenax, 232
Cololejeunea minutissima, 184
Colt's-foot, 137
Columbine, 89
Colwyn Brook, 16

Index

Comfrey, Common, 123
COMPOSITAE, 135
Concrete, 219
CONIFERS, 78
Coniocybe furfuracea, 232
Conium maculatum, 117
Conocephalum conicum, 175
Conopodium majus, 117
Convallaria majalis, 147
CONVOLVULACEAE, 122
Convolvulus arvensis, 123
Coriscium viride, 232
CORNACEAE, 116
Corncockle, 86
Cornflower, 140
 , Perennial, 139
Cornsalad, Common, 133
 , Keeled-fruited, 133
 , Narrow-fruited, 133
Cornus sanguinea, 116
 sericea, 116
Coronopus didymus, 94
 squamatus, 94
Cors y Llyn, 28, 31, 55, 63
Corydalis, Climbing, 90
Corydalis claviculata, 90
 lutea, 90
CORYLACEAE, 80
Corylus avellana, 80
Cotoneaster, Himalayan, 102
 , Small-leaved, 102
 , Wall, 102
Cotoneaster horizontalis, 102
 microphyllus, 102
 simonsii, 102
Cotteswold Club, 67
Cottongrass, Broad-leaved, 157
 , Common, 157
 , Hare's-tail, 157
Couch, Bearded, 153
 , Common, 153
Cow parsley, 116
Cow-wheat, Common, 130
Cowbane, 118
Cowberry, 119
Cowslip, 120
Coxhead, 31
Craig Cwm Clyd, 10
Craig Fawr, 218
Craig Goch Dam, 3
Craig Goch Resr., 163
Craig y Bwch, 11
Cranberry, 119
Crane's-bill, Bloody, 107
 , Cut-leaved, 108
 , Dove's-foot, 107
 , Dusky, 107
 , Hedgerow, 107
 , Long-stalked, 108
 , Meadow, 107
 , Shining, 108
 , Small-flowered, 107
 , Wood, 107
Crash barrier posts, 219
CRASSULACEAE, 95
Crataegus monogyna, 102
Cratoneuron commutatum var.
 commutatum, 206
 commutatum var. falcatum, 207
 filicinum, 206
Creeping-jenny, 120
Crepis capillaris, 143
 paludosa, 143
 vesicaria ssp. haenseleri, 143
Crosswort, 122
Crowberry, 119
Crowfoot, Common Water-, 89
 , Ivy-leaved, 88
 , Pond Water-, 88
 , River Water-, 89
 , Round-leaved, 88
 , Stream Water-, 89
 , Thread-leaved Water-, 89
 , Three-lobed, 88
 , Water, 88
Cruciata laevipes, 122
CRUCIFERAE, 90
Crucigenia, 164

Crychell, 31
Cryphaea heteromalla, 205
Cryptogramma crispa, 74
CRYPTOGRAMMACEAE, 74
Cryptomeria japonica, 78
Cryptothallus mirabilis, 176
Ctenidium molluscum, 212
Cuckooflower, 92
CUCURBITACEAE, 114
Cudweed, Common, 135
 , Heath, 136
 , Marsh, 136
 , Small, 135
Cumminsiella mirabilissima, 261
CUPRESSACEAE, 78
Currant, Black, 97
 , Flowering, 97
 , Mountain, 97
 , Red, 97
Cuscuta epithymum, 122
Cwm Coel, 26, 28
Cwm Elan, 11
Cwm Elan Mine, 218
Cwm Gilla, 27
Cwm Gwynllyn, 26, 218
Cwmystwyth, 11
Cyanobacteria 163
Cyclamen, 120
Cyclamen hederifolium, 120
Cyclotella meneghiniana, 163
Cymbalaria muralis, 129
Cynodontium bruntonii, 188
Cynoglossum officinale, 124
Cynosurus cristatus, 152
CYPERACEAE, 156
Cyphelium inquinans, 232
Cypress, Lawson, 78
Cystocoleus ebeneus, 232
Cystopteris fragilis, 76
Cytisus scoparius, 103
 striatus, 103
Dactylis glomerata, 152
Dactylorhiza majalis ssp. purpurella, 162
 fuchsii, 162
 incarnata, 162
 maculata, 162
 majalis ssp. praetermissa, 162
Daffodil, Wild, 148
Daisy, 135
Dame's-violet, 91
Dandelions, 141-143
Danthonia decumbens, 155
Daphne laureola, 111
 mezereum, 111
Datura stramonium, 127
Daucus carota, 118
Davy Morgan's Dingle, 12
Deergrass, 157
Dermatocarpon luridum, 233
 meiophyllizum, 233
 miniatum, 233
Descampsia cespitosa ssp. cespitosa, 154
 flexuosa, 154
Desmazeria rigida, 151
Devil's Garden, 65
Devil's Gulch, 3
Dewberry, 99
Dianthus barbatus, 86
 deltoides, 86
Diatoma vulgare, 164
Dichodontium flavescens, 189
 pellucidum, 188
DICOTYLEDONES, 79
Dicranella cerviculata, 189
 heteromalla, 189
 palustris, 189
 rufescens, 189
 schreberana, 189
 staphylina, 189
 subulata, 189
 varia, 189
Dicranodontium denudatum, 190
Dicranoweissia cirrata, 189
Dicranum bonjeanii, 189
 fuscesens, 190
 majus, 189
 montanum, 190
 scoparium, 189

Digitalis purpurea, 129
Dimerella diluta, 233
 lutea, 233
Dinobryon sertularia, 163
DIOSCOREACEAE, 148
Diphasiastrum alpinum, 73
 foliosum, 187
Diploicia canescens, 233
Diplophyllum albicans, 180
 obtusifolium, 180
Diploschistes scruposus, 233
Diplotaxis muralis, 94
Diplotomma alboatrum, 233
DIPSACACEAE, 134
Dipsacus fullonum, 134
 pilosus, 134
Dirina massiliensis f. sorediata, 233
Discelium nudum, 197
Ditrichum cylindricum, 188
 flexicaule, 188
 heteromallum, 188
 lineare, 188
Dock, Broad-leaved, 83
 , Clustered, 83
 , Curled, 83
 , Water, 83
 , Wood, 83
Dodder, 122
Dog's-tail, Crested, 152
 , Common, 113
 , Early, 113
 , Heath, 113
Dogwood, 116
 , Red-osier, 116
Dol y Fan, 13
Dol-y-fan Hill, 8
Dolau, 2, 34, 35
Doldowlod, 8
Dolyhir, 2, 11, 13, 29, 173, 218, 220
Dolymynach Resr., 39, 163
Doronicum pardalianches, 138
 pantagineum var. excelsum, 138
Douglas Fir, 1
Draba muralis, 93
Drepanocladus aduncus, 207
 exannulatus, 207
 fluitans, 207
 revolvens, 207
 sendtneri, 207
 uncinatus, 208
 vernicosus, 208
Drewett, D.R., 70
Drosera rotundifolia, 95
DROSERACEAE, 95
Druce, G.Claridge, 67
Dry stone walls, 39
DRYOPTERIDACEAE, 76
Dryopteris affinis, 76
 carthusiana, 76
 dilatata, 76
 filix-mas, 76
 oreades, 76
Drysgol, 16
Duckweed, Common, 156
 , Ivy-leaved, 156
Dyer's greenweed, 104
Dyfed, 217
Dyffryn Wood, 64
Echium vulgare, 123
Elan, 168, 217, 218
Elan Valley, 2, 5, 163, 166-169, 213
ELATINACEAE, 113
Elatine hexandra, 113
Elder, 133
 , Dwarf, 133
Elderberry, Red-berried, 133
ELEAGNACEAE, 111
Elecampane, 136
Elenydd, 63, 218
Eleocharis acicularis, 157
 multicaulis, 157
 palustris, 157
 quinqueflora, 157
Ellis, R.G., 69
Elm, English, 81
 , Wych, 81
Elodea canadensis, 145
Elymus caninus, 153

Index

Elymus repens, 153
Empetrum nigrum, 119
Encalypta ciliata, 192
Encalypta streptocarpa, 192
    vulgaris, 192
Enchanter's-nightshade, 114
    , Alpine, 114
    , Upland, 114
Endococcus rugulosus, 234
Enterographa crassa, 234
Enteromorpha flexuosa, 164
Entyloma ficariae, 265
Environmentally Sensitive Areas, 63
Ephebe lanata, 234
Ephemerum serratum, 198
Epigloea soleiformis, 234
Epilobium angustifolium, 114
    brunnescens, 115
    ciliatum, 115
    hirsutum, 114
    lanceolatum, 115
    montanum, 115
    obscurum, 115
    palustre, 115
    parviflorum, 115
    roseum, 115
    tetragonum, 115
Epipactis helleborine, 161
    palustris, 161
Epipterygium tozeri, 199
EQUISETACEAE, 73
Equisetum arvense, 73
    fluviatile, 73
    palustre, 73
    sylvaticum, 73
    telmateia, 73
Eranthis hyemalis, 87
Erica cinerea, 119
    tetralix, 118
ERICACEAE, 118
Erigeron acer, 135
Erinus alpinus, 129
Eriophorum angustifolium, 157
    latifolium, 157
    vaginatum, 157
Erodium cicutarium, 108
Erophila verna, 93
Erwood, 2, 163, 167
Erysimum cheiranthoides, 91
Euastrum ansatum, 163
Eucladium verticillatum, 195
Euglena, 164
Euglenophyta, 164
Eunotia, 164
Euonymus europaeus, 110
Eupatorium cannabinum, 135
Euphorbia amygdaloides, 109
    exigua, 109
    helioscopia, 108
    lathyris, 109
    peplus, 109
EUPHORBIACEAE, 108
Euphrasia anglica, 131
    arctica ssp. borealis, 131
    confusa, 131
    micrantha, 131
    nemorosa, 131
    officinalis agg., 130
    roskoviana ssp. roskoviana, 131
    scottica, 131
Eurhynchium praelongum, 210
    pumilum, 210
    schleicheri, 211
    striatum, 210
    swartzii, 210
Evans, John, 65
Evening Primrose, Large-flowered, 114
Everlasting, Mountain, 136
    , Pearly, 136
Evernia prunastri, 234
Eyebright, 130
FAGACEAE, 80
Fagus sylvatica, 80
Fallopia convolvulus, 82
Farnoldia jurana, 234
Fat-hen, 83
Fedw Turbary, 15
Feg, 155

Felindre, 3
Fen-sedge Great, 158
Fens, 44
Fenugreek, 105
Fern, Beech, 75
    , Brittle Bladder-, 76
    , Broad Buckler-, 76
    , Dwarf Male, 76
    , Filmy, 74, 75
    , Hard-, 77
    , Hard Shield-, 77
    , Hart's-tongue, 76
    , Killarney, 74
    , Lady-, 76
    , Limestone, 77
    , Male, 76
    , Marsh, 75
    , Mountain, 75
    , Narrow Buckler-, 76
    , Oak, 77
    , Parsley, 74
    , Royal, 74
    , Rusty-back, 76
    , Scaly Male, 76
    , Soft Shield-, 76
    , Wilson's Filmy, 75
Fern-grass, 151
FERNS, 73
Fescue, Giant, 150
    , Meadow, 150
    , Rat's-tail, 151
    , Red, 150
    , Sheep's, 150
    , Squirreltail, 151
    , Tall, 150
    , Wood, 150
Festuca altissima, 150
    arundinacea, 150
    gigantea, 150
    ovina agg., 150
    pratensis, 150
    rubra, 150
Feverfew, 137
Fferllys, 1
Ffordd-fawr, 4, 28
Ffos Trosol, 16, 30
Figwort, Common, 128
    , Water, 128
Filaginella uliginosa, 136
Filago minima, 135
    vulgaris, 135
FILICOPSIDA, 74
Filipendula ulmaria, 97
Filmy Fern, 74, 75
    , Wilson's, 75
Fir, Douglas, 78
    , European Silver-, 78
Fissidens adianthoides, 192
    algarvicus, 191
    bryoides, 191
    celticus, 191
    crassipes, 191
    cristatus, 192
    curnovii, 191
    exilis, 191
    incurvus, 191
    osmundoides, 191
    pusillus, 191
    rivularis, 191
    rufulus, 191
    taxifolius ssp. taxifolius, 191
    viridulus, 191
Flag, Sweet, 155
Flax, 108
    , Fairy, 108
    , Pale, 108
Fleabane, Blue, 135
    , Common, 136
Fluellen, Round-leaved, 129
    , Sharp-leaved, 129
Fontinalis antipyretica, 204
    squamosa, 205
Foraminella ambigua, 234
Forest Nature Reserve, 63
Forestry, 57
Forget-me-not, Changing, 123
    , Creeping, 124
    , Early, 123

Forget-me-not, Field, 123
    , Tufted, 124
    , Water, 124
    , Wood, 124
Fossombronia pusilla var. pusilla, 177
    wondraczekii, 177
Fox and Cubs, 143
Foxglove, 129
    , Fairy, 129
Foxtail, Marsh, 155
    , Meadow, 155
Fragaria moschata, 101
    vesca, 101
Fragilaria pinnata, 164
    sp., 163
Frangula alnus, 110
Frank's Bridge, 5, 218
Fraxinus excelsior, 121
Fringe-cups, 96
Frommea obtusa, 261
Fron Barn Dingle, 28
Frullania dilatata, 184
    fragilifolia, 184
    tamarisci, 183
Fumaria muralis, 90
    officinalis, 90
    purpurea, 90
FUMARIACEAE, 90
Fumitory, Common, 90
    , Yellow, 90
Funaria attenuata, 198
    fascicularis, 198
    hygrometrica, 197
    muhlenbergii, 197
    obtusa, 198
    pulchella, 197
Fuscidea cyathoides, 234
    kochiana, 234
    lightfootii, 234
    praeruptarum, 234
    recensa, 234
    tenebrica, 234
Gagea bohemica, 147
Galanthus nivalis, 148
Galeobdolon luteum ssp. montanum, 126
Galeopsis angustifolia, 125
    speciosa, 125
    tetrahit agg., 125
Galium aparine, 122
    mollugo agg., 122
    odoratum, 122
    palustre agg., 122
    saxatile, 122
    uliginosum, 122
    verum, 122
Gardens, 40
Garreg Ddu, 11
Gelli-Cadwgan, 10
Genista anglica, 104
    tinctoria, 104
Gentian, Field, 121
GENTIANACEAE, 121
Gentianella campestris, 121
GERANIACEAE, 107
Geranium columbinum, 108
    dissectum, 108
    lucidum, 108
    molle, 107
    phaeum, 107
    pratense, 107
    pusillum, 107
    pyrenaicum, 107
    robertianum, 108
    sanguineum, 107
    sylvaticum, 107
Geum rivale, 100
    urbanum, 100
Gibbin, O., 69
Gilfach Farm, 2, 63
Gilwern Brook, 8, 16
Gipsywort, 127
Glanalder, 26, 28
Glannau, 26
Glasbury, 2, 3, 13
Glascwm, 12
Glaslyn, 8
Glechoma hederacea, 126
Globe flower, 87

Index

Glog Fawr, 11
Glyceria declinata, 152
    fluitans, 152
    maxima, 152
    plicata, 152
Goat's-beard, 140
Gogerddan, 19
Golden-saxifrage, Alternate-leaved, 97
    , Opposite-leaved, 97
Goldenrod, 135
    , Canadian, 135
Gooseberry, 97
Goosefoot, Many-seeded, 83
    , Red, 83
Gore Quarry, 8
Gorse, 104
    , French, 104
    , Western, 104
Gorsgoch, 16, 30, 31
GRAMINEAE, 150
Graphis elegans, 234
    scripta, 235
Grass, Disco, 155
Grasslands, Acidic, 43
    , Calcareous, 44
    , Mesotrophic, 44
Great Creigiau, 12
Grimmia decipiens, 196
    donniana, 196
    hartmanii, 196
    laevigata, 196
    montana, 196
    orbicularis, 196
    ovalis, 196
    pulvinata, 196
    retracta, 196
    trichophylla, 196
GROSSULARIACEAE, 97
Ground-elder, 117
Groundsel, 138
    , Heath, 138
    , Sticky, 138
Gumma, 27
GUTTIFERAE, 111
Gwastedyn Hill, 10
Gwaunceste Hill, 12, 31, 218
Gwernaffel Dingle, 26, 27
Gwyn Llyn, 15
Gwynllyn, 36, 37, 38, 167
Gwynllyn Wood, 64
Gwystre, 15
Gyalecta jenensis, 235
    truncigena, 235
Gyalidea subscutellaris, 235
Gyalideopsis anastomosans, 235
Gymnadenia conopsea, 161
Gymnocarpium dryopteris, 77
    robertianum, 77
Gymnocolea inflata var. inflata, 178
GYMNOSPERMAE, 78
Gymnostomum aeruginosum, 194
Haematomma caesium, 235
    elatinum, 235
    chroleucum, 235
    ventosum, 235
Hair-grass, Early, 154
    , Silver, 154
    , Tufted, 154
    , Wavy, 154
Hairy-brome, Lesser, 153
HALORAGACEAE, 115
Hammarbya paludosa, 162
Hanter, 7
Hard-fern, 77
Harebell, 134
Hargreaves, D.P., 70
Harley Dingle, 12
Harrison, S.G., 69
Hawk's-beard, Beaked, 143
    , Marsh, 143
    , Smooth, 143
Hawkbit, Autumn, 140
    , Lesser, 140
    , Rough, 140
Hawkweed, Mouse-ear, 143
    spp., 143
Hawthorn, 102
Hay on Wye, 163

Hazel, 80
Heath, Cross-leaved, 118
Heath-grass, 155
Heather, 119
Heather, Bell, 119
Heaths, 42
Heavy metal, 217, 218
Hedera colchica, 116
    helix, 116
    hibernica, 116
Hedge-parsley, Knotted, 118
    , Spreading, 118
    , Upright, 118
Hedwigia ciliata, 204
Helianthemum nummularium, 113
Helianthus annuus, 136
Heliotrope, Winter, 138
Hellebore, Green, 87
    , Stinking, 87
Helleborine, Marsh, 161
    , Broad-leaved, 161
Helleborus foetidus, 87
    viridis, 87
Hemlock, 117
    , Western, 78
Hemp, Indian, 81
Henbane, 127
Henllyn Mawr, 34
HEPATICAE, 175
Heracleum mantegazzianum, 118
    sphondylium, 118
Herefordshire, 213, 217
Hesperis matronalis, 91
Heterocladium heteropterum, 206
Hieracium acuminatum, 143
    argenteum, 144
    calcaricola, 145
    carneddorum, 144
    cinderella, 143
    diaphanoides, 144
    diaphanum, 144
    lasiophyllum, 144
    perpropinquum, 144
    placerophylloides, 144
    reticulatum, 144
    scabrisetum, 144
    schmidtii, 144
    spp., 143
    stenstroemii, 143
    strictiforme, 144
    subamplifolium, 143
    subcrocatum, 144
    sublepistoides, 143
    submutabile, 143
    substrigosum, 144
    uiginskyense, 144
    umbellatum ssp. bichlorophyllum, 144
    umbellatum ssp. umbellatum, 144
    vagense, 144
    vagum, 144
    vulgatum, 143
Hindwell Pool, 38
HIPPOCASTANACEAE, 110
Hippophae rhamnoides, 111
Hirschfeldia incana, 94
Hoel-y-gaer, 4
Hogweed, 118
    , Giant, 118
Holcus lanatus, 154
    mollis, 154
Holly, 110
Homalia trichomanoides, 205
Homalothecium lutescens, 209
    sericeum, 209
Honesty, 93
Honeysuckle, 133
    , Perfoliate, 133
Hookera lucens, 206
Hop, 81
Hordeum distichon, 153
    murinum, 154
    sp., 153
Horehound, Black, 126
Hormidium, 163, 164
Hornbeam, 80
Horns Adder's, 156
Hornwort, Rigid, 87

HORNWORTS, 175
Horse-radish, 92
Horses, 56
Horsetail, Common, 74
    , Great, 74
    , Marsh, 73
    , Water, 73
    , Wood, 73
HORSETAILS, 73
Hound's Tongue, 124
House-leek, 95
Hullia albocaerulescens, 235
    crustulata, 235
    hydrophilia, 235
    macrocarpa, 235
    platycarpoides, 235
    soredizodes, 236
    tuberculosa, 236
Humphreys, D.R., 70
Humulus lupulus, 81
Huperzia selago, 73
Hyacinthoides hispanica, 147
    non-scripta, 147
Hyalopsora polypodii, 261
Hyde, H.A., 68
HYDROCHARITACEAE, 145
Hydrocotyle vulgaris, 116
Hydrodictyon reticulatum, 165
Hygrohypnum luridum, 208
    ochraceum, 208
Hylocomium brevirostre, 212
    splendens, 212
Hymenelia lacustris, 236
HYMENOPHYLLACEAE, 74
Hymenophyllum wilsonii, 75
Hyocomium armoricum, 212
Hyophila stanfordensis, 193
Hyoscyamus niger, 127
Hypericum androsaemum, 111
    calycinum, 111
    elodes, 112
    hirsutum, 111
    humifusum, 112
    linariifolium, 112
    maculatum, 112
    montanum, 112
    perforatum, 112
    pulchrum, 112
    tetrapterum, 112
Hyperphyscia adglutinata, 236
Hypnum cupressiforme var. cupressiforme, 212
    cupressiforme var. lacunosum, 212
    cupressiforme var. resupinatum, 212
    jutlandicum, 212
    lindbergii, 212
    mammillatum, 212
Hypocenomyce caradocensis, 236
    friesii, 236
    scalaris, 236
Hypochaeris radicata, 140
Hypogymnia physodes, 236
    tubulosa, 236
HYPOLEPIDACEAE, 74
Iberis sempervirens, 93
    umbellata, 93
Ilex aquifolium, 110
Impatiens glandulifera, 110
    parviflora, 110
Inquisition, Elizabethan, 57
Inula conyza, 136
    helenium, 136
IRIDACEAE, 148
Iris, Stinking, 148
    , Yellow, 148
Iris foetidissima, 148
    pseudacorus, 148
ISOETACEAE, 73
Isoetes lacustris, 73
Isopterygium elegans, 211
    pulchellum, 211
Isothecium holtii, 209
    myosuroides, 209
    myurum, 209
Ithon Valley, 15, 19, 35, 38
Ivy, 116
    , Atlantic, 116
    , Persian, 116

Ivy, Ground, 126
Jacob's Ladder, 122
Jamesoniella autumnalis, 178
Jasione montana, 135
Jones, Edward W.H., 67
Jubula hutchinsiae, 184
JUNCACEAE, 148
JUNCAGINACEAE, 145
Juncus acutiflorus, 149
　　articulatus, 149
　　bufonius, 149
　　bulbosus, 149
　　conglomeratus, 149
　　effusus, 148
　　inflexus, 148
　　squarrosus, 149
　　subnodulosus, 149
　　tenuis, 149
Jungermannia atrovirens, 178
　　caespiticia, 179
　　exsertifolia ssp.cordifolia, 179
　　gracillima, 179
　　hyalina, 179
　　obovata, 179
　　pumila, 178
Juniper, 78
Juniperus communis, 78
Kickxia elatine, 129
　　spuria, 129
Killarney fern, 61
Kington, 2
Knapweed, 139
　　, Greater, 139
Knautia arvensis, 134
Knawel, Annual, 85
　　, Perennial, 85
Knighton, 2, 39
Knotgrass, 81
Knotweed, Giant, 82
　　, Himalayan, 82
　　, Japanese, 82
Knucklas, 7
Kurzia pauciflora, 182
LABIATAE, 125
Laburnum, 103
　　, Scotch, 103
Laburnum alpinum, 103
　　anagyroides, 103
Lady's-mantle, 101
　　, Hairy, 101
　　, Intermediate, 101
　　, Smooth, 102
Lady-fern, 76
Lagarosiphon major, 145
Lakeside Wood, 28, 218
Lambs, 56
Lambsear, 126
Lamium album, 125
　　maculatum, 125
　　purpurem, 125
Langerheimia, 164
Lapsana communis ssp. communis, 143
Larch, 1, 78
　　, Japanese, 1, 78
Larix decidua, 78
　　kaempferi, 78
Lasallia pustulata, 236
Lathraea squamaria, 132
Lathyrus latifolius, 105
　　montanus, 105
　　pratensis, 105
　　sylvestris, 105
Laurel, Cherry, 103
Lea Hall, 26
Lecanactis abietina, 236
　　amylacea, 236
　　dilleniana, 236
　　premnea, 236
Lecania cyrtella, 237
　　erysibe, 237
　　nylanderiana, 237
Lecanora aitema, 237
　　albescens, 237
　　atra, 237
　　badia, 237
　　campestris, 237
　　carpinea, 237
　　chlarotera, 237

Lecanora confusa, 237
　　conizaeoides, 237
　　crenulata, 237
　　dispersa, 237
　　epanora, 237
　　expallens, 237
　　gangaleoides, 238
　　grumosa, 238
　　intricata, 238
　　intumescens, 238
　　jamesii, 238
　　muralis, 238
　　polytropa, 238
　　pulicaris, 238
　　quercicola, 238
　　rupicola, 238
　　saligna, 238
　　soralifera, 239
　　subcarnea, 239
　　symmicta, 239
　　varia, 239
Lecidea carrollii, 239
　　chalybeioides, 239
　　erratica, 239
　　fuliginosa, 239
　　furvella, 239
　　fuscoatra, 239
　　hypnorum, 239
　　hypopta, 239
　　insularis, 239
　　lactea, 239
　　lapicida, 239
　　leucophaea, 240
　　lithophila, 240
　　monticola, 240
　　orosthea, 240
　　ecidea phaeops, 240
　　speirea, 240
　　sublivescens, 240
　　sulphurea, 240
　　valentior, 240
　　vernalis, 240
Lecidella elaeochroma, 240
　　scabra, 240
　　stigmatea, 240
　　subincongrua, 240
LEGUMINOSAE, 103
Lejeunea cavifolia, 184
　　lamacerina, 184
　　patens, 184
　　ulicina, 184
Lemanea fluviatilis, 164
Lemna minor, 156
　　trisulca, 156
LEMNACEAE, 156
LENTIBULAREACEAE, 132
Leontodon autumnalis, 140
　　hispidus, 140
　　taraxacoides, 140
Leopard's-bane, 138
　　, Plantain-leaved, 138
Lepidium campestre, 93
　　heterophyllum, 94
　　virginicum, 94
Lepidozia cupressina, 182
　　reptans, 182
Lepiocinclis, 164
Lepraria caesioalba, 241
　　incana, 241
　　lesdainii, 241
　　neglecta, 241
　　nivalis, 241
　　umbricola 241
Leprocaulon microscopicum, 241
Leproloma angardianum, 241
　　diffusum var. chrysodetoides, 241
　　diffusum var. diffusum, 241
　　membranaceum, 241
　　spp., 241
　　vouauxii, 241
Leproplaca chrysodeta, 241
Leptodontium flexifolium, 195
Leptogium cyanescens, 241
　　gelatinosum, 241
　　lichenoides, 241
　　palmatum, 241
　　teretiusculum, 242
　　turgidum, 242

Leptorhaphis epidermidis, 242
Leskea polycarpa, 206
Lettuce, Wall, 141
Leucobryum glaucum, 190
　　juniperoideum, 190
Leucodon sciuroides, 205
Ley, Rev. Augustin, 66
Ligustrum ovalifolium, 121
　　vulgare, 121
Lilac, 121
LILIACEAE, 146
Lilium pyrenaicum, 147
Lily, Pyrenean, 147
　　, Radnor, 147
Lily-of-the-Valley, 147
Lime, 111
　　, Large-leaved, 110
　　, Small-leaved, 110
Lime-mortared walls, 39
LIMNANTHACEAE, 107
Limnanthes douglasii, 107
Limosella aquatica, 128
LINACEAE, 108
Linaria purpurea, 128
　　repens, 128
Linum bienne, 108
　　catharticum, 108
　　usitatissimum, 108
Listera ovata, 161
Little Creigiau, 12
Little Lodge, 4
Littorella uniflora, 132
LIVERWORTs, 175
Llaithdu, 33
Llanbadarn Fynydd, 11, 29
Llanbedr Hill, 1, 12
Llanbister, 29, 33
Llanbwchllyn, 15, 36, 38, 65, 173
Llanddewi Ystradenni, 3, 11
Llandegley, 8, 10, 32
Llandegley Rhos, 15
Llandegley Rocks, 213
Llandeilo Graban, 34
Llandeilo Hill, 1, 12, 37
Llandrindod Lake, 36, 164
Llandrindod Wells, 2, 19, 66, 166, 167, 213, 218
Llanelwedd 2, 8, 10, 29, 32, 37
Llanelwedd Rocks, 218
Llanfawr, 10
Llangunllo, 2, 38
Llansantffraed Cwmdeuddwr, 5, 167
Llansantffraed-in-Elvel, 10
Llanstephan, 2, 7, 12, 164, 167
Llanwrthl, 164
Llanyre, 33, 166
Lloyd, Rev. John B., 68
Llyn, 15
Llyn Gwyn, 15, 38, 67
Llyn Heilyn, 15, 36, 38
Llyncerrigllwydion Isaf, 39
Llyncerrigllwydion Uchaf, 39
Lobaria amplissima, 242
　　pulmonaria, 242
Lobelia, Garden, 135
　　, Water, 135
Lobelia dortmanna, 135
　　erinus, 135
Lobularia maritima, 93
Lolium multiflorum, 150
　　perenne, 150
London Pride, 96
Lonicera caprifolium, 133
　　periclymenum, 133
Loosestrife, Purple, 114
　　, Dotted, 120
　　, Fringed, 120
　　, Yellow, 120
Lopadium disciforme, 242
Lophocolea bidentata, 180
　　cuspidata, 180
　　fragans, 180
　　heterophylla, 180
Lophozia bicrenata, 178
　　excisa var. excisa, 178
　　incisa, 178
　　sudetica, 178
　　ventricosa, 177

Index

LORANTHACEAE, 81
Lords-and-Ladies, 156
Lotus corniculatus, 106
    uliginosus, 106
Lousewort, 131
    , Marsh, 131
Lover's Leap, 166
Lower Woodhouse, 28
Lucerne, 105
Lunaria annua, 93
Lungwort, 123
Lunularia cruciata, 175
Lupin, Garden, 104
Lupinus polyphyllus, 104
Luzula campestris, 149
    forsteri, 150
    multiflora, 149
    pilosa, 150
    sylvatica, 149
Lychnis flos-cuculi, 86
    viscaria, 86
Lycium barbarum, 127
Lycopersicon esculentum, 127
LYCOPODIACEAE, 73
Lycopodium clavatum, 73
LYCOPSIDA, 73
Lycopus europaeus, 127
Lyngbya limnetica, 164
    putealis, 164
Lyonshall, 19
Lysimachia ciliata, 120
    nemorum, 120
    nummularia, 120
    punctata, 120
    vulgaris, 120
LYTHRACEAE, 114
Lythrum portula, 114
    salicaria, 114
Macentina stigonemoides, 242
Madder, Field, 121
Maesyfed, 13
Mahonia aquifolium, 89
Maize, 155
Mallow, Common, 111
    , Dwarf, 111
    , Musk, 111
Malomonas longiseta, 163
Malus domestica, 102
    sylvestris, 102
Malva moschata, 111
    neglecta, 111
    sylvestris, 111
MALVACEAE, 111
Manchester, 167
Maple, Field, 109
    , Norway, 109
Marchantia polymorpha, 175
Marcheini Uplands, 218
Marchesinia mackaii, 184
Marigold, Corn, 137
    , Pot, 138
Maritime, 217
Marjoram, 126
Marsh-marigold, 87
Marshwort, Lesser, 118
MARSILEACEAE, 77
Marsupella emarginata, 179
    funckii, 179
Marteg, 2, 5, 29
Massalongia carnosa, 242
Mat-grass, 155
Matricaria matricarioides, 137
    recutita, 136
Mawn Pools, 5, 15
Mayweed, Scented, 136
    , Scentless, 136
Meadow-grass, Annual, 151
    , Broad-leaved, 151
    , Flattened, 151
    , Narrow-leaved, 151
    , Rough, 151
    , Smooth, 151
    , Spreading, 151
    , Wood, 151
Meadow-rue, Lesser, 89
Meadowsweet, 97
Meconopsis cambrica, 90
Medicago lupulina, 105

Medicago sativa, 105
Medick, Black, 105
Melampsora euphorbiae, 261
    hypericorum, 262
    larici-populina, 262
    lini, 262
    populnea, 262
Melampsoridium betulinum, 262
Melampyrum pratense, 130
Melica nutans, 152
    uniflora, 152
Melick, Mountain, 152
    , Wood, 152
Melilot, White, 105
Melilotus alba, 105
Melissa officinalis, 126
Mentha aquatica, 127
    arvensis, 127
    spicata, 127
    suaveolens, 127
    x gentilis, 127
    x piperata, 127
    x smithiana, 127
    x verticillata, 127
    x villosa, 127
MENYANTHACEAE, 121
Menyanthes trifoliata, 121
Mercurialis annua, 108
    perennis, 108
Mercury, Annual, 108
    , Dog's, 108
Meridion circulare, 163
Methan, 11
Metzgeria conjugata, 176
    fruticulosa, 176
    furcata, 176
    temperata, 176
Mezereon, 111
Micarea adnata, 242
    bauschiana, 242
    botryoides, 242
    cinerea, 242
    denigrata, 242
    leprosula, 242
    lignaria, 242
    lutulata, 242
    melaena, 243
    nitschkeana, 243
    peliocarpa, 243
    prasina, 243
    subnigrata, 243
    sylvicola, 243
Michaelchurch-on-Arrow, 36
Michaelmas-daisy, 135
Micractinium, 164
Micrasterias spp., 163
Microcystis aeruginosa, 164
Microglaena muscorum, 243
Microspora, 164
Mid-Wales Railway, 2
Mignonette, Wild, 95
Milesina blechni, 262
    dieteliana, 262
    kriegeriana, 262
    scolopendrii, 262
    whitei, 262
Milium effusum, 155
Milkwort, Common, 109
    , Heath, 109
Millet, Wood, 155
Mimulus guttatus, 128
    moschatus, 128
Mint, Bushy, 127
    , Corn, 127
    , Large Apple, 127
    , Pepper, 127
    , Red, 127
    , Round-leaved, 127
    , Water, 127
    , Whorled, 127
Mires, 42
Mistletoe, 81
Miyagia pseudosphaeria, 262
Mnium hornum, 201
    stellare, 201
Moehringia trinervia, 84
Moelleropsis nebulosa, 243
Moenchia erecta, 85

Moity, 27
Molinea caerulea, 155
Monaughty, 7
Monk's-hood, 87
Monkeyflower, 128
Monoraphidium tortile, 163
Montia fontana, 84
    sibirica, 84
Moonwort, 74
Moor-grass Purple, 155
Moschatel, 133
MOSSES, 184
Mougeotia, 164
Mouse-ear, Common, 85
    , Field, 85
    , Little, 85
    , Sea, 85
    , Sticky, 85
Mudwort, 128
Muellerella pygmaea, 243
Mugwort, 137
Mullein, Dark, 128
    , Great, 128
MUSCI, 184
Musk, 128
Mustard, Black, 94
    , Garlic, 91
    , Hedge, 91
    , Hoary, 94
    , Treacle, 91
    , White, 94
Mycelis muralis, 141
Mycoblastus sanguinarius, 243
    sterilis, 243
Mycoporum quercus, 243
Mylia anomala, 178
    taylorii, 178
Myosotis arvensis, 123
    discolor, 123
    laxa ssp. caespitosa, 124
    ramosissima, 123
    scorpioides, 124
    secunda, 124
    sylvatica, 124
Myosoton aquaticum, 85
Myrinia pulvinata, 206
Myriophyllum alterniflorum, 116
    spicatum, 115
    verticillatum, 115
Myrrhis odorata, 117
Nannerth, 11, 218
Nantmel, 13
Narcissus pseudonarcissus, 148
Nardia compressa, 179
    scalaris, 179
Nardus stricta, 155
Narthecium ossifragum, 146
Nasturtium microphyllum, 92
Nasturtium officinale, 92
National Nature Reserves, 63
National Vegetation Classification, 26
Nature Conservancy Council, 70
Navelwort, 95
Navicula cryptocephala, 163
    pupula var. capitata, 163
Neckera complanata, 205
    crispa, 205
    pumila, 205
Neidium iridis, 163
Neottia nidus-avis, 161
Nephroma laevigatum, 243
    parile, 244
Nettle, Common, 81
    , Small, 81
    , Common Hemp-, 125
    , Large-Flowered Hemp-, 125
    , Red Dead-, 125
    , Red Hemp-, 125
    , Spotted Dead-, 125
    , White Dead-, 125
New Radnor, 2, 7, 39
Newbridge on Wye, 2, 8, 214
Newchurch, 7
Newmead Farm, 10
Nightshade, Black, 127
Nipplewort, 143
Nitella flexilis var flexilis, 165

Index

Nitrate, 218
Nitzschia dissipata, 163
Nitzschia spp., 164
Normandina pulchella, 244
Norton, 13
Nothofagus obliqua, 81
    procera, 81
Nowellia curvifolia, 182
Nuphar lutea, 87
Nymphaea alba, 87
NYMPHAEACEAE, 87
Oak, Hybrid, 80
    , Pedunculate, 80
    , Red, 80
    , Sessile, 80
    , Turkey, 80
Oat, Cultivated, 154
Oat-grass, Downy, 154
    , False, 154
    , Yellow, 154
Ochrolechia androgyna, 244
    parella, 244
    subviridis, 244
    tartarea, 244
    tuneri, 244
Odontites verna, 131
Odontoschisma denudatum, 181
    sphagni, 181
Oedogonium, 165
Oenanthe crocata, 117
    fistulosa, 117
Oenothera erythrosepala, 114
Old Radnor, 7, 8, 32
OLEACEAE, 121
Oligotrichum hercynicum, 187
Omalotheca sylvatica, 136
Omphalina cupulatoides, 244
    ericetorum, 244
    velutina, 244
ONAGRACEAE, 114
Onion, Wild, 147
Ononis repens, 105
Oocystis, 164
    crassa, 163
Opegrapha atra, 244
    gyrocarpa, 244
    herbarum, 245
    lithyrga, 245
    lyncea, 245
    ochrocheila, 245
    rufescens, 245
    saxatilis, 245
    sorediifera, 245
    varia, 245
    vermicellifera, 245
    vulgata, 245
    zonata, 245
OPHIOGLOSSACEAE, 74
Ophioglossum vulgatum, 74
Orache, Common, 83
    , Spear-leaved, 83
Orchid, Bird's-nest, 161
    , Bog, 162
    , Common Spotted-, 162
    , Early Marsh-, 162
    , Early-purple, 162
    , Fragant, 161
    , Frog, 161
    , Greater Butterfly-, 161
    , Green-winged, 162
    , Heath Spotted-, 162
    , Lesser Butterfly-, 161
    , Northern Marsh-, 162
    , Southern Marsh-, 162
ORCHIDACEAE, 161
Orchis mascula, 162
    morio, 162
Ordovices, 8
Oregon-grape, 89
Oreopteris limbosperma, 75
Origanum vulgare, 126
Ornithopus perpusillus, 107
OROBANCHACEAE, 132
Orobanche rapum-genistae, 132
Orpine, 95
Orthodontium lineare, 198
Orthothecium intricatum, 211
Orthotrichum affine, 203

Orthotrichum anomalum, 204
    cupulatum, 204
    diaphanum, 204
    lyellii, 203
    pulchellum, 204
    rivulare, 203
    rupestre, 203
    sprucei, 204
    stramineum, 204
    striatum, 203
    tenellum, 204
Oscillatoria redekii, 164
Osier, 79
Osmunda regalis, 74
OSMUNDACEAE, 74
Oswestry, 213
Ox-tongue, Bristly, 140
    , Hawkweed, 140
Oxalidaceae, 107
Oxalis acetosella, 107
    corniculata, 107
    europaea, 107
Oxidised nitrogen, 217
Oxystegus tenuirostris, 195
Painscastle, 5, 15
Pannaria conoplea, 245
    leucophaea, 245
Pansy, Field, 113
    , Mountain, 113
    , Wild, 113
Pant y Dwr, 15, 16, 19
Pantllwyd, 5
Papaver argemone, 89
    dubium, 89
    lecoqii, 89
    rhoeas, 89
    somniferum, 89
PAPAVERACEAE, 89
Parietaria judaica, 81
Paris, Herb, 148
Paris quadrifolia, 148
Parmelia britannica, 245
    caperata, 245
    conspersa, 246
    discordans, 246
    disjuncta, 246
    elegantula, 246
    exasperata, 246
    exasperatula, 246
    glabratula, 246
    laciniatula, 246
    laevigata, 246
    loxodes, 246
    mougeotii, 246
    omphalodes, 246
    pastillifera, 246
    perlata, 246
    reddenda, 246
    revoluta, 247
    saxatillis, 247
    stygia, 247
    subaurifera, 247
    subrudecta, 247
    sulcata, 247
    taylorensis, 247
    verruculifera, 247
Parmeliella jamesii, 247
    triptophylla, 247
Parmeliopsis aleurites, 247
Parsley, Fool's, 117
    , Garden, 118
Parsley-piert, 102
Parsnip, Wild, 118
Pastinaca sativa, 118
Pea, Broad-leaved Everlasting-, 105
    , Garden, 105
    , Narrow-leaved Everlasting-, 105
Pear, Wild, 102
Pearlwort, Fringed, 85
    , Knotted, 85
    , Procumbent, 85
Peatland, 16
Pediastrum, 164
Pedicularis palustris, 131
    sylvatica, 131
Pellia endiviifolia, 177
    epiphylla, 176
    neesiana, 176

Pellitory-of-the-wall, 81
Peltigera collina, 248
    didactyla, 248
    horizontalis, 248
    lactucifolia, 248
    leucophlebia, 248
    membranacea, 248
    praetexta, 248
    rufescens, 248
Pen y Berth, 28
Pen-y-banc, 10
Pen-y-Garreg, 3, 11, 28
Pencerrig Lake, 37
Penddol, 67
Pennines, 215
Penny-cress, Alpine, 93
    , Field, 93
Pennywort, Marsh, 116
Pentaglottis sempervirens, 123
Pentrosfa Bog, 37
Penybont, 31
Pepper-saxifrage, 117
Pepperwort, Field, 93
    , Poor-man's, 94
    , Smith's, 94
Peridinium willei, 163
Periwinkle, Greater, 121
    , Lesser, 121
Perring, F.H., 69
Persicaria, Pale, 82
Pertusaria albescens, 248
    amara, 248
    coccodes, 248
    corallina, 249
    coronata, 249
    dealbescens, 249
    excludens, 249
    flavicans, 249
    flavida, 249
    hemisphaerica, 249
    hymenea, 249
    lactea, 249
    leioplaca, 249
    multipuncta, 249
    pertusa, 249
    pseudocorallina, 249
    pupillaris, 250
Petasites hybridus, 137
Petroselinum crispum, 118
Petty whin, 104
Pezizella epithallina, 250
Phaeoceros laevis ssp.laevis, 175
Phaeographis dendritica, 250
    smithii, 250
Phaeophyscia orbicularis, 250
Phaeopyxis varia, 250
Phalaris arundinacea, 155
    canariensis, 155
Phascum cuspidatum, 193
Phascus, 164
Phegopteris connectilis, 75
Philonotis arnellii, 202
    calcarea, 202
    fontana, 202
    rigida, 202
Phleum pratense ssp. bertolonii, 155
    pratense ssp. pratense, 155
Phlyctis argena, 250
Phragmidium bulbosum, 262
    fragariae, 262
    mucronatum, 262
    violaceum, 262
Phragmites australis, 155
Phragmonaevia fuckelii, 250
Phyllitis scolopendrium, 76
Phyllopsora rosei, 250
Physcia adscendens, 250
    aipolia, 250
    caesia, 250
    dubia, 250
    tenella, 250
    tribacia, 250
    wainioi, 251
Physcomitrium pyriforme, 198
Physconia distorta, 251
    enteroxantha, 251
    grisea, 251
Picea abies, 78

Index

Picea sitchensis, 78
Picris echioides, 140
 hieracioides, 140
Pignut, 117
Pigweed, 84
Pillwort, 61, 77
Pilosella aurantiaca ssp. carpathicola, 143
 officinarum, 143
 praealta, 143
Pilularia globulifera, 61, 77
Pimpernel, Bog, 120
 , Scarlet, 121
 , Yellow, 120
Pimpinella saxifraga, 117
PINACEAE, 78
Pine, Lodge-pole, 78
 , Scots, 78
Pineappleweed, 137
Pingos, 15
Pinguicula vulgaris, 132
Pink, Maiden, 86
Pinus contorta, 78
 nigra, 78
 peuce, 78
 strobus, 78
 sylvestris, 78
 wallichiana, 78
Pisum sativum, 105
Placopsis gelida, 251
 lambii, 251
Placynthiella icmalea, 251
 uliginosa, 251
Placynthium nigrum, 251
Plagiobryum zieri, 199
Plagiochila asplenoides, 179
 britannica, 179
 porelloides, 179
 punctata, 180
 spinulosa, 180
Plagiomnium affine, 201
 cuspidatum, 201
 latum, 201
 ellipticum, 201
 rostratum, 201
 undulatum, 201
Plagiopus oederi, 202
Plagiothecium curvifolium, 211
 denticulatum, 211
 laetum, 211
 latebricola, 211
 nemorale, 211
 ruthei, 211
 succulentum, 211
 undulatum, 211
PLANTAGINACEAE, 132
Plantago lanceolata, 132
 major, 132
 media, 132
Plantain, Greater, 132
 , Hoary, 132
 , Ribwort, 132
 , Water, 145
Platanthera bifolia, 161
 chlorantha, 161
Platismatia glauca, 251
Pleuridium acuminatum, 187
 subulatum, 187
Pleurotaenium trabecula, 163
Pleurozium schreberi, 212
Ploughman's-spikenard, 136
Plum, 103
Poa angustifolia, 151
 annua, 151
 chaixii, 151
 compressa, 151
 nemoralis, 151
 pratensis s.s., 151
 subcaerulea, 151
 trivialis, 151
Poached-egg-flower, 107
Pogonatum aloides, 187
 nanum, 187
 urnigerum, 187
Pohlia annotina agg., 198
 bulbifera, 198
 camptotrachela, 199
 carnea, 199
 cruda, 198

Pohlia elongata ssp.elongata, 198
 lutescens, 199
 nutans, 198
 proligera, 199
 wahlenbergii, 199
POLEMONIACEAE, 122
Polemonium caeruleum, 122
Pollen remains, 55
Polyblastia allobata, 251
Polychidium muscicola, 251
Polygala serpyllifolia, 109
 vulgaris, 109
POLYGALACEAE, 109
POLYGONACEAE, 81
Polygonatum multiflorum, 148
Polygonum amphibium, 82
 arenastrum, 82
 aviculare, 81
 bistorta, 82
 hydropiper, 82
 lapathifolium, 82
 persicaria, 82
 polystachyum, 82
POLYPODIACEAE, 77
Polypodium cambricum, 77
 interjectum, 77
 vulgare, 77
Polypody, 77
Polysporina simplex, 252
Polystichum aculeatum, 77
 setiferum, 76
Polytrichum alpinum, 186
 commune, 186
 formosum, 186
 juniperinum, 187
 longisetum, 186
 piliferum, 186
Pondweed, Blunt-leaved, 146
 , Bog, 145
 , Broad-leaved, 145
 , Canadian, 145
 , Curled, 146
 , Fennel, 146
 , Flat-stalked, 146
 , Horned, 146
 , Long-stalked, 146
 , Perfoliate, 146
 , Red, 146
 , Shining, 146
 , Small, 146
Poplar, Black, 80
 , White, 79
Poppy, Common, 89
 , Long-headed, 89
 , Opium, 89
 , Prickly, 89
 , Welsh, 90
 , Yellow-juiced, 89
Populus alba, 79
 gileadensis, 80
 nigra, 80
 tremula, 79
 trichocarpa, 80
 x canescens, 80
Porella arboris-vitae var. arboris-vitae, 183
 cordeana var. cordeana, 183
 pinnata, 183
 platyphylla, 183
Porina aenea, 252
 ahlesiana, 252
 chlorotica, 252
 guentheri, 252
 lectissima, 252
 leptalea, 252
 linearis, 252
PORTULACEAE, 84
Potamogeton alpinum, 146
 berchtoldii, 146
 crispus, 146
 friesii, 146
 lucens, 146
 natans, 145
 obtusifolius, 146
 pectinatus, 146
 perfoliatus, 146
 polygonifolius, 145
 praelongus, 146

POTAMOGETONACEAE, 145
Potato, 127
Potentilla anglica, 101
 anserina, 101
 erecta, 101
 palustris, 101
 reptans, 101
 rupestris, 101
 sterilis, 101
Pottia crinata, 193
 truncata, 193
Powell, Miss A.C., 69
Powell, Rev. A.Wentworth, 68
Pragmidium tuberculatum, 262
Precambrian, 7
Preissia quadrata, 175
Presteigne, 2, 13, 39
Primrose, 120
Primula veris, 120
 vulgaris, 120
PRIMULACEAE, 120
Privet, Garden, 121
 , Wild, 121
Protoblastenia calva, 252
 rupestris, 252
Prunella vulgaris, 126
Prunus avium, 103
 cerasus, 103
 domestica, 103
 laurocerasus, 103
 padus, 103
 spinosa, 103
Pseudephemerum nitidum, 188
Pseudevernia furfuracea, 252
Pseudoscleropodium purum, 210
Pseudotsuga menziesii, 78
Psilolechia clavulifera, 252
 leprosa, 252
 lucida, 252
Pteridium aquilinum, 74
PTERIDOPHYTA, 73
Pterogonium gracile, 205
Ptilidium ciliare, 183
 pulcherrimum, 183
Ptychographa xylographoides, 252
Ptychomitrium polyphyllum, 197
Puccinia acetosae, 262
 adoxae, 262
 aegopodii, 262
 albescens, 262
 annularis, 262
 antirrhini, 262
 arenariae, 263
 behenis, 263
 betonicae, 263
 brachypodii, 263
 buxi, 263
 calcitrapae, 263
 calthicola, 263
 caricina, 263
 chaerophylli, 263
 circaeae, 263
 cnici, 263
 cnici-oleracei, 263
 coronata, 263
 crepidicola, 263
 deschampsiae, 263
 galii-verni, 263
 glechomatis, 263
 hieracii, 263
 lagenophorae, 263
 lapsanae, 264
 maculosa, 264
 menthae, 264
 obscura, 264
 pimpinellae, 264
 porri, 264
 pulverulenta, 264
 punctata, 264
 punctiformis, 264
 sessilis, 264
 tumida, 264
 umblilici, 264
 veronicae, 264
 violae, 264
 vaccinii, 264
Pulicaria dysenterica, 136

Index

Pulmonaria officinalis, 123
Purslane, Pink, 84
Pycnothelia papillaria, 252
Pylaisia polyantha, 212
Pyrenopsis rhodosticta, 252
Pyrenula chlorospila, 252
　　macrospora, 253
　　nitidia var. nitidella, 252
　　occidentalis, 253
Pyrrhospora quernea, 253
Pyrus pyraster, 102
Quaking-grass, 152
Quercus cerris, 80
　　petraea, 80
　　robur, 80
　　rubra, 80
　　x rosacea, 80
Quickthorn, 102
Quill-wort, 73
Racodium rupestre, 253
Racomitrium aciculare, 196
　　affine, 197
　　aquaticum, 196
　　canescens group, 197
　　fasciculare, 197
　　heterostichum, 197
　　lanuginosum, 197
Radish, Wild, 95
Radnor Forest, 5, 12, 17, 173
Radnorshire Wildlife Trust, 63
Radula complanata, 183
Ragged-robin, 86
Ragwort, Common, 138
　　, Marsh, 138
　　, Oxford, 138
Railways, 2
Rainfall, 21
Ramalina calicaris, 253
　　farinacea, 253
　　fastigiata, 253
　　fraxinea, 253
　　pollinaria, 253
　　subfarinacea, 253
Ramping-fumitory, 90
　　, Common, 90
　　, Purple, 90
Ramsons, 147
RANUNCULACEAE, 87
Ranunculus acris, 88
　　aquatilis, 88, 89
　　arvensis, 88
　　auricomus, 88
　　bulbosus, 88
　　ficaria, 88
　　flammula, 88
　　fluitans, 89
　　hederaceus, 88
　　lingua, 88
　　omiophyllus, 88
　　parviflorus, 88
　　peltatus, 88
　　penicillatus, 89
　　repens, 88
　　sceleratus, 88
　　trichophyllus, 89
　　tripartitus, 88
Rape, 94
Raphanus raphanistrum, 95
Raspberry, 97
Rattle, Yellow, 131
Reboulia hemisphaerica, 175
Redshank, 82
Redwoods, 78
Reed, Common, 155
Reseda lutea, 95
　　luteola, 95
RESEDACEAE, 95
Restharrow, Common, 105
Reynoutria japonica, 82
　　sachalinensis, 82
Rhabdoweisia crenulata, 188
　　crispata, 188
　　fugax, 188
RHAMNACEAE, 110
Rhamnus catharticus, 110
Rhayader, 2, 213, 218
Rhayader Quarries, 8
Rhinanthus minor, 131

Rhizocarpon concentricum, 253
　　distinctum, 253
　　geminatum, 253
　　geographicum agg., 254
　　hochstetteri, 254
　　lavatum, 254
　　obscuratum, 254
　　oederi, 254
　　polycarpum, 254
　　riparium ssp. lindsayanum, 254
　　viridiatrum, 254
Rhizoclonium hieroglyphicum, 165
Rhizomnium pseudopunctatum, 201
　　punctatum, 201
Rhodobryum roseum, 201
Rhododendron, 119
Rhododendron ponticum, 119
Rhoicosphenia curvata, 163
Rhosgoch, 16, 28, 30, 36, 37, 38, 55, 60, 63, 65, 173
Rhosgoch Common, 173
Rhynchospora alba, 158
Rhynchostegiella teesdalei, 211
　　tenella, 211
Rhynchostegium confertum, 210
　　lusitancium, 210
　　murale, 210
　　ripariodes, 210
Rhytiadelphus loreus, 212
　　squarrosus, 212
　　triquetrus, 212
Ribes alpinum, 97
　　nigrum, 97
　　rubrum, 97
　　sanguineum, 97
　　uva-crispa, 97
Riccardia chamaedrifolia, 176
　　latifrons, 176
　　multifida, 176
　　palmata, 176
Riccia beyrichiana, 175
　　fluitans, 175
　　glauca, 175
　　nigrella, 176
　　sorocarpa, 176
　　subbifurca, 175
Riddelsdell, Rev. H.J., 67
Ridley, H.N., 67
Rinodina exigua, 254
　　griseosoralifera, 254
　　lecideina, 254
　　roboris, 254
　　sophodes, 254
River Arrow, 7
River Edw, 5, 15
River Elan, 163, 164
River Irfon, 163
River Ithon, 5, 163, 164
River Lugg, 7, 164
River Teme, 7, 167
River Wye, 4, 5, 163, 167, 172, 173, 218
Rock-cress, Bristol, 62, 92
　　, Hairy, 92
Rock-rose, Common, 113
Rocket, Eastern, 90
　　, London, 90
　　, Tall, 90
Rogers, Rev. W.Moyle, 66
Rorippa palustris, 92
　　sylvestris, 91
Rosa afzeliana, 100
　　arvensis, 99
　　canina, 100
　　coriifolia, 100
　　dumetorum, 100
　　micrantha, 100
　　obtusifolia, 100
　　pimpinellifolia, 99
　　rubiginosa, 100
　　rugosa, 100
　　sherardii, 100
　　stylosa, 100
　　tomentosa, 100
　　villosa, 100
ROSACEAE, 97
Rose, Burnet, 99
　　, Close-styled, 100
　　, Dog, 100

Rose, Downy, 100
　　, Field, 99
　　, Japanese, 100
Rose, Guelder, 133
Rose of Sharon, 111
Rosemary, Bog, 119
Round F.E., 163
Rowan, 102
Royal Society for the Protection of Birds, 64
RUBIACEAE, 121
Rubus acclivitatum, 98
　　amplificatus, 98
　　babingtonii, 99
　　bartonii, 98
　　bertrami, 98
　　biloensis, 98
　　caesius, 99
　　cardiophyllus, 98
　　dasyphyllus, 99
　　dumnoniensis, 98
　　echinatus, 99
　　euryanthemus, 99
　　hylocharis, 99
　　idaeus, 97
　　incurvatus, 98
　　laciniatus, 98
　　leyanus, 99
　　lindebergii, 98
　　lindleianus, 98
　　longithyrsiger, 99
　　longus, 98
　　ludensis, 98
　　merlini, 99
　　moylei, 98
　　mucronatoides, 98
　　nemorales, 98
　　pallidus, 99
　　pascuorum, 99
　　perdigitatus, 98
　　platyacanthus, 98
　　plicatus, 98
　　polyanthemus, 98
　　prolongatus, 98
　　pruinosus, 99
　　purchasianus, 99
　　pyramidalis, 98
　　raduloides, 98
　　rhombifolius, 98
　　rubritinctus, 98
　　saxatilis, 97
　　scissus, 98
　　septentrionalis, 98
　　silurum, 98
　　tuberculatus, 99
　　ulmifolius, 98
　　vestitus, 98
　　vigorosus, 98
　　wirralensis, 98
Rumex acetosa, 83
　　acetosella, 82
　　conglomeratus, 83
　　crispus, 83
　　hydrolapathum, 83
　　obtusifolius, 83
　　sanguineus, 83
Rush, Blunt-flowered, 149
　　, Bulbous, 149
　　, Compact, 149
　　, Flowering, 145
　　, Hard, 148
　　, Heath, 149
　　, Jointed, 149
　　, Sharp-flowered, 149
　　, Slender, 149
　　, Soft, 148
　　, Toad, 149
Rye-grass, Italian, 150
　　, Perennial, 150
Saccogyna viticulosa, 180
Saffron, Meadow, 147
Sage, Wood, 125
Sagina apetala, 85
　　nodosa, 85
　　procumbens, 85
SALICACEAE, 79
Salix alba, 79
　　aurita, 79

Index

Salix caprea, 79
   cinerea, 79
   cinerea ssp. oleifolia, 79
   fragilis, 79
   purpurea, 79
   repens, 79
   triandra, 79
   viminalis, 79
Sambucus ebulus, 133
   nigra, 133
   racemosa, 133
Sandwort, Mossy, 84
   , Three-nerved, 84
   , Thyme-leaved, 84
Sanguisorba minor, 100
   officinalis, 100
Sanicle, 116
Sanicula europaea, 116
Saponaria officinalis
Sarcogyne privigna, 254
   regularis, 254
Saw-wort, 139
Saxifraga cymbalaria, 96
   granulata, 96
   hypnoides, 96
   rosacea, 96
   tridactylites, 96
   urbium, 96
SAXIFRAGACEAE, 96
Saxifrage, Celandine, 96
   , Irish, 96
   , Meadow, 96
   , Mossy, 96
   , Rue-leaved, 96
Scabious, Devil's-bit, 134
   , Field, 134
Scandinavia, 167, 168
Scandix pecten-veneris, 116
Scapania aspera, 181
   compacta, 181
   gracilis, 181
   irrigua, 181
   nemorea, 180
   scandica, 180
   subalpina, 181
   umbrosa, 180
   undulata, 181
Scenedesmus spp., 163, 164
Schaereria cinereorufa, 254
   tenebrosa, 255
Schismatomma cretaceum, 255
   decolorans, 255
Schistidium alpicola, 196
   apocarpum, 196
Schistostega pennata, 198
Scirpus cespitosus ssp.germanicus,157
   fluitans, 157
   setaceus, 156
   sylvaticus, 156
Scleranthus annuus, 85
   perennis, 85
Sclerococcum sphaerale, 255
Scleropodium cespitans, 210
   tourettii, 210
Scoliciosporum chlorococcum, 255
   umbrinum, 255
Scorpidium scorpioides, 208
Scotland, 167, 170, 215
Scrophularia auriculata, 128
   nodosa, 128
SCROPHULARIACEAE, 128
Scrubland, 41
Scutellaria galericulata, 125
   minor, 125
Sedge, Bladder, 159
   , Bottle, 159
   , Broom, 158
   , Carnation, 159
   , Common, 160
   , Common Yellow-, 160
   , Dioecious, 158
   , False Fox-, 158
   , Flea, 161
   , Glaucous, 159
   , Greater Pond-, 159
   , Green-ribbed, 160
   , Grey, 158
   , Hairy, 158

Sedge, Lesser Pond-, 159
   , Long-stalked Yellow-, 160
   , Oval, 158
   , Pale, 160
   , Pendulous, 159
   , Pill, 160
   , Prickly, 158
   , Remote, 158
   , Sand, 158
   , Slender Tufted-, 160
   , Smooth-stalked, 160
   , Spiked, 158
   , Spring, 160
   , Star, 158
   , Tawny, 160
   , Thin-spiked Wood-, 159
   , White, 158
   , Wood, 159
Sedum acre, 96
   album, 96
   anglicum, 96
   forsterianum, 95
   reflexum, 95
   sexangulare, 96
   spurium, 95
   telephium, 95
Selfheal, 126
Seligeria pusilla, 188
   recurvata, 188
Sempervivum tectorum, 95
Senecio aquaticus, 138
   jacobaea, 138
   squalidus, 138
   sylvaticus, 138
   viscosus, 138
   vulgaris spp. vulgaris, 138
Sequoia sempervirens, 78
Sequoiadendron giganteum, 78
Serratula tinctoria, 139
Service, Wild, 102
Severn, 215
Shaky Bridge, 166
Sheep, 56
Sheep's-bit, 135
Shepherd's needle, 116
Shepherd's-cress, 93
Shepherd's-purse, 93
Sherardia arvensis, 121
Shield-fern, Hard, 77
   , Soft, 76
Shoreweed, 132
Shrewsbury, 213
Silaum silaus, 117
Silene dioica, 86
   gallica, 86
   pratensis, 86
   vulgaris, 86
Silverweed, 101
Sinapis alba, 94
   arvensis, 94
Sisymbrium altissimum, 90
   irio, 90
   officinale, 91
   orientale, 90
Sites of Special Scientific Interest, 63
Sketty, 2
Skullcap, 125
   , Lesser, 125
Slater, F.M., 70
Snapdragon, 128
Sneezewort, 136
Snow-in-summer, 84
Snowberry, 133
Snowdrop, 148
Snowfall, 24
Soane, I.D., 70
Soapwort, 86
Soft-grass, Creeping, 154
SOLANACEAE, 127
Solanum dulcamara, 127
   tuberosum, 127
Solidago canadensis, 135
   virgaurea, 135
Solomon's-seal, 148
Sonchus arvensis, 140
   asper, 140
   oleraceus, 140
Sorbus anglica, 102

Sorbus aucuparia, 102
   porrigentiformis, 102
   rupicola, 102
   torminalis, 102
Sorrel, Common, 83
   , Sheep's, 82
Sow-thistle, Blue, 141
   , Perennial, 140
   , Prickly, 140
   , Smooth, 140
SPARGANIACEAE, 156
Sparganium emersum, 156
   erectum, 156
Spearmint, 127
Spearwort, Greater, 88
   , Lesser, 88
Speedwell, Blue Water-, 129
   , Common Field-, 130
   , Germander, 129
   , Green Field-, 130
   , Grey Field-, 130
   , Heath, 129
   , Ivy-leaved, 130
   , Marsh, 129
   , Slender, 130
   , Spiked, 130
   , Thyme-leaved, 129
   , Wall, 130
   , Wood, 129
Spergula arvensis, 86
Spergularia rubra, 86
Sphaerophorus fragilis, 255
   globosus, 255
Sphagnum auriculatum, 185
   auriculatum var. inundatum, 185
   capillifolium, 185
   compactum, 185
   contortum, 186
   cuspidatum, 186
   fimbriatum, 185
   girghensohnii, 185
   palustre, 184
   quinquefarium, 185
   recurvum, 186
   russowii, 185
   squarrosum, 184
   subnitens, 185
   subsecundum, 185
   tenellum, 186
   teres, 184
   warnstorfii, 185
SPHENOPSIDA, 73
Sphinctrina turbinata, 255
Spike-rush, Common, 157
   , Few-flowered, 157
   , Many-stalked, 157
   , Needle, 157
Spindle, 110
Spiraea, 97
Spirogyra, 164, 165
Splachnum ampullaceum, 198
   sphaericum, 198
Spleenwort, Black, 75
   , Forked, 75
   , Green, 75
   , Maidenhair, 75
Spruce, Norway, 78
   , Sitka, 1, 78
Spurge, Caper, 109
   , Dwarf, 109
   , Petty, 109
   , Sun, 108
   , Wood, 109
Spurge Laurel, 111
Spurrey, Corn, 86
   , Sand, 86
Squitch, 153
St George's Land, 12
St Harmon, 5, 19
St John's-wort, Hairy, 111
   , Imperforate, 112
   , Marsh, 112
   , Pale, 112
   , Perforate, 112
   , Slender, 112
   , Square-stalked, 112
   , Toadflax-leaved, 112
Stachys arvensis, 126

Index

Stachys byzantina, 126
  officinalis, 126
  palustris, 126
  sylvatica, 126
Stanage, 13, 38
Stanner, 2, 13, 29, 213, 218
Stanner Rocks, 7, 32, 63, 166, 167, 168, 169, 172, 173
Starwort, Blunt-fruited Water-, 124
  , Common Water-, 124
  , Intermediate Water-, 124
  , Pedunculate, 124
  , Various-leaved Water-, 124
Staurastrum anatinum, 163
  bibrachiatum, 164
  irregulare, 164
  paradoxum, 163
  pseudotetracerum, 164
  tetracerum, 164
Staurothele fissa, 255
Steinia geophana, 255
Stellaria graminea, 84
  holostea, 84
  media, 84
  neglecta, 84
  nemorum, 84
  pallida, 84
  uliginosa, 84
Stenocybe pullulata, 255
Stereocaulon evolutum, 256
  leucophaeopsis, 256
  nanodes, 256
  vesuvianum, 256
Sticta fuliginosa, 256
  limbata, 256
Stigeoclonium, 164
Stitchwort, Bog, 84
  , Greater, 84
  , Lesser, 84
  , Wood, 84
Stonecrop, Biting, 96
  , Caucasian, 95
  , English, 96
  , Reflexed, 95
  , Rock, 95
  , Tasteless, 96
  , White, 96
Stoneworts, 165
Stork's-bill, Common, 108
Strangospora moriformis, 256
  ochrophora, 256
Straurodesmus cuspidatus, 164
Strawberry, Barren, 101
  , Hautbois, 101
  , Wild, 101
Succisa pratensis, 134
Sulphur dioxide, 217
Summergil Brook, 7, 13, 15, 16
Sundew, Common, 95
Sunflower, Common, 136
Sunshine, 21
Surirella brebissonii, 164
Swamps, 44
Swansea Field Naturalists Society, 68
Swdge, Great Fen-, 158
Swede, 94
Sweet Briar, 100
Sweet Cicely, 117
Sweet William, 86
Sweet-grass, Floating, 152
  , Plicate, 152
  , Reed, 152
  , Small, 152
Swine-cress, 94
  , Lesser, 94
Sycamore, 109
Symphoricarpos albus, 133
Symphytum officinale, 123
  x uplandicum, 123
Syringa vulgaris, 121
Tabellaria fenestrata, 163, 164
  flocculosa, 163, 164
Tachyspora intrusa, 264
Tamus communis, 148
Tanacetum parthenium, 137
  vulgare, 137
Tansy, 137
Taraxacum alatum, 142

Taraxacum ancistrolobum, 142
  argutum, 141
  atactum, 141
  brachyglossum, 141
  bracteatum, 141
  canulum, 141
  cordatum, 142
  croceiflorum, 142
  degelii, 141
  dilaceratum, 142
  ekmanii, 142
  euryphyllum, 141
  exacutum, 142
  expallidiforme, 142
  faeroense, 141
  fasciatum, 142
  fulviforme, 141
  gelertii, 141
  hamatiforme, 142
  hamatum, 142
  hamiferum, 142
  hemicyclum, 142
  hesperium, 142
  horridifrons, 142
  insigne, 142
  lacistophyllum, 141
  laeticolor, 142
  laticordatum, 142
  lingulatum, 142
  maculosum, 141
  marklundii, 142
  necessarium, 142
  nordstedii, 141
  oblongatum, 142
  oxoniense, 141
  pachymerum, 142
  pannucium, 142
  pannulatiforme, 142
  pannulatum, 142
  pectinatiforme, 142
  piceatum, 142
  proximum, 141
  pseudohamatum, 142
  quadrans, 142
  raunkiaeri, 141
  rhamphodes, 142
  richardsianum, 141
  rubicundum, 141
  sagittipotens, 142
  sellandii, 143
  subbracteatum, 141
  sublaeticolor, 143
  sublongisquameum, 143
  undulatum, 143
  unguilobum, 141
  xanthostigma, 143
Tare, Hairy, 104
  , Smooth, 104
Targionia hypophylla, 175
TAXACEAE, 78
Taxiphyllum wissgrillii, 211
TAXODIACEAE, 78
Taxus baccata, 78
Teaplant, Duke of Argyll's, 127
Teasel, 134
  , Small, 134
Teesdalia nudicaulis, 93
Tellima grandiflora, 96
Tetra desmus, 164
Tetraedron, 164
Tetraphis pellucida, 186
Tetraplodon mnioides, 198
Teucrium scorodonia, 125
Thalassiosira fluvialilis, 163
Thale-cress, 91
Thalictrum minus, 89
Thamnobryum alopecuroides, 205
Thelidium minutulum, 256
  pyrenophorum, 256
Thelocarpon epibolum, 256
  intermediellum, 256
  laureri, 256
  lichenicola, 256
Thelopsis rubella, 256
Thelotrema lepadinum, 256
THELYPTERIDACEAE, 75
Thelypteris thelypteroides, 75
Thistle, Carline, 138

Thistle, Creeping, 139
  , Marsh, 139
  , Meadow, 139
  , Melancholy, 139
  , Musk, 139
  , Spear, 139
  , Welted, 139
  , Woolly, 139
Thlaspi alpestre, 93
  arvense, 93
Thorn Apple, 127
Thorow-wax, False, 117
Three Cocks, 2
Three Wells, 10
Thrombium epigaeum, 256
Thuidium delicatulum, 206
  philibertii, 206
  tamariscinum, 206
Thuja plicata, 78
Thyme, Wild, 127
THYMELAEACEAE, 111
Thymus praecox ssp. articus, 127
Tilia cordata, 110
  platyphyllos, 110
  x vulgaris, 111
TILIACEAE, 110
Tilletia sphaerococca, 265
Timberhill, 26, 27
Timothy, 155
Tir-uched Farm, 4
Titley, 2
Toadflax, Common, 129
  , Ivy-leaved, 129
  , Pale, 128
  , Purple, 128
  , Small, 128
Tolypothrix distorta, 164
Tomasellia gelatinosa, 256
Tomato, 127
Toninia aromatica, 257
Toothwort, 132
Tor-grass, 153
Torilis arvensis, 118
  japonica, 118
  nodosa, 118
Tormentil, 101
  , Trailing, 101
Tortella nitida, 195
  tortuosa, 195
Tortula canescens, 192
  intermedia, 192
  laevipila, 192
  latifolia, 193
  muralis, 193
  papillosa, 193
  ruralis ssp. ruralis, 192
  subulata, 193
Trachelomonas, 164
Tragopogon pratensis ssp. minor, 140
Trapelia coarctata, 257
  coarctata agg., 257
  corticola, 257
  involuta, 257
  obtegens, 257
  placodioides, 257
Trapeliopsis flexuosa, 257
  gelatinosa, 257
  glaucolepidea, 257
  granulosa, 257
  pseudogranulosa, 257
  wallrothii, 257
Traveller's Joy, 87
Treboeth, 17
Tree, Wayfaring, 133
Trefoil, Common Bird's-foot, 106
  , Greater Bird's-foot, 106
  , Hop, 106
  , Lesser, 106
  , Slender, 106
Tregoed, 3
Tremolecia atrata, 257
Trichocolea tomentella, 183
Trichomanes speciosum, 74
Trichostomum brachydontium, 195
  crispulum, 195
Trifolium arvense, 106
  campestre, 106
  dubium, 106

Index

Trifolium hybridum, 106
  medium, 106
  micranthum, 106
  ornithopodioides, 105
  pratense, 106
  repens, 105
  scabrum, 106
  striatum, 106
  strictum, 105
Triglochin palustris, 145
Triphragmium ulmariae, 264
Tripleurospermum inodorum, 136
Trisetum flavescens, 154
Triticum aestivum, 153
Tritomaria exsectiformis, 178
  quinquedentata, 178
Trollius europaeus, 87
Trumau, 16, 30
Tsuga heterophylla, 78
Turnip, 94
Tussilago farfara, 137
Tussock-sedge, Greater, 158
  , Lesser, 158
Tutsan, 111
Twayblade, Common, 161
Twyi anticline, 5
Tylothallia biformigera, 258
Tylwch, 2
Tyn-y-berth, 15
Tyncoed, 28
Typha angustifolia, 156
  latifolia, 156
TYPHACEAE, 156
Ulex europaeus, 104
  gallii, 104
ULMACEAE, 81
Ulmus glabra, 81
  procera, 81
Ulota crispa, 204
  phyllantha, 204
Ulothrix, 163, 164
UMBELLIFERAE, 116
Umbilicaria deusta, 258
  polyphylla, 258
  polyrrhiza, 258
Umbilicus rupestris, 95
Urocystis ranunculi, 265
  violae, 265
Uromyces acetosae, 264
  dactylidis, 264
  fallens, 264
  ficariae, 264
  muscari, 264
  pisi-sativi, 264
  polygoni-aviculare, 264
  valerianae, 264
  vicia-fabae, 265
Urtica dioica, 81
  urens, 81
URTICACEAE, 81
Usnea articulata, 258
  filipendula, 258
  florida, 258
  inflata, 258
  rubicunda, 258
  subfloridana, 258
Ustilago anomala, 265
  avenae, 265
  longissima. 265
  striiformis, 265
  succisae, 265
  tragopogonis-pratensis, 265
  violacea, 265
Utricularia australis, 132
  minor, 132
Vaccinium myrtillus, 119
  oxycoccus, 119
  vitis-idaea, 119
Valerian, Common, 133
  , Marsh, 133
  , Red, 134
Valeriana dioica, 133
  officinalis, 133
VALERIANACEAE, 133
Valerianella carinata, 133
  dentata, 133
  locusta, 133
Vélindre, 19

Velvetleaf, 111
Verbascum nigrum, 128
  thapsus, 128
Verbena officinalis, 124
VERBENACEAE, 124
Vernal-grass, Sweet, 154
Veronica agrestis, 130
  anagallis-aquatica s.s., 129
  arvensis, 130
  beccabunga, 129
  chamaedrys, 129
  filiformis, 130
  hederifolia, 130
  montana, 129
  officinalis, 129
  persica, 130
  polita, 130
  scutellata, 129
  serpyllifolia ssp. serpyllifolia, 129
  spicata, 129
Verrucaria aethiobola, 258
  baldensis, 258
  bryoctona, 258
  caerulea, 258
  dolosa, 258
  glaucina, 258
  hochstetteri, 259
  hydrela, 259
  margacea, 259
  muralis, 259
  mutabilis, 258
  nigrescens, 259
  praetermissa, 259
  rheitrophila, 259
  viridula, 259
Vervain, 124
Vetch, Bitter, 105
  , Bush, 104
  , Common, 104
  , Kidney, 106
  , Spring, 105
  , Tufted, 104
  , Wood, 104
Vetchling, Meadow, 105
Vezdaea acicularis, 259
  aestivalis, 259
  leprosa, 259
  retigera, 259
  rheocarpa, 259
Viburnum lantana, 133
  opulus, 133
Vicia cracca, 104
  faba, 105
  hirsuta, 104
  lathyroides, 105
  orobus, 104
  sativa, 104
  sepium, 104
  sylvatica, 104
  tetrasperma, 104
Vinca major, 121
  minor, 121
Viola arvensis, 113
  canina, 113
  hirta, 113
  lutea, 113
  odorata, 112
  palustris, 113
  reichenbachiana, 113
  riviniana, 113
  tricolor, 113
VIOLACEAE, 112
Violet, Hairy, 113
  , Marsh, 113
  , Sweet, 112
Viper's-bugloss, 123
Viscum album, 81
Vulpia bromoides, 151
  myuros, 151
Wade, A.E., 68
Wahlenbergia hederacea, 135
Wall-rocket, Annual, 94
Wall-rue, 75
Walton, 37
Wash, 215
Water-break-its-neck, 12, 65, 167, 172
Water-cress, 92
  , Fool's, 117

Water-crowfoot, Pond, 89
  , River, 89
  , Stream, 89
  , Thread-leaved, 89
Water-dropwort, Hemlock, 117
  , Tubular, 117
Water-lily, White, 87
  , Yellow, 87
Water-milfoil, Alternate, 116
  , Spiked, 115
  , Whorled, 115
Water-pepper, 82
Water-purslane, 114
Water-thyme, Curly, 145
Waterwort, Six-stamened, 113
Watson, H.C., 67
Watsonian vice-counties, 3
Webb, J.A., 68
Weedkillers, 56
Weissia controversa, 195
  longifolia, 195
  microstoma, 195
  rutilans, 195
Weld, 95
Welsh names, 46
Wenallt, 8
Westcombe, T., 65
Wheat, Bread, 153
Whinberry, 119
Whinyard Rocks, 12
Whitebeam, 61, 102
  , English, 102
Whitlow-grass, Common, 93
  , Wall, 93
Whitty, 102
Whorl-grass, 152
Willow, Almond, 79
  , Crack, 79
  , Creeping, 79
  , Eared, 79
  , Goat, 79
  , Grey, 79
  , Purple, 79
  , Rusty, 79
  , White, 79
Willowherb, American, 115
  , Broad-leaved, 115
  , Great, 114
  , Hoary, 115
  , Marsh, 115
  , New Zealand, 115
  , Pale, 115
  , Rosebay, 114
  , Short-fruited, 115
  , Spear-leaved, 115
  , Square-stalked, 115
Wimberry, 119
Wind, 24
Winter-cress, 91
  , American, 91
  , Intermediate, 91
Wood-rush, Field, 149
  , Great, 149
  , Hairy, 150
  , Heath, 149
  , Southern, 150
Wood-sorrel, 107
Woodland Trust, 64
Woodland type, 26
Woodlands, 40
Woodlands, Ancient, 58
  Broad-leaved, 58
Woodruff, 122
Woolhope Naturalists Field Club, 65
Wormwood, 137
Worsell, 7, 13, 27
Woundwort, Field, 126
  , Hedge, 126
  , Marsh, 126
Wye Valley, 5, 168, 169, 215
Wyloer, 11, 30
Xanthidium antilopaeum, 163
  armatum, 163
Xanthoria calcicola, 259
  candelaria, 259
  elegans, 259
  parietina, 259
  polycarpa, 260